Lyme Borreliosis
Biology, Epidemiology and Control

Lyme Borreliosis
Biology, Epidemiology and Control

Edited by

J. Gray
Department of Environmental Resource Management
University College Dublin
Republic of Ireland

O. Kahl
Institut für Angewandte Zoologie
Freie Universität Berlin
Germany

R.S. Lane
Department of Environmental Science, Policy and Management
University of California at Berkeley
USA

and

G. Stanek
Institute of Hygiene
University of Vienna
Austria

CABI *Publishing*

CABI *Publishing* is a division of CAB *International*

CABI Publishing
CAB International
Wallingford
Oxon OX10 8DE
UK

CABI Publishing
10 E 40th Street
Suite 3203
New York, NY 10016
USA

Tel: +44 (0)1491 832111
Fax: +44 (0)1491 833508
E-mail: cabi@cabi.org
Web site: www.cabi-publishing.org

Tel: +1 212 481 7018
Fax: +1 212 686 7993
E-mail: cabi-nao@cabi.org

© CAB *International* 2002. All rights reserved. No part of this publication may be reproduced in any form or by any means, electronically, mechanically, by photocopying, recording or otherwise, without the prior permission of the copyright owners.

A catalogue record for this book is available from the British Library, London, UK.
A catalogue record for this book is available from the Library of Congress, Washington DC, USA.

ISBN 0 85199 632 9

Typeset by Wyvern 21 Ltd, Bristol
Printed and bound in the UK by Cromwell Press, Trowbridge

Contents

Contributors   vii

Preface   ix

1. History and Characteristics of Lyme Borreliosis   1
   G. Stanek, F. Strle, J. Gray and G.P. Wormser

2. Ecological Research on *Borrelia burgdorferi* sensu lato: Terminology and Some Methodological Pitfalls   29
   O. Kahl, L. Gern, L. Eisen and R.S. Lane

3. Molecular and Cellular Biology of *Borrelia burgdorferi* sensu lato   47
   S. Bergström, L. Noppa, Å. Gylfe and Y. Östberg

4. Vectors of *Borrelia burgdorferi* sensu lato   91
   L. Eisen and R.S. Lane

5. *Borrelia burgdorferi* sensu lato in the Vertebrate Host   117
   K. Kurtenbach, S.M. Schäfer, S. de Michelis, S. Etti and H.-S. Sewell

6. Ecology of *Borrelia burgdorferi* sensu lato in Europe   149
   L. Gern and P.-F. Humair

7. Ecology of *Borrelia burgdorferi* sensu lato in Russia   175
   E.I. Korenberg, N.B. Gorelova and Y.V. Kovalevskii

8. Ecology of *Borrelia burgdorferi* sensu lato in Japan and East Asia   201
   K. Miyamoto and T. Masuzawa

9. Ecology of *Borrelia burgdorferi* sensu lato in North America   223
   J. Piesman

10. Epidemiology of Lyme Borreliosis   251
    D.T. Dennis and E.B. Hayes

11. Vaccination against Lyme Borreliosis   281
    E.B. Hayes and M.E. Schriefer

12. Environmental Management for Lyme Borreliosis Control   301
    K.C. Stafford and U. Kitron

Index   335

Contributors

S. Bergström, *Department of Molecular Biology, Umeå University, SE-901 87 Umeå, Sweden.*

D.T. Dennis, *Division of Vector-borne Infectious Diseases, National Center for Infectious Diseases, Centers for Disease Control and Prevention, Public Health Service, US Department of Health and Human Services, PO Box 2087, Fort Collins, CO 80522, USA.*

L. Eisen, *Department of Environmental Science, Policy and Management, University of California, Berkeley, CA 94720, USA.*

S. Etti, *Department of Infectious Disease Epidemiology, Imperial College of Science, Technology and Medicine, St Mary's Campus, Norfolk Place, London W2 1PG, UK.*

L. Gern, *Institut de Zoologie, University of Neuchâtel, Rue Emile-Argand 11, Case Postale 2, CH 2007 Neuchâtel 7, Switzerland.*

N.B. Gorelova, *Gamaleya Research Institute for Epidemiology and Microbiology, Russian Academy of Medical Sciences, 18 Gamaleya Street, Moscow, 123098 Russia.*

J. Gray, *Department of Environmental Resource Management, University College Dublin, Ireland.*

Å. Gylfe, *Department of Molecular Biology, Umeå University, SE-901 87 Umeå, Sweden.*

E.B. Hayes, *Division of Vector-borne Infectious Diseases, National Center for Infectious Diseases, Centers for Disease Control and Prevention, Public Health Service, US Department of Health and Human Services, PO Box 2087, Fort Collins, CO 80522, USA.*

P.-F. Humair, *Institute of Zoology, University of Neuchâtel, Chantemerle 22, 2000 Neuchâtel, Switzerland.*

O. Kahl, *Institute of Applied Zoology/Animal Ecology, Free University of Berlin, 12163 Berlin, Germany. Present address: Blackwell Verlag, Kurfürstendamm 57, 10707 Berlin, Germany.*

U. Kitron, *Department of Veterinary Pathobiology, College of Veterinary Medicine, University of Illinois, 2001 S. Lincoln, Urbana, IL 61802, USA.*

E.I. Korenberg, *Gamaleya Research Institute for Epidemiology and Microbiology, Russian Academy of Medical Sciences, 18 Gamaleya Street, Moscow, 123098 Russia.*

Y.V. Kovalevskii, *Gamaleya Research Institute for Epidemiology and Microbiology, Russian Academy of Medical Sciences, 18 Gamaleya Street, Moscow, 123098 Russia.*

K. Kurtenbach, *Department of Infectious Disease Epidemiology, Imperial College of Science, Technology and Medicine, St Mary's Campus, Norfolk Place, London W2 1PG, UK.*

R.S. Lane, *Department of Environmental Science, Policy and Management, University of California, Berkeley, CA 94720, USA.*

T. Masuzawa, *Department of Microbiology, School of Pharmaceutical Sciences, University of Shizuoka, 52-1 Yada, Shizuoka 422-8526, Japan.*

S. de Michelis, *Department of Infectious Disease Epidemiology, Imperial College of Science, Technology and Medicine, St Mary's Campus, Norfolk Place, London W2 1PG, UK.*

K. Miyamoto, *Department of Parasitology, Asahikawa Medical College, Asahikawa, 078-8510, Japan.*

L. Noppa, *Department of Molecular Biology, Umeå University, SE-901 87 Umeå, Sweden.*

Y. Östberg, *Department of Molecular Biology, Umeå University, SE-901 87 Umeå, Sweden.*

J. Piesman, *Division of Vector-borne Infectious Diseases, National Center for Infectious Diseases, Centers for Disease Control and Prevention, Public Health Service, US Department of Health and Human Services, PO Box 2087, Fort Collins, CO 80522, USA.*

S.M. Schäfer, *Department of Infectious Disease Epidemiology, Imperial College of Science, Technology and Medicine, St Mary's Campus, Norfolk Place, London W2 1PG, UK.*

M.E. Schriefer, *Division of Vector-borne Infectious Diseases, National Center for Infectious Diseases, Centers for Disease Control and Prevention, Public Health Service, US Department of Health and Human Services, PO Box 2087, Fort Collins, CO 80522, USA.*

H.-S. Sewell, *Department of Infectious Disease Epidemiology, Imperial College of Science, Technology and Medicine, St Mary's Campus, Norfolk Place, London W2 1PG, UK.*

K.C. Stafford, *Connecticut Agricultural Experiment Station, PO Box 1106, New Haven, CT 06540, USA.*

G. Stanek, *Clinical Institute of Hygiene and Medical Microbiology, University of Vienna, Austria.*

F. Strle, *Department of Infectious Diseases, University Medical Centre, Ljubljana, Slovenia.*

G.P. Wormser, *Division of Infectious Diseases, Department of Medicine, New York Medical College and Westchester Medical Center, Valhalla, New York, USA.*

Preface

Lyme borreliosis, often referred to as Lyme disease, emerged during the mid-1970s when an unusual form of juvenile arthritis was observed in the small town of Lyme, Connecticut, USA. Since then a wealth of literature on the subject has been published and eight international conferences as well as several smaller-scale specialist or local meetings have taken place. Most publications, including several textbooks, have concentrated on clinical aspects, which reflects the importance of the disease in addition to the well-documented difficulties associated with diagnosis and treatment.

The complex ecology of the causative agent, *Borrelia burgdorferi*, has also attracted much attention. Awareness of this complexity increased markedly when it became evident during the early 1990s that *B. burgdorferi* occurs as several distinct genospecies, referred to collectively as *B. burgdorferi* sensu lato, with apparently differing host preferences and pathological significance. In fact, most recent advances in our knowledge of Lyme borreliosis have concerned aspects of the basic biology and ecology of the causative spirochaetes and their tick vectors, leading to the development of vaccines and ecological preventive measures. However, the literature on these topics is widely scattered in research papers, reviews and meeting proceedings, and there is now a need for such information to be made available in book form.

In this book, the first chapter outlines historical and clinical aspects including descriptions of the disease in humans and domestic animals, and of diagnosis and treatment. The second chapter deals with ecological methods and terminology and Chapters 3–5 describe the biology of the spirochaetes and their behaviour in vectors and vertebrates. The next four chapters (6–9) concern the ecology of *B. burgdorferi* s.l. in Europe, Russia, Southeast Asia and North America, and the last three chapters (10–12) deal with the application of the biological and ecological attributes of the pathogens to disease epidemiology, vaccine development and the ecological management of Lyme borreliosis.

The intended readership includes microbiologists, immunologists, zoologists, ecologists, epidemiologists, specialist clinicians, public health workers, specialist veterinarians, teachers, research students and all those interested in zoonotic or vector-borne infectious diseases. It is intended that the book will not only inform but will also stimulate much needed further research on the complex biology of *B. burgdorferi* s.l.

Acknowledgements

We are especially grateful to Janet Robertson for her editorial assistance. Thanks are also due to the following authorities in the subject who gave generously of their time and expertise as reviewers: Lise Gern, Thomas Jaenson, Andreas Krause, Klaus Kurtenbach, Michael Levin, Jim Oliver, Eva Olsson, Joe Piesman, Danièle Postic, Sarah Randolph, Janet Robertson, Markus Simon, Susan Shaw, Dan Sonenshine, Reinhardt Straubinger, Franc Strle and Gary Wormser. We are also grateful to Bernard Kaye for creating some of the figures and for his assistance in the formatting of most of them.

History and Characteristics of Lyme Borreliosis

Gerold Stanek,[1] Franc Strle,[2] Jeremy Gray[3] and Gary P. Wormser[4]

[1]*Clinical Institute of Hygiene and Medical Microbiology, University of Vienna, Austria;* [2]*Department of Infectious Diseases, University Medical Centre, Ljubljana, Slovenia;* [3]*Department of Environmental Resource Management, University College Dublin, Ireland;* [4]*Division of Infectious Diseases, Department of Medicine, New York Medical College and Westchester Medical Center, Valhalla, New York, USA*

History

Before the first description of Lyme disease

When the first cases were observed in Connecticut, USA, of a condition now known as Lyme disease (or Lyme borreliosis) (Steere *et al.*, 1977a), it was 92 years after the first description in Europe of a chronic skin disorder (Buchwald, 1883) that was later named acrodermatitis chronica atrophicans (ACA) (Herxheimer and Hartmann, 1902). Thus, ixodid tick-borne borreliosis was present in Europe and emerged into clinical awareness at the end of the 19th century. In 1909, Afzelius presented a patient to the Swedish Dermatological Society in Stockholm with a slowly expanding skin lesion, which he called erythema migrans (EM) (Afzelius, 1910). A few years later the dermatologist Lipschütz from Vienna described a similar lesion in a 29-year-old woman, which expanded over 7 months from the back of the knee to the entire back and the neck (Lipschütz, 1913). Lipschütz was already aware of the transmission of some agent by the bite of a tick and he suggested that '... attention should be directed towards microscopic/bacteriologic investigations of the intestinal tract and of the salivary gland secretions of the tick' (Lipschütz, 1923). However, no one followed his suggestions in Europe. The histological appearance of another skin manifestation of the disease, today called borrelial lymphocytoma, was first described in 1911 (Burckhardt, 1911). An extended description and definition of solitary lymphocytoma benigna cutis based on about 150 cases was given by Bäferstedt (1943).

A tick bite was also the link to what is considered the first description of borrelial meningo-polyradiculoneuritis by the French neurologists Garin and

Bujadoux (1922). The case report was an exciting story that elucidated the dramatic changes that may accompany borrelial infection of the central nervous system and/or the nerve roots. A relationship between tick bite, EM and disorders of the nervous system was first considered by Hellerström (1930). Bannwarth from Munich performed a careful analysis of several patients with chronic lymphocytic meningitis with the clinical syndrome of neuralgia and neuritis. He identified three groups, one with intense radicular pain, another with facial palsy and the third with chronic lymphocytic meningitis and cerebral symptoms such as severe headache and vomiting (Bannwarth, 1941). From today's viewpoint it is evident that Bannwarth's valuable work led to incorrect conclusions. He missed the relationship between tick bite and skin and nervous system manifestations and thought that the conditions were due to a rheumatic-allergic cause. Despite this, Bannwarth's name is now synonymous with meningo-polyradiculoneuritis, the full picture of neuroborreliosis.

Following the introduction of antibiotics after the Second World War, large studies were undertaken on the skin manifestations ACA and EM. A Swedish physician treated 57 patients with ACA with antibiotics and followed them for 2 years. He reported complete recovery and considerable improvement in about 70% (Thyresson, 1949). Successful antibiotic treatment stimulated the search for a causative agent of ACA and also renewed interest in the clinical course and pathogenicity of this chronic skin disease. More than 200 patients in Germany were evaluated in detail and associations between EM and lymphocytomas described by Hauser (1955). In a spectacular inoculation experiment, skin from an ACA lesion was successfully transplanted to the investigator and to three colleagues. An erythema-like rash developed on the skin of the volunteers around the borders of the inoculum; all skin changes disappeared after antibiotic therapy (Götz, 1954, 1955). Among the many studies performed at this time, one observation should be mentioned. Studies of ACA patients showed that peripheral neuropathy of the distal type was part of the clinical picture in a considerable proportion of patients and that the condition responded well to penicillin (Hopf, 1966). Electron microscope treatment studies were performed by Hollström (1958) in which it was reported that the EM disappeared within 2 weeks on average after the initiation of low-dose penicillin therapy.

It was evident that an agent sensitive to antibiotics must play a role in the aetiology of these disorders. This was also suggested by observations of neurological disorders after EM and tick bite (Schaltenbrand, 1967). The most common neurological signs and symptoms after tick bites were painful radiculomeningitis, neuritis, myelitis and cranial nerve palsies that developed within a few weeks to 8 months (Sälde, 1946), and severe pain constituted the hallmark of the disease that lasted for 3 weeks on average before neurological deficit occurred (Hörstrup and Ackermann, 1973). In 1974, Weber reported on a patient who developed meningitis after EM, who responded very well to intravenous penicillin (Weber, 1974). Weber got very close to the 'real' aetiologic agent as he systematically and logically ruled out viruses and many tick-transmitted bacteria known at that time, and mentioned borreliae as possible causative agents. Unfortunately, he could not follow this track, and the enigmatic aetiology of these tick-borne clinical conditions persisted until the early 1980s.

A new clinical entity

In October 1975, two mothers from Lyme, Connecticut, reported to the Connecticut State Health Department that a number of children who lived close together were thought to have juvenile rheumatoid arthritis.

A surveillance system was organized in order to identify all children with inflammatory joint disease in the communities of Old Lyme, Lyme and East Haddam, Connecticut. The arthritis cases found were of a similar type, characterized by brief but recurrent attacks of swelling and pain in a few large joints, especially the knee, over a period of years. Parents and neighbours of the affected children were found to have a similar type of arthritis. Moreover, the frequency of this type of arthritis was 100 times greater than juvenile rheumatoid arthritis. The majority of patients noted the onset of arthritis in summer or early autumn. Furthermore, a certain proportion of patients noted a skin rash approximately 4 weeks before the onset of arthritis. This skin lesion was found to be compatible with EM, described in Europe many decades before (Afzelius, 1910; Lipschütz, 1913). However, the disease in Lyme residents was thought to be a previously unrecognized clinical entity and was called Lyme arthritis (Steere *et al.*, 1977a,b). In the following years, the Yale investigators identified patients with the skin lesion EM and followed the patients prospectively. Many of them subsequently developed arthritis, and some also acquired neurological and cardiac abnormalities. Thus, a complex, multi-system disorder was recognized and the name was changed from Lyme arthritis to Lyme disease. Furthermore, the vector of the disease was identified as the hard tick *Ixodes scapularis* (then called *Ixodes dammini*). In the following years, Lyme disease was found to be present along the northeastern coast of the USA and also in Wisconsin, California and Oregon; there was strong evidence that ixodid ticks were the vectors (Steere and Malawista, 1979). The agent was apparently a non-pyrogenic, penicillin-sensitive bacterium since penicillin treatment shortened the duration of EM and appeared to prevent subsequent arthritis (Steere *et al.*, 1980). But the agent was still unknown.

Discovery of the Lyme disease agent

The discovery of the Lyme disease agent came unexpectedly in conjunction with a tick and rickettsial survey of Long Island, New York (Burgdorfer *et al.*, 1982). Unable to find the Rocky Mountain spotted fever agent in *Dermacentor variabilis* ticks, the researchers speculated that possibly other ticks such as *I. scapularis* (*dammini*), which were there in abundance, might carry the agent. No rickettsiae were found in the midguts of many dissected ticks but poorly stained and rather long, irregularly coiled spirochaetes were observed. The bacteria were isolated and cultivated in a modified Kelly's medium, which is now known as BSKII medium and the organisms were linked with Lyme disease after they were observed to react with sera of Lyme disease patients in immunofluorescence assays (Burgdorfer *et al.*, 1982; Barbour, 1984). Shortly thereafter the spirochaete was isolated from blood, skin and cerebrospinal fluid of a few Lyme disease patients (Benach *et al.*, 1983; Steere *et al.*, 1983).

A worldwide tick-borne disease

Very soon after the successful isolation of a new borrelia from hard ticks and patients in the USA, researchers in Europe isolated spirochaetes from ticks, and from skin and cerebrospinal fluid of patients in Sweden, Germany, Switzerland and Austria (Preac-Mursic *et al.*, 1984; Asbrink and Hovmark, 1985; Stanek *et al.*, 1985). With the methods available initially, the spirochaetes appeared to be the same as the USA isolates. The bacterium was identified as a new species among borreliae and named *Borrelia burgdorferi* (Johnson *et al.*, 1984). Barbour and Schrumpf (1986) raised monoclonal antibodies to the outer surface proteins (Osp) A and B, and to flagellin proteins of the spirochaetes, which resulted in the detection of substantial differences in reactivity. An OspA typing system was developed in Munich (Wilske *et al.*, 1986) and the resulting findings in conjunction with molecular biological studies (Baranton *et al.*, 1998) provided evidence that two new *Borrelia* genospecies, *Borrelia afzelii* and *Borrelia garinii*, were also human pathogens. These genospecies occur in Europe and Asia but not in North America. The USA type was called *B. burgdorferi* sensu stricto (s.s.), and the overall term for the genospecies complex is *B. burgdorferi* sensu lato (s.l.).

The discovery of a new tick-transmitted agent sparked off a spate of research in various branches of science, including human and veterinary medicine, molecular biology, and tick and reservoir host biology and ecology. Lyme disease, or Lyme borreliosis (LB) as it is more widely known, is now recognized as a tick-borne spirochaetal infection with a wide geographical distribution in the northern hemisphere. However, the story is still incomplete and new findings continue to emerge, including the discovery of additional genospecies.

Scientific publications and international conferences on tick-borne diseases

Up to mid-July 2001, the PubMed service of the National Library of Medicine listed 5592 scientific publications under the search words 'Lyme disease', and research is still proceeding vigorously. Besides these publications there are numerous others available in English and in other languages, and many theses and several books on LB have been written and published since the early 1980s. An international conference on LB has been held every 2 or 3 years since 1983, and the number of smaller international symposia occurring in different places in the world is increasing (Table 1.1). Topics have included clinical manifestations, treatment regimens, concepts on prophylaxis, vectors, the still poorly understood changes of the agent during its infectious cycle, and studies on pathogenicity of the disease. The European Union Concerted Action on Lyme Borreliosis (EUCALB) (Gray and Stanek, 1998) was the first continent-wide approach to attempt risk assessment in relation to exposure and to improve diagnostic criteria for both case management and the quality of epidemiological data.

Lyme borreliosis as a social phenomenon

Contrary to the success in identifying the aetiology of multi-system LB and in introducing proper treatment, people in endemic areas of the USA and subsequently in Europe have exhibited a high degree of anxiety concerning this disease (Barbour and Fish, 1993). Lyme disease/borreliosis self-help groups have been formed with 'tick-victims' as their members. These activities appear to be acceptable or even truly helpful as long as they form an information desk that guides persons to trained physicians who follow the therapeutic concepts of evidence-based medicine. Problems arise when such groups develop their own pictures of the disease. Moreover, certain physicians have increased the fear of patients with their private opinion that long-term treatment (over many months) is obligatory in order to get rid of the pathogens hidden in the body. Other physicians have developed obscure diagnostic procedures such as identification of borreliae in the peripheral blood by dark-field microscopy and adhere to this approach although it has been discredited by other methods including nucleic acid amplification techniques. Some consequences of such activities were described in an article published in the *New York Times* of June 2001 (Grann, 2001).

What follows is a brief review of the salient features of the disease, its diagnosis and treatment.

Clinical Features

Three stages of LB have been described, but the disease itself does not necessarily develop in stages. Neuroborreliosis, for example, is not infrequently preceded by EM but neurological disorders may also develop without it. Chronic manifestations such as ACA usually develop unrecognized initially and are not preceded by EM. The concept of localized versus disseminated infection is open to question because up to 40% of US patients with a single EM have microbiological evidence of haematogenous dissemination. The main clinical manifestations of LB are listed in Table 1.2 and Table 1.3 provides a simplified outline of the differences between the clinical features in Europe/Eurasia and those in North America. In the following paragraphs the various clinical manifestations will be introduced briefly.

Table 1.1. International conferences on Lyme disease/Lyme borreliosis.

Year	Location	Conference	Publication
1983	New Haven, Connecticut, USA	I International Conference on Lyme Disease	*Yale J. Biol. Med.* 57 1984
1985	Vienna, Austria	II International Conference on Lyme Disease and Related Disorders	Stanek et al., 1986
1987	Baden, Austria	1st European Symposium on Lyme Borreliosis	Stanek et al., 1989
1987	New York, USA	III International Conference on Lyme Disease and Related Disorders	Benach and Bosler, 1988
1990	Stockholm, Sweden	IV International Conference on Lyme Borreliosis	Sköldenberg and Stiernsted, 1991
1992	Arlington, USA	V International Congress on Lyme Borreliosis	Abstracts only
1993	London, UK	2nd European Symposium on Lyme Borreliosis	Axford and Rees, 1994
1994	Bologna, Italy	VI International Conference on Lyme Borreliosis	Cevenini et al., 1994
1994	Shizuoka, Japan	1st International Symposium on Lyme Disease in Japan	Yanagihara and Masuzawa, 1994
1995	Portoroz, Slovenia	International Symposium on the Therapy and Prophylaxis for Lyme Borreliosis	*Infection* 24 (1996) 60–102 24 (1996) 170–212
1995	Warsaw, Poland	WHO Workshop on Lyme Borreliosis Diagnosis and Surveillance	Anon., 1995
1996	San Francisco, USA	VII International Congress on Lyme Borreliosis	Abstracts only
1997	Berlin, Germany	4th International Potsdam Symposium on Tick-borne Diseases	Süss and Kahl, 1997
1997	Shizuoka, Japan	2nd International Symposium on Lyme Disease in Japan	Yanagihara and Masuzawa, 1997
1998	Vienna, Austria	International Symposium on the Pathogenesis and Management of Tick-borne Diseases	*Wien. klin. Wochenschr.* 110 (1998) 847–910 111 (1999) 909–1004
1999	Berlin, Germany	5th International Potsdam Symposium on Tick-borne Diseases	Süss and Kahl, 1999
1999	Munich, Germany	VIII International Conference on Lyme Borreliosis and Other Emerging Tick-borne Diseases	Abstracts only
2001	Berlin, Germany	6th International Potsdam Symposium on Tick-borne Diseases	Süss et al. (2002)
2001	Ljubljana, Slovenia	International Symposium on Tick-transmitted Diseases	*Wien. klin. Wochenschr.*, in press
2002	New York, USA	IX International Conference on Lyme Borreliosis and Other Tick-borne Diseases	

Table 1.2. Main clinical manifestations of Lyme borreliosis.

Organ system	Clinical feature
Skin	Erythema (chronicum) migrans Erythema migrans multiple lesions Borrelial lymphocytoma Acrodermatitis chronica atrophicans
Nervous system	Meningo-radiculoneuritis (Bannwarth syndrome) Cranial nerve palsy Meningitis Meningoencephalitis Encephalomyelitis Cerebral vasculitis Peripheral neuropathy
Musculoskeletal	Arthritis Myositis
Heart	Carditis
Eye	Conjunctivitis, keratitis, endophthalmitis, panophthalmitis

Skin manifestations

Erythema (chronicum) migrans

EM is the most frequent clinical manifestation of LB. It occurs at any age and in both sexes. EM is most frequently diagnosed in late spring and early summer but can be seen in individual cases throughout the year. It develops at the site of a tick bite, though the bites are often not detected by patients. The incubation period ranges from a few days to more than 1 month, but is usually about 10 days. The initial EM is commonly a homogeneous lesion, which may either be bright red or turn bluish-red, and the homogeneous nature may not change in some patients until treatment or spontaneous healing. In cases of long duration, the lesion will fade or clear in the central part and a peripheral bright red erythematous ring will further expand, again until treatment or spontaneous healing. How the lesion expands depends partly on the site of the initial infection. The border of EM is usually demarcated. In adult European patients, constitutional symptoms accompanying EM (most often fatigue, malaise, arthralgia, myalgia and headache) are present in fewer than 50% and are regularly mild, while local symptoms such as itching, burning and/or mild local pain are reported by about 50% of patients (Strle *et al.*, 1993; Logar *et al.*, 1999). Constitutional symptoms are more frequent in the USA (up to 15% of cases).

The different shapes that can be observed with EM make a definition of the skin lesion a challenging task. Generally, EM is an enlarging erythematous to

Table 1.3. Lyme borreliosis in Eurasia and North America. (Modified after Steere, 2001.)

Organ system	Europe/Eurasia (*Borrelia afzelii*, *B. garinii*, *B. burgdorferi* s.s.)	North America (*B. burgdorferi* s.s.)
Skin		
Acute phase	Erythema (chronicum) migrans slower spreading, less intensely inflamed, of longer duration; less frequent haematogenous dissemination, possible regional or contiguous spread to other sites. Borrelial lymphocytoma rarely and predominantly seen in children and scarcely in adults.	Erythema migrans faster spreading, more intensely inflamed, of briefer duration; frequent, possibly widespread haematogenous dissemination.
Chronic phase	Acrodermatitis chronica atrophicans (ACA), caused primarily by *B. afzelii*. Circumscribed scleroderma and Lichen sclerosus et atrophicus; reports on the isolation of *B. afzelii* from lesional skin, improvement after antibiotic therapy in single cases.	ACA rarely reported. None reported.
Nervous system		
Acute phase	Meningo-polyradiculoneuritis with severe radicular pain, pleocytosis and intrathecal antibody production, caused primarily by *B. garinii*.	Meningitis, less prominent radiculoneuritis.
Chronic phase	Subtle sensory neuropathy often within areas affected by acrodermatitis. Severe encephalomyelitis, spasticity, cognitive abnormalities, marked intrathecal antibody production.	Subtle sensory polyneuropathy without acrodermatitis. Subtle encephalopathy, cognitive disturbance, slight intrathecal antibody production.
Heart		
Acute phase	Atrioventricular block and subtle myocarditis.	Atrioventricular block and subtle myocarditis.
Chronic phase	Dilated cardiomyopathy; isolation of *B. burgdorferi* s.l. from endomyocardial biopsies in only a single case.	None reported.
Musculoskeletal		
Acute phase	Less frequent oligoarticular arthritis, less intense joint inflammation.	More frequent oligoarticular arthritis, more intense joint inflammation.
Chronic phase	Persistent arthritis rare.	Treatment-resistant arthritis in about 10% of patients, probably due to autoimmune mechanism.

bluish-red patch with an advancing border, with or without central clearing. However, there are conditions with which these lesions may be confused and which should be considered. Differential diagnosis comprises non-specific insect-bite reactions, erythema annulare centrifugum, urticaria, contact dermatitis, folliculitis, vasculitis, granuloma annulare, tinea corporis and fixed drug eruption.

The histology of EM-lesional skin is usually unimpressive. Histological findings of skin samples from the central and peripheral parts of EM show a superficial perivascular infiltrate with lymphocytes, histiocytes, a few plasma cells, eosinophils and neutrophils. Blood vessels may show endothelial swelling, and extravasation of erythrocytes is frequently found.

Borrelial lymphocytoma

Lymphocytoma is a term that denotes cutaneous proliferation of lymphoid cells without detectable extracutaneous involvement. Because of its benign course, the lesion has also been called lymphadenosis benigna cutis. Borrelial lymphocytoma was the term introduced by Weber *et al.* (1985) in order to specify the aetiology of the condition (Hovmark *et al.*, 1986a). Borrelial lymphocytoma presents as a bluish-red tumour-like skin infiltrate, up to a few centimetres in diameter, with lymphoreticular proliferation in the dermis and/or subcutis. The lesion may develop several weeks to months after a tick bite. Predilection sites are the ear lobe, ear helix, nipple or scrotum in children, and nipple in adults. The lesion may be preceded by EM or occur simultaneously with it. Borrelial lymphocytoma heals spontaneously but can persist and even grow for several months (Strle *et al.*, 1992). This skin manifestation is not seen in the USA.

The histological picture of borrelial lymphocytoma is impressive. It presents either as a diffuse, dense dermal polyclonal lymphocytic infiltrate or a follicular lesion with germinal centres.

Acrodermatitis chronica atrophicans

ACA usually begins on the extensor sites of the extremities, most commonly on the lower leg with initial involvement of one foot. Besides this, the soles and palms, toes, fingers and knees can be involved, and, in rare cases, ACA may also develop on the trunk. The lesion may persist over a long period of time and expand on one extremity. With time, however, the lesions will usually become symmetric but in some cases may be asymmetric, restricted to one extremity. The first signs are localized, livid-red colour changes and a doughy swelling of the involved skin. The swelling may be pronounced, which might cause the patient to seek medical attention. ACA does not heal spontaneously. Within years of onset, fibrous thickening may develop over, or close to, joints and after years of progression the lesional skin will gradually become atrophic, hyperpigmentation and sometimes depigmentation will occur and the veins may appear prominent because of the thin and wrinkled, 'cigarette paper-like' skin. Chronic ulceration may develop and form the basis for squamous or basal cell carcinomas. Sclerotic changes of the skin may develop, resembling morphoea.

The borrelial aetiology of ACA was confirmed by culture and serology (Asbrink and Hovmark, 1985). *B. afzelii* is predominantly isolated from samples of lesional skin, but the other pathogenic genospecies have been isolated from cases of ACA (Picken *et al.*, 1998). ACA rarely, if ever, occurs in the USA.

Associated with long-standing ACA is peripheral neuropathy (Kristoferitsch *et al.*, 1988) and joint manifestations such as partial or complete dislocations of the small joints of hands and feet, clawed toes, periosteal thickening, joint effusion and hyperpathia, and periarticular manifestations (Asbrink *et al.*, 1986).

Chronic sclerodermic skin manifestations

B. afzelii strains have been isolated from the lesional skin of some cases of morphoea (circumscribed scleroderma) (Breier *et al.*, 1999, 2001).

Neuroborreliosis

Early neurological manifestations

The full picture of neuroborreliosis is defined as meningo-polyradiculoneuritis, also named Bannwarth syndrome. All age groups may be affected, although painful radiculoneuritis has only very rarely been observed in children. Several studies have reported that the median age for Bannwarth syndrome is the fifth decade, ranging from 6 to 80 years. The incubation period for neuroborreliosis is on average 7 weeks in adults and 4 weeks in children. In adults, this disorder often presents as a triad, which includes radicular pain, peripheral pareses, most frequently facial palsy, and lymphoplasmocytic pleocytosis in the cerebrospinal fluid with a cell count of up to 1000×10^6 l^{-1}. The majority of cases of early neuroborreliosis occur between July and November, with most cases observed in August (Kristoferitsch, 1989). Nevertheless, the disorder is diagnosed throughout the year.

The clinical picture of Bannwarth syndrome begins with pain because of radiculoneuritis. The character of the pain varies, is usually severe, and patients describe it as burning, tearing and migrating. A characteristic of neuroborreliosis pain is its aggravation at night. Also remarkable is that this pain is refractory to normal dosages of analgesics. Furthermore, the skin at the site of the pain may be dysaesthetic and hyperpathic. The slightest pressure by contact of cloth, for example, may cause severe burning pain. Pain constituted the only symptom in up to 20% of patients. Because of its severity, about one-third of patients become depressed, agitated, restless, suffer from sleeplessness and are full of anxiety. Even autoaggressive behaviour has been observed. Patients with pain syndrome alone have been certified insane as a result of personality changes and misdiagnosis. In most cases, however, the onset of pareses, especially cranial nerve involvement 1 or 2 weeks after onset of pain, leads to the correct diagnosis. Before the discovery of the agent of LB and the introduction of specific antibiotic treatment, the syndrome was observed in its spontaneous course, which lasted for up to 25 weeks until complete recovery or recovery with incomplete resolution of nerve function occurred.

Neuroborreliosis in children presents predominantly as acute facial palsy and/or acute aseptic meningitis. Bannwarth syndrome was observed in fewer than 4% of all cases studied, and mostly in older children (Christen et al., 1993).

Chronic neuroborreliosis

The major manifestations of chronic neuroborreliosis are encephalitis, radiculomyelitis, transverse myelitis, stroke-like disorders and cranial nerve deficits. The condition is very rare, and diagnosis has to fulfil strict diagnostic criteria with other diagnostic considerations including neurosyphilis, fungal meningoencephalitis and brain tumour (Wokke et al., 1987; Ackermann et al., 1988).

Joint manifestations

In the USA, arthritis was the predominant manifestation that led to the discovery of this complex borrelial infection. Episodes of frank arthritis occurred in about 50% of patients who did not receive antibiotic treatment for EM (Steere et al., 1987). In Europe, the situation is different and arthritis is not one of the major complications. Lyme arthritis may occur within several months after an unrecognized primary infection and the resolution of an untreated EM. It occurs predominantly in the fourth decade and, if in childhood, older children are more often affected. The joints most frequently involved are the knee in about half of all cases, and the ankle, wrist, elbow and rarely smaller joints (Herzer, 1991). Lyme arthritis is caused by an intra-articular infection (Priem et al., 1998). A small number of bacteria are liable to provoke severe arthritis by inducing mechanisms (including the induction of cytokines and chemokines) that amplify the inflammatory response (Franz et al., 1999).

Cardiac manifestations

Acute cardiac involvement is a rarely observed manifestation that may present as conduction or rhythm disturbance, and occasionally as acute myopericarditis or mild left ventricular dysfunction (Steere, 2001). The condition occurs mostly in the course of an EM or within a few weeks after onset of infection, and appears to be transient and self-limiting. In Europe, however, *B. burgdorferi* s.s. was isolated from the endo-myocardium of a patient suffering from longstanding dilated cardiomyopathy (Stanek et al., 1990).

Lyme Borreliosis in Domestic Animals

So far this review has considered LB only in humans. LB is categorized as a zoonotic disease in humans because the infection is maintained in nature by animals; humans are infected accidentally and are dead-end hosts for *B. burgdorferi* s.l. The animal reservoirs of *B. burgdorferi* s.l. are many and varied (see Kurtenbach *et al.*, Chapter 5) and the majority are wild small mammals and birds. In common with many animal reservoirs of zoonotic infections, these animals do not seem to show signs of disease as a result of infection with *B. burgdorferi* s.l., but other animals, particularly domestic animals, can become victims of this abnormal host/pathogen relationship and may show similar manifestations to those that occur in humans.

Most data on LB in domestic animals concern dogs and horses, with isolated reports of infection in cattle, sheep and cats. Anti-borrelial antibodies have been detected in a wide range of domestic animal species, but it is probable that most infections are subclinical and self-limiting. It is also possible that infections are missed or misdiagnosed because of the non-specificity of clinical symptoms. Domestic animals do not show any pathognomic symptoms such as EM and there are no case definitions for LB in animals. There are also considerable difficulties involved in laboratory diagnosis. In contrast with diagnostic methods for human LB, little quality assurance has been applied to the laboratory diagnosis of animal LB. Furthermore, it should be borne in mind that the tick vectors of LB transmit other pathogens, which may confound both clinical and laboratory diagnosis of LB (Gray, 1999).

Lyme borreliosis in dogs

Canine LB was first reported in the original LB focus and neighbouring regions in the USA (Lissman *et al.*, 1984; Kornblatt *et al.*, 1985). Typical signs are lameness combined with malaise. Fever may accompany the malaise, which manifests as listlessness and inappetance, and the lameness may be extreme claudication with localized painful joints or a general stiff gait without clear pain localization. Kidney involvement (Dambach *et al.*, 1997) and heart block (Levy and Duray, 1988) have also been reported.

Diagnosis of canine LB requires evidence of exposure to ticks and the presence of typical clinical signs together with antibodies to *B. burgdorferi* s.l. As with all LB serology, problems arise with specificity and cross-reactions with other spirochaetal organisms such as *Treponema* spp. Positive serology is not diagnostic on its own since most infections are evidently subclinical, though serosurveys have shown that prevalence of *B. burgdorferi* s.l. antibodies is usually higher in symptomatic dogs compared with healthy ones (Hovius, 1999). A positive response to antibiotics (e.g. tetracyclines) is also regarded as supportive of the diagnosis, but there is evidence of persistence of clinical signs despite treatment (Straubinger *et al.*, 1998). Increasing use is being made of PCR to support diagnosis (Appel *et al.*, 1993; Hovius *et al.*, 1999) but definitive demonstration of the organism by culture is rarely successful.

In Europe serosurveys suggest widespread infection (Hovius, 1999) and there are several reports where typical symptoms have been accompanied by elevated antibody titres (Font *et al.*, 1992; McKenna *et al.*, 1995; Overduin and van den Bogaard, 1997; Hovius *et al.*, 2000). *B. burgdorferi* s.l. can be detected in symptomatic dogs by PCR (Hovius *et al.*, 1999), but so far only one study has reported diagnosis supported by isolation and identification of *B. burgdorferi* s.l. (Liebisch and Liebisch, 1999). More recently *B. afzelii* was isolated from the blood of a dog with neurological abnormalities, but the dog was euthanased before any definite association between the infection and the symptoms could be established (Speck *et al.*, 2001).

Despite the difficulties associated with unequivocal diagnosis, clinical canine LB is considered sufficiently prevalent in the USA and Europe for the marketing of both inactivated whole-cell and recombinant vaccines (Wormser, 1999).

Lyme borreliosis in cats

LB has not been reported in cats exposed to natural infection though anti-*B. burgdorferi* antibodies have been detected (Magnarelli *et al.*, 1990; Oliver *et al.*, 1999). In one study, experimental infections resulted in seroconversion but not clinical manifestations (Burgess, 1992), whereas, in another, clinical manifestations were reported that apparently resembled those in humans sufficiently closely for this feline model to be suggested for vaccine testing (Gibson *et al.*, 1995). There are insufficient data for evaluation of the significance of LB in cats but the latter observation suggests that these animals may be affected where they are exposed to a heavy tick challenge.

Lyme borreliosis in horses

Suspected equine LB was reported in endemic areas in the USA a few years after identification of *B. burgdorferi* as the causal agent of the human disease (Burgess *et al.*, 1986; Burgess and Mattison, 1987; Magnarelli *et al.*, 1988). Serosurveys have shown that the infection is widespread, and that seroprevalence may be as high as 45% in some areas (Magnarelli *et al.*, 2000). Serosurveys conducted in Europe have also shown high seroprevalence in some countries (Käsbohrer and Schönberg, 1990; Gerhards and Wollanke, 1996; Egenvall *et al.*, 2001; Müller *et al.*, 2001).

Clinical signs of LB in horses consist mainly of malaise, low grade fever, stiffness, laminitis and swollen joints, and there are also several reports of eye involvement such as uveitis (Burgess *et al.*, 1986; Parker and White, 1992; Hahn *et al.*, 1996), two of cardiac problems (Browning *et al.*, 1993; Liebisch *et al.*, 1999) and at least one of encephalitis (Burgess and Mattison, 1987). However, clinical cases appear to be uncommon and, although some may be missed because of the non-specific nature of symptoms, it is apparent that in the vast majority of cases infection is asymptomatic. No clinical signs arose even when infection was induced experimentally with infected nymphal ticks in immunosuppressed animals (Chang *et al.*, 2000). There is no evidence that horses can be a source of infection for ticks.

Despite increased awareness, improvement of serological diagnostics and the advent of molecular methods (Manion *et al.*, 1998), unequivocal diagnosis of equine LB is still very difficult. It is only recently that modified Koch's postulates have been satisfied in the diagnosis of equine LB, in this case involving the presence of appropriate clinical signs in four animals exposed to ticks in Germany, successful isolation of spirochaetes and their identification by PCR as *B. burgdorferi* s.l. (Liebisch *et al.*, 1999). However, the PCR product was apparently not sequenced and the evident rapidity of the growth of the isolated spirochaete culture has led to some doubts about its identity.

There is some interest in the development of vaccines against equine LB, particularly in the USA (Chang *et al.*, 1999), but, in view of the diagnostic difficulties and uncertain risk of disease in horses, justification for this preventive measure may require further attention.

Lyme borreliosis in cattle

Several cases of clinical LB in cattle reported in the USA in the late 1980s and early 1990s were diagnosed by association of symptoms such as lameness, weight loss and abortion, with positive serology, isolation of *B. burgdorferi* from colostrum, serum, urine, milk or synovial fluids, apparent response to tetracyclines and seroconversion in mice and a cat after exposure to bovine urine and milk, respectively (Burgess *et al.*, 1987; Burgess, 1988; Post *et al.*, 1988; Wells *et al.*, 1993). The apparent infectivity of milk could have profound implications for human health, but, despite continued interest in this topic, no more evidence has been forthcoming and it must be concluded that milk from infected animals (or humans) is not infective. In fact, very few cases of LB in cattle have been diagnosed subsequently, though isolated reports appeared from Australia (Rothwell *et al.*, 1989), Japan (Isogai *et al.*, 1992) and the UK (Blowey *et al.*, 1994). Doubts concerning the specificity of the serology have been reinforced by studies implicating *Treponema* spp. as the causal agent of *B. burgdorferi* seropositivity of cattle with digital dermatitis (Choi *et al.*, 1997; Demirkan *et al.*, 1999) and cross-reactions with other spirochaetes have also been recorded (Rogers *et al.*, 1999).

When cattle were experimentally infected with three pathogenic genospecies of *B. burgdorferi* s.l. (*B. afzelii*, *B. burgdorferi* s.s. and *B. garinii*) no clinical symptoms resulted (Tuomi *et al.*, 1998), though infection by tick bite might have produced a different result. It appears likely that *B. burgdorferi* s.l. is usually non-pathogenic in cattle. However, a recent report strongly suggests that at least two genospecies may be pathogenic under certain circumstances (Lischer *et al.*, 2000). In this

Swiss study of two dairy cows, the clinical signs included erythema on the udder, loss of weight, swollen legs, swollen joints, stiff gait, acute laminitis and oligoarthritis. In order to decrease the possibility of erroneous results arising through contamination, real-time PCR was used as a diagnostic tool and *B. burgdorferi* s.s. DNA was detected in synovial fluid and milk of one cow and *B. afzelii* DNA in synovial fluid of the other.

LB in cattle is probably an infrequent disease that is difficult to diagnose due to the non-specific nature of symptoms and doubt surrounding the specificity of serology. The application of PCR technology as a diagnostic aid may result in more certain identification of cases in the future.

Cattle are not regarded as reservoir hosts of *B. burgdorferi* s.l. (Gern *et al.*, 1998) and the association of infected ticks with cattle pastures (Borko and Bole-Hribovsek, 1998) is probably due to the presence of other hosts in the same habitat (Gray *et al.*, 1995).

Lyme borreliosis in sheep

In the only report of possible LB in sheep, the suspected disease manifested in seropositive Norwegian lambs as pronounced lameness in one leg (Fridriksdottir *et al.*, 1992). In the absence of other reports it must be concluded that *B. burgdorferi* s.l. infection of sheep rarely results in disease. There are several reports of the detection of anti-borrelial antibodies in the blood of sheep (e.g. Hovmark *et al.*, 1986b; Fridriksdottir *et al.*, 1992; Ciceroni *et al.*, 1996), but, as discussed above, the same caveats regarding the specificity of the tests apply here.

The susceptibility of sheep to systemic infection appears to be low (Stuen and Fridriksdottir, 1991), which limits the ability of *B. burgdorferi* s.l. to cause disease and also reduces the proportion of infected ticks in the immediate environment (Gray *et al.*, 1995), thus limiting the risk to farmers. There is some evidence that adult ticks may become infected as the result of the transfer of the pathogens from infected to uninfected nymphal ticks co-feeding on sheep (Ogden *et al.*, 1997), but the impact that this might have on the ecology and epidemiology of the disease is unknown.

Lyme borreliosis in other livestock

Antibodies to *B. burgdorferi* s.l. have been detected in the blood of goats (Doby and Chevrier, 1990; Ciceroni *et al.*, 1996) but there are no reports of LB in these animals. Seroprevalence in domestic pigs has not been investigated, though wild boar have been found to be seropositive (Doby *et al.*, 1991). Studies on the susceptibility of deer to *B. burgdorferi* s.l. suggest that infected animals do not exhibit clinical signs though seroconversion occurs (Lane *et al.*, 1994; Luttrell *et al.*, 1994).

Chickens have been infected experimentally and it was found that these animals quickly became immune to *B. burgdorferi* s.s. and did not show any clinical symptoms (Piesman *et al.*, 1996). More recent studies have shown that pheasants can function as reservoir hosts of *B. garinii* and *Borrelia valaisiana* in the UK (Kurtenbach *et al.*, 1998), but no symptoms of disease in infected birds have been reported.

In summary, there is no doubt that LB occurs in some domestic animals, and serological studies suggest that infection with *B. burgdorferi* s.l. is common. However, at present there is little evidence that LB is responsible for frequent and significant morbidity in any domestic animal species, with the possible exception of the dog. Diagnosis of LB in domestic animals is particularly problematic because of the lack of case definitions and of quality-assured laboratory tests.

Diagnostic Problems

Clinical diagnosis of Lyme borreliosis

Clinical signs and symptoms should always form a basis for diagnosis of human LB (Strle, 1999). Thus, a solid knowledge of clinical features is essential and case definitions (Stanek *et al.*, 1996; Anon., 1997; Nadelman and Wormser, 1998) could be of help in everyday clinical practice and especially for comparison of the findings of different authors or groups. Caution should be used when clinical and laboratory experience from one side of the Atlantic Ocean is applied to patients on the other (Strle *et al.*, 1999).

A characteristic sign that enables diagnosis of LB in clinical practice is the typical EM skin lesion, and highly suggestive manifestations are ear lobe lymphocytoma, ACA and Bannwarth syndrome. The majority of the many other signs and symptoms are only suggestive and when expressed individually may have a very limited value for clinical diagnosis.

Laboratory evidence of borrelial infection

Laboratory confirmation of borrelial infection is, as a rule, needed for all manifestations of LB with the exception of typical EM (Tugwell *et al.*, 1997; Nadelman and Wormser, 1998; Strle *et al.*, 1999). In clinical practice indirect laboratory methods are widely used to support the diagnosis of borrelial infection. Determination of borrelial IgM and IgG antibodies by immunofluorescence assays or ELISA has not been standardized, and the correlation of results from different laboratories and/or different commercial tests may be poor. Immunoblotting may solve some of the many dilemmas but could (especially in Europe) add additional questions to the field in which many uncertainties still exist (Robertson *et al.*, 2000).

A clinician should understand the sensitivity and specificity of these tests, and be aware of the predictive values for certain clinical manifestations. However, in general, there have not been clear-cut answers to such questions. In addition, even if a clinician relied on the results of serological tests, several important problems remain with their interpretation. There is no doubt that we need (but do not have) standardized and dependable tests to ascertain borrelial infection, with the ability to distinguish between recent and long-lasting infection, and also between active and inactive infection.

T-cell proliferation assay results are also difficult to interpret, and apparently the specificity and sensitivity of T-lymphocyte recognition by *B. burgdorferi* s.l. antigens remain controversial (Strle *et al.*, 1999). There have also been several problems with tests for direct detection of borrelial infection. The reliability of methods other than culture to detect spirochaetes in tissue specimens is open to question. Although *B. burgdorferi* s.l. grows relatively well under laboratory conditions, the spirochaete is not easily recovered from clinical specimens other than skin biopsy samples of EM. Culture results are as a rule available only several weeks after the material has been obtained (Wilske and Preac-Mursic, 1993), which is not very desirable for clinicians or for their patients. Moreover, the procedure itself is time-consuming and relatively expensive. However, for a new or unusual clinical manifestation to be proclaimed as a sign of LB, borrelial infection should be firmly ascertained – isolation of *B. burgdorferi* s.l. from the tissue (organ) involved is the best possible approach to such cases. Sometimes the question arises as to which specimen has the highest chance for successful culture. For example, for isolation of *B. burgdorferi* s.l. from EM, a skin biopsy on the periphery of the lesion is usually performed (Strle *et al.*, 1996). However, in reviewing previous work on this topic, conclusive evidence was not found that the peripheral aspect of the lesion represents the optimal site for biopsy. Results of a prospective study in which the culture isolation rate of *B. burgdorferi* s.l. from the centre and the margin of typical EM skin lesions was compared (Jurca *et al.*, 1998) did not reveal any significant difference in the isolation rate. However, in a continuation of the study, a significant difference was found in favour of central biopsy sites in patients with homogeneous EM lesions (Jurca *et al.*, 1998).

Detection of borrelial DNA by PCR has been acclaimed for its high sensitivity and specificity. However, the value of detection of borreliae by this method in human tissues and fluids has not been unequivocally established (Tugwell *et al.*, 1997; Nadelman and Wormser, 1998). In a 2-year prospective study undertaken in Ljubljana, the detection rate of borreliae by culture and PCR was compared using skin specimens from patients with untreated or treated skin lesions of EM and ACA. Results for 758 typical EM specimens showed positive results in 36% for culture in modified Kelly-Pettenkofer medium, 25% for PCR and 24% for culture in BSKII medium. The differences are statistically significant. The overall positivity rate for all methods combined was 54%, but few specimens were positive by all three methods (only 6%). Results for typical ACA were similar (Picken *et al.*, 1997). These findings suggest that the distribution of spirochaetes in skin biopsies is not homogeneous and that, although possessing the potential to provide a rapid diagnosis, PCR is not more sensitive than culture for direct detection of borreliae in skin (Picken *et al.*, 1997).

Treatment

Basic principles of treatment

A well-established principle in medicine maintains that a reliable diagnosis forms a basis for rational treatment. However, in LB, with its lack of clinical specificity compounded by the lack of standardized serological tests and direct detection procedures, a reliable diagnosis is usually not an easily achievable goal.

In general, treatment with antibiotics is efficacious in all stages of LB and for all clinical manifestations; however, it has been most effective early in the course of the illness (Steere, 2001). Choice of antibiotic depends upon many factors including the efficacy, pharmacokinetic profile, side effects, expected compliance and price. The most effective antibiotic, the optimal dosage and the most appropriate duration of treatment have not been precisely determined for any of the many clinical manifestations of the disease. However, those uncertainties should not and cannot be the reason for the extent of the disparities in treatment recommendations that have been evident in the last few years. Some physicians treat only patients who fulfil all rigorous criteria for established LB. Others treat all poorly defined 'chronic' LB patients with repeated or prolonged courses of antibiotics, or sometimes even with combinations of antibiotics that are ineffective *in vitro*.

In the last few years a trend of prolongation of treatment of LB has been observed. The reasons for this have been predominantly based on disappointments with the results of treatment according to current recommendation, sometimes on the basis of reports of anecdotal experiences and not on the results of controlled studies. However, it is reasonable to expect that the recommendations for treatment will continue to change and there are indications that shorter treatment regimens rather than longer ones will prove to be the correct approach (Klempner *et al.*, 2001). Current shortcomings in therapy should be a stimulus for well-designed studies to answer these unresolved questions.

There are several potential reasons for antibiotic treatment failure in LB. One of the possible causes for treatment failure is persistence of borreliae in tissues. There are clinical and experimental data, predominantly in European literature, showing that, after presumed adequate (that is, recommended) treatment, borreliae can persist in tissue (Preac-Mursic *et al.*, 1996; Straubinger *et al.*, 1997); however, it is difficult to assess the magnitude of this problem. In some cases a possible explanation for treatment failure could be irreversible tissue damage caused during active borrelial infection, or inflammation in association with this infection, or the induction of autoimmune mechanisms; in the last two cases treatment with an antibiotic will be ineffective.

An important and probably a common cause for treatment failure is misdiagnosis. It is quite possible that treating a patient with non-specific symptoms such as arthralgia and myalgia and with apparently detectable borrelial serum antibodies will fail, because this may not be treatment of LB but of patients with certain serological test results.

No reliable data support the use of antibiotics in asymptomatic persons with positive borrelial serological test results. A decision to treat with an antibiotic should be based on the presence of at least suggestive clinical signs and symptoms. Since IgM and IgG antibodies to *B. burgdorferi* s.l. may persist for years after clinical recovery (Wilske *et al.*, 2000), the role of serology in measuring response to treatment is dubious (Nadelman and Wormser, 1998). According to present knowledge, repeated courses of (prolonged) antibiotic treatment are not indicated. In general, the main indication for intravenous therapy (usually with ceftriaxone) should be based on proven or suspected involvement of the central nervous system; for the large majority of other manifestations, oral therapy is adequate (Nadelman and Wormser, 1998). All antibiotics have potential adverse effects and some patients have had serious or fatal complications during treatment of poorly substantiated LB.

Treatment of Cases

Two recent reviews have addressed this topic in some detail (Strle *et al.*, 1999; Wormser *et al.*, 2000). In central Europe, various antibiotics and antimicrobial chemotherapeutics are used for the therapy of the different manifestations of LB. The recommended duration of treatment ranges from 2 to 4 weeks. However, the necessary duration of the treatment of LB could be shorter and we have to await results of further studies on this subject. Patients with solitary EM and borrelial lymphocytoma are treated with amoxicillin, azithromycin, doxycycline, cefuroxime-axetil or phenoxymethylpenicillin but only very exceptionally with erythromycin. The usual duration of treatment is 14 days. Nervous system involvement and severe Lyme carditis are ordinarily treated with ceftriaxone or penicillin G for 2–3 weeks and only exceptionally orally with doxycycline or amoxicillin. Ceftriaxone, penicillin G, doxycycline or amoxicillin are used for the therapy of ACA and arthritis. Parenteral therapy is used when there is concomitant or peripheral central nervous system involvement and may also be useful for treatment of EM in pregnancy (Maraspin *et al.*, 1996) and in immunodeficient patients. Duration of treatment is usually 3 weeks for ACA, 2 weeks for arthritis and 4 weeks in the case of oral therapy for arthritis. In most cases, Lyme arthritis may be cured by antibiotic therapy, but about 10% of Lyme arthritis patients do not respond satisfactorily (Franz *et al.*, 1999).

Prophylaxis of Lyme borreliosis

Chemoprophylaxis of LB, that is, administration of antibiotics after a tick bite to prevent infection, is not standard practice for several reasons. Firstly, the infection rate after a tick bite is low, between 1 and 2%, and, secondly, the rate of preventing eventual LB is not predictable (Stanek and Kahl, 1999). However, in a prospective study with adults bitten by ticks in endemic areas of North America, a single dose of 200 mg doxycycline was effective for the prevention of LB, provided the antimicrobial was given within 72 h after the tick bite (Nadelman *et al.*, 2001).

Immunoprophylaxis of LB with a recombinant OspA vaccine was introduced in the USA in the mid-1990s. This vaccine is designed to prevent infection by destroying *B. burgdorferi* s.s. in the midgut of the feeding tick. Other vaccines are being designed and studied that would be appropriate in preventing infections with the more heterogeneous agents of LB in Europe (see Hayes and Schrider, Chapter 11).

References

Ackermann, R., Rehse-Küpper, B., Gollmer, E. and Schmidt, R. (1988) Chronic neurologic manifestations of erythema migrans borreliosis. *Annals of the New York Academy of Sciences* 539, 16–23.

Afzelius, A. (1910) Verhandlungen der Dermatologischen Gesellschaft zu Stockholm, 28 Oct 1909. *Archives of Dermatology and Syphilis* 101, 404.

Anon. (1995) *Report of WHO Workshop on Lyme Borreliosis Diagnosis and Surveillance.* WHO/CDS/VPH/ 95.14-1, Warsaw, Poland, 220 pp.

Anon. (1997) Case definitions for infectious conditions under public health surveillance: Lyme disease (revised 9/96). *Mortality and Morbidity Weekly Report* 46 (suppl. RR-10), 20–21.

Appel, M.J., Allan, S., Jacobson, R.H., Lauderdale, T.L., Chang, Y.F., Shin, S.J., Thomford, J.W., Todhunter, R.J. and Summers, B.A. (1993) Experimental Lyme disease in dogs produces arthritis and persistent infection. *Journal of Infectious Diseases* 167, 651–664.

Asbrink, E. and Hovmark, A. (1985) Successful cultivation of spirochetes from skin lesions of patients with erythema chronicum migrans Afzelius and acrodermatitis chronica atrophicans. *Acta Pathologica Microbiologica et Immunologica Scandinavica (B)* 93, 161–163.

Asbrink, E., Hovmark, A. and Olsson, I. (1986) Clinical manifestations of acrodermatitis chronica atrophicans in 50 Swedish patients. *Zentralblatt für Bakteriologie und Hygiene A* 263, 253–261.

Axford, J.S. and Rees, D.H.E. (eds) (1994) *Lyme Borreliosis.* Life Sciences Vol. 260, Plenum Press, New York, 344 pp.

Bäferstedt, B. (1943) Über Lymphadenosis benigna cutis. *Acta Dermato-Venereologica (Stockholm) Suppl.* 24, 1–202.

Bannwarth, A. (1941) Chronische lymphozytäre Meningitis, entzündliche Polyneuritis und 'Rheumatismus'. *Archiv Psychiatrischer Nerven krankheiten* 113, 284–376.

Baranton, G., Marti Ras, N.N. and Postic, D. (1998) Molecular epidemiology of the aetiological agents of Lyme borreliosis. *Wiener Klinische Wochenschrift* 110, 850–855.

Barbour, A.G. (1984) Isolation and cultivation of Lyme disease spirochetes. *Yale Journal of Biology and Medicine* 57, 521–525.

Barbour, A.G. and Fish, D. (1993) The biological and social phenomenon of Lyme disease. *Science* 260, 1610–1616.

Barbour, A.G. and Schrumpf, M.E. (1986) Polymorphisms of major surface proteins of *Borrelia burgdorferi*. *Zentralblatt für Bakteriologie und Hygiene A* 263, 83–91.

Benach, J.L. and Bosler, E.M. (eds) (1988) Lyme disease and related disorders. *Annals of the New York Academy of Sciences* 539, 1–513.

Benach, J.L., Bosler, E.M., Hanrahan, J.P., Coleman, J.L., Bast, T.F., Habicht,

G.S., Cameron, D.J., Ziegler, J.L., Burgdorfer, W., Barbour, A.G., Edelman, R. and Kaslow, R.A. (1983) Spirochetes isolated from the blood of two patients with Lyme disease. *New England Journal of Medicine* 308, 740–742.

Blowey, R.W., Carter, S.D., White, A.G. and Barnes, A. (1994) *Borrelia burgdorferi* infections in UK cattle: a possible association with digital dermatitis. *Veterinary Record* 135, 577–578.

Borko, C. and Bole-Hribovsek, V. (1998) Infection rate of the tick *Ixodes ricinus* (Ixodidae, Acarina) with bacteria *Borrelia burgdorferi* sensu lato and in connection with the infection rate in grazing cattle. *Zbornik Veterinarske Fakultete Univerza Ljubljana* 36, 173–180.

Breier, F.H., Aberer, E., Stanek, G., Khanakah, G., Schlick, A. and Tappeiner, G. (1999) Isolation of *Borrelia afzelii* from circumscribed scleroderma. *British Journal of Dermatology* 140, 925–930.

Breier, F., Khanakah, G., Stanek, G., Kunz, G., Aberer, E., Schmidt, B. and Tappeiner, G. (2001) Isolation and polymerase chain reaction typing of *Borrelia afzelii* from a skin lesion in a seronegative patient with generalized ulcerating bullous lichen sclerosus et atrophicus. *British Journal of Dermatology* 144, 387–392.

Browning, A., Carter, S.D., Barnes, A., May, C. and Bennett, D. (1993) Lameness associated with *Borrelia burgdorferi* infection in the horse. *Veterinary Record* 132, 610–611.

Buchwald, A. (1883) Ein Fall von diffuser idiopathischer Haut Atrophie. *Archives of Dermatology and Syphilis* 10, 553–556.

Burckhardt, J.L. (1911) Zur Frage der Follikel- und Keimzentrenbildung in der Haut. *Frankfurter Zeitschrift für Pathologie* 6, 352–359.

Burgdorfer, W., Barbour, A.G., Hayes, S.F., Benach, J.L., Grunwaldt, E. and Davis, J.P. (1982) Lyme disease – a tick-borne spirochetosis? *Science* 216, 1317–1319.

Burgess, E.C. (1988) *Borrelia burgdorferi* infection in Wisconsin horses and cows. *Annals of the New York Academy of Sciences* 539, 235–243.

Burgess, E.C. (1992) Experimentally induced infection of cats with *Borrelia burgdorferi*. *American Journal of Veterinary Research* 53, 1507–1511.

Burgess, E.C. and Mattison, M. (1987) Encephalitis associated with *Borrelia burgdorferi* infection in a horse. *Journal of the American Veterinary Medical Association* 191, 1457–1458.

Burgess, E., Gillette, D. and Pickett, J.P. (1986) Arthritis and panuveitis as manifestations of *Borrelia burgdorferi* infection in a Wisconsin pony. *Journal of the American Veterinary Medical Association* 186, 1340–1342.

Burgess, E.C., Gendron-Fitzpatrick, A. and Wright, W.O. (1987) Arthritis and systemic disease caused by *Borrelia burgdorferi* infection in a cow. *Journal of the American Veterinary Medical Association* 191, 1468–1470.

Cevenini, R., Sambri, V. and La Placa, M. (eds) (1994) *Advances in Lyme Borreliosis Research*. Societa Editrice Esculapio, Bologna, 239 pp.

Chang, Y., Novosol, V., McDonough, S.P., Chang, C.F., Jacobson, R.H., Divers, T., Quimby, F.W., Shin, S. and Lein, D.H. (1999) Vaccination against Lyme disease with recombinant *Borrelia burgdorferi* outer-surface protein A (rOspA) in horses. *Vaccine* 18, 540–548.

Chang, Y.E., Nosovol, V., McDonough, S.P., Chang, C.F., Jacobson, R.H., Divers, T., Quimby, F.W., Shin, S. and Lein, D.H. (2000) Experimental infection of ponies with *Borrelia burgdorferi* by exposure to ixodid ticks. *Veterinary Pathology* 37, 68–76.

Choi, B.K., Natterman, H., Grund, S., Haider, W. and Gobel, U.B. (1997) Spirochetes from digital dermatitis lesions in cattle closely related to treponemes associated with human periodontitis. *International Journal of Systematic Bacteriology* 47, 175–181.

Christen, H.J., Hanefeld, F., Eiffert, H. and Thomssen, R. (1993) Epidemiology and clinical manifestations of Lyme borreliosis in childhood. *Acta Paediatrica* 82 (suppl. 386), 1–76.

Ciceroni, L., Simeoni, J., Pacetti, A.I., Ciarrocchi, S. and Cacciapuoti, B. (1996) Antibodies to *Borrelia burgdorferi* in sheep and goats. Alto Adige-South Tyrol, Italy. *Microbiologica* 19, 171–174.

Dambach, D.M., Smith, C.A., Lewis, R.M. and van Winkle, T.J. (1997) Morphologic, immunohistochemical, and ultrastructural characterization of a distinctive renal-lesion in dogs putatively associated with *Borrelia burgdorferi* infection: 49 cases (1987–1992). *Veterinary Pathology* 34, 85–96.

Demirkan, I., Walker, R.I., Murray, R.D., Blowey, R.W. and Carter, S.D. (1999) Serological evidence of spirochaetal infections associated with digital dermatitis in dairy cattle. *Veterinary Journal* 157, 69–77.

Doby, J.M. and Chevrier, S. (1990) Recherche des anticorps anti-*Borrelia burgdorferi* chez les caprins en Bretagne. Bilan serologique sur 602 chevres. *Recueil de Médecine Vétérinaire* 166, 799–804.

Doby, J.M., Betremieux, C., Rolland, C. and Barrat, J. (1991) Les grands mammifères forestiers, réservoirs de germe pour *Borrelia burgdorferi*, agent de la maladie de Lyme? Recherche des anticorps chez 543 cervides et suides. *Recueil de Médecine Vétérinaire* 167, 55–61.

Egenvall, A., Franzen, P., Gunnarsson, A., Engvall, E.O., Vagsholm, I., Wikstrom, U.B. and Artursson, K. (2001) Cross-sectional study of the seroprevalence to *Borrelia burgdorferi* sensu lato and granulocytic *Ehrlichia* spp. and demographic, clinical and tick-exposure factors in Swedish horses. *Preventive Veterinary Medicine* 49, 191–208.

Font, A., Closa, J.M. and Mascort, J. (1992) Lyme disease in dogs in Spain. *Veterinary Record* 130, 227–228.

Franz, J.K., Priem, S., Rittig, M.G., Burmester, G.R. and Krause, A. (1999) Studies on the pathogenesis and treatment of Lyme arthritis. *Wiener Klinische Wochenschrift* 111, 981–984.

Fridriksdottir, V., Overnes, G. and Stuen, S. (1992) Suspected Lyme borreliosis in sheep. *Veterinary Record* 130, 323–324.

Garin, C. and Bujadoux, R. (1922) Paralysie par les tiques. *Journal de Médecine de Lyon* 71, 765–767.

Gerhards, H. and Wollanke, B. (1996) Antikorpertiter gegen Borrelien bei Pferden im Serum und im Auge und Vorkommen der equinen rezidivierenden Uveitis (ERU). *Berliner und Münchener Tierarztliche Wochenschrift* 109, 273–278.

Gern, L., Estrada-Pena, A., Frandsen, F., Gray, J.S., Jaenson, T., Jongejan, F., Kahl, O., Korenberg, E., Mehl, R. and Nuttall, P. (1998) European reservoir hosts of *Borrelia burgdorferi* s.l. *Zentralblatt für Bakteriologie* 283, 196–204.

Gibson, M.D, Omran, M.T. and Young, C.R. (1995) Experimental feline Lyme borreliosis as a model for testing *Borrelia burgdorferi* vaccines. *Advances in Experimental Medicine and Biology* 383, 73–82.

Götz, H. (1954) Die Acrodermatitis chronica atrophicans Herxheimer als Infektionskrankheit. *Hautarzt* 5, 491–504.

Götz, H. (1955) Die Acrodermatitis chronica atrophicans Herxheimer als Infektionskrankheit. *Hautarzt* 6, 294–252.

Grann, D. (2001) Stalking Dr. Steere over Lyme disease. *The New York Times*, Magazine Desk, 17 June 2001.

Gray, J.S. (1999) Tick-borne pathogen interactions. *Infectious Disease Review* 1, 117–121.

Gray, J.S. and Stanek, G. (eds) (1998) Risk assessment in Lyme Borreliosis, a 'concerted action' in the EU Biomedicine and Health Programme (1993–1996). *Zentralblatt für Bakteriologie und Hygiene* 287, 175–269.

Gray, J.S., Kahl, O, Janetzki, C, Stein, J. and Guy, E. (1995) The spatial distribution of *Borrelia burgdorferi*-infected *Ixodes ricinus* in the Connemara region of Co. Galway, Ireland. *Experimental and Applied Acarology* 19, 163–172.

Hahn, C.N., Mayhew, I.G., Whitwell, K.E., Smith, K.C., Carey, D., Carter, S.D. and Read, R.A. (1996) A possible case of Lyme borreliosis in a horse in the UK. *Equine Veterinary Journal* 28, 84–88.

Hauser, W. (1955) Zur Kenntnis der Akrodermatitis chronica atrophicans. *Archives of Dermatology and Syphilis* 199, 350–393.

Hellerström, S. (1930) Erythema chronicum migrans Afzelii. *Acta Dermato-Venereologica (Stockholm)* 11, 315–321.

Herxheimer, K. and Hartmann, K. (1902) Über Acrodermatitis chronica atrophicans. *Archives of Dermatology and Syphilis* 61, 57–76.

Herzer, P. (1991) Joint manifestations of Lyme borreliosis in Europe. *Scandinavian Journal of Infectious Diseases* 77, 55–63.

Hollström, E. (1958) Penicillin treatment of erythema chronicum migrans Afzelius. *Acta Dermato-Venereologica (Stockholm)* 38, 285–289.

Hopf, H.C. (1966) Acrodermatitis chronica atrophicans (Herxheimer) und Nervensystem. In: *Monographien aus dem Gesamtgebiet der Neurologie und Psychiatrie*, Vol. 114, Springer, Berlin, pp. 1–130.

Hörstrup, P. and Ackermann, R. (1973) Durch Zecken übertragene Meningopolyneuritis (Garin-Bujadoux, Bannwarth). *Fortschritte der Neurologie Psychiatrie* 41, 583–606.

Hovius, J.W., Hovius, K.E., Oei, A., Houwers, D.J. and van Dam, A.P. (2000) Antibodies against specific proteins of and immobilizing activity against three strains of *Borrelia burgdorferi* sensu lato can be found in symptomatic but not in infected asymptomatic dogs. *Journal of Clinical Microbiology* 38, 2611–2621.

Hovius, K.E. (1999) Borrelia infections in dogs: epidemiological, clinical and diagnostic aspects. PhD thesis, University of Utrecht, The Netherlands.

Hovius, K.E., Stark, L.A., Bleumink-Pluym, N.M., van de Pol, I., Verbeek-de Kruif, N., Rijpkema, S.G., Schouls, L.M. and Houwers, D.J. (1999) Presence and distribution of *Borrelia burgdorferi* sensu lato species in internal organs and skin of naturally infected symptomatic and asymptomatic dogs, as detected by polymerase chain reaction. *Veterinary Quarterly* 21, 54–58.

Hovmark, A., Asbrink, E. and Olsson, I. (1986a) The spirochetal etiology of lymphadenosis benigna cutis solitaria Afzelius. *Acta Dermato-Venereologica (Stockholm)* 66, 479–484.

Hovmark, A., Asbrink, E., Schwan, O., Hederstedt, B. and Christensson, D. (1986b) Antibodies to *Borrelia* spirochetes in sera from Swedish cattle and sheep. *Acta Veterinaria Scandinavica* 27, 479–485.

Isogai, H., Isogai, E., Masuzawa, T., Yanagihara, Y., Matsubara, M., Shimanuki, M., Seta, T., Fukai, K., Kurosawa, N. and Enokidani, M. (1992) Seroepidemiological survey for antibody to *Borrelia burgdorferi* in cows. *Microbiology and Immunology* 36, 1029–1039.

Johnson, R.C., Schmid, G.P., Hyde, F.W., Steigerwaldt, A.G. and Brenner, D.J. (1984) *Borrelia burgdorferi* sp.nov.: etiologic agent of Lyme disease. *International Journal of Systematic Bacteriology* 34, 496–497.

Jurca, T., Ruzic-Sabljic, E., Lotric-Furlan, S., Maraspin, V., Cimperman, J. and Picken R.N. (1998) Comparison of peripheral and central biopsy sites for the isolation of *Borrelia burgdorferi* sensu lato from erythema migrans skin lesions. *Clinical Infectious Disease* 27, 636–638.

Käsbohrer, A. and Schönberg, A. (1990) Serologische Untersuchungen zum Vorkommen von *Borrelia burgdorferi* bei

Haustieren in Berlin (West). *Berliner und Munchener Tierarztliche Wochenschrift* 103, 374–378.

Klempner, M.S., Hu, L.T., Evans, J., Schmid, C.H., Johnson, G.M., Trevino, R.P., Norton, D., Levy, L., Wall, D., McCall, J., Kosinski, M. and Weinstein, A. (2001) Two controlled trials of antibiotic treatment in patients with persistent symptoms and a history of Lyme disease. *New England Journal of Medicine* 345, 85–92.

Kornblatt, A.N., Urband, P.H. and Steere, A.C. (1985) Arthritis caused by *Borrelia burgdorferi* in dogs. *Journal of the American Veterinary Medicine Association* 186, 960–964.

Kristoferitsch, W. (1989) *Neuropathien bei Lyme-Borreliose.* Springer Verlag, Vienna, 207 pp.

Kristoferitsch, W., Sluga, E., Graf, M., Partsch, H., Neumann, R., Stanek, G. and Budka, H. (1988) Neuropathy associated with acrodermatitis chronica atrophicans: clinical and morphological features. *Annals of the New York Academy of Sciences* 539, 35–45.

Kurtenbach, K., Peacey, M., Rijpkema, S.G.T., Hoodless, A.N., Randolph, S.E. and Nuttall, P.A. (1998) Differential transmission of genospecies of *Borrelia burgdorferi* sensu lato by game birds and small rodents in England. *Applied and Environmental Microbiology* 64, 1169–1174.

Lane, R.S., Berger, D.M., Casher, L.E. and Burgdorfer, W. (1994) Experimental infection of Columbian black-tailed deer with the Lyme disease spirochete. *Journal of Wildlife Diseases* 30, 20–28.

Levy, S.A. and Duray, P.H. (1988) Complete heart block in a dog seropositive for *Borrelia burgdorferi*. Similarity to human Lyme carditis. *Journal of Veterinary International Medicine* 2, 138–144.

Liebisch, G. and Liebisch, A. (1999) Lyme-borreliosis in dogs: risk of infection, interpretation of laboratory diagnosis and vaccination. *Praktische Tierarzt* 80, 404.

Liebisch, G., Assman, G. and Liebisch, A. (1999) Infection with *Borrelia burgdorferi* s.l. causes disease (Lyme borreliosis) in horses in Germany. *Praktische Tierarzt* 80, 498.

Lipschütz, B. (1913) Über eine seltene Erythemform (Erythema chronicum migrans). *Archives of Dermatology and Syphilis* 118, 349–356.

Lipschütz, B. (1923) Weiterer Beitrag zur Kenntnis der 'Erythema chronicum migrans'. *Archives of Dermatology and Syphilis* 143, 365–374.

Lischer, C.J., Leutenegger, C.M., Braun, U. and Lutz, H. (2000) Diagnosis of Lyme disease in two cows by the detection of *Borrelia burgdorferi* DNA. *Veterinary Record* 146, 497–499.

Lissman, B.A., Bosler, E.M., Camay, H., Ormiston, B.G. and Benach, J.L. (1984) Spirochete-associated arthritis (Lyme disease) in a dog. *Journal of the American Veterinary Medicine Association* 185, 219–220.

Logar, M., Lotric-Furlan, S., Maraspin, V., Cimperman, J., Jurca, T., Ruzic-Sabljic, E. and Strle, F. (1999) Has the presence or absence of *Borrelia burgdorferi* sensu lato as detected by skin culture any influence on the course of erythema migrans? *Wiener Klinische Wochenschrift* 111, 945–950.

Luttrell, M.P., Nakagaki, K., Howerth, E.W., Stallknecht, D.E. and Lee, K.A. (1994) Experimental infection of *Borrelia burgdorferi* in white-tailed deer. *Journal of Wildlife Diseases* 30, 146–154.

Magnarelli, L.A., Anderson, J.F., Shaw, E., Post, J.E. and Palka, F.C. (1988) Borreliosis in northeastern United States. *American Journal of Veterinary Research* 49, 359–362.

Magnarelli, L.A., Anderson, J.F., Levine, H.R. and Levy, S.A. (1990) Tick parasitism and antibodies to *Borrelia burgdorferi* in cats. *Journal of the American Veterinary Medical Association* 197, 63–66.

Magnarelli, L.A., Ijdo, J.W., van Andel, A.E., Wu, C., Padula, S.J. and Fikrig, E. (2000) Serologic confirmation of *Erhrlichia equi* and *Borrelia burgdorferi* infections in horses from the northeastern United States. *Journal of the American Veterinary Medical Association* 217, 1045–1050.

Manion, T.B., Khan, M.I., Dinger, J. and Bushmich, S.L. (1998) Viable *Borrelia burgdorferi* in the urine of two clinically normal horses. *Journal of Veterinary Diagnostic Investigation* 10, 196–199.

Maraspin, V., Cimperman, J., Lotric-Furlan, S., Pleterski-Rigler, D. and Strle, F. (1996) Treatment of erythema migrans in pregnancy. *Clinical Infectious Disease* 22, 788–793.

McKenna, P., Clement, J., van Dijck, D., Lauwerys, M., Carey, D., van den Bogaard, T. and Bigaignon, G. (1995) Canine Lyme disease in Belgium. *Veterinary Record* 136, 244–247.

Müller, I., Khanakah, G., Kundi, M. and Stanek, G. (2002) Horses and *Borrelia*: immunoblot patterns with five *Borrelia burgdorferi* sensu lato strains and sera from horses of various stud farms in Austria and from the Spanish Riding School in Vienna. *International Journal of Medical Microbiology* 291, Suppl. 33 pp. 80–87.

Nadelman, R.B. and Wormser, G.P. (1998) Lyme borreliosis. *Lancet* 352, 557–565.

Nadelman, R.B., Nowakowski, J., Fish, D., Falco, R.C., Freeman, K., McKenna, D., Welch, P., Marcus, R., Aguero-Rosenfeld, M.E., Dennis, D.T. and Wormser, G.P. (2001) Prophylaxis with single-dose doxycycline for the prevention of Lyme disease after an *Ixodes scapularis* tick bite. *New England Journal of Medicine* 345, 79–84.

Ogden, N.H., Nuttall, P.A. and Randolph, S.E. (1997) Natural Lyme disease cycles maintained via sheep by co-feeding ticks. *Parasitology* 115, 591–599.

Oliver, J.H. Jr, Magnarelli, L.A., Hutcheson, H.J. and Anderson, J.F. (1999) Ticks and antibodies to *Borrelia burgdorferi* from mammals at Cape Hatteras, NC and Assateague Island, MD and VA. *Journal of Medical Entomology* 36, 578–587.

Overduin, L.M. and van den Bogaard, A.E. (1997) Lyme borreliosis in dogs. *Tijdschrift voor Diergeneeskunde* 122, 7–9.

Parker, J.L. and White, K.K. (1992) Lyme borreliosis in cattle and horses: a review of the literature. *Cornell Veterinarian* 82, 253–274.

Picken, M.M., Picken, R.N., Han, D., Cheng, Y., Ruzic-Sabljic, E., Cimperman, J., Maraspin, V., Lotric-Furlan, S. and Strle, F. (1997) A two year prospective study to compare culture and polymerase chain reaction amplification for the detection and diagnosis of Lyme borreliosis. *Molecular Pathology* 50, 186–193.

Picken, R.N., Strle, F., Picken, M.M., Ruzic-Sabljic, E., Maraspin, V., Lotric-Furlan, S. and Cimperman, J. (1998) Identification of three species of *Borrelia burgdorferi* sensu lato (*B. burgdorferi* sensu stricto, *B. garinii*, and *B. afzelii*) among isolates from acrodermatitis chronica atrophicans lesions. *Journal of Investigative Dermatology* 110, 211–214.

Piesman, J., Dolan, M.C., Schriefer, M.E. and Burkot, T.R. (1996) Ability of experimentally infected chickens to infect ticks with the Lyme disease spirochete, *Borrelia burgdorferi*. *American Journal of Tropical Medicine and Hygiene* 54, 294–298.

Post, J.E., Shaw, E.E. and Wright, S.D. (1988) Suspected borreliosis in cattle. *Annals of the New York Academy of Sciences* 539, 488.

Preac-Mursic, V., Wilske, B., Schierz, G., Pfister, H.W. and Einhäupl, K. (1984) Repeated isolation of spirochetes from the cerebrospinal fluid of a patient with meningoradiculitis Bannwarth. *European Journal of Clinical Microbiology* 3, 564–565.

Preac-Mursic, V., Marget, W., Busch, U., Pleterski-Rigler, D. and Hagl, S. (1996) Kill kinetics of *Borrelia burgdorferi* and

bacterial findings in relation to treatment of Lyme borreliosis. *Infection* 24, 9–16.
Priem, S., Burmester, G.R., Kamradt, T., Wolbart, K., Rittig, M.G. and Krause, A. (1998) Detection of *Borrelia burgdorferi* by polymerase chain reaction in synovial membrane, but not in synovial fluid from patients with persisting Lyme arthritis after antibiotic therapy. *Annals of Rheumatic Diseases* 57, 118–121.
Robertson, J., Guy, E., Andrews, N., Wilske, B., Anda, P., Granstrom, M., Hauser, U., Moosmann, Y., Sambri, V., Schellekens, J., Stanek, G. and Gray, J. (2000) A European multicenter study of immunoblotting in serodiagnosis of Lyme borreliosis. *Journal of Clinical Microbiology* 38, 2097–2102.
Rogers, A.B., Smith, R.D. and Kakoma, I. (1999) Serologic cross-reactivity of antibodies against *Borrelia theileri*, *Borrelia burgdorferi*, and *Borrelia coriaceae* in cattle. *American Journal of Veterinary Research* 60, 694–697.
Rothwell, J.T., Christie, B.M., Williams, C. and Walker, K.H. (1989) Suspected Lyme disease in a cow. *Australian Veterinary Journal* 66, 296–298.
Sälde, H. (1946) Erythema chronicum migrans Afzelius with meningitis. *Acta Dermato-Venereologica (Stockholm)* 31, 227–234.
Schaltenbrand G. (1967) Durch Arthropoden übertragene Erkrankung der Haut und des Nervensystems. *Verhandlungen der Deutschen Gesellschaft für Kreislaufforschung* 72, 975–1006.
Sköldenberg, B. and Stiernsted, G. (eds) (1991) *Lyme Borreliosis 1990*. Scandinavian Journal of Infectious Diseases, suppl. 77, 157 pp.
Speck, S., Reiner, D. and Wittenbrink, M.M. (2001) Isolation of *Borrelia afzelii* in a dog. *Veterinary Record* 149, 19–20.
Stanek, G. and Kahl, O. (1999) Chemoprophylaxis for Lyme borreliosis? *Zentralblatt für Bakteriologie* 289, 655–665.

Stanek, G., Wewalka, G., Groh, G. and Neumann, R. (1985) Isolation of spirochetes from the skin of patients with erythema chronicum migrans in Austria. *Zentralblatt für Bakteriologie und Hygiene A* 260, 88–90.
Stanek, G., Flamm, H., Barbour, A.G. and Burgdorfer, W. (eds) (1986) *Lyme Borreliosis*. Gustav Fischer Verlag, Stuttgart, 501 pp.
Stanek, G., Kristoferitsch, W., Pletschette, M., Barbour, A.G. and Flamm, H. (eds) (1989) *Lyme Borreliosis II*. Zentralblatt für Bakteriologie, suppl. 18, 365 pp.
Stanek, G., Klein, J., Bittner, R. and Glogar, D. (1990) Isolation of *Borrelia burgdorferi* from the myocardium of a patient with longstanding cardiomyopathy. *New England Journal of Medicine* 322, 249–252.
Stanek, G., O'Connell, S., Cimmino, M., Aberer, E., Kristoferitsch, W., Granstrom, M., Guy, E. and Gray, J. (1996) European Union Concerted Action on Risk Assessment in Lyme Borreliosis: clinical case definitions for Lyme borreliosis. *Wiener Klinische Wochenschrift* 108, 741–747.
Steere, A.C. (2001) Lyme disease. *New England Journal of Medicine* 345, 115–125.
Steere, A.C. and Malawista, S.E. (1979) Cases of Lyme disease in the United States: locations correlated with distribution of *Ixodes dammini*. *Annals of Internal Medicine* 91, 730–733.
Steere, A.C., Malawista, S.E., Snydman, D.R., Shope, R.E., Andiman, W.A., Ross, M.R. and Steele, F.W. (1977a) Lyme arthritis: an epidemic of oligoarticular arthritis in children and adults in three Connecticut communities. *Arthritis and Rheumatism* 20, 7–17.
Steere, A.C., Malawista, S.E., Hardin, J.A., Ruddy, S., Askenase, P.W. and Andiman, W.A. (1977b) Erythema chronicum migrans and Lyme arthritis: the enlarging clinical spectrum. *Annals of Internal Medicine* 86, 685–698.
Steere, A.C., Malawista, S.E., Newman,

J.H., Spieler, P.N. and Bartenhagen, N.H. (1980) Antibiotic therapy in Lyme disease. *Annals of Internal Medicine* 93, 1–8.

Steere, A.C., Grodzicki, R.L., Kornblatt, A.N., Craft, J.E., Barbour, A.G., Burgdorfer, W., Schmid, G.P., Johnson, E. and Malawista, S.E. (1983) The spirochetal etiology of Lyme disease. *New England Journal of Medicine* 308, 733–740.

Steere, A.C., Schoen, R.T. and Taylor, E. (1987) The clinical evolution of Lyme arthritis. *Annals of Internal Medicine* 107, 725–731.

Straubinger, R.K., Summers, B.A., Chang, Y.F. and Appel, M.J. (1997) Persistence of *Borrelia burgdorferi* in experimentally infected dogs after antibiotic treatment. *Journal of Clinical Microbiology* 35, 111–116.

Straubinger, R.K., Straubinger, A.F., Summers, B.A., Jacobson, R.H. and Erb, H.N. (1998) Clinical manifestations, pathogenesis, and effect of antibiotic treatment on Lyme borreliosis in dogs. *Wiener Klinische Wochenschrift* 110, 874–881.

Strle, F. (1999) Principles of the diagnosis and treatment of Lyme borreliosis. *Wiener Klinische Wochenschrift* 111, 911–915.

Strle, F., Pleterski-Rigler, D., Stanek, G., Pejovnik-Pustinek, A., Ruzic, E. and Cimperman, J. (1992) Solitary borrelial lymphocytoma: report of 36 cases. *Infection* 20, 201–206.

Strle, F., Preac-Mursic, V., Cimperman, J., Ruzic, E., Maraspin, V. and Jereb, M. (1993) Azithromycin versus doxycycline for treatment of erythema migrans: clinical and microbiological findings. *Infection* 21, 83–88.

Strle, F., Nelson, J.A., Ruzic-Sabljic, E., Cimperman, J., Maraspin, V., Lotric-Furlan, S., Cheng, Y., Picken, M.M., Trenholme, G.M. and Picken, R.N. (1996) European Lyme borreliosis: 231 culture-confirmed cases involving patients with erythema migrans. *Clinical Infectious Disease* 23, 61–65.

Strle, F., Nadelman, B.N., Cimperman, J., Nowakowski, J., Picken, R.N., Schwartz, I., Maraspin, V., Aguero-Rosenfeld, M.E., Varde, S., Lotric-Furlan, S. and Wormser, G.P. (1999) Comparison of culture-confirmed erythema migrans caused by *Borrelia burgdorferi* sensu stricto in New York State and by *Borrelia afzelii* in Slovenia. *Annals of Internal Medicine* 130, 32–36.

Stuen, S. and Fridriksdottir, V. (1991) Experimental inoculation of sheep with *Borrelia burgdorferi*. *Veterinary Record* 129, 315.

Süss, J. and Kahl, O. (eds) (1997) *Tick-borne Encephalitis and Lyme Borreliosis*. 4th International Potsdam Symposium on Tick-borne Diseases, Pabst Science Publishers, Lengerich, Germany, 285 pp.

Süss, J. and Kahl, O. (eds) (1999) Tick-borne encephalitis and Lyme borreliosis. 5th International Potsdam Symposium on Tick-borne Diseases. *Zentralblatt für Bakteriologie* 289, 491–769.

Süss, J., Kahl, O. and Dautel (eds) (2002) Tick-borne encephalitis and Lyme borreliosis. 5th International Potsdam Symposium on Tick-borne Diseases. *Zentralblatt für Bakteriologie*.

Thyresson, N. (1949) The penicillin treatment of acrodermatitis chronica atrophicans (Herxheimer). *Acta Dermato-Venereologica (Stockholm)* 29, 572–621.

Tugwell, P., Dennis, D.T., Weinstein, A., Wells, G., Shea, B., Nichol, G., Hayward, R., Lightfoot, R., Baker, P. and Steere, A.C. (1997) Laboratory evaluation in the diagnosis of Lyme disease. *Annals of Internal Medicine* 127, 1109–1123.

Tuomi, J., Rantamaki, L.K. and Tanskanen, R. (1998) Experimental infection of cattle with several *Borrelia burgdorferi* sensu lato strains; immunological heterogeneity of strains as revealed in serological tests. *Veterinary Microbiology* 60, 27–43.

Weber, K. (1974) Erythema-chronicum-migrans-Meningitis – eine bakterielle Infektions-krankheit? *München Medizinische Wochenschrift* 116, 1993–1998.

Weber, K. and Neubert, U. (1986) Clinical features of early erythema migrans disease and related disorders. *Zentralblatt für Bakteriologie und Hygiene A* 263, 209–228.

Weber, K., Schierz, G., Wilske, B. and Preac-Mursic, V. (1985) Das Lymphozytom – eine Borreliose. *Zeitschrift für Hautkrankheiten* 69, 1585–1598.

Wells, S.J., Trent, A.M., Robinson, R.A., Knutson, K.S. and Bey, R.F. (1993) Association between clinical lameness and *Borrelia burgdorferi* antibody in dairy cows. *American Journal of Veterinary Research* 54, 398–405.

Wilske, B. and Preac-Mursic, V. (1993) Microbiological diagnosis of Lyme Borreliosis. In: Weber, K., Burgdorfer, W. and Schierz, G. (eds) *Aspects of Lyme Borreliosis*. Springer, Berlin, pp. 267–299.

Wilske, B., Preac-Mursic, V., Schierz, G. and von Busch, K. (1986) Immunochemical and immunological analysis of European *Borrelia burgdorferi* strains. *Zentralblatt für Bakteriologie und Hygiene A* 263, 92–102.

Wilske, B., Zöller, L., Brade, V., Eiffert, H., Göbel, U.B. and Stanek, G. (2000) *Qualitätsstandards in der mikrobiologischen Diagnostik, MIQ 12/2000 Lyme-Borreliose.* Urban & Fischer, Munich, Jena, 59 pp.

Wokke, J.H.J., van Gijn, J., Elderson, A. and Stanek, G. (1987) Chronic forms of *Borrelia burgdorferi* infection of the nervous system. *Neurology* 37, 1031–1034.

Wormser, G.P. (1999) Vaccination as a modality to prevent Lyme disease. A status report. *Infectious Diseases Clinic of North America* 13, 135–148.

Wormser, G.P., Nadelman, R.B., Dattwyler, R.J., Dennis, D.T., Shapiro, E.D., Steere, A.C., Rush, T.J., Rahn, D.W., Coyle, P.K., Persing, D.H., Fish, D. and Luft, B.J. (2000) Practice guidelines for the treatment of Lyme disease. *Clinical Infectious Disease* 31 (suppl. 1), S1–S14.

Yanagihara, Y. and Masuzawa, T. (eds) (1994) *Present Status of Lyme Borreliosis and Biology of Lyme Borrelia.* University of Shizuoka Press, 230 pp.

Yanagihara, Y. and Masuzawa, T. (eds) (1997) *Emerging and Re-emerging Diseases Transmitted by Arthropod Vectors and Rodents.* University of Shizuoka Press, 409 pp.

Ecological Research on *Borrelia burgdorferi* sensu lato: Terminology and Some Methodological Pitfalls

Olaf Kahl,[1a] **Lise Gern,**[2] **Lars Eisen**[3] **and Robert S. Lane**[3]

[1]*Institute of Applied Zoology/Animal Ecology, Free University of Berlin, 12163 Berlin, Germany;* [2]*Institut de Zoologie, University of Neuchâtel, Emile Argand 11, 2007 Neuchâtel, Switzerland;* [3]*Department of Environmental Science, Policy and Management, University of California, Berkeley, CA 94720, USA*

Introduction

Among arthropod-borne human diseases in the cool-temperate regions of the northern hemisphere Lyme borreliosis (LB) is the most commonly reported. The causative agents, *Borrelia afzelii*, *Borrelia burgdorferi* sensu stricto (s.s.) and *Borrelia garinii*, are spirochaetes belonging to the *B. burgdorferi* species complex, or *B. burgdorferi* sensu lato (s.l.). Almost all known *Borrelia* spp. circulate between ticks (Arachnida, Acari, Ixodoidea) and a wide variety of vertebrates, which may include different orders of mammals (mainly insectivores, rodents, lagomorphs and ungulates), some bird species and a few reptiles. Tick-borne relapsing fever borreliae are usually transmitted by soft ticks (Argasidae, genus *Ornithodoros*), whereas hard ticks (Ixodidae) transmit *B. burgdorferi* s.l. (Burgdorfer and Hayes, 1989). The epidemiologically most important ticks or transmitting *B. burgdorferi* s.l. to humans are *Ixodes persulcatus* (northern mid Asia), *Ixodes scapularis* (eastern North America), *Ixodes pacificus* (western North America) and *Ixodes ricinus* (Europe and some adjacent areas) (see Eisen and Lane, Chapter 4).

Each of the three components of the ecological system: the pathogens, pathogen-transmitting ticks (**vectors**) and pathogen-transmitting tick hosts (usually called **reservoir hosts**), and also their interrelationships, are dependent on external influence by various biotic and abiotic factors (Fig. 2.1). For example, vector ticks are dependent upon the availability of suitable host individuals for

---

[a]Present address: Blackwell Verlag, Kurfürstendamm 57, 10707 Berlin, Germany.

each life stage and also need high microclimatic relative humidities during the often extended off-host phases. Host-seeking in ticks is usually a seasonal activity and is strongly dependent on environmental factors. The existence of *B. burgdorferi* s.l. requires the occurrence of vector ticks and also vertebrate populations that must be able to transmit borreliae efficiently to feeding ticks. Although humans often become infected with *B. burgdorferi* s.l. and subsequently may fall ill, they are dead-end hosts and they are not involved in the natural circulation of the pathogen. The main ecological influence of humans on *B. burgdorferi* s.l. is indirect, by their impact on landscape ecology and on the distribution and abundance of some tick hosts, for example deer.

It became clear soon after the discovery of *B. burgdorferi* (Burgdorfer *et al.*, 1982) that this organism was an important cause of human disease. Although specific ecological research began immediately in the early 1980s both in North America and Europe and shortly thereafter in the former USSR and Japan, many key ecological and epidemiological questions still remain unanswered, at least regionally and locally. The main variables in the ecological puzzle of LB are: (i) the different vector species, most of them occurring in regions with strongly varying macroclimatic conditions and differing host spectra; and (ii) the existence of several closely related *Borrelia* spp., referred to as genospecies, that can only be differentiated with molecular methods. Research in tick-borne diseases seeks

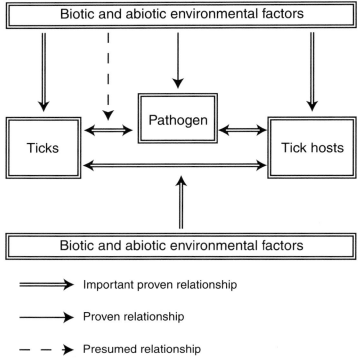

Fig. 2.1. The ecological system of Lyme borreliosis.

to determine what environmental and behavioural factors influence the risk of human infection and disease. Answering these and other questions is of both major ecological and epidemiological relevance.

Although it is undoubtedly a principal task to identify those ticks and vertebrates that maintain the natural transmission cycle and those ticks capable of transmitting pathogens to humans, it is also essential to have a standardized and plausible set of terms for describing the different kinds of functional involvement of ticks and vertebrates in the ecology of LB and other tick-borne diseases. In this chapter, a terminology is suggested that emphasizes clarity, consistency and simplicity, and is largely based on terms introduced by Pavlovsky (1966). Our concept tries to be broad and logical by also paying attention to the minor and even antagonistic roles of certain ticks and vertebrates in the circulation of tick-borne pathogens. Thus, the ecological effect rather than the mechanism involved is emphasized, a distinction that is crucial. This chapter also evaluates borrelia detection methods and outlines different experimental approaches for determining the functional involvement of ticks and their hosts. Some major pitfalls in ecological research and some of the more common errors and faulty interpretations occurring in the literature are identified. For example, not all naturally infected ticks and vertebrates are capable of transmitting a given agent, nor does the presence of specific antibodies against the agent in vertebrates confirm their infectiousness for feeding ticks.

Ecological Terminology

A large variety of qualitative and quantitative terms and phrases have been used to describe the multiple functions and attributes of ticks and their hosts in the ecology (epizootiology) of tick-borne zoonoses (e.g. Pavlovsky, 1966; Balashov, 1972; Ginsberg, 1993; Nuttall and Labuda, 1994; Sonenshine and Mather, 1994; Randolph and Craine, 1995). However, there has been inconsistency in the use of these terms, and a standardized and logical terminology would increase the clarity of communication. The system of terms presented here is intended to draw attention to the present confused situation and, in particular, to provide the basis for future discussions on this topic.

The terms **tick**, **tick host** and **vertebrate** hereafter usually refer to species but, depending on the context, they can also be used in relation to populations and individuals.

Qualitative terms (categories)

To transmit a given pathogen and to be considered a **vector**, a tick species must: (i) feed on infectious vertebrates; (ii) be able to acquire the pathogen during the blood meal; (iii) maintain it through one or more life stages (transstadial passage); and (iv) pass it on to other hosts when feeding again. Otherwise it is a

non-vector tick. In some cases tick infection may be acquired from another tick, if a male infects a female during mating (Alekseev and Dublinina, 1996) or females transovarially infect their offspring (Burgdorfer and Varma, 1967). These four successive crucial steps in the transmission of pathogens by vectors are considered in a system of qualitative categories (Fig. 2.2).

Ticks that feed on infectious hosts may be categorized as **pathogen-exposed ticks**. Most pathogen detection methods used in ticks identify **carrier ticks**, which harbour a particular pathogen without necessarily being infective for a host. These methods are unable to differentiate vector ticks from carrier ticks. Numerous species of ixodid ticks have been shown to harbour *B. burgdorferi* s.l. naturally (see Chapter 4). Detection of spirochaetes in unfed nymphal or adult ticks clearly indicates that borreliae were passed transstadially from the larval or nymphal stage. At least seven such carrier tick species are unable to transmit spirochaetes to the host (see Eisen and Lane, Chapter 4) and may therefore be considered **non-vector ticks**. Infected **vector ticks**, however, are able to transmit the given pathogen to a host during their blood meal and are involved in the agent's natural maintenance cycle.

A comparable system of qualitative categories is presented for vertebrates (Fig. 2.3). To be considered a **reservoir host**, the vertebrate species concerned must be a source of infection for ticks. It must also fulfil similar successive transmission criteria that define a vector: (i) it must be fed on by infected vector ticks, at least occasionally; (ii) it must take up a critical number of infectious agents during an infectious tick bite; (iii) it must allow the pathogen to multiply and to

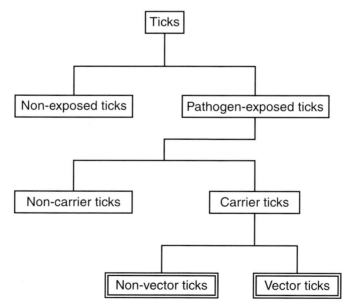

Fig. 2.2. Categories describing the functional involvement of ticks in tick-borne zoonoses (key ecological terms are in a double frame).

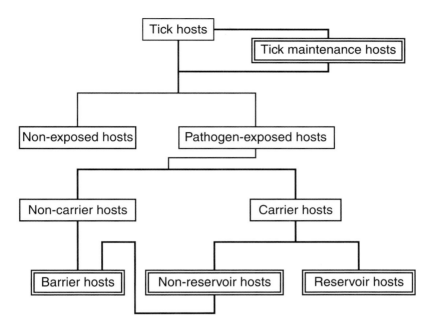

Fig. 2.3. Categories describing the functional involvement of tick hosts in tick-borne zoonoses (key ecological terms are in a double frame).

survive for some time in at least certain parts of its body; and last but not least (iv) it must allow the pathogen to find its way into other feeding ticks. Each of these four successive steps is crucial to make a tick host an active link in the chain of transmission and will therefore be considered in the following system of qualitative categories (Fig. 2.3).

Tick hosts include all vertebrates that ticks feed on in nature. Local and regional tick abundance (or even tick occurrence) largely depends on the presence of one or more **tick maintenance hosts** (sometimes called 'amplifying hosts', a term also confusingly applied occasionally to vertebrates that transmit pathogens). The availability of suitable hosts (medium-sized and large mammals) for adults may be a limiting factor or bottle-neck for *Ixodes* spp. in some areas, for example in city parks or some suburban environments with an impoverished vertebrate fauna (Dautel and Kahl, 1999). Empirical studies have found correlations between local vector tick density and local abundance of deer both in Europe (Gray *et al.*, 1992) and North America (reviewed by Wilson, 1998).

Pathogen-exposed hosts are fed on by infected vector ticks, at least occasionally. Most pathogen detection methods used in tick hosts determine the functional category **carrier hosts**, i.e. hosts that harbour a particular pathogen for at least a short period but are not necessarily infective for ticks. These same methods do not distinguish carrier hosts from reservoir hosts, and at least in the case of *B. burgdorferi* s.l. the unwarranted assumption is occasionally made that these categories are synonymous. 'Susceptible hosts' harbour, at least temporarily, live agents in a density suggesting that they could be infectious for feeding ticks but with experimental proof lacking. **Reservoir hosts** are proven natural hosts of vector ticks, and ticks may become infected while feeding on them. It is therefore common to all reservoir hosts that they increase the number of infected ticks in a particular area and thereby exert a positive ecological effect on the pathogen concerned. Infectivity lasts either lifelong, as in many small mammalian species when infected with *B. afzelii* or *B. burgdorferi* s.s., or for only some shorter period, as in certain small mammalian species infected with tick-borne encephalitis (TBE) virus for only a few days (Randolph *et al.*, 1996).

Uninfected ticks can also acquire an infection when feeding together with infected vector ticks on a 'co-feeding transmission host'. Although the vertebrate is used only as a short-term vehicle by the agent, without showing an overt infection, co-feeding transmission may be ecologically very important for some tick-borne pathogens, as postulated for TBE virus by Randolph *et al.* (1999). Because of the short-term infection after an infectious tick bite, a co-feeding transmission host's significance depends strongly on the frequency with which infective and non-infected vector ticks co-feed on it (Randolph *et al.*, 2000). Although the pathogen/host relationships in co-feeding transmission strongly differ from those in 'classical' reservoir host transmission, the same ecological effect of increasing the number of infected ticks in an area may justify the inclusion of 'co-feeding transmission hosts' under the well-established term 'reservoir hosts'. Note that the term co-feeding transmission refers essentially to a transmission mechanism that may or may not be ecologically significant.

Non-reservoir hosts may have contact with infected ticks and may or may not develop a long-lasting infection but they are incapable of transmitting the infection to ticks. Consequently, non-reservoir hosts do not directly support the circulation of the agent. Nevertheless, infected ticks may retain their infection when feeding on such a host and disseminate it later to another host or other ticks. As an example, white-tailed deer, *Odocoileus virginianus*, although apparently a non-reservoir host of *B. burgdorferi* s.s. (Telford *et al.*, 1988), seem to be strongly implicated in the recent spread of that agent in parts of eastern North America, probably because of their function as maintenance hosts of the vector, *I. scapularis* (Spielman *et al.*, 1993).

Barrier hosts decrease the number of infected ticks, sometimes only after multiple contacts with vector ticks or the agent. Barrier host is a collective term describing an ecological effect rather than a particular mechanism. The mechanism by which the negative effect on the agent's perpetuation is exerted can vary, for example, refractoriness to infection and elimination of infection in feeding ticks (zooprophylactic host), or destruction of large numbers of vector ticks (during or after feeding) because of, for example, efficient grooming or an acquired anti-tick immunity. Thus, it is important to know about tick performance (survival, developmental success, fecundity) after feeding on a given host species to assess its role in LB ecology. Confirmed barrier hosts for *B. burgdorferi* s.s. include two North American lizard species, *Sceloporus occidentalis* and *Elgaria multicarinata* (Lane and Quistad, 1998; Wright *et al.*, 1998).

Qualitative terms not included in Figs 2.2 and 2.3 include bridging vectors, bridging hosts and experimental vectors/reservoir hosts. **Bridging vectors** transmit the agent to humans without necessarily being important vectors for the maintenance of the agent in nature (example: *I. pacificus* for *B. burgdorferi* s.s. in western North America, the non-anthropophilic *Ixodes spinipalpis* being the vector maintaining the enzootic cycle in small mammals). The term 'bridging vector' may relate only to a certain life stage, the one actually transmitting an infection to humans (example: female *I. persulcatus* for the TBE virus, for *B. afzelii* and *B. garinii*). **Bridging hosts** are naturally infested by more than one vector tick species in nature and may therefore form a link between distinct enzootic cycles of infection. Examples include the European hedgehog (*Erinaceus europaeus*) fed upon by *I. ricinus* and *Ixodes hexagonus* (Gern *et al.*, 1997). Bridging hosts may be the source of infection for some bridging vectors, for example, for dusky-footed woodrats, *Neotoma fuscipes*, in California, which can be parasitized by the human-biting *I. pacificus* and the non-human-biting *I. spinipalpis* in the same area (Lane and Keirans, 1999).

An **experimental vector** (**tick**) must have been shown to be capable of acquiring a particular agent while feeding, maintaining it until the following blood meal and subsequently transmitting it to a host via the next life stage, but is not necessarily involved in its natural circulation (e.g. the hard tick *Rhipicephalus appendiculatus* for the TBE virus (Labuda *et al.*, 1993a,b)). If a tick species carrying *B. burgdorferi* s.l. in the field is shown to transmit borreliae in a feeding experiment there is strong evidence that it is a vector tick. Similarly, **experimental reservoir hosts** are capable of passing on the agent to feeding ticks in the laboratory but are not necessarily involved in its natural circulation, for example, the Mongolian gerbil, *Meriones unguiculatus*, for *B. afzelii* and *B. burgdorferi* s.s. (Kahl *et al.*, 1998a; O. Kahl *et al.*, unpublished data). If a vertebrate species carrying *B. burgdorferi* s.l. in the field is shown to transmit borreliae in a tick feeding experiment it is very likely to be a reservoir host. In both cases, experiments have to show whether pathogen transmission is possible for only one or a few highly adapted genotypes of a particular genospecies or whether this is also a common phenomenon when feral borreliae are used.

The meanings of the terms 'vector/reservoir competence' have been diverse and often misleading in the literature and it might be extremely difficult to introduce new standardized definitions for such popular terms. We suggest that these terms should be used only qualitatively, describing the proven function of a particular species or population. Vector and reservoir competence are definitely not synonymous merely with the experimentally proven capability of transmission of a particular agent during tick feeding. The situation in nature must also be taken into account. The adjective 'competent' is superfluous, and instead of describing vectors or reservoirs as having high or low competence, we suggest describing them to be of high or low 'importance' on the basis of their general significance in the particular ecosystem concerned. This importance is a function of their abundance in the habitat, their infectivity and their 'capacity' as described under 'quantitative terms' below. It is appropriate to distinguish primary versus secondary vectors and reservoirs, the former being ecologically important, the latter not. Ticks or vertebrates incapable of transmitting a pathogen are by definition not vectors or reservoirs of that pathogen. 'Incompetent vector' and 'incompetent reservoir' are contradictions in terms.

Quantitative terms

There is even more confusion about the use of quantitative than of qualitative terms. Quantitative terms refer to the epizootiological and epidemiological consequences that arise from certain physiological capabilities as well as ecological and behavioural traits of ticks or their hosts in connection with the natural circulation of infectious agents.

Quantitative terms relating to questing ticks and ticks on hosts have been reviewed by Wilson (1994). Briefly, **abundance** of questing ticks is a dimensionless relative measure, whereas **density** is an absolute measure describing the number of ticks per area unit. With regard to ticks on hosts, **prevalence of infestation** describes the percentage of examined hosts that are infested. **Mean density** of ticks per host is calculated as the total number of ticks observed divided by the total number of hosts examined, whereas **intensity of infestation** is the total number of ticks observed divided by the number of infested hosts. However, the distribution of ticks is typically highly aggregated within host populations, with most hosts harbouring a few ticks and a few hosts harbouring high numbers. Therefore the median value of infesting ticks may better describe the typical infestation level for an individual host. Finally, **prevalence of infection** refers to the percentage of examined ticks infected with a pathogen and **intensity of infection** describes the average number of pathogens per infected tick.

Tick infectivity denotes the efficiency with which the infection is transmitted from a tick population or species to hosts. It can vary between different life stages and populations and also in relation to different vertebrate hosts. **Specific tick infectivity** denotes the efficiency with which the infection is transmitted from a tick population or species to a particular host species. The specific tick infectivity also may differ depending on the tick stage and phenotype of the agent involved (Dolan *et al.*, 1998). **Host infectivity** denotes the efficiency with which the infection is transmitted from a tick host population or species to feeding ticks. The infectivity of a vertebrate may vary during its lifetime and between individuals due to physiological factors associated with ageing or immunological phenomena such as prior experience with ticks and tick-borne agents. **Specific host infectivity** denotes the proportion of ticks of a particular species and life stage that becomes infected during feeding. Also, the specific host infectivity may differ depending on the phenotype of the agent involved.

Vector capacity describes the *absolute* contribution made by a particular vector species to the natural prevalence of infection with a given pathogen in vertebrates in a certain area. Determination of the vector capacity of a tick should include quantitative data on the proportion of reservoir hosts naturally infested by the tick, the percentage of infected ticks, and the likelihood that one infected tick transmits the pathogen to a reservoir host (Randolph and Craine, 1995). Accordingly, **relative vector capacity** describes the *proportional* contribution made by a particular vector to the natural prevalence of infection in vertebrates in a certain area in relation to that of other possible vectors.

Reservoir capacity describes the *absolute* contribution made by a particular reservoir or co-feeding transmission host to the natural prevalence of infection with a given pathogen in vector ticks in a certain area. Determination of the reservoir capacity of a vertebrate should include quantitative data on its density, its infectivity for feeding ticks, the duration of the infective period as well as vector tick infestation, vector tick feeding success and the subsequent tick moulting success (Mather *et al.*, 1989; Manelli *et al.*, 1993; Randolph and Craine, 1995). Accordingly, **relative reservoir capacity** (synonymous with reservoir potential, as used by Mather *et al.* (1989)) describes the *proportional* contribution made by a particular reservoir or co-feeding transmission host to the natural prevalence of infection with a given pathogen in vector ticks in a certain area in relation to that of other possible reservoir hosts. Empirical assessment of the relative reservoir capacity of various vertebrates in an area is an arduous task that few researchers have attempted, with few species considered (e.g. Mather *et al.*, 1989; Tälleklint and Jaenson, 1994; Kurtenbach *et al.*, 1995).

Vector potential is a subjective assessment of the putative vector capacity of a given tick based on the results of incomplete laboratory experiments or insufficient field observations on, for example, its abundance, its tendency to feed on reservoir hosts, its feeding performance on reservoir hosts including its subsequent developmental success and/or its infectivity for different host taxa. Vector potential has some similarity with the collective term 'vectorial capacity' as used by Spielman *et al.* (1984), including all those vector-related variables affecting stability of pathogen transmission. **Reservoir potential** is a subjective assessment of the putative reservoir capacity of a given vertebrate based on the limited results of incomplete laboratory experiments or insufficient field observations on, for example, its abundance, its levels of infestation by different vector tick stages, its infectivity for vector ticks, and/or the duration of the infective period.

Pathogen Detection: Advantages and Limitations of Different Methods

Care must be taken in choosing an adequate experimental approach to demonstrate that a particular tick or vertebrate belongs to a certain functional category in tick-borne zoonoses, or to answer a specific key ecological epidemiology question. In addition to the isolation of an agent from field-collected ticks and evidence of blood-feeding contact between suspected vectors and reservoir hosts, *in vivo* transmission experiments are crucial for proving the capability of a vector or of a reservoir host (DeFoliart *et al.*, 1987).

Identification of borrelia carriers

The first step in an ecological epidemiology investigation is usually to identify one or more carriers of a given pathogen in a certain area. The next step typically is to assess the prevalence of infection in different parts of a tick population (i.e. different life stages, host-seeking or feeding individuals) over time. In addition to a proper spatial and temporal sampling pattern, reliable identification of the causative agents in ticks and their hosts is an important prerequisite. The detection method or combination of methods used must be sufficiently sensitive and specific. Several methods that can be used to identify carriers of *B. burgdorferi* s.l. offer different combinations of advantages and limitations and are summarized in Table 2.1.

Table 2.1. The strengths and limitations of *Borrelia burgdorferi* s.l. detection methods.

Target	Detection method	Remarks[a]
Living (motile) cells	DFM PCM Culture	Detection of live spirochaetes, low specificity
Dead cells	IFA ELISA FISH	Detection of marked surface structures (IFA, ELISA) or genetic material (FISH), quantification of infection, high specificity possible
Genetic material	FISH PCR Sequencing	Detection of genetic material, high specificity
Specific antibodies in the vertebrate host	ELISA IFA Immunoblot	Determination of exposure, low to moderate specificity

[a]The various methods may differ in their sensitivity, with culture, DFM and PCM probably being less sensitive, and IFA, PCR and FISH being more sensitive. However, there is no single method that is intrinsically more sensitive than the others.
DFM, dark-field microscopy; PCM, phase-contrast microscopy; IFA, immunofluorescence assay; ELISA, enzyme-linked immunosorbent assay; FISH, fluorescence *in situ* hybridization; PCR, polymerase chain reaction.

Ticks

There are three detection methods that demonstrate the viability and motility of borreliae: dark-field microscopy (DFM), phase-contrast microscopy (PCM) and culture. All three methods have a low specificity, so that it is not possible to differentiate between *Borrelia* species. Culture is too costly to be used for routine detection of borreliae but it affords a large number of cells that facilitates subsequent identification and characterization by PCR, monoclonal antibodies with differential diagnostic value and other sophisticated methods. However, culture is probably not possible with every *B. burgdorferi* s.l. isolate and may also result in a reduced genetic diversity after an increasing number of passages, due to selection for specific culture-adaptive genotypes (Norris *et al.*, 1997). To save expense and time, DFM and culture can be combined so that only those ticks having a visible spirochaetal infection are cultivated. The remaining detection methods (see Table 2.1) detect only certain parts of the borrelial cell such as outer surface proteins, other proteins or genetic material. In doing so, they demonstrate infection but, strictly speaking, not the presence of living agents, although the detection of RNA by PCR may indicate living organisms since RNA degrades very rapidly. The immunofluorescence assay (IFA) may also indicate living borreliae by showing the typical helical morphology of the spirochaete. It is also possible to combine culture and IFA to obtain clear evidence of living borreliae. IFA is a simple, relatively inexpensive technique that can be as sensitive as PCR for detecting *B. burgdorferi* s.l. (e.g. Kahl *et al.*, 1998b) and provides data on the intensity of infection, although some skill is required to eliminate subjectivity. If large numbers of ticks have to be examined quickly, IFA and PCR are the methods of choice because the samples can be fixed before investigation. However, differentiation of *B. burgdorferi* s.l. genospecies with IFA in a large number of ticks is, although possible, not practical.

Ticks and their hosts, especially in Europe and Asia, may be co-infected with multiple genospecies of *B. burgdorferi* s.l. In central Europe, for example, *I. ricinus* populations harbour *B. afzelii*, *B. garinii*, *B. burgdorferi* s.s. and *Borrelia valaisiana* to varying degrees, and identification of mixed infections in ticks is possible with PCR (e.g. Rijpkema *et al.*, 1995). Pooling groups of ticks, although cost-saving and offering some advantages from a biometric view (Abel *et al.*, 1999), makes it impossible to determine the proportion of mixed infections.

Because the tick moult seems to be a critical physiological challenge for borreliae in ticks (Piesman *et al.*, 1990), unfed ticks found to be infected with borreliae under natural conditions are promising candidates as vectors. Partially or fully fed ticks removed from hosts that contain borreliae may or may not be vectors because nearly all haematophagous arthropods feeding on reservoir hosts of *B. burgdorferi* s.l. are likely to ingest some spirochaetes with the blood meal. Detection of infection in a particular arthropod or vertebrate is not indicative of infectiousness. This can only be proved by transmission experiments, as mentioned above.

Vertebrates

Detection of infection in vertebrates can be based on host seroconversion, detection of spirochaetal DNA in host fluids or tissue, or recovery of live borreliae from cultured host tissue, e.g. ear-punch biopsies (Sinsky and Piesman, 1989). Host seroconversion demonstrates that the vertebrate was exposed to spirochaetes but does not confirm that a viable infection was established, nor does seropositivity indicate infectiousness, but it may be a first step towards describing the presence of a microorganism in an area. Likewise, detection of spirochaetal DNA in host tissue or fluids, although allowing detection of the genospecies involved, is not necessarily indicative of a viable infection. As previously suggested, detection of spirochaetal RNA might indicate living organisms, but this approach does not seem to have been adopted for detection of *Borrelia* spp.

Xenodiagnosis offers an alternative to culture for certain hosts (Donahue *et al.*, 1987) and is the method of choice for proving host infectivity to feeding ticks. The result is positive if spirochaetes are detected in previously spirochaete-free ticks after they have fed on either naturally or experimentally exposed hosts.

Identification of borrelia vectors and reservoir hosts

None of the static tests capable of identifying carriers of borreliae is able to predict with certainty whether a given tick or vertebrate can transmit the agent, i.e. act as a vector or reservoir host. This can only be determined by carrying out additional transmission experiments with a well-defined pathogen. As noted by Lane (1994) and Randolph and Nuttall (1994), there are several difficulties in conducting biologically meaningful transmission studies. The variation sometimes observed in specific tick or host infectivity for different isolates of the same *B. burgdorferi* genospecies indicates that the use of more than one isolate in transmission studies would be preferable. Furthermore, because of the enlarging spectrum of *B. burgdorferi* genospecies (see Bergström *et al.*, Chapter 3), it is especially difficult to draw the right conclusions from transmission experiments with a negative result. A vertebrate may be a reservoir host for one *B. burgdorferi* genospecies but not for another, as the example of different types of European and Asian *B. garinii* impressively demonstrates even within a particular genospecies (see Miyamoto and Masuzawa, Chapter 8). As one would expect, studies based on a single isolate that performs well in an experimental model may not always be representative of the specific tick or host infectivity for the whole genospecies.

Problems related to the use of long-term cultured spirochaetes in studies (Schwan *et al.*, 1988) can be diminished or avoided when using borreliae from a quasi-natural maintenance system in an experimental reservoir host species and a vector tick species in the laboratory. During feeding, nymphal vector ticks infect laboratory hosts, which subsequently transmit the infection to larval ticks.

A new transmission cycle (or vertebrate passage) begins with the feeding of the resultant hungry nymphs. Although it is possible to establish such a system by introducing an isolate from culture into a tick or a vertebrate, the feeding of field-collected vector ticks on a naive experimental reservoir host may form an interesting alternative, especially in areas where only one genospecies exists or where certain hosts only carry a single genospecies. Using this approach, a number of feral *B. afzelii* strains have been maintained for transmission experiments and one strain has completed its sixth gerbil passage after 5 years in the laboratory using *I. ricinus* as the vector (O. Kahl *et al.*, unpublished data). These kinds of experimental studies can also begin with naturally infected feral hosts and xenodiagnostic vector ticks.

Vectors

The capability of a particular tick species to transmit a given agent must be shown by: (i) infecting uninfected ticks – usually larvae – during feeding on an infective host; (ii) allowing them to moult to the next life stage; (iii) feeding the resultant ticks – usually nymphs – on naive hosts; and (iv) demonstrating infection in the hosts after tick-feeding.

Infecting ticks by artificial feeding (e.g. on a capillary tube containing living borreliae in culture medium) followed by feeding the ticks on a host is an unnatural situation. Agent transmission by *in vitro*-infected ticks must be confirmed by an *in vivo* approach. Transstadial survival is a critical challenge for a tick-borne agent and passing through all the different developmental phases (engorged phase, moult, unfed phase and feeding again) of the tick may allow the agent to synchronize its own physiology with that of the vector tick (Schwan *et al.*, 1995; de Silva *et al.*, 1996), and therefore should be part of any basic transmission experiment. Although survival of a pathogen from one life stage to another is commonly referred to as **transstadial transmission** there is in fact no transmission involved and a more accurate term would be **transstadial passage** since the tick simply stays infected through the moult.

Reservoir hosts

Xenodiagnosis is the classical method of choice for determining the host infectivity or specific host infectivity of a particular vertebrate species for a given agent. Feral animals captured in the field and caged for collection of detached ticks can be used and such an *in vivo* approach appears to be superior to the use of needle-infected animals (Gern *et al.*, 1993; Piesman, 1993; Kurtenbach *et al.*, 1994). Xenodiagnosis on feral animals may be problematic if infectivity in reservoir hosts and especially in co-feeding transmission hosts is short-lived, as in small mammals infected with TBE virus.

It is very difficult in most geographical areas to determine the whole spectrum of reservoir hosts or the relative reservoir capacity of one particular or several vertebrate species. Blood-meal identification promises to be an elegant and effective tool for the identification of important reservoir hosts of *B. burgdorferi* s.l. and other tick-borne agents (Kirstein and Gray, 1996). Briefly, each field-collected unfed nymphal vector tick is examined both for its *B. burgdorferi* s.l. infection status to genospecies level and for the source of its larval blood meal in order to identify the larval host. The latter can be done by amplifying PCR products of the cytochrome b gene from the remnants of the larval blood meal. It is thus possible to identify the vertebrate host source of the pathogen detected in the nymphal tick. This approach also takes account of the developmental success of vector ticks after they have fed on different host species, a factor whose importance can easily be underestimated in laboratory studies. By using this method, a list of candidate reservoir hosts can be identified with minimal trapping and handling of feral animals.

Concluding Remarks

As is common for other tick-borne pathogens, the causative agents of LB, *B. burgdorferi* s.l., circulate between species of ticks and several tick hosts. A number of different terms exist for the description of the various functions and properties of ticks and their hosts in tick-borne zoonoses, but there has been inconsistency in their usage. A simple scheme has been presented here in an attempt to establish a logical set of well-defined ecological terms describing the various roles of ticks and tick hosts in LB and other tick-borne zoonoses, using previously established terms wherever possible. Vector or non-vector, reservoir host or non-reservoir host are the key qualitative questions in relation to the role played by ticks and their hosts, respectively, in the circulation of a particular zoonotic agent in nature. A further set of terms has been suggested that might help quantify the contribution of a given species or population to the natural circulation of a given zoonotic agent and to its ecological epidemiology importance. On the basis of all these definitions, the advantages and limitations of different experimental approaches for determining what particular functional category a tick or tick host belongs to have been outlined. Serodiagnosis in vertebrates only determines exposure to a pathogen. Although carrier ticks and carrier hosts may not be key functional categories, the great majority of detection methods identify just these categories but not infectiousness of the carrier. Only transmission experiments can help to clarify whether or not a given tick is a vector or non-vector and a given tick host is a reservoir host or non-reservoir host. It must be emphasized that the finding in a laboratory experiment that a particular tick or vertebrate species transmits a given agent does not mean that it is a vector or reservoir host under natural conditions. The species under investigation must be, at least sporadically, in contact with the agent in nature, usually by feeding on reservoir hosts in the case of vector ticks or by being fed on by vector ticks in the case of reservoir hosts.

Acknowledgement

We are especially grateful to Jeremy Gray (Dublin, Irish Republic) for his contribution to the development and writing of this chapter.

References

Abel, U., Schosser, R. and Süss, J. (1999) Estimating the prevalence of infectious agents using pooled samples: biometrical considerations. *Zentralblatt für Bakteriologie* 289, 550–563.

Alekseev, A.N. and Dubinina, H.V. (1996) Exchange of *Borrelia burgdorferi* between *Ixodes persulcatus* (Acari: Ixodidae) sexual partners. *Journal of Medical Entomology* 33, 351–354.

Balashov, Y.S. (1972) Bloodsucking ticks (Acari, Ixodidae) – vectors of diseases of man and animals. *Miscellaneous Publications of the Entomological Society of America* 8, 163–376.

Burgdorfer, W. and Hayes, S.F. (1989) Vector–spirochete relationships in louse-borne and tick-borne borreliosis with emphasis on Lyme disease. In: Harris, K.F. (ed.) *Advances in Disease Vector Research*. Springer-Verlag, New York, pp. 127–150.

Burgdorfer, W. and Varma, M.G.R. (1967) Trans-stadial and transovarial development of disease agents in arthropods. *Annual Review of Entomology* 12, 347–376.

Burgdorfer, W., Barbour, A.G., Hayes, S.F., Benach, J.L., Grunwaldt, E. and Davis, J.P. (1982) Lyme disease – a tick-borne spirochetosis? *Science* 216, 1317–1319.

Dautel, H. and Kahl, O. (1999) Ticks (Acari: Ixodoidea) and their medical importance in the urban environment. In: Robinson, W.H., Rettich, F. and Rambo, G.W. (eds) *Proceedings of the 3rd International Conference on Urban Pests*, Vol. 3. Grafické závody Hronov, Czech Republic, pp. 73–82.

DeFoliart, G.R., Grimstad, P.R. and Watts, D.M. (1987) Advances in mosquito-borne arbovirus/vector research. *Annual Review of Entomology* 32, 479–505.

Dolan, M.C., Piesman, J., Mbow, M.L., Maupin, G.O., Péter, O., Brossard, M. and Golde, W.T. (1998) Vector competence of *Ixodes scapularis* and *Ixodes ricinus* (Acari, Ixodidae) for three genospecies of *Borrelia burgdorferi*. *Journal of Medical Entomology* 35, 465–470.

Donahue, J.G., Piesman, J. and Spielman, A. (1987) Reservoir competence of white-footed mice for Lyme disease spirochetes. *American Journal of Tropical Medicine and Hygiene* 36, 92–96.

Gern, L., Schaible, U.E. and Simon, M.M. (1993) Mode of inoculation of the Lyme disease agent *Borrelia burgdorferi* influences infection and immune responses in inbred strains of mice. *Journal of Infectious Diseases* 167, 971–975.

Gern, L., Rouvinez, E., Toutoungi, L.N. and Godfroid, E. (1997) Transmission cycles of *Borrelia burgdorferi* sensu lato involving *Ixodes ricinus* and/or *I. hexagonus* ticks and the European hedgehog, *Erinaceus europaeus*, in suburban and urban areas in Switzerland. *Folia Parasitologica* 44, 309–314.

Ginsberg, H.S. (ed.) (1993) *Ecology and Environmental Management of Lyme Disease*. Rutgers University Press, New Brunswick, New Jersey, 224 pp.

Gray, J.S., Kahl, O., Janetzki, C. and Stein, J. (1992) Studies on the ecology of Lyme disease in a deer forest in Co. Galway, Ireland. *Journal of Medical Entomology* 29, 915–920.

Kahl, O., Janetzki-Mittmann, C., Gray, J.S., Jonas, R., Stein, J. and de Boer, R.

(1998a) Risk of infection with *Borrelia burgdorferi* sensu lato for a host in relation to the duration of nymphal *Ixodes ricinus* feeding and the method of tick removal. *Zentralblatt für Bakteriologie* 287, 41–52.

Kahl, O., Gern, L., Gray, J.S., Guy, E.C., Jongejan, F., Kirstein, F., Kurtenbach, K., Rijpkema, S.G.T. and Stanek, G. (1998b) Detection of *Borrelia burgdorferi* sensu lato in ticks: immunofluorescence assay versus polymerase chain reaction. *Zentralblatt für Bakteriologie* 287, 205–210.

Kirstein, F. and Gray, J.S. (1996) A molecular marker for the identification of the zoonotic reservoirs of Lyme borreliosis by analysis of the blood meal in its European vector *Ixodes ricinus*. *Applied and Environmental Microbiology* 62, 4060–4065.

Kurtenbach, K., Dizij, A., Seitz, H.M., Margos, G., Moter, S.E., Kramer, M.D., Wallich, R., Schaible, U.E. and Simon, M.M. (1994) Differential immune responses to *Borrelia burgdorferi* in European wild rodent species influence spirochete transmission to *Ixodes ricinus* L. (Acari, Ixodidae). *Infection and Immunity* 62, 5344–5352.

Kurtenbach, K., Kampen, H., Dizij, A., Arndt, S., Seitz, H.M., Schaible, U.E. and Simon, M.M. (1995) Infestations of rodents with larval *Ixodes ricinus* (Acari, Ixodidae) is an important factor in the transmission cycle of *Borrelia burgdorferi* s.l. in German woodlands. *Journal of Medical Entomology* 32, 807–817.

Labuda, M., Danielova, V., Jones, L.D. and Nuttall, P.A. (1993a) Amplification of tick-borne encephalitis virus infection during co-feeding of ticks. *Medical and Veterinary Entomology* 7, 339–342.

Labuda, M., Jones, L.D., Williams, T., Danielova, V. and Nuttall, P.A. (1993b) Efficient transmission of tick-borne encephalitis virus between cofeeding ticks. *Journal of Medical Entomology* 30, 295–299.

Lane, R.S. (1994) Competence of ticks as vectors of microbial agents with an emphasis on *Borrelia burgdorferi*. In: Sonenshine, D.E. and Mather, T.N. (eds) *Ecological Dynamics of Tick-borne Zoonoses*. Oxford University Press, New York, pp. 45–67.

Lane, R.S. and Keirans, J.E. (1999) *Ixodes spinipalpis*: a probable enzootic vector of *Borrelia burgdorferi* in California. In: Needham, G.R., Mitchell, R., Horn, D.J. and Welbourn, W.C. (eds) *Acarology IX*: Vol. 2, *Symposia*. Ohio Biological Survey, Columbus, Ohio, pp. 395–399.

Lane, R.S. and Quistad, G.B. (1998) Borreliacidal factor in the blood of the western fence lizard (*Sceloporus occidentalis*). *Journal of Parasitology* 84, 29–34.

Manelli, A., Kitron, U., Jones, C.J. and Slajchert, T.L. (1993) *Ixodes dammini* (Acari: Ixodidae) infestation on medium-sized mammals and blue jays in northwestern Illinois. *Journal of Medical Entomology* 30, 950–952.

Mather, T.N., Wilson, M.L., Moore, S.I., Ribeiro, J.M.C. and Spielman, A. (1989) Comparing the relative potential of rodents as reservoirs of the Lyme disease spirochete (*Borrelia burgdorferi*). *American Journal of Epidemiology* 130, 143–150.

Norris, D.E., Johnson, B.J.B., Piesman, J., Maupin, G.O., Clark, J.L. and Black, W.W. IV (1997) Culturing selects for specific genotypes of *Borrelia burgdorferi* in an enzootic cycle in Colorado. *Journal of Clinical Microbiology* 35, 2359–2364.

Nuttall, P.A. and Labuda, M. (1994) Tick-borne encephalitis subgroup. In: Sonenshine, D.E. and Mather, T.N. (eds) *Ecological Dynamics of Tick-borne Zoonoses*. Oxford University Press, New York, pp. 351–391.

Pavlovsky, E.N. (1966) *Natural Nidality of Transmissible Diseases with Special Reference to the Landscape Epidemiology of Zooanthroponoses*. Levine, N.D. (ed.), University of Illinois Press, Urbana, 261 pp. (translated from Russian).

Piesman, J. (1993) Standard system for

infecting ticks (Acari, Ixodidae) with the Lyme disease spirochete, *Borrelia burgdorferi*. *Journal of Medical Entomology* 30, 199–203.

Piesman, J., Oliver, J.R. and Sinsky, R.J. (1990) Growth kinetics of the Lyme disease spirochete (*Borrelia burgdorferi*) in vector ticks (*Ixodes dammini*). *American Journal of Tropical Medicine and Hygiene* 42, 352–357.

Randolph, S.E. and Craine, N.G. (1995) General framework for comparative quantitative studies on transmission of tick-borne diseases using Lyme borreliosis in Europe as an example. *Journal of Medical Entomology* 32, 765–777.

Randolph, S.E. and Nuttall, P.A. (1994) Nearly right or precisely wrong? Natural versus laboratory studies of vector-borne diseases. *Parasitology Today* 10, 458–462.

Randolph, S.E., Gern, L. and Nuttall, P.A. (1996) Co-feeding ticks: epidemiological significance for tick-borne pathogen transmission. *Parasitology Today* 12, 472–479.

Randolph, S.E., Miklisová, D., Lysy, J., Rogers, D.J. and Labuda, M. (1999) Incidence from coincidence: patterns of tick infestations on rodents facilitate transmission of tick-borne encephalitis virus. *Parasitology* 118, 177–186.

Randolph, S.E., Green, R.M., Peacey, M.F. and Rogers, D.J. (2000) Seasonal synchrony: the key to tick-borne encephalitis foci identified by satellite data. *Parasitology* 121, 15–23.

Rijpkema, S.G.T., Molkenboer, M.J.C.H., Schouls, L.M., Jongejan, F. and Schellekens, J.F.P. (1995) Simultaneous detection and genotyping of three genomic groups of *Borrelia burgdorferi* s.l. in Dutch *Ixodes ricinus* ticks by characterization of the amplified intergenic spacer region between 5S and 23S rRNA genes. *Journal of Clinical Microbiology* 33, 3091–3095.

Schwan, T.G., Burgdorfer, W. and Garon, C.F. (1988) Changes in infectivity and plasmid profile of the Lyme disease spirochete, *Borrelia burgdorferi*, as a result of *in vitro* cultivation. *Infection and Immunity* 56, 1831–1836.

Schwan, T.G., Piesman, J., Golde, W.T., Dolan, M.C. and Rosa, P.A. (1995) Induction of an outer surface protein on *Borrelia burgdorferi* during tick feeding. *Proceedings of the National Academy of Sciences, USA* 92, 2909–2913.

de Silva, A., Telford, S.R., Brunet, L.R., Barthold, S.W. and Fikrig, E. (1996) *Borrelia burgdorferi* OspA is an arthropod-specific transmission-blocking Lyme disease vaccine. *Journal of Experimental Medicine* 183, 271–275.

Sinsky, R.J. and Piesman, J. (1989) Ear punch biopsy method for detection and isolation of *Borrelia burgdorferi* from rodents. *Journal of Clinical Microbiology* 27, 1723–1727 (Erratum: (1989) *Journal of Clinical Microbiology* 27, 2401).

Sonenshine, D.E. and Mather, T.N. (eds) (1994) *Ecological Dynamics of Tick-borne Zoonoses*. Oxford University Press, New York, 447 pp.

Spielman, A., Levine, J.F. and Wilson, M.L. (1984) Vectorial capacity of North American *Ixodes* ticks. *Yale Journal of Biology and Medicine* 57, 507–513.

Spielman, A., Telford, S. III and Pollack, R.J. (1993) The origins and course of the present outbreak of Lyme disease. In: Ginsberg, H.S. (ed.) *Ecology and Environmental Management of Lyme Disease*. Rutgers University Press, New Brunswick, New Jersey, pp. 83–96.

Tälleklint, L. and Jaenson, T.G. (1994) Transmission of *Borrelia burgdorferi* s.l. from mammal reservoirs to the primary vector of Lyme Borreliosis, *Ixodes ricinus* (Acari, Ixodidae), in Sweden. *Journal of Medical Entomology* 31, 880–886.

Telford, S.R. III, Mather, T.N., Moore, S.I., Wilson, M.L. and Spielman, A. (1988) Incompetence of deer as reservoirs of the Lyme disease spirochete. *American Journal of Tropical Medicine and Hygiene* 39, 105–109.

Wilson, M.L. (1994) Population ecology of tick vectors: interaction, measurement, and analysis. In: Sonenshine, D.E. and Mather, T.N. (eds) *Ecological Dynamics of Tick-borne Zoonoses*. Oxford University Press, New York, pp. 20–44.

Wilson, M.L. (1998). Distribution and abundance of *Ixodes scapularis* (Acari, Ixodidae) in North America: ecological processes and spatial analysis. *Journal of Medical Entomology* 35, 446–457.

Wright, S.A., Lane, R.S. and Clover, J.R. (1998) Infestation of the southern alligator lizard (Squamata: Anguidae) by *Ixodes pacificus* (Acari: Ixodidae) and its susceptibility to *Borrelia burgdorferi*. *Journal of Medical Entomology* 35, 1044–1049.

Molecular and Cellular Biology of *Borrelia burgdorferi* sensu lato

3

Sven Bergström, Laila Noppa, Åsa Gylfe, and Yngve Östberg

Department of Molecular Biology, Umeå University, SE-901 87 Umeå, Sweden

Introduction

The genus *Borrelia* is a member of the spirochaete phylum, which is an ancient evolutionary branch only remotely related to Gram-negative and Gram-positive bacteria (Woese, 1987). The genus is divided into three major pathogenic groups of organisms, all transmitted by arthropods: the relapsing fever borreliae, Lyme disease (borreliosis) borreliae and the aetiological agent of avian spirochaetosis, *Borrelia anserina*. The Lyme disease borreliae of the *Borrelia burgdorferi* sensu lato (s.l.) species complex have been divided into at least 11 different genomic species (Table 3.1). Amongst these, the human pathogens include *B. burgdorferi* sensu stricto (s.s.), *Borrelia afzelii, Borrelia garinii* (Baranton *et al.*, 1992; Marconi and Garon, 1992a; Canica *et al.*, 1993) and possibly *Borrelia bissettii* (Picken *et al.*, 1996; Strle *et al.*, 1997). The strains designated to these genomic species are characterized by similar protein patterns, reactivity with monoclonal antibodies, amino acid sequence of outer surface proteins, 16S rRNA analysis, Southern blot and PCR analysis (Adam *et al.*, 1991; Jonsson *et al.*, 1992; Baranton *et al.*, 1992), and 16S rRNA, *flaB* and 5S–23S rRNA intergenic spacer sequences (Wang, G. *et al.*, 1999a).

The distribution of Lyme borreliosis (LB) globally is closely tied to the presence of ticks of the *Ixodes ricinus* species complex in the northern hemisphere (see Eisen and Lane, Chapter 4). There are no confirmed cases of LB in the southern hemisphere apart from cases where it may have been acquired in Europe (Hudson *et al.*, 1998), although *B. garinii* DNA has been amplified from the seabird tick *Ixodes uriae* in the southern hemisphere (Olsen *et al.*, 1995).

Members of the genus *Borrelia* have a very complex lifestyle as they are host-associated spirochaetes that shuttle between haematophagous arthropods and

Table 3.1. *Borrelia burgdorferi* sensu lato genospecies and their distribution.

Borrelia species	Reference	Geographical distribution	Reference
B. burgdorferi s.s.	Baranton et al. (1992)	Europe and North America	(Baranton et al., 1998)
B. garinii	Baranton et al. (1992)	Europe and parts of Asia	(Hubálek and Halouzka, 1997; Baranton et al., 1998; Li et al., 1998)
B. afzelii	Canica et al. (1993)	Europe and parts of Asia	(Hubálek and Halouzka, 1997; Baranton et al., 1998; Li et al., 1998)
B. valaisiana	Wang et al. (1997)	Central Europe, Ireland, Great Britain and the Netherlands	(Hubálek and Halouzka, 1997; Kurtenbach et al., 1998)
B. lusitaniae	Levy et al. (1993)	Portugal, Tunisia, rarely in Central and Eastern Europe	(Le Fleche et al., 1997; de Michelis et al., 2000)
B. bissettii	Postic et al. (1998)	Slovenia and North America	(Picken et al., 1996; Strle et al., 1997; Postic et al., 1998)
B. japonica	Kawabata et al. (1993)	Japan	(Kawabata et al., 1993; Fukunaga et al., 1996a; Masuzawa et al., 1996)
B. tanuki	Fukunaga et al. (1996a)	Japan	(Kawabata et al., 1993; Fukunaga et al., 1996a; Masuzawa et al., 1996)
B. sinica	Masuzawa et al. (2001)	China	(Masuzawa et al., 2001)
B. turdii	Fukunaga et al. (1996a)	Japan	(Kawabata et al., 1993; Fukunaga et al., 1996a; Masuzawa et al., 1996)
B. andersoni	Marconi et al. (1995)	North America	(Marconi et al., 1995)

vertebrates. When the spirochaetes move from an arthropod to a vertebrate, they are changing from an environment with no antibody-based immune system to an environment where the immunological pressure is high; thus the biology of these organisms requires a high degree of adaptive capacity. In order to increase the probability of borreliae being transmitted by an infrequently feeding arthropod from one host to another, borreliae must persist in the host as long as possible. The antigens, which evoke immune reactions in the host, are mainly the surface molecules of the microorganism. Thus, a good way to escape the immune reaction is to change the molecules on the surface. This strategy is used by many microorganisms and may be accomplished in a number of ways, most of which involve DNA rearrangements (Borst, 1991). In addition, molecules on the bacterial surface mediate interaction between bacteria and host cells.

These are usually protein or glycolipid molecules. Adherence of *B. burgdorferi* s.l. to cultured epithelium, endothelium, differentiated neural cells, rat brain and glial cells has been shown (Garcia-Monco *et al.*, 1989; Thomas and Comstock, 1989; Shoberg and Thomas, 1993; Peters and Benach, 1997). The ability of *B. burgdorferi* to infect different organ and tissue types might also be receptor-mediated, analogous to the tissue tropism described for *Escherichia coli*. Thus, it is obvious that the characteristics of the surface-exposed proteins and their ability to vary their expression or undergo specific changes is important for the biology of this organism.

Morphology, Shape and Membrane Structure

Borreliae are thin, elongated, motile, wave-like bacteria (Fig. 3.1). A general feature of the spirochaetes, including *B. burgdorferi* s.l., is that they possess a fragile outer membrane surrounding the protoplasmic cylinder. This cylinder consists of a peptidoglycan layer, a cytoplasmic membrane (inner membrane) and the enclosed cytoplasmic contents (Johnson *et al.*, 1984; Barbour and Hayes, 1986).

The outer cell membrane is fluid and consists of 45–62% protein, 23–50% lipid and 3–4% carbohydrate (Barbour and Hayes, 1986). The borreliae have 7–11 bipolar flagella located in the periplasmic space where they are attached to the poles and wrapped around the cell cylinder, giving the bacterium its characteristic flat wave shape (Barbour and Hayes, 1986; Goldstein *et al.*, 1994; Motaleb *et al.*, 2000). Unlike other bacterial flagella, those of the spirochaetes are endoflagella, located in the periplasmic space between the protoplasmic cylinder and the outer cell membrane and attached to the poles of the bacterial cell.

More than 100 polypeptides have been identified in the outer cell membrane of *B. burgdorferi* s.l. (Luft *et al.*, 1989). The proteins most extensively studied are the outer surface proteins (Osps). To date, six Osps have been characterized, OspA–F. They are all surface-exposed and anchored by a lipid moiety to the fluid outer surface membrane (Brandt *et al.*, 1990).

A more complete understanding of the membrane constituents, their function and the overall membrane structure of *B. burgdorferi* s.l. will provide insight into the strategies that enable the bacterium to establish and maintain the infection during LB.

Flagella, motility and chemotaxis

The periplasmic flagella (PFs), located in the periplasmic space between the cytoplasm–peptidoglycan layer and the outer membrane, have an essential role in cell morphology and motility (Sadziene *et al.*, 1991; Goldstein *et al.*, 1994, 1996; Motaleb *et al.*, 2000). The distinct cell morphology of this bacterium is caused by the flagella that are wrapped around the protoplasmic cylinder (Burgdorfer *et al.*, 1982; Hovind Hougen, 1984; Barbour and Hayes, 1986;

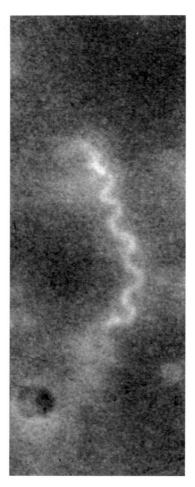

Fig. 3.1. *Borrelia burgdorferi* sensu lato stained by indirect immunofluorescence using the flagellin (FlaB) antibody H9724. (Photo courtesy of Björn Olsen.)

Goldstein *et al.*, 1996). Each PF is attached at only one end and in *Borrelia* spp. they are long enough to overlap in the centre of the cell (Johnson *et al.*, 1984). The genes encoding the flagellar proteins are *flaA* (Ge and Charon, 1997a), *flaB* (Wallich *et al.*, 1990), *flgE* (Jwang *et al.*, 1995), *fliH* and *fliI* (Ge *et al.*, 1996), and the *flgK* operons (Ge *et al.*, 1997), and are located on the chromosome. These genes have homologies to various genes in the flagellar apparatus of both animal and plant pathogens. This suggests that the flagellar apparatus and its protein export pathways are well conserved. Swimming *B. burgdorferi* s.l. cells have a flattened wave-like shape similar to that of eukaryotic flagella (Goldstein *et al.*, 1994)

A striking feature of *B. burgdorferi* s.l. is its capacity to swim efficiently in a viscous medium such as connective tissue where other bacteria are slowed down or immobilized (Kimsey and Spielman, 1990; Goldstein et al., 1994). Motility in B. burgdorferi s.l. requires an environment similar to interstitial fluid (e.g. pH 7.6, 0.15 M NaCl and viscous) (Shi et al., 1998). The relative immobility of B. burgdorferi s.l. spirochaetes in less viscous media suggests active dissemination via the skin, as optimal mobility of borreliae is seen in collagen-like viscosity (Kimsey and Spielman, 1990). Furthermore, virulence studies on a spontaneously occurring non-flagellated, non-motile, mutant strain indicate that motility is important for the pathogenesis of this organism (Sadziene et al., 1991, 1996).

The spirochaete motility apparatus is similar to that of other bacteria in that it has a filament, hook and basal body (Holt, 1978; Hovind Hougen, 1984; Barbour et al., 1986; Brahamsha and Greenberg, 1988; Charon et al., 1992). However, the PFs of spirochaetes differ from the flagella of most other bacteria as they are composed of two classes of proteins instead of one class (Cockayne et al., 1987; Norris et al., 1988; Koopman et al., 1992; Trueba et al., 1992; Rosey et al., 1995; Ge et al., 1998). These two classes are the flagellar outer sheath protein FlaA and the core protein FlaB. FlaA is unique to spirochaetes whereas FlaB shows homology to flagellar proteins of other bacteria (Magnarelli et al., 1987; Wallich et al., 1990). Until recently it was a general belief that the PFs of B. burgdorferi differed from those of other spirochaetes since the flagellar core protein FlaB was not thought to be surrounded by an outer sheath protein. However, using a modified extraction technique, Ge and Charon identified a FlaA homologue, showing 54–58% similarity to FlaA proteins from other spirochaetes (Ge and Charon, 1997a; Ge et al., 1998). In other spirochaetes, the FlaA and FlaB proteins are among the most abundant cell proteins (Koopman et al., 1992; Li et al., 1993; Norris, 1993). Depending on species, the PFs consist of one to two different FlaA proteins and two to four different FlaB proteins (Brahamsha and Greenberg, 1988; Norris et al., 1988; Charon et al., 1992; Koopman et al., 1992; Trueba et al., 1992; Li et al., 1993; Ge et al., 1998). In B. burgdorferi, the PFs are composed of only FlaA (39 kDa) and FlaB (41 kDa), where FlaB is one of the most abundant proteins and FlaA is expressed in lower amounts.

Chemotaxis studies have revealed that rabbit serum acts as an attractant while ethanol and butanol act as repellents (Shi et al., 1998). Interestingly, B. burgdorferi does not exhibit chemotaxis towards common sugars or amino acids (Shi et al., 1998). It has also been noted that spirochaetes grown in tissue chambers implanted in rats (host-adapted bacteria) show more vigorous motility compared with spirochaetes grown in vitro (Akins et al., 1998). The chemotaxis genes cheA, cheE (Trueba et al., 1997), cheW and cheY (Fraser et al., 1997; Ge and Charon, 1997b; Trueba et al., 1997) have been identified in B. burgdorferi s.l. In E. coli, chemotaxis is achieved by regulating the direction of rotation of the flagellar bundle. In B. burgdorferi and other spirochaetes, this is not easily achieved since they have two flagellar bundles, one at each cell end. In order to switch the direction of swimming, they have to rotate in different directions and then switch synchronously. This type of motility is observed as smooth swimming, used when an attractant is present. Another type of observed swimming pattern, a flexing movement, is observed in the presence of repellents. The flexing movement is the result of two bundles rotating in the same direction (Charon *et al.*, 1992; Goldstein *et al.*, 1994; Shi *et al.*, 1998).

Structure of the outer membrane

Since *B. burgdorferi* s.l. possesses both an outer membrane and a cytoplasmic membrane, the surface of this bacterium is analogous to enteric Gram-negative bacteria. However, its membrane composition differs from that of Gram-negative bacteria and has several distinct features (Fig. 3.2). For example, *B. burgdorferi* s.l. has an extraordinarily high abundance of membrane proteins covalently modified with lipids (Brandt *et al.*, 1990; Fraser *et al.*, 1997). Another characteristic is the absence of phosphatidylethanolamine (Belisle *et al.*, 1994) and lipopolysaccharide (LPS) (Takayama *et al.*, 1987), and the presence of glycolipid antigens other than LPS (Eiffert *et al.*, 1991; Wheeler *et al.*, 1993; Belisle *et al.*, 1994). Glycolipids represent about 50% of the total lipids and comprise only galactose as monosaccharide constituents (Berg, 2001; Hossain *et al.*, 2001). The existence of LPS on the surface of *B. burgdorferi* s.l. has been controversial, although most evidence currently points to its absence (Beck *et al.*, 1985; Takayama *et al.*, 1987; Cinco *et al.*, 1991; Schoenfeld *et al.*, 1992; de Souza *et*

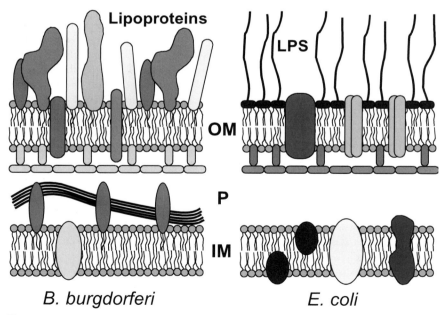

Fig. 3.2. Comparison of *Borrelia burgdorferi* sensu lato and *Escherichia coli* cell envelopes. IM, inner membrane; LPS, lipopolysaccharide; OM, outer membrane; P, protein.

al., 1992). The *B. burgdorferi* outer membrane contains a relatively low density of transmembrane-spanning proteins as determined by freeze–fracture electron microscopy studies (Walker *et al.*, 1991; Radolf *et al.*, 1994; Jones *et al.*, 1995). This may explain why *B. burgdorferi* is more susceptible to disruption by routine physical manipulations such as centrifugation and resuspension and is more susceptible to detergents compared with Gram-negative bacteria. It is very likely that the unusual surface of this bacterium contributes to the unique properties that *B. burgdorferi* s.l. possesses, e.g. the ability to survive in both the mammalian host and the tick and to evade the mammalian immune system. Thus, a comprehensive analysis of the membrane architecture and constituents would yield insight into the unique physiological mechanisms by which the bacterium survives in diverse environments and may also elucidate pathogenic mechanisms operative during LB.

Blebs

B. burgdorferi s.l. is capable of producing extracellular vesicles (blebs) when grown in laboratory culture media (Barbour and Hayes, 1986). From earlier studies on the aetiologic agent of relapsing fever, it is evident that a number of different chemical and physical agents can disturb the bacteria, i.e. when the borrelial cells are treated with penicillin, freeze–thawing, prolonged cultivation, or the addition of a specific antibody together with a complement source, the spirochaetes form outer membrane blebs which detach from the parent organism (Barbour and Hayes, 1986). In the case of *B. burgdorferi*, the exposure to stress from the mammalian milieu, even in the absence of anti-*B. burgdorferi* antibodies, is enough to trigger the production of membrane-bound and possibly secreted blebs (Shoberg and Thomas, 1995). All borrelial blebs appear to contain an outer membrane, while some also contain cytoplasmic membranes (Radolf *et al.*, 1994). However, all vesicle preparations made to date have no detectable quantities of the periplasmic 41 kDa flagellin, investigated by Western blotting (Dorward *et al.*, 1991; Shoberg and Thomas, 1995). These blebs contain plasmid DNA (Garon *et al.*, 1989), including *osp* DNA (Persing *et al.*, 1994), and in their membranes the OspA and OspB proteins and another low molecular weight lipoprotein, Lp6.6, have been found (Dorward *et al.*, 1991; Katona *et al.*, 1992; Lahdenne *et al.*, 1997). The bleb membranes are also thought to contain the borrelial adhesin for human endothelial cells since they are capable of binding to these cells (Shoberg and Thomas, 1993). As an anecdote, in the first half of the 20th century an alternative viewpoint was that these large blebs or 'gemmae' represented another stage of the life cycle of spirochaetes and this encouraged the early classification of borreliae as protozoans (Barbour and Hayes, 1986).

The recently described phenomenon of cyst formation is another sign of disturbance of cell architecture. Cyst formation involves changes in membrane integrity and cell shape (Alban *et al.*, 2000). It is apparently a starvation response, which can be induced by cultivation in serum-free BSKII medium. It is a slow but active process requiring protein synthesis and is reversible in up to 53% of the cells. It is not known whether cyst formation is important for persistence of borreliae in the host and the explanation for survival during antibiotic therapy.

The outer membrane proteins

There is much evidence that *B. burgdorferi* s.l. contains an unusually high number of proteins covalently modified by lipids (e.g. lipoproteins). The recently published genome sequence of *B. burgdorferi* s.s. revealed 132 open reading frames (corresponding to about 4.9% of the coding sequence) that contained the signal sequence characteristic for lipoproteins and 32 additional open reading frames that had other similarities to lipoproteins (Hayashi and Wu, 1990; Fraser *et al.*, 1997; Casjens *et al.*, 2000). Typically, proteins involved in interaction with the host are encoded on plasmids and, on the *B. burgdorferi* s.l. plasmids, putative lipoproteins make up at least 14.5% of the open reading frames (Casjens *et al.*, 2000). The complete sequence of other related bacterial genomes revealed lower proportions of putative lipoproteins. For example, in *Treponema pallidum* and *Helicobacter pylori*, only 2.1% and 1.3%, respectively, of the coding sequences were potential lipoproteins (Tomb *et al.*, 1997; Fraser *et al.*, 1998; Casjens *et al.*, 2000). Direct evidence for a large number of lipoproteins in *B. burgdorferi* was obtained by metabolic labelling of *in vitro*-growing cultures of *B. burgdorferi* with [^{3}H]-palmitic acid, which revealed an abundance of metabolically labelled lipoproteins (Brandt *et al.*, 1990). The function of the lipid moiety attached to lipoproteins is to anchor these polypeptides to the appropriate membrane. These modifications can then ensure membrane localization of otherwise hydrophilic proteins. Illustrating this, three *T. pallidum* lipoproteins are hydrophilic when synthesized in an *in vitro* transcription–translation system where acylation does not occur (Chamberlain *et al.*, 1989; Swancutt *et al.*, 1989) and a recombinant non-lipidated form of OspA is soluble in aqueous solutions (Dunn *et al.*, 1990).

The modification of lipoproteins occurs in several steps. First, the signal sequence LXYC specifies the addition of diacylglycerol in a thioether linkage to the sulphur of the cysteine side chain. Then signal peptidase II cleaves at the amino side of the cysteine residue and a third fatty acid is added in an amide linkage to the new amino terminus (Hayashi and Wu, 1990). The enzymes responsible for these reactions were first discovered in *E. coli* (Sankaran *et al.*, 1995). It is likely that *B. burgdorferi* lipoproteins are generated in a similar fashion since *B. burgdorferi* encodes enzymes highly homologous to the *E. coli* enzymes that are required for lipoprotein maturation (Fraser *et al.*, 1997).

It appears that the immunopotentiating activities of bacterial lipoproteins are mostly conferred to their lipid constituents since the lipid moiety can elicit an inflammatory response (Bessler *et al.*, 1985; Lex *et al.*, 1986; Hoffmann *et al.*, 1988; Deres *et al.*, 1989; Erdile *et al.*, 1993). Further studies have revealed that synthetic lipohexapeptides of varying peptide composition, but sharing a tripalmitoyl-cysteine residue, were equally effective (Radolf *et al.*, 1995b).

However, non-lipidated recombinant OspA is still recognized by antibodies from LB patients (Dunn et al., 1990).

Prokaryotic lipoproteins are secreted through the Sec-dependent general secretory pathway and, in general, the periplasmic space is the final destination for their polypeptide portions (Pugsley, 1993). This hypothesis is supported by the fact that lipoproteins are rarely surface-exposed in Gram-negative bacteria and that lipidated OspA expressed in *E. coli* is associated with the cytoplasmic membrane (Dunn et al., 1990; Pugsley, 1993). In *T. pallidum*, the agent for venereal syphilis, all lipoproteins appear to be located beneath the surface (Cox et al., 1992; Radolf, 1994).

In *B. burgdorferi* s.l. there is evidence for outer-membrane association of many lipoproteins, and in some cases lipoproteins appear to be located in both the outer and inner (cytoplasmic) membranes. Many borrelial proteins are surface-exposed based on experimental evidence, e.g. proteinase K accessibility assays of intact cells and immunogold labelling (Barbour et al., 1984; Bunikis et al., 1995; Carroll et al., 2001; Noppa et al., 2001). Except for minor differences in relative abundance, the lipoprotein constituents in outer membranes from borreliae labelled with [^{3}H]-palmitic acid were essentially identical to those of whole cells (Radolf et al., 1995a). However, not all lipoproteins in *B. burgdorferi* s.l. are located in the outer membrane since the protoplasmic cylinders contained lipoproteins not observed in the outer membranes (Radolf et al., 1995a). Several studies suggest that the predominant location of the lipoproteins OspA and OspB is in the periplasmic space, but these proteins are also found in the fluid outer membrane (Brusca et al., 1991; Bledsoe et al., 1994; Radolf et al., 1995a; Skare et al., 1995; Carroll and Gherardini, 1996; Cox et al., 1996).

In contrast to the highly abundant lipoproteins in the borrelial outer membrane, the abundance of transmembrane-spanning proteins in the outer-membrane proteins of *B. burgdorferi* s.l. is very low compared with that of Gram-negative bacteria. However, when compared with another spirochaete, *T. pallidum*, the membrane-spanning proteins are approximately tenfold more abundant in *B. burgdorferi* s.l., as determined by freeze–fracture electron microscopy studies (Walker et al., 1991; Radolf et al., 1994).

The major Osp lipoproteins

Several lipoproteins have been characterized in *B. burgdorferi* s.l. Many of these proteins are called outer surface proteins, Osps, and most of these proteins do not have any significant sequence or structural homology to proteins of known function. When various *B. burgdorferi* s.l. isolates are compared, the Osps show considerable heterogeneity within and among the different species.

Both OspA and OspB are encoded by the same operon on a 49 kb linear plasmid (Barbour and Garon, 1988; Bergström et al., 1989). The first gene that was cloned and sequenced in *B. burgdorferi*, *ospA*, encodes the most studied protein in these bacteria (Howe et al., 1986; Bergström et al., 1989). In an attempt to reveal the function of OspA, the crystal structure of a complex of recombinant unlipidated OspA and the Fab fragment of an anti-OspA monoclonal antibody was elucidated (Li *et al.*, 1997). The structure was not consistent with that of an enzyme but it had a hydrophobic cavity buried in a positively charged cleft in the carboxyterminal domain, which might be a binding site for an unknown ligand. This presence of conserved charged groups in a pronounced cleft is typical of proteins that bind small or linear polymeric ligands. OspA could thus recognize either a soluble or host surface molecule and possibly transmit this information to another membrane component, suggesting that OspA could act as a receptor (Li *et al.*, 1997). The structure of OspA also indicated that the conserved NH_2-terminus could interact with another membrane protein (Li *et al.*, 1997). OspA has been shown to be expressed by the spirochaete in the unfed tick's midgut and mediate adhesion to the midgut epithelium (Pal *et al.*, 2000). This preferential expression in the midgut suggests that OspA may have an important function in the vector. Thus, this OspA-mediated attachment to the tick midgut epithelium provides a possible mechanism for how stage-specific protein expression can contribute to pathogenesis during the natural cycle of borrelial spirochaetes (Pal *et al.*, 2000).

OspA and OspB have, in a large number of *in vivo* and *in vitro* experiments, been shown to play important roles in the pathogenesis of LB. Studies have shown that these two major proteins are involved in adherence to and penetration of mammalian host cells. Thus, anti-OspA monoclonal antibodies inhibit adhesion of *B. burgdorferi* to endothelial cell surfaces, and during subculturing in BSKII medium the spirochaetes lose OspB and their infectivity (Schwan and Burgdorfer, 1987; Schwan *et al.*, 1988; Comstock *et al.*, 1993). Furthermore, mutants lacking OspB do not penetrate into endothelial cells and their infectivity is from 30- to 300-fold lower than that of wild-type spirochaetes (Sadziene *et al.*, 1993b).

Development of what has been termed treatment-resistant Lyme arthritis has been shown to be associated with particular major histocompatibility complex class II alleles and immunoreactivity to the OspA protein (Chen *et al.*, 1999). Most of the patients with treatment-resistant Lyme arthritis have HLA-DRB1*0401 or related alleles, and the severity and duration of their arthritis correlate with cellular and humoral immune responses to the *B. burgdorferi* s.s. OspA protein. In a gene bank homology search, one human protein, the lymphocyte function-associated antigen-1 (hLFA-1), which had sequence homology with $OspA_{165-173}$ and predicted binding in the DRB1*0401 molecule, was identified (Gross *et al.*, 1998). Only individuals with treatment-resistant Lyme arthritis, in contrast to other forms of arthritis, generated these responses to OspA, hLFA-1 and the highly related peptide epitopes. From these results, Steere and co-workers proposed a model of molecular mimicry affecting genetically susceptible individuals to explain the autoimmune mechanisms in antibiotic treatment-resistant Lyme arthritis (Steere *et al.*, 2001; Trollmo *et al.*, 2001).

OspC is homologous to proteins in both the relapsing fever species *Borrelia hermsii* and other *Borrelia* spp. Because of this, these proteins were suggested as defining a genus-wide protein family (Carter *et al.*, 1994). OspC is very heterogeneous and a large degree of molecular polymorphism of the *ospC* gene has been shown when analysing the different *Borrelia* genospecies (Jauris-Heipke *et al.*, 1995; Livey *et al.*, 1995). So far 19 major OspC groups have been identified (Wang, I.N. *et al.*, 1999). Molecular analysis of the *ospC* sequences from different *Borrelia* genospecies suggests that recombination between different *ospC* alleles can occur frequently and that this genetic exchange is mediated by lateral transfer between *ospC* sequences. This can be achieved either by lateral transfer of *ospC* genes from the same genospecies or between different *Borrelia* spp. (Livey *et al.*, 1995; Wang, G. *et al.*, 1999b).

There is an inverse relationship between the expression of OspA and OspC *in vitro*: strains with low expression of OspC have strong expression of OspA and vice versa (Margolis and Rosa, 1993). During tick feeding, the spirochaete alters its expression of Osps. In non-engorged ticks, the spirochaete expresses and produces OspA and probably OspB on the surface. Furthermore, OspA has been shown to mediate adhesion to the midgut epithelium in the unfed tick's midgut (Pal *et al.*, 2000). The production of OspA and OspB proteins decreases at the time of feeding, and instead the spirochaete promotes production of OspC (Schwan *et al.*, 1995; Ohnishi *et al.*, 2001). OspC expression is induced 36–48 h into the blood meal and probably mediates the escape of the spirochaete from the tick midgut via the haemolymph to the salivary glands from where it enters the new vertebrate host (Anderson *et al.*, 1989; Schwan *et al.*, 1995; Schwan and Piesman, 2000; Seshu and Skare, 2000; Ohnishi *et al.*, 2001). The OspC protein is encoded on a 27 kb circular plasmid. Recent studies have suggested that the expression of the *ospC* gene can be influenced, in an inverse relationship, by the expression of the *ospA* and *ospB* genes (Sadziene *et al.*, 1993a; Jonsson and Bergström, 1995; Schwan *et al.*, 1995; Schwan and Piesman, 2000).

Antibodies against OspA in the blood of vertebrate hosts kill spirochaetes in the tick midgut before dispersal to the salivary glands and thereby block transmission to the vertebrate host (de Silva *et al.*, 1996, 1999). Antibodies against OspC are also efficient, probably blocking the transmission by preventing migration of spirochaetes to the salivary glands (Gilmore and Piesman, 2000). Recombinant OspA is an efficient vaccine against LB, and immunization with OspC confers protective immunity in animal studies (Gilmore *et al.*, 1996; Sigal *et al.*, 1998; Steere *et al.*, 1998). However, the use of these vaccines may be limited by the apparent variability of these proteins (Wilske *et al.*, 1996).

OspD is an abundant 28 kDa surface-exposed component of *B. burgdorferi* s.l. (Norris *et al.*, 1992). The *ospD* gene is located on a 38 kb linear plasmid present in most low-passage strains of *B. burgdorferi*. Molecular cloning and sequence analysis of the *ospD* gene locus revealed an open reading frame encoding a 28,436 Da polypeptide with a putative signal peptidase II leader sequence (Norris *et al.*, 1992). Sequence analysis has demonstrated the presence of various numbers of 17 bp repeated sequences in the upstream control (promoter) region of the gene (Norris *et al.*, 1992; Marconi *et al.*, 1994). However, only one transcription initiation site was active as determined by primer extension analysis (Norris *et al.*, 1992).

OspE and OspF are the founding members of a protein family known as the **O**sp**EF-r**elated **p**roteins (Erps) (Lam *et al.*, 1994b; Stevenson *et al.*, 1996). Individual bacteria carry multiple *erp* operons, each located on a different plasmid of the cp32 family (Stevenson *et al.*, 1996; Casjens *et al.*, 1997a; Akins *et al.*, 1999; Stevenson *et al.*, 2000a). Clonal *B. burgdorferi* with as many as nine different *erp* loci per cell have been characterized (Stevenson *et al.*, 1996, 1998; Casjens *et al.*, 1997a; Akins *et al.*, 1999; El-Hage *et al.*, 1999; Casjens *et al.*, 2000). Since several Erps are immunogenic, these lipoproteins are surface-exposed and may therefore be important for the virulence of *B. burgdorferi*. Surface localization of these infection-associated proteins indicates the potential for interactions of Erps with vertebrate tissues. At least some Erps can bind complement factor H, suggesting that these proteins might be important for the borreliae during mammalian infection by inhibition of complement activation. Recombination of the genes encoding the Erp antigens might contribute to the evasion of the mammalian immune response and could play roles in the establishment and persistence of *B. burgdorferi* s.l. infections in mammalian hosts (Hellwage *et al.*, 2001). All LB spirochaetes contain *erp* genes, but these genes have not been found in other species of the genus *Borrelia*, indicating that Erps perform a function unique to the biology of LB borreliae (Stevenson *et al.*, 2000a,b).

Recently, an immunodominant Osp of *B. afzelii*, Osp17, was identified (Jauris-Heipke *et al.*, 1999). Moreover, the *B. burgdorferi* genome sequence defined several hypothetical lipoproteins, which might increase the number of possible Osp proteins in the future.

The decorin-binding proteins

Microbial adhesion to and colonization of host tissues are an important event in the bacterial infection process. Most pathogenic bacteria express adhesins that can participate in attachment mechanisms that might result in tissue colonization. Spirochaetes injected in the skin by an ixodid tick are found associated with collagen fibres in the extracellular matrix (ECM). However, *B. burgdorferi* s.l. do not attach directly to collagen but to decorin, a small leucine-rich proteoglycan (SLRP) that is associated with collagen fibres (Guo *et al.*, 1995, 1998). Decorin is a collagen-associated ECM proteoglycan found in the skin (the site of entry for the spirochaete) and in many other tissues. There are two borrelial-encoded adhesins that can bind to decorin, DbpA and DbpB (**d**ecorin-**b**inding **p**rotein) (Guo *et al.*, 1995; Hanson *et al.*, 1998). DbpA and DbpB were also isolated as antigens that were recognized by sera from mice infected with *B. burgdorferi* s.s. (Feng *et al.*, 1998; Hagman *et al.*, 1998). Both DbpA and DbpB are outer membrane-associated and may function as adhesins to facilitate bacterial adherence to mammalian tissue. However, alternative adhesion mechanisms may also be used by the spirochaetes depending on the tissue target and stage of dissemination (Duray, 1992; van Mierlo *et al.*, 1993). To further investigate the role of Dbps in the pathogenesis of LB, a decorin-deficient (Dcn–/–) mouse was developed and used in experimental LB (Brown *et al.*, 2001). In that study it was demonstrated that Dcn−/− mice were more resistant to LB, which suggests that decorin could be a limiting factor as a substrate for adherence of *B. burgdorferi* s.l. Interestingly, this resistance to *B. burgdorferi* s.l. in Dcn−/− could be reduced by increasing the infection dose, which indicates that additional adhesion ligands can participate in the disease process *in vivo* (Brown *et al.*, 2001).

Another adhesin, the fibronectin-binding protein BBK32, has been identified in all LB genospecies (Probert and Johnson, 1998). Molecular analysis of the *bbk32* gene and gene product defined its fibronectin-binding domain, which shared sequence homology with a fibronectin-binding peptide in *Streptococcus pyogenes*, indicating a common mechanism of adhesion for *B. burgdorferi* s.l. and *S. pyogenes* (Probert *et al.*, 2001).

The oligopeptide permease (Opp)

Homologues of all five subunits of the peptide-binding component OppA (*oppA*I–V) as well as the other components of oligopeptide permease (OppB, OppC, OppD, OppF) were identified and characterized in *B. burgdorferi* s.l. (Das *et al.*, 1996; Bono *et al.*, 1998; Kornacki and Oliver, 1998). It is not known if oligopeptide permease has a sensing capability in *B. burgdorferi* s.l. similar to other bacteria although an interesting feature to note is that Opp-like proteins are involved in expression of virulence determinants in other bacteria (Payne and Gilvarg, 1968; Goodell and Higgins, 1987; Finlay and Falkow, 1989; Gominet *et al.*, 2001). The Opp system is particularly interesting in *B. burgdorferi* s.l. since it was evident from the genome sequence analysis that only a rudimentary machinery for synthesis or transport of amino acids exists. In fact, only a single intact peptide transporter operon could be found. From the genome analysis of *B. burgdorferi* s.l. it is also apparent that the *opp* operon differs from that of *E. coli* as it has three separate substrate-binding proteins, OppA-1, 2 and 3. Interestingly, all of the *B. burgdorferi* s.l. OppA proteins were found to complement *E. coli* OppBCDF to form a functional peptide transport system for nutritional purposes (Lin *et al.*, 2001).

The *vmp*-like sequence, Vls, of *B. burgdorferi* s.l.

To increase their chances of transmission by an arthropod from one host to another, borreliae persist in the host for as long as possible. Persistence can be achieved by a mechanism called antigenic variation, and this mechanism has been studied for several years in relapsing fever borreliosis where new serotypes appear during relapses as a change in novel surface proteins called variable major proteins (Vmps). In the relapsing fever spirochaete *B. hermsii*, at least 26 different surface-exposed lipoproteins encoded by genes situated on linear plasmids are alternately expressed and presented, one at a time, on the bacterial surface

(Barbour, 1990, 1993; Burman *et al.*, 1990; Kitten and Barbour, 1990; Barbour *et al.*, 1991a). Only one *vmp* gene, located at the expression site, is expressed whereas the other genes are kept silent at storage sites. Antigenic variation occurs when a silent *vmp* gene replaces the expressed *vmp* gene by homologous recombination (Plasterk *et al.*, 1985; Barbour *et al.*, 1991b; Restrepo and Barbour, 1994; Restrepo *et al.*, 1994).

In *B. burgdorferi* s.s a 28 kb linear plasmid was found to have a *vmp*-like sequence (*vls* locus) that resembles the Vmp system in *B. hermsii* (Zhang *et al.*, 1997). Fifteen non-expressed *vls* cassettes can be recombined into a functional and expressed *vlsE* region resulting in antigenic variation of the expressed lipoprotein (Zhang *et al.*, 1997) and involving partial gene conversion of the expressed *vls* gene by parts of the *vls* genes at silent loci in a unidirectional process (Zhang and Norris, 1998a,b). The mechanism of Vls antigenic variation in LB *Borrelia* spp. is similar to the mechanism that is responsible for the recombination between the expressed pilin gene *pilE* and silent copies of *pilS* in *Neisseria gonorrhoeae* (Hagblom *et al.*, 1985; Haas and Meyer, 1986; Zhang *et al.*, 1992). VlsE contains two invariable domains and one variable domain that includes six variable and six invariable regions. Among molecules that undergo antigenic variation, VlsE is unusual in that more than 75% of its primary structure is invariable (Liang and Philipp, 1999). The relevance of the antigenic variation of VlsE in the pathogenesis of *B. burgdorferi* s.l. remains unclear. Even though there is a correlation between the presence of the plasmid that encodes the Vls lipoprotein and infectivity, the Vls proteins appear to be a minor constituent of the bacterial surface. However, it is likely that the antigenic variation expressed by the *vls* gene locus of *B. burgdorferi* s.l. is used in order to persist in mammals and to evade the host immune response.

Little is known about the signals that trigger recombination events at the *vls* locus. Recombination occurs *in vivo* as early as 4 days after experimental infection of mice but not *in vitro*, suggesting that the mammalian host provides the signal for *vls* recombination (Zhang and Norris, 1998a). The role of the mammalian host is important since the *vlsE* recombination rates by *B. burgdorferi* s.l. were found to be lower in IFN-γ R-deficient mice than in control animals. These results suggest that the murine immune response can promote the *in vivo* adaptation of *B. burgdorferi* s.l. (Anguita *et al.*, 2001).

In addition to the importance of VlsE in the immune evasion of *B. burgdorferi* s.l., the *ospE* family of genes is able to undergo mutational changes during infection, leading to the generation of antigenically distinct OspE protein variants. This adds OspE to the repertoire of antigens that can be important for adaptation and immune evasion during LB (Sung *et al.*, 2000).

Integral membrane proteins

Bacteria other than *B. burgdorferi* s.l. commonly ensure correct localization of the membrane proteins by including transmembrane-spanning domains in their proteins. Proteins containing such domains appear to be rare in *B. burgdorferi* s.l.

This is based on the observation that biosynthetic labelling of borrelia cultures with [^{14}C]-amino acids identified only a few amphiphilic polypeptides that do not co-migrate with lipoproteins (Brandt *et al.*, 1990). It was also shown that these putative transmembrane proteins were in relatively low abundance when visualized by freeze–fracture electron microscopy (Walker *et al.*, 1991; Radolf *et al.*, 1994). Virtually none of the proteins were recognized by sera from patients with chronic LB (Brandt *et al.*, 1990). Thus, *B. burgdorferi* s.l. contains two classes of integral membrane proteins: abundant lipoproteins and rare transmembrane-spanning proteins.

Three characterized membrane proteins have been predicted to contain transmembrane-spanning domains, P66 (Oms66), Oms28 (Oms refers to **o**uter **m**embrane **s**panning) and P13 (Bunikis *et al.*, 1995; Skare *et al.*, 1996, 1997; Noppa *et al.*, 2001). P66 and Oms28 have also been shown to exhibit porin activities (Skare *et al.*, 1996, 1997). In Gram-negative bacteria, porins are known to form large water-filled channels that allow the diffusion of hydrophilic molecules into the periplasmic space, thus facilitating transport of molecules across the bacterial membrane (Cowan *et al.*, 1992). The third protein, P13, contains transmembrane-spanning domains and was shown to be surface-exposed by immunofluorescence assays, immunoelectron microscopy and protease sensitivity assays (Sadziene *et al.*, 1995; Noppa *et al.*, 2001). The deduced sequence of the P13 peptide revealed possible signal peptidase type I cleavage sites and the computer analysis predicted that P13 has three transmembrane-spanning domains. Mass spectrometry, *in vitro* translation and N- and C-terminal amino acid sequencing analyses have indicated that P13 is very unusual as it is post-translationally processed at both ends and modified by an unknown mechanism (Noppa *et al.*, 2001).

Other surface-exposed membrane proteins

We do not intend to cover all aspects and importance of every membrane protein of *B. burgdorferi* s.l. However, we will add some comments below on the characteristics of the outer surface-associated proteins and protein families P22, P35, Bmp, Mlp, RevA and Epp. Many of these proteins are probably important both as virulence determinants and as possible antigens for prophylactic and diagnostic use.

P22

P22 is a 21.8 kDa protein encoded from the linear chromosome and is homologous to the IpLA7 protein (98.5% homology) (Simpson *et al.*, 1991a; Wallich *et al.*, 1993). The P22 protein is a surface-exposed lipoprotein, as demonstrated by [^{3}H]-palmitate and immunogold labelling (Lam *et al.*, 1994a).

P35

This is yet another *B. burgdorferi* s.l. lipoprotein that is recognized by sera from *B. burgdorferi* s.l.-infected mammals. Molecular analysis revealed that the *p35* gene was selectively expressed *in vivo* and is upregulated or initiated during the post-logarithmic bacterial growth phase *in vitro* (Indest *et al.*, 1997; Indest and Philipp, 2000).

Bmp (P39)

BmpA, BmpB, BmpC and BmpD constitute a family of homologous lipoproteins that are immunogenic during infection in vertebrate hosts. The genetic organization of the *bmp* region has been characterized for a variety of *B. burgdorferi* s.l. strains by Southern hybridization, PCR amplification and DNA sequencing (Simpson *et al.*, 1991b; Gorbacheva *et al.*, 2000).

Mlp

Mlp is the designation of a *B. burgdorferi* s.l. family of multicopy lipoproteins encoded in at least seven versions on the circular cp32 and cp18 plasmids (Porcella *et al.*, 1996; Yang *et al.*, 1999). The *mlp* transcription has an *ospC*-like temperature regulation as the *mlp* genes are not expressed at low temperature but expression is increased when shifted to higher temperatures (Porcella *et al.*, 2000; Yang *et al.*, 2000).

RevA

RevA is another *B. burgdorferi* s.l. surface-exposed protein that is expressed at high temperature in the mammalian host. The level of RevA expression was also found to be dependent upon the pH of the culture (Carroll *et al.*, 2001).

EppA

The exported plasmid protein A (*eppA*) genes are only expressed during infection in the mammalian host (Champion *et al.*, 1994). The *eppA* genes are located on either the plasmid cp9-1 or cp9-2 (Miller *et al.*, 2000). The widespread occurrence of *eppA* genes and 9 kb circular plasmids suggests that they contribute to the survival of *B. burgdorferi* s.l. in nature, and possibly to the pathogenesis of LB (Champion *et al.*, 1994; Miller *et al.*, 2000).

Genome and Genome Organization

B. burgdorferi s.l. has a low guanine and cytosine content of 28–30.5 mole % (Schmid *et al.*, 1984; Bergström *et al.*, 1989; Fraser *et al.*, 1997). The low DNA homology and the low guanine and cytosine content separate the genus *Borrelia* from the genera of *Treponema* and *Leptospira* (Hyde and Johnson, 1984).

The genetic organization of *Borrelia* spp. is unusual among prokaryotes in the sense that they contain a small linear chromosome approximately 1 Mb in size (Baril *et al.*, 1989; Ferdows and Barbour, 1989; Bergström *et al.*, 1992; Davidson *et al.*, 1992; Casjens and Huang, 1993; Fraser *et al.*, 1997) and numerous linear and circular plasmids (Barbour and Garon, 1987; Barbour, 1988; Casjens *et al.*, 1997a). Most of the genome of the *B. burgdorferi* s.s. strain B31 was published in 1997 (Fraser *et al.*, 1997) and it was completed in 2000 (Casjens *et al.*, 2000). The sequencing of the *B. burgdorferi* s.s. B31 genome revealed a total genome size of 1.52 Mbp: 910,725 bp on the linear chromosome, and 611 kbp divided on 12 linear and nine circular plasmids (Fig. 3.3). There were only 1283 putative genes

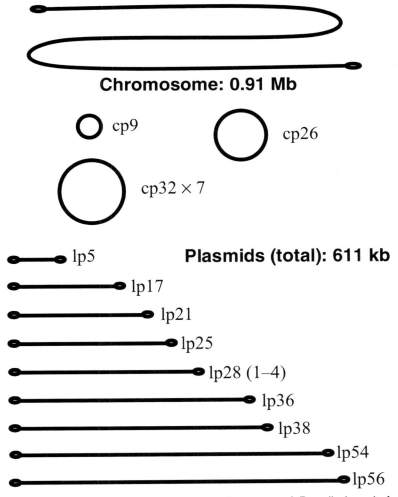

Fig. 3.3. Schematic representation of the total genome of *Borrelia burgdorferi* sensu stricto B31.

(Fraser *et al.*, 1997), compared with the 4405 genes present in *E. coli*. The small number of genes in *B. burgdorferi* s.s is similar to the number of genes in the related spirochaete *T. pallidum* (Fraser *et al.*, 1998) and the obligate intracellular bacterium *Mycoplasma genitalium* (Fraser *et al.*, 1995). As in *T. pallidum* and *M. genitalium*, a number of genes encoding cellular biosynthetic enzymes are missing in *B. burgdorferi* s.s. B31. Since *Mycoplasma* spp. and spirochaetes are only distantly related, the limited metabolic capacity probably reflects convergent evolution by the loss of genes, allowed by the parasitic lifestyle of these bacteria.

A linear chromosome has been described in only two other eubacteria, *Streptomyces lividans* (Lin *et al.*, 1993) and *Agrobacterium tumefaciens* (Allardet-Servent *et al.*, 1993), while linear plasmids have been observed in *Streptomyces* spp., *Rhodococcus fascians* and *Thiobacillus versutus* (Kinashi *et al.*, 1987; Wlodarzyk and Nowicka, 1988; Crespi *et al.*, 1992). The vast majority of bacterial species have a circular chromosome and circular plasmids. The ends of the *Borrelia* spp. chromosome and linear plasmids are covalently closed loops similar to eukaryotic telomeres (Hinnebusch *et al.*, 1990; Casjens *et al.*, 1997b). Replicons with covalently closed hairpin ends have also been described in several other organisms including poxviruses, African swine fever virus, *Chlorella* viruses, the *E. coli* N15 prophage, certain mitochondrial DNAs, mitochondrial plasmids and plastid DNAs (Meinhardt *et al.*, 1997; Nosek *et al.*, 1998; Rybchin and Svarchevsky, 1999; Volff and Altenbuchner, 2000).

The problem of the replication of *B. burgdorferi* s.l. DNA molecules has been discussed ever since the discovery of its linear DNA replicons. The mechanisms by which the hairpin telomeres are processed during replication have now been investigated by Chaconas *et al.* (2001). They showed that a synthetic 140 bp sequence, having the predicted structure of a replicated telomere, could function as a viable substrate for telomere resolution *in vivo*, and as such was sufficient to convert a circular replicon to a linear form. They suggested that the final step in the replication of linear *B. burgdorferi* s.l. replicons is a site-specific DNA breakage and reunion event to regenerate covalently closed hairpin ends (Chaconas *et al.*, 2001).

The linear chromosome (the maxi chromosome)

The genomic sequence of the *B. burgdorferi* linear chromosome contains 853 chromosomal genes encoding proteins involved in fundamental processes such as replication, transcription, translation, energy metabolism and transport across the membranes. The average size of the open reading frames is 992 bp, and 93% of the genome is within coding sequences. The GC content of only 28.6% of the *B. burgdorferi* s.s. chromosome has been confirmed and all 61 possible codon triplets are used, with a marked bias towards the AU-rich triplets (Hyde and Johnson, 1984; Johnson *et al.*, 1984; Schmid *et al.*, 1984; Fraser *et al.*, 1997).

Linear plasmids (mini chromosomes) and circular plasmids

The genome sequence of the entire *B. burgdorferi* s.s. B31 isolate MI was partially published in 1997 and completed in 2000. From those studies it was shown that *B. burgdorferi* s.s. B31 carries 21 extrachromosomal DNA elements, which is the largest number known for any bacterium (Fraser *et al.*, 1997; Casjens *et al.*, 2000). Among these are 12 linear and nine circular plasmids, in total 610,694 bp. Each linear plasmid of *B. burgdorferi* s.l. has a copy number of approximately one per chromosome (Hinnebusch and Barbour, 1992; Casjens and Huang, 1993), and many plasmids appear to contain homologous regions of DNA (Simpson *et al.*, 1990; Barbour *et al.*, 1996; Zuckert and Meyer, 1996; Casjens *et al.*, 1997a).

Prolonged *in vitro* cultivation (10–17 passages) results in the loss of some plasmids, both linear and circular. It seems likely that the lost plasmid-encoded genes are non-essential for growth *in vitro*. However, loss of some plasmids mediates changes in protein expression and a lost ability to infect laboratory animals (Schwan *et al.*, 1988; Norris *et al.*, 1995; Xu *et al.*, 1996; Purser and Norris, 2000). Low-passage strains are therefore often more virulent than high-passage strains, and thus these plasmids may encode key virulence determinants and/or proteins involved in infection and immune evasion. This hypothesis is supported by the observation that *B. burgdorferi* s.l. contains plasmid-encoded genes that are selectively expressed during infection of mammalian hosts (Champion *et al.*, 1994; Akins *et al.*, 1995; Suk *et al.*, 1995). The plasmids mainly contain unique genes with unknown functions, 14.5% of the plasmid genes being putative lipoprotein genes that may be involved in host–parasite interactions. Some plasmids have a high fraction of pseudogenes, possibly reflecting an ongoing rapid evolution (Casjens *et al.*, 2000).

Infectivity-associated plasmids have been identified in *B. burgdorferi* s.s. B31 using prolonged cultivation resulting in clones with high, low and intermediate infectivity phenotypes based on their frequency of isolation from needle-inoculated C3H/HeN mice. Thus, the plasmid lp25 had a direct correlation with infectivity as it was always present in clones of high or intermediate infectivity and was absent in all low-infectivity clones. In addition, the plasmid lp28-1, encoding the *vlsE* locus, also correlated with infectivity. In contrast, the plasmids cp9, cp32-3, lp21, lp28-2, lp28-4 and lp56 apparently are not required for infection in this model because clones lacking these plasmids exhibited a high-infectivity phenotype. Plasmids cp26, cp32-1, cp32-2 and/or cp32-7, cp32-4, cp32-6, cp32-8, cp32-9, lp17, lp28-3, lp36, lp38 and lp54 were consistently present in all clones examined. These results clearly indicate that the plasmids lp25 and lp28-1 encode virulence factors important for the pathogenesis of *B. burgdorferi* s.s. B31 (Purser and Norris, 2000).

Bacteriophages

Using electron microscopic studies of *B. burgdorferi* s.s., the first structures resembling bacteriophages were observed and identified by Hayes *et al.* (1983). Earlier studies also suggested that *B. burgdorferi* s.l. can carry prophages, which can be lysogenized upon treatment with ciprofloxacin (Neubert *et al.*, 1993). However, it was not until recently that a bacteriophage designated ΦBB-1of *B. burgdorferi* s.l. was molecularly characterized. In that study, a DNase-protected, chloroform-resistant molecule of DNA from the cell-free supernatant of a *B. burgdorferi* s.l. culture was recovered (Eggers and Samuels, 1999). This bacteriophage was found to be a temperate prophage derived from the 32 kb circular plasmids (cp32s) of the *B. burgdorferi* s.l. genome. Electron microscopy of samples from which the 32 kb DNA molecule was purified revealed bacteriophage particles (Eggers *et al.*, 2000, 2001a,b). These authors also successfully demonstrated that ΦBB-1 could package and transduce DNA, which is the first direct evidence of a mechanism for lateral gene transfer in *B. burgdorferi* s.l.

Genetic tools

Although sequencing of the complete genome of *B. burgdorferi* s.s., including all its various plasmids, has been achieved, most of the advances in pathogenesis research will come from experimental observations in the laboratory. This will not be possible unless good and reliable methods to mutagenize specific virulence-associated genes have been developed. Despite the fact that spirochaetes are important pathogens, spirochaete genetics are in their early stages and new genetic tools for studying these important pathogens are under way. Gene inactivation in spirochaetes has been achieved in *B. burgdorferi* s.l., *Brachyspira hyodysenteriae* and *Treponema denticola* (Tilly *et al.*, 2000). A gene of interest can be inactivated by allelic exchange or by plasmid integration and both of these methods have been used in *B. burgdorferi* s.l. Random insertional mutagenesis using a transposable element has yet to be developed for spirochaetes (Berg and Berg, 1996).

Until recently, genetic studies in *Borrelia* spp. have been hindered by the lack of an exogenous selectable marker, low transformation frequency and difficulties in growing *B. burgdorferi* s.l. on solid medium. The first selectable marker used for *B. burgdorferi* s.l. was $gyrB^r$, a mutated form of the chromosomal *gyrB* gene, which encodes the B subunit of DNA gyrase and confers resistance to the antibiotic coumermycin A_1 (Samuels *et al.*, 1994a,b). The use of coumermycin as a selectable marker is limited by a high frequency of recombination with the endogenous *gyrB* gene. A major improvement was made when Bono *et al.* (2000) developed an efficient marker for mutant selection in *B. burgdorferi* s.l. By linking the *B. burgdorferi flaA* or *flgB* promoter with the *kan* gene from Tn903, transformants were resistant to high levels of kanamycin. With this innovation the pace of gene inactivation will increase, and genes that have been inactivated in *B. burgdorferi* s.l. and published are shown in Table 3.2. However, gene inactivation has not been achieved in an infectious background and therefore analysis of phenotypes with regard to infection has not been possible. The challenges associated with inactivating genes in infectious *B. burgdorferi* s.l. are not surpris-

Table 3.2. Genes that have been inactivated in *B. burgdorferi*.

Gene	Marker	Reference
ospC	*gyrB^r*	Tilly *et al.* (1997)
*opp*AIV	*gyrB^r*	Bono *et al.* (1998)
guaA	*gyrB^r*	Tilly *et al.* (1998)
rpoS	*gyrB^r*	Elias *et al.* (2000)
gac	*gyrB^r*	Knight *et al.* (2000)
*opp*AV	kan	Bono *et al.* (2000)
flaB	*gyrB^r*	Motaleb *et al.* (2000)
chbB	*gyrB*	Tilly *et al.* (2001)

ing when considering that *Borrelia* spp. are fastidious bacteria, with limited metabolic capacity, and infectious clones generally grow more slowly and display lower plating efficiency. Surface characteristics of the bacteria, which often correlate with infectivity, are among the obstacles to effective transformation by electroporation. Other possible reasons for reduced transformation frequency could be that a restriction/modification system is present in infectious *B. burgdorferi* s.l. but absent or inactive in culture-adapted bacteria. Homologous recombination may be more tightly regulated in low-passage infectious strains and recombination with a targeted locus may take place at a lower rate. Shuttle vectors have also been developed for *B. burgdorferi* s.l. and a broad-host-range plasmid functioning as a *B. burgdorferi*–*E. coli* shuttle vector has been reported by Sartakova *et al.* (2000). Stewart *et al.* (2001) successfully transformed infectious *B. burgdorferi* s.s. using a shuttle vector including a 3.3 kb region of the *B. burgdorferi* s.l. circular plasmid 9, and this vector could be maintained in a stable form in the spirochaete (Stewart *et al.*, 2001). Optimal transformation frequencies are obtained with log-phase bacteria, large amounts of DNA (up to 50 µg per transformation) and high field strength (12.5–37.5 kV cm^{-1}). Infectious *B. burgdorferi* isolates transform with frequencies 100-fold lower than those found for high-passage, non-infectious strains. A summary of the transformation and plating procedure is shown in Fig. 3.4.

Classification and Evolution

Of the 11 known members of the *B. burgdorferi* s.l. species complex (Table 3.1), only three, *B. burgdorferi* s.s., *B. afzelii* and *B. garinii*, are known to be pathogenic in humans, but there are also unknown types of *B. burgdorferi* s.l. isolated from LB patients and the species defining process is still going on (Picken *et al.*, 1996; Strle *et al.*, 1997; Wang, G. *et al.*, 1999a). Classification of the relapsing fever spirochaetes was based on the different argasid tick vectors involved: the 'one tick vector species – one *Borrelia* species' concept. This approach for the classification of LB spirochaetes transmitted by ixodid ticks of the *I. ricinus* species complex (see Eisen and Lane, Chapter 4) was abandoned when it became

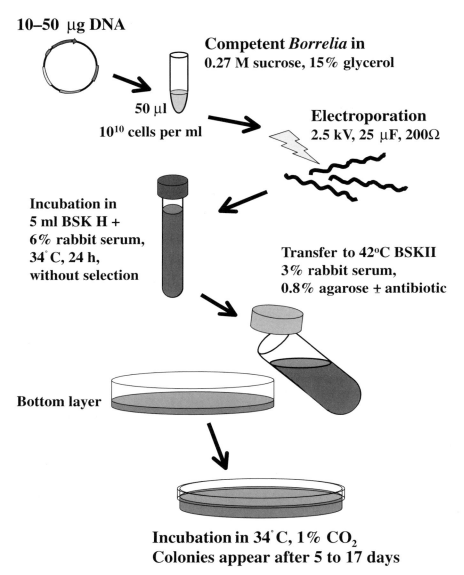

Fig. 3.4. Summary of transformation and plating methodology for *Borrelia burgdorferi* sensu lato.

apparent that there were many exceptions (Felsenfeld, 1971; Barbour and Hayes, 1986). For example, *I. ricinus* transmits at least five genospecies (Hubálek and Halouzka, 1997) and *B. garinii* has been isolated from at least four tick species: *I. ricinus* (Baranton *et al.*, 1992), *Ixodes persulcatus* (Kryuchechnikov *et al.*, 1988), *I. uriae* (Olsen *et al.*, 1993) and *Ixodes hexagonus* (Gern *et al.*, 1997).

Each species of *B. burgdorferi* s.l. can be characterized to a certain extent by its vectors, host spectrum, organotropism and geographical distribution. However, speciation is mainly based on molecular methods such as phylogenetic analysis of the 5S–23S rRNA intergenic spacer sequence, 16S rRNA gene sequence, whole genome DNA reassociation experiments, analysis of rRNA gene restriction patterns, protein electrophoresis patterns and differences in reactivity to specific murine monoclonal antibodies.

Molecular characterization

Numerous *B. burgdorferi* s.l. strains have been isolated and even more PCR fragments have been amplified from various sources. In order to determine the *Borrelia* species present in the samples several genotypic and phenotypic methods have been developed.

Among the phenotypic methods, serotyping using monoclonal antibodies against OspA and OspC is commonly performed (Wilske *et al.*, 1993, 1995; Wang, G. *et al.*, 1999c). The drawback is that not all strains express all of the investigated proteins *in vitro*. Solely determining the protein profile on SDS-PAGE is not sufficiently reliable.

Genotypic methods are not only applicable to cultivated strains but also to PCR-amplified DNA. PCR-based methods include species-specific PCR, randomly amplified polymorphic DNA (RAPD) fingerprinting or arbitrary primed (AP)-PCR, restriction fragment length polymorphism of PCR-amplified fragments (PCR-RFLP) and nucleotide sequencing (Wang, G. *et al.*, 1999c). Methods requiring cultivation of the spirochaetes are, for example, ribotyping by hybridization with rDNA-directed probes to RFLP-generated fragments of genomic DNA, pulse field gel electrophoresis (PFGE) and the labour-intensive DNA–DNA reassociation analyses. Many methods have high congruence but differing resolution at the species and subspecies levels (Wang, G. *et al.*, 1999c).

DNA–DNA reassociation analysis is one of the reference methods for bacterial species delineation (Wayne *et al.*, 1987). The general rule in bacterial taxonomy, that less than 70% similarity between genomes constitutes a different species, applies to *B. burgdorferi* s.l. (Postic *et al.*, 1994). DNA–DNA reassociation analysis is, however, laborious to perform and therefore rarely used.

Various genes have been sequenced in order to delineate and characterize *B. burgdorferi* s.l. In particular, the 16S rDNA, the 5S–23S rDNA intergenic spacer and the *fla* gene supply data representative for the whole genome (Fukunaga *et al.*, 1996b; Wang *et al.*, 1997). These sequences are very conserved, and in particular the 16S rRNA gene sequences in *B. burgdorferi* s.l are highly homologous, indicating a recent speciation (Le Fleche *et al.*, 1997). Other gene sequences, for example *p66* (Jonas Bunikis, personal communication) or the *hbb* gene (Valsangiacomo *et al.*, 1997), are more variable and may contain more phylogenetic information. The outer surface protein genes *ospA* and *ospC* are less reliable in phylogeny as they are highly variable (Theisen *et al.*, 1995; Wang *et al.*,

2000) and subject to lateral gene transfer (Rosa *et al.*, 1992; Dykhuizen *et al.*, 1993; Jauris-Heipke *et al.*, 1995; Livey *et al.*, 1995; Wang, G. *et al.*, 1999b). However, it is usually possible to determine the species correctly from *ospA* and *ospC* sequences (Wang *et al.*, 2000).

Variations in the rDNA and its intergenic spacer have been used to develop different typing systems for *Borrelia* strains and DNA. These ribotyping methods are commonly used since they are easy to perform on both cultivated strains and *B. burgdorferi* s.l. DNA isolated from various sources. Analysis of restriction fragment length polymorphisms and sequencing of 16S rDNA, 23S rDNA and the intergenic spacer between the duplicated 23S and 5S rRNA genes have also been very useful in phylogenetic studies of *B. burgdorferi* s.l. (Marconi and Garon, 1992b; Postic *et al.*, 1994).

Molecular epidemiology

Molecular typing of *B. burgdorferi* s.l. DNA and strains from ticks and animals are important tools in epidemiology. For example, we have studied the importance of seabirds for transport of *B. burgdorferi* s.l. over long distances, and using molecular methods have identified *B. burgdorferi* s.l. DNA in seabird ticks, *I. uriae* (Olsen *et al.*, 1995). The seabird tick, *I. uriae*, has a circumpolar distribution in both the northern and southern hemispheres and in this study identical *B. garinii* flagellin gene (*flaB*) sequences were detected in *I. uriae* from these hemispheres, indicating a transequatorial transport of *B. garinii*. Phylogenetic analysis of *I. uriae* ITS1 and 16S rDNA sequences suggested that northern and southern *I. uriae* might be reproductively separated. Therefore, passive transport of infected ticks between the polar regions is unlikely and instead seabirds probably carry an active *B. burgdorferi* s.l. infection during their migration.

Furthermore, in another study we have demonstrated that *B. burgdorferi* s.l. can persist for several months in passerine birds and the infection in redwing thrushes can be reactivated in response to migration (Gylfe *et al.*, 2000). Thus, birds may be more infectious to ticks during their migration and therefore important long-range disseminators of *B. burgdorferi* s.l. This work shows that migrating seabirds and passerine birds are probably important for the long-range dispersal of *B. burgdorferi* s.l. and that this mechanism of dispersal could be important for the distribution of human Lyme disease.

Using similar approaches, other researchers have demonstrated the relative importance of different animals as reservoir hosts for different *B. burgdorferi* s.l species in Europe. For example, since *Borrelia valaisiana* has only been isolated from birds, birds have been implicated as being the only reservoir for this *Borrelia* species (Kurtenbach *et al.*, 1998). *B. afzelii*, on the other hand, is more frequently isolated from small rodents and ticks in areas where rodents are abundant, suggesting that rodents are the main reservoir hosts for *B. afzelii* (Hu *et al.*, 1997). In areas where birds are important as tick hosts, *B. garinii* and *B. valaisiana* predominate (Kurtenbach *et al.*, 1998; Gray *et al.*, 2000). There seem to be at least two distinct groups of *B. garinii*: the eastern type, which is mainly infectious to mice, and the western type, which is mainly infectious to birds (see Kurtenbach *et al.*, Chapter 5).

Evolution of *B. burgdorferi* s.l.

Since the 1980s, bacterial genomics have been performed in order to investigate the evolution of the members of the bacterial domain. Using very conserved genes, it has also been possible to compare the evolution between all the three domains: Bacteria, Archae and Eukarya. Spirochaetes have been separated into eight different genera based on habitat, pathogenicity, rRNA sequences and morphological and physiological characters. They belong to a very disparate group of microorganisms that is found in a variety of ecological niches such as soil, marine sediments, in arthropods or as parasites of vertebrates. The spirochaetes constitute a major bacterial grouping that by multicellular eukaryotic standards would be considered at least a division or phylum (Woese, 1987; Olsen *et al.*, 1994). In a recent review by Carl Woese *et al.* (2000), the aminoacyl-tRNA synthetases were used very elegantly to construct a picture of the evolutionary process in bacteria. This was done to obtain a clearer concept of how modern cells have evolved and to identify various stages in the process of bacterial evolution (Woese *et al.*, 2000). The genome sequence of two pathogenic spirochaetes, *B. burgdorferi* s.s. and *T. pallidum*, was recently published. The sequenced genomes from these two members of the spirochaetal division do not cover the full phylogenetic extent of this division but they are well separated phylogenetically. These two spirochaetes contain 19 of the aminoacyl-tRNA synthetases (they have no GlnRS), and from the comparison of these genes it was evident that in only two cases, threonine and proline, are the spirochaete enzymes clearly evolutionarily separated from one another. Interestingly, it was also apparent from this analysis that the spirochaetes have seven out of 19 aminoacyl-tRNA synthetases that are of both archaeal and eukaryotic origin.

Although the treponemes and borreliae exhibit a lot of similarities, a striking difference is their genome organization and structure. The members of the genus *Borrelia* have a very unusual genome organization among prokaryotic organisms as they have a polyploid genome that is mostly linear. The small linear replicons in borreliae have been called plasmids but, as in the genus *Trypanosoma*, there is justification for designating them mini chromosomes instead. A comparison of the ends of these replicons, i.e. the telomeres, with telomeric sequences of other linear double-stranded DNA replicons revealed sequence similarities with poxviruses, especially with the iridovirus agent of African swine fever. Interestingly, the latter virus and a relapsing fever *Borrelia* sp., *Borrelia duttoni*, share the same tick vector. These findings suggest that the linear replicons of *B. burgdorferi* s.l. originated through a horizontal gene transfer across the kingdom. Since this structural feature is found throughout the genus *Borrelia*, the evolutionary change must have occurred a very long time ago (Hinnebusch and Barbour, 1992).

An interesting hypothesis on the evolution of *Borrelia* spp. spirochaetes is that they have co-evolved together with their different arthropod vectors. The concept of the 'one vector – one *Borrelia* species' has been accepted in relapsing fever *Borrelia* spp. as its original species designation was related to its specific tick vector. Despite the fact that, as mentioned previously, this concept cannot be applied to LB borreliae, several phylogenetic analyses of *B. burgdorferi* s.l. and its different ixodid tick vectors have indicated that these organisms might have co-evolved. In a study by Fukunaga *et al.* (2000), an internal transcribed spacer (ITS2) sequence between the 5.8S and 28S rRNA genes was used to determine the phylogenetic relationships among *Ixodes* spp. vectors of LB-causing *Borrelia* spp. spirochaetes. From these analyses it was concluded that the *Borrelia* spp. associated with LB are found mainly in ticks of the *I. ricinus* species complex, whereas other closely related tick species, e.g. *I. uriae*, not known to transmit the *Borrelia* spp. that cause LB in humans, appear to have a specific association with other closely related *Borrelia* spp. Thus, from these studies it was found that there is a large similarity in the phylogeny of the *Borrelia* taxa and the phylogenetic relationships among *Ixodes* spp. ticks.

The pattern of distribution of LB is comparable with a multifocal worldwide epidemic with the endemic regions extending over the continents. However, most species of the *B. burgdorferi* s.l. complex have a relatively limited distribution, the only exception being *B. burgdorferi* s.s., which is found both in Europe and in the USA. The question is, from where and how have the LB *Borrelia* spp. migrated? The larger genetic variability and the clinical polymorphism seen in the Old World suggest that the *B. burgdorferi* s.l. complex emanated from this geographical area. However, in a series of experiments Baranton and co-workers compared the genetic polymorphism of *B. burgdorferi* s.s. strains isolated in Europe or the USA, using several molecular methods (Foretz *et al.*, 1997; Marti Ras *et al.*, 1997). A genetic distance method and a parsimony method were used to perform phylogenetic analysis of each of the two sets of data. The results were consistent and showed that the American *B. burgdorferi* s.s. strains were more heterogeneous than the European ones; thus, it might be possible that this species evolved in America and was introduced to Europe, and that the different populations have evolved separately.

In contrast to the distribution of *B. burgdorferi* s.s., *B. afzelii* appears to be more localized and stable within regions in Europe and Asia where it does not spread over large distances. This is probably due to its vector and host relationships. *B. garinii*, on the other hand, was found to be spread over large geographical regions in Europe and was associated with both the ticks within the *I. ricinus* complex and the seabird tick *I. uriae* involved in the maintenance of a novel, seabird-related, enzootic cycle of LB (Olsen *et al.*, 1993). In an attempt to delineate the expansion and molecular polymorphism of *B. burgdorferi* s.l. in the north of Europe, *Borrelia* strains recovered from different tick species on islands in the northern Baltic Sea and North Atlantic were genotypically and phenotypically characterized. Molecular polymorphism of *B. garinii* was shown to be a result of interactions among several regional enzootic foci of this genospecies.

Concluding Comments

The aim of this chapter has been to describe the current status of the molecular and cellular biology of LB spirochaetes. Much of this recent knowledge is directly coupled to the published genome sequence of *B. burgdorferi* s.s. B31. The new genomics and proteomics era that we have entered will be instrumental for future understanding of *B. burgdorferi* s.l. In addition, the combination of data from other microorganisms also has the potential of further increasing knowledge of the biology of *Borrelia* spp. spirochaetes in general and their pathogenesis in particular. The approach in the future will be to identify the central regulators of common host responses and the manner in which *Borrelia* spp. spirochaetes attempt to defeat this process. This should lead to the development of valuable tools for designing novel prophylactic and diagnostic methods.

References

Adam, T., Gassmann, G.S., Rasiah, C. and Gobel, U.B. (1991) Phenotypic and genotypic analysis of *Borrelia burgdorferi* isolates from various sources. *Infection and Immunity* 59, 2579–2585.

Akins, D.R., Porcella, S.F., Popova, T.G., Shevchenko, D., Baker, S.I., Li, M., Norgard, M.V. and Radolf, J.D. (1995) Evidence for *in vivo* but not *in vitro* expression of a *Borrelia burgdorferi* outer surface protein F (OspF) homologue. *Molecular Microbiology* 18, 507–520.

Akins, D.R., Bourell, K.W., Caimano, M.J., Norgard, M.V. and Radolf, J.D. (1998) A new animal model for studying Lyme disease spirochetes in a mammalian host-adapted state. *Journal of Clinical Investigation* 101, 2240–2250.

Akins, D.R., Caimano, M.J., Yang, X., Cerna, F., Norgard, M.V. and Radolf, J.D. (1999) Molecular and evolutionary analysis of *Borrelia burgdorferi* 297 circular plasmid-encoded lipoproteins with OspE- and OspF-like leader peptides. *Infection and Immunity* 67, 1526–1532.

Alban, P.S., Johnson, P.W. and Nelson, D.R. (2000) Serum-starvation-induced changes in protein synthesis and morphology of *Borrelia burgdorferi*. *Microbiology* 146, 119–127.

Allardet-Servent, A., Michaux-Charachon, S., Jumas-Bilak, E., Karayan, L. and Ramuz, M. (1993) Presence of one linear and one circular chromosome in the *Agrobacterium tumefaciens* C58 genome. *Journal of Bacteriology* 175, 7869–7874.

Anderson, J.F., Magnarelli, L.A., LeFebvre, R.B., Andreadis, T.G., McAninch, J.B., Perng, G.C. and Johnson, R.C. (1989) Antigenically variable *Borrelia burgdorferi* isolated from cottontail rabbits and *Ixodes dentatus* in rural and urban areas. *Journal of Clinical Microbiology* 27, 13–20.

Anguita, J., Thomas, V., Samanta, S., Persinski, R., Hernanz, C., Barthold, S.W. and Fikrig, E. (2001) *Borrelia burgdorferi*-induced inflammation facilitates spirochete adaptation and variable major protein-like sequence locus recombination. *Journal of Immunology* 167, 3383–3390.

Baranton, G., Postic, D., Saint Girons, I., Boerlin, P., Piffaretti, J.C., Assous, M. and Grimont, P.A. (1992) Delineation of *Borrelia burgdorferi* sensu stricto, *Borrelia garinii* sp. nov., and group VS461 associated with Lyme borreliosis. *International Journal of Systematic Bacteriology* 42, 378–383.

Baranton, G., Marti Ras, N. and Postic, D.

(1998) *Borrelia burgdorferi*, taxonomy, pathogenicity and spread. *Annales de Médecine Interne (Paris)* 149, 455–458 (in French).

Barbour, A.G. (1988) Plasmid analysis of *Borrelia burgdorferi*, the Lyme disease agent. *Journal of Clinical Microbiology* 26, 475–478.

Barbour, A.G. (1990) Antigenic variation of a relapsing fever *Borrelia* species. *Annual Reviews in Microbiology* 44, 155–171.

Barbour, A.G. (1993) Linear DNA of *Borrelia* species and antigenic variation. *Trends in Microbiology* 1, 236–239.

Barbour, A.G. and Garon, C.F. (1987) Linear plasmids of the bacterium *Borrelia burgdorferi* have covalently closed ends. *Science* 237, 409–411.

Barbour, A.G. and Garon, C.F. (1988) The genes encoding major surface proteins of *Borrelia burgdorferi* are located on a plasmid. *Annals of the New York Academy of Sciences* 539, 144–153.

Barbour, A.G. and Hayes, S.F. (1986) Biology of *Borrelia* species. *Microbiological Reviews* 50, 381–400.

Barbour, A.G., Tessier, S.L. and Hayes, S.F. (1984) Variation in a major surface protein of Lyme disease spirochetes. *Infection and Immunity* 45, 94–100.

Barbour, A.G., Hayes, S.F., Heiland, R.A., Schrumpf, M.E. and Tessier, S.L. (1986) A *Borrelia*-specific monoclonal antibody binds to a flagellar epitope. *Infection and Immunity* 52, 549–554.

Barbour, A.G., Burman, N., Carter, C.J., Kitten, T. and Bergstrom, S. (1991a) Variable antigen genes of the relapsing fever agent *Borrelia hermsii* are activated by promoter addition. *Molecular Microbiology* 5, 489–493.

Barbour, A.G., Carter, C.J., Burman, N., Freitag, C.S., Garon, C.F. and Bergstrom, S. (1991b) Tandem insertion sequence-like elements define the expression site for variable antigen genes of *Borrelia hermsii*. *Infection and Immunity* 59, 390–397.

Barbour, A.G., Carter, C.J., Bundoc, V. and Hinnebusch, J. (1996) The nucleotide sequence of a linear plasmid of *Borrelia burgdorferi* reveals similarities to those of circular plasmids of other prokaryotes. *Journal of Bacteriology* 178, 6635–6639.

Baril, C., Richaud, C., Baranton, G. and Saint Girons, I.S. (1989) Linear chromosome of *Borrelia burgdorferi*. *Research in Microbiology* 140, 507–516.

Beck, G., Habicht, G.S., Benach, J.L. and Coleman, J.L. (1985) Chemical and biologic characterization of a lipopolysaccharide extracted from the Lyme disease spirochete (*Borrelia burgdorferi*). *Journal of Infectious Diseases* 152, 108–117.

Belisle, J.T., Brandt, M.E., Radolf, J.D. and Norgard, M.V. (1994) Fatty acids of *Treponema pallidum* and *Borrelia burgdorferi* lipoproteins. *Journal of Bacteriology* 176, 2151–2157.

Berg, C.M. and Berg, D.E. (1996) Transposable element tools for microbial genetics. In: Neidhardt, F.C., Curtis, R.C. III, Ingraham, J.L., Lin, E.C.C., Low, K.B. *et al.* (eds) *Escherichia coli and Salmonella*. ASM Press, Washington, DC, pp. 2588–2612.

Berg, S. (2001) The glucolipid pathway in *Acholeplasma laidlawii*. Functional and structural properties of three consecutively-acting enzymes and similarities to orthologs in prokaryotes. PhD thesis, Umeå University, Umeå, Sweden. ISBN 91-7305-025-3.

Bergström, S., Bundoc, V.G. and Barbour, A.G. (1989) Molecular analysis of linear plasmid-encoded major surface proteins, OspA and OspB, of the Lyme disease spirochaete *Borrelia burgdorferi*. *Molecular Microbiology* 3, 479–486.

Bergström, S., Garon, C.F., Barbour, A.G. and MacDougall, J. (1992) Extrachromosomal elements of spirochetes. *Research in Microbiology* 143, 623–628.

Bessler, W.G., Cox, M., Lex, A., Suhr, B., Wiesmuller, K.H. and Jung, G. (1985) Synthetic lipopeptide analogs of bacterial lipoprotein are potent polyclonal activators for murine B lymphocytes. *Journal of Immunology* 135, 1900–1905.

Bledsoe, H.A., Carroll, J.A., Whelchel, T.R., Farmer, M.A., Dorward, D.W. and Gherardini, F.C. (1994) Isolation and partial characterization of *Borrelia burgdorferi* inner and outer membranes by using isopycnic centrifugation. *Journal of Bacteriology* 176, 7447–7455.

Bono, J.L., Tilly, K., Stevenson, B., Hogan, D. and Rosa, P. (1998) Oligopeptide permease in *Borrelia burgdorferi*: putative peptide-binding components encoded by both chromosomal and plasmid loci. *Microbiology* 144, 1033–1044.

Bono, J.L., Elias, A.F., Kupko, J.J. III, Stevenson, B., Tilly, K. and Rosa, P. (2000) Efficient targeted mutagenesis in *Borrelia burgdorferi*. *Journal of Bacteriology* 182, 2445–2452.

Borst, P. (1991) Molecular genetics of antigenic variation. *Immunology Today* 12, A29–A33.

Brahamsha, B. and Greenberg, E.P. (1988) Biochemical and cytological analysis of the complex periplasmic flagella from *Spirochaeta aurantia*. *Journal of Bacteriology* 170, 4023–4032.

Brandt, M.E., Riley, B.S., Radolf, J.D. and Norgard, M.V. (1990) Immunogenic integral membrane proteins of *Borrelia burgdorferi* are lipoproteins. *Infection and Immunity* 58, 983–991.

Brown, E.L., Wooten, R.M., Johnson, B.J., Iozzo, R.V., Smith, A., Dolan, M.C., Guo, B.P., Weis, J.J. and Hook, M. (2001) Resistance to Lyme disease in decorin-deficient mice. *Journal of Clinical Investigation* 107, 845–852.

Brusca, J.S., McDowall, A.W., Norgard, M.V. and Radolf, J.D. (1991) Localization of outer surface proteins A and B in both the outer membrane and intracellular compartments of *Borrelia burgdorferi*. *Journal of Bacteriology* 173, 8004–8008.

Bunikis, J., Noppa, L. and Bergström, S. (1995) Molecular analysis of a 66-kDa protein associated with the outer membrane of Lyme disease *Borrelia*. *FEMS Microbiology Letters* 131, 139–145.

Burgdorfer, W., Barbour, A.G., Hayes, S.F., Benach, J.L., Grunwaldt, E. and Davis, J.P. (1982) Lyme disease – a tickborne spirochetosis? *Science* 216, 1317–1319.

Burman, N., Bergström, S., Restrepo, B.I. and Barbour, A.G. (1990) The variable antigens Vmp7 and Vmp21 of the relapsing fever bacterium *Borrelia hermsii* are structurally analogous to the VSG proteins of the African trypanosome. *Molecular Microbiology* 4, 1715–1726.

Canica, M.M., Nato, F., du Merle, L., Mazie, J.C., Baranton, G. and Postic, D. (1993) Monoclonal antibodies for identification of *Borrelia afzelii* sp. nov. associated with late cutaneous manifestations of Lyme borreliosis. *Scandinavian Journal of Infectious Diseases* 25, 441–448.

Carroll, J.A. and Gherardini, F.C. (1996) Membrane protein variations associated with *in vitro* passage of *Borrelia burgdorferi*. *Infection and Immunity* 64, 392–398.

Carroll, J.A., El-Hage, N., Miller, J.C., Babb, K. and Stevenson, B. (2001) *Borrelia burgdorferi* RevA antigen is a surface-exposed outer membrane protein whose expression is regulated in response to environmental temperature and pH. *Infection and Immunity* 69, 5286–5293.

Carter, C.J., Bergstrom, S., Norris, S.J. and Barbour, A.G. (1994) A family of surface-exposed proteins of 20 kilodaltons in the genus *Borrelia*. *Infection and Immunity* 62, 2792–2799.

Casjens, S. and Huang, W.M. (1993) Linear chromosomal physical and genetic map of *Borrelia burgdorferi*, the Lyme disease agent. *Molecular Microbiology* 8, 967–980.

Casjens, S., van Vugt, R., Tilly, K., Rosa, P.A. and Stevenson, B. (1997a) Homology throughout the multiple 32-kilobase circular plasmids present in

Lyme disease spirochetes. *Journal of Bacteriology* 179, 217–227.

Casjens, S., Murphy, M., DeLange, M., Sampson, L., van Vugt, R. and Huang, W.M. (1997b) Telomeres of the linear chromosomes of Lyme disease spirochaetes: nucleotide sequence and possible exchange with linear plasmid telomeres. *Molecular Microbiology* 26, 581–596.

Casjens, S., Palmer, N., van Vugt, R., Huang, W.M., Stevenson, B., Rosa, P., Lathigra, R., Sutton, G., Peterson, J., Dodson, R.J., Haft, D., Hickey, E., Gwinn, M., White, O. and Fraser, C.M. (2000) A bacterial genome in flux: the twelve linear and nine circular extrachromosomal DNAs in an infectious isolate of the Lyme disease spirochete *Borrelia burgdorferi*. *Molecular Microbiology* 35, 490–516.

Chaconas, G., Stewart, P.E., Tilly, K., Bono, J.L. and Rosa, P. (2001) Telomere resolution in the Lyme disease spirochete. *EMBO Journal* 20, 3229–3237.

Chamberlain, N.R., DeOgny, L., Slaughter, C., Radolf, J.D. and Norgard, M.V. (1989) Acylation of the 47-kilodalton major membrane immunogen of *Treponema pallidum* determines its hydrophobicity. *Infection and Immunity* 57, 2878–2885.

Champion, C.I., Blanco, D.R., Skare, J.T., Haake, D.A., Giladi, M., Foley, D., Miller, J.N. and Lovett, M.A. (1994) A 9.0-kilobase-pair circular plasmid of *Borrelia burgdorferi* encodes an exported protein: evidence for expression only during infection. *Infection and Immunity* 62, 2653–2661.

Charon, N.W., Greenberg, E.P., Koopman, M.B. and Limberger, R.J. (1992) Spirochete chemotaxis, motility, and the structure of the spirochetal periplasmic flagella. *Research in Microbiology* 143, 597–603.

Chen, J., Field, J.A., Glickstein, L., Molloy, P.J., Huber, B.T. and Steere, A.C. (1999) Association of antibiotic treatment-resistant Lyme arthritis with T cell responses to dominant epitopes of outer surface protein A of *Borrelia burgdorferi*. *Arthritis and Rheumatism* 42, 1813–1822.

Cinco, M., Banfi, E., Balanzin, D., Godeas, C. and Panfili, E. (1991) Evidence for (lipo) oligosaccharides in *Borrelia burgdorferi* and their serological specificity. *FEMS Immunology and Medical Microbiology* 3, 33–38.

Cockayne, A., Bailey, M.J. and Penn, C.W. (1987) Analysis of sheath and core structures of the axial filament of *Treponema pallidum*. *Journal of General Microbiology* 133, 1397–1407.

Comstock, L.E., Fikrig, E., Shoberg, R.J., Flavell, R.A. and Thomas, D.D. (1993) A monoclonal antibody to OspA inhibits association of *Borrelia burgdorferi* with human endothelial cells. *Infection and Immunity* 61, 423–431.

Cowan, S.W., Schirmer, T., Rummel, G., Steiert, M., Ghosh, R., Pauptit, R.A., Jansonius, J.N. and Rosenbusch, J.P. (1992) Crystal structures explain functional properties of two *E. coli* porins. *Nature* 358, 727–733.

Cox, D.L., Chang, P., McDowall, A.W. and Radolf, J.D. (1992) The outer membrane, not a coat of host proteins, limits antigenicity of virulent *Treponema pallidum*. *Infection and Immunity* 60, 1076–1083.

Cox, D.L., Akins, D.R., Bourell, K.W., Lahdenne, P., Norgard, M.V. and Radolf, J.D. (1996) Limited surface exposure of *Borrelia burgdorferi* outer surface lipoproteins. *Proceedings of the National Academy of Sciences, USA* 93, 7973–7978.

Crespi, M., Messens, E., Caplan, A.B., van Montagu, M. and Desomer, J. (1992) Fasciation induction by the phytopathogen *Rhodococcus fascians* depends upon a linear plasmid encoding a cytokinin synthase gene. *EMBO Journal* 11, 795–804.

Das, S., Shraga, D., Gannon, C., Lam,

T.T., Feng, S., Brunet, L.R., Telford, S.R., Barthold, S.W., Flavell, R.A. and Fikrig, E. (1996) Characterization of a 30-kDa *Borrelia burgdorferi* substrate-binding protein homologue. *Research in Microbiology* 147, 739–751.

Davidson, B.E., MacDougall, J. and Saint Girons, I. (1992) Physical map of the linear chromosome of the bacterium *Borrelia burgdorferi* 212, a causative agent of Lyme disease, and localization of rRNA genes. *Journal of Bacteriology* 174, 3766–3774.

Deres, K., Schild, H., Wiesmuller, K.H., Jung, G. and Rammensee, H.G. (1989) In vivo priming of virus-specific cytotoxic T lymphocytes with synthetic lipopeptide vaccine. *Nature* 342, 561–564.

Dorward, D.W., Schwan, T.G. and Garon, C.F. (1991) Immune capture and detection of *Borrelia burgdorferi* antigens in urine, blood, or tissues from infected ticks, mice, dogs, and humans. *Journal of Clinical Microbiology* 29, 1162–1170.

Dunn, J.J., Lade, B.N. and Barbour, A.G. (1990) Outer surface protein A (OspA) from the Lyme disease spirochete, *Borrelia burgdorferi*: high level expression and purification of a soluble recombinant form of OspA. *Protein Expression and Purification* 1, 159–168.

Duray, P.H. (1992) Target organs of *B. burgdorferi* infections: functional responses and histology. In: Schutzer, S.E. (ed.) *Lyme Disease: Molecular and Immunologic Approaches*. Cold Spring Harbor Laboratory Press, Cold Spring Harbor, New York, pp. 11–30.

Dykhuizen, D.E., Polin, D.S., Dunn, J.J., Wilske, B., Preac-Mursic, V., Dattwyler, R.J. and Luft, B.J. (1993) *Borrelia burgdorferi* is clonal: implications for taxonomy and vaccine development. *Proceedings of the National Academy of Sciences, USA* 90, 10163–10167.

Eggers, C.H. and Samuels, D.S. (1999) Molecular evidence for a new bacteriophage of *Borrelia burgdorferi*. *Journal of Bacteriology* 181, 7308–7313.

Eggers, C.H., Casjens, S., Hayes, S.F., Garon, C.F., Damman, C.J., Oliver, D.B. and Samuels, D.S. (2000) Bacteriophages of spirochetes. *Journal of Molecular Microbiology and Biotechnology* 2, 365–373.

Eggers, C.H., Casjens, S. and Samuels, D.S. (2001a) Bacteriophages of *B. burgdorferi* and other spirochetes. In: Saier, M.H. Jr and García-Lara, J. (eds) *The Spirochetes: Molecular and Cellular Biology*. Horizon Press, Wymondham, UK, p. 222.

Eggers, C.H., Kimmel, B.J., Bono, J.L., Elias, A.F., Rosa, P. and Samuels, D.S. (2001b) Transduction by phiBB-1, a bacteriophage of *Borrelia burgdorferi*. *Journal of Bacteriology* 183, 4771–4778.

Eiffert, H., Lotter, H., Jarecki-Khan, K. and Thomssen, R. (1991) Identification of an immunoreactive non-proteinaceous component in *Borrelia burgdorferi*. *Medical Microbiology and Immunology* 180, 229–237.

El-Hage, N., Lieto, L.D. and Stevenson, B. (1999) Stability of *erp* loci during *Borrelia burgdorferi* infection: recombination is not required for chronic infection of immunocompetent mice. *Infection and Immunity* 67, 3146–3150.

Elias, A.F., Bono, J.L., Carroll, J.A., Stewart, P., Tilly, K. and Rosa, P. (2000) Altered stationary-phase response in a *Borrelia burgdorferi rpoS* mutant. *Journal of Bacteriology* 182(10), 2909–2918.

Erdile, L.F., Brandt, M.A., Warakomski, D.J., Westrack, G.J., Sadziene, A., Barbour, A.G. and Mays, J.P. (1993) Role of attached lipid in immunogenicity of *Borrelia burgdorferi* OspA. *Infection and Immunity* 61, 81–90.

Felsenfeld, O. (1971) Borrelia: *Strains, Vectors, Human and Animal Borreliosis*. Warren H. Green Inc., St Louis, Missouri, 180 pp.

Feng, S., Hodzic, E., Stevenson, B. and Barthold, S.W. (1998) Humoral immunity to *Borrelia burgdorferi* N40 decorin binding proteins during infection of laboratory mice. *Infection and Immunity* 66, 2827–2835.

Ferdows, M.S. and Barbour, A.G. (1989) Megabase-sized linear DNA in the bacterium *Borrelia burgdorferi*, the Lyme disease agent. *Proceedings of the National Academy of Sciences, USA* 86, 5969–5973.

Finlay, B.B. and Falkow, S. (1989) Common themes in microbial pathogenicity. *Microbiology Reviews* 53, 210–230.

Foretz, M., Postic, D. and Baranton, G. (1997) Phylogenetic analysis of *Borrelia burgdorferi* sensu stricto by arbitrarily primed PCR and pulsed-field gel electrophoresis. *International Journal of Systematic Bacteriology* 47, 11–18.

Fraser, C.M., Gocayne, J.D., White, O., Adams, M.D., Clayton, R.A., Fleischmann, R.D., Bult, C.J., Kerlavage, A.R., Sutton, G., Kelley, J.M. et al. (1995) The minimal gene complement of *Mycoplasma genitalium*. *Science* 270, 397–403.

Fraser, C.M., Casjens, S., Huang, W.M., Sutton, G.G., Clayton, R., Lathigra, R., White, O., Ketchum, K.A., Dodson, R., Hickey, E.K., Gwinn, M., Dougherty, B., Tomb, J.F., Fleischmann, R.D., Richardson, D., Peterson, J., Kerlavage, A.R., Quackenbush, J., Salzberg, S., Hanson, M., van Vugt, R., Palmer, N., Adams, M.D., Gocayne, J., Weidman, J., Utterback, T., Watthey, L., McDonald, L., Artiach, P., Bowman, C., Garland, S., Fujii, C., Cotton, M.D., Horst, K., Roberts, K., Hatch, B., Smith, H.O. and Venter, J.C. (1997) Genomic sequence of a Lyme disease spirochaete, *Borrelia burgdorferi*. *Nature* 390, 580–586.

Fraser, C.M., Norris, S.J., Weinstock, G.M., White, O., Sutton, G.G., Dodson, R., Gwinn, M., Hickey, E.K., Clayton, R., Ketchum, K.A., Sodergren, E., Hardham, J.M., McLeod, M.P., Salzberg, S., Peterson, J., Khalak, H., Richardson, D., Howell, J.K., Chidambaram, M., Utterback, T., McDonald, L., Artiach, P., Bowman, C., Cotton, M., Fujii, C., Garland, S., Hatch, B., Horst, K., Roberts, K., Sandusky, M., Weidman, J., Smith, H.O. and Venter, J.C. (1998) Complete genome sequence of *Treponema pallidum*, the syphilis spirochete. *Science* 281, 375–388.

Fukunaga, M., Hamase, A., Okada, K. and Nakao, M. (1996a) *Borrelia tanukii* sp. nov. and *Borrelia turdae* sp. nov. found from ixodid ticks in Japan: rapid species identification by 16S rRNA gene-targeted PCR analysis. *Microbiology and Immunology* 40, 877–881.

Fukunaga, M., Okada, K., Nakao, M., Konishi, T. and Sato, Y. (1996b) Phylogenetic analysis of *Borrelia* species based on flagellin gene sequences and its application for molecular typing of Lyme disease borreliae. *International Journal of Systematic Bacteriology* 46, 898–905.

Fukunaga, M., Yabuki, M., Hamase, A., Oliver, J.H. Jr and Nakao, M. (2000) Molecular phylogenetic analysis of ixodid ticks based on the ribosomal DNA spacer, internal transcribed spacer 2, sequences. *Journal of Parasitology* 86, 38–43.

Garcia-Monco, J.C., Fernandez-Villar, B. and Benach, J.L. (1989) Adherence of the Lyme disease spirochete to glial cells and cells of glial origin. *Journal of Infectious Diseases* 160, 497–506.

Garon, C.F., Dorward, D.W. and Corwin, M.D. (1989) Structural features of *Borrelia burgdorferi* – the Lyme disease spirochete: silver staining for nucleic acids. *Scanning Microscopy* 3 (suppl.), 109–115.

Ge, Y. and Charon, N.W. (1997a) An unexpected *flaA* homolog is present and expressed in *Borrelia burgdorferi*. *Journal of Bacteriology* 179, 552–556.

Ge, Y. and Charon, N.W. (1997b) Molecular characterization of a flagellar/chemotaxis operon in the spirochete

Borrelia burgdorferi. FEMS Microbiology Letters 153, 425–431.

Ge, Y., Old, I., Saint Girons, I., Yelton, D.B. and Charon, N.W. (1996) FliH and FliI of *Borrelia burgdorferi* are similar to flagellar and virulence factor export proteins of other bacteria. *Gene* 168, 73–75.

Ge, Y., Old, I.G., Girons, I.S. and Charon, N.W. (1997) The *flgK* motility operon of *Borrelia burgdorferi* is initiated by a sigma 70-like promoter. *Microbiology* 143, 1681–1690.

Ge, Y., Li, C., Corum, L., Slaughter, C.A. and Charon, N.W. (1998) Structure and expression of the FlaA periplasmic flagellar protein of *Borrelia burgdorferi*. *Journal of Bacteriology* 180, 2418–2425.

Gern, L., Rouvinez, E., Toutoungi, L.N. and Godfroid, E. (1997) Transmission cycles of *Borrelia burgdorferi* sensu lato involving *Ixodes ricinus* and/or *I. hexagonus* ticks and the European hedgehog, *Erinaceus europaeus*, in suburban and urban areas in Switzerland. *Folia Parasitologica* 44, 309–314.

Gilmore, R.D. and Piesman, J. (2000) Inhibition of *Borrelia burgdorferi* migration from the midgut to the salivary glands following feeding by ticks on OspC-immunized mice. *Infection and Immunity* 68, 411–414.

Gilmore, R.D. Jr, Kappel, K.J., Dolan, M.C., Burkot, T.R. and Johnson, B.J. (1996) Outer surface protein C (OspC), but not P39, is a protective immunogen against a tick-transmitted *Borrelia burgdorferi* challenge: evidence for a conformational protective epitope in OspC. *Infection and Immunity* 64, 2234–2239.

Goldstein, S.F., Charon, N.W. and Kreiling, J.A. (1994) *Borrelia burgdorferi* swims with a planar waveform similar to that of eukaryotic flagella. *Proceedings of the National Academy of Sciences, USA* 91, 3433–3437.

Goldstein, S.F., Buttle, K.F. and Charon, N.W. (1996) Structural analysis of the *Leptospiraceae* and *Borrelia burgdorferi* by high-voltage electron microscopy. *Journal of Bacteriology* 178, 6539–6545.

Gominet, M., Slamti, L., Gilois, N., Rose, M. and Lereclus, D. (2001) Oligopeptide permease is required for expression of the *Bacillus thuringiensis plcR* regulon and for virulence. *Molecular Microbiology* 40, 963–975.

Goodell, E.W. and Higgins, C.F. (1987) Uptake of cell wall peptides by *Salmonella typhimurium* and *Escherichia coli*. *Journal of Bacteriology* 169, 3861–3865.

Gorbacheva, V.Y., Godfrey, H.P. and Cabello, F.C. (2000) Analysis of the *bmp* gene family in *Borrelia burgdorferi* sensu lato. *Journal of Bacteriology* 182, 2037–2042.

Gray, J.S., Robertson, J.N. and Key, S. (2000) Limited role of rodents as reservoirs of *Borrelia burgdorferi* sensu lato in Ireland. *European Journal of Epidemiology* 16, 101–103.

Gross, D.M., Forsthuber, T., Tary-Lehmann, M., Etling, C., Ito, K., Nagy, Z.A., Field, J.A., Steere, A.C. and Huber, B.T. (1998) Identification of LFA-1 as a candidate autoantigen in treatment-resistant Lyme arthritis. *Science* 281, 703–706.

Guo, B.P., Norris, S.J., Rosenberg, L.C. and Hook, M. (1995) Adherence of *Borrelia burgdorferi* to the proteoglycan decorin. *Infection and Immunity* 63, 3467–3472.

Guo, B.P., Brown, E.L., Dorward, D.W., Rosenberg, L.C. and Hook, M. (1998) Decorin-binding adhesins from *Borrelia burgdorferi*. *Molecular Microbiology* 30, 711–723.

Gylfe, Å., Olsen, B., Strasevicius, D., Marti Ras, N., Weihe, P., Noppa, L., Östberg, Y., Baranton, G. and Bergström, S. (1999) Isolation of Lyme disease *Borrelia* from puffins (*Fratercula arctica*) and seabird ticks (*Ixodes uriae*) on the Faeroe islands. *Journal of Clinical Microbiology* 37, 890–896.

Gylfe, Å., Bergström, S., Lundström, J. and Olsen, B. (2000) Reactivation of

Borrelia infection in birds. *Nature* 403, 724–725.

Haas, R. and Meyer, T.F. (1986) The repertoire of silent pilus genes in *Neisseria gonorrhoeae*: evidence for gene conversion. *Cell* 44, 107–115.

Hagblom, P., Segal, E., Billyard, E. and So, M. (1985) Intragenic recombination leads to pilus antigenic variation in *Neisseria gonorrhoeae*. *Nature* 315, 156–158.

Hagman, K.E., Lahdenne, P., Popova, T.G., Porcella, S.F., Akins, D.R., Radolf, J.D. and Norgard, M.V. (1998) Decorin-binding protein of *Borrelia burgdorferi* is encoded within a two-gene operon and is protective in the murine model of Lyme borreliosis. *Infection and Immunity* 66, 2674–2683.

Hanson, M.S., Cassatt, D.R., Guo, B.P., Patel, N.K., McCarthy, M.P., Dorward, D.W. and Hook, M. (1998) Active and passive immunity against *Borrelia burgdorferi* decorin binding protein A (DbpA) protects against infection. *Infection and Immunity* 66, 2143–2153.

Hayashi, S. and Wu, H.C. (1990) Lipoproteins in bacteria. *Journal of Bioenergetics and Biomembranes* 22, 451–471.

Hayes, S.F., Burgdorfer, W. and Barbour, A.G. (1983) Bacteriophage in the *Ixodes dammini* spirochete, etiological agent of Lyme disease. *Journal of Bacteriology* 154, 1436–1439.

Hellwage, J., Meri, T., Heikkila, T., Alitalo, A., Panelius, J., Lahdenne, P., Seppala, I.J. and Meri, S. (2001) The complement regulator factor H binds to the surface protein OspE of *Borrelia burgdorferi*. *Journal of Biological Chemistry* 276, 8427–8435.

Hinnebusch, J. and Barbour, A.G. (1992) Linear- and circular-plasmid copy numbers in *Borrelia burgdorferi*. *Journal of Bacteriology* 174, 5251–5257.

Hinnebusch, J., Bergstrom, S. and Barbour, A.G. (1990) Cloning and sequence analysis of linear plasmid telomeres of the bacterium *Borrelia burgdorferi*. *Molecular Microbiology* 4, 811–820.

Hoffmann, P., Heinle, S., Schade, U.F., Loppnow, H., Ulmer, A.J., Flad, H.D., Jung, G. and Bessler, W.G. (1988) Stimulation of human and murine adherent cells by bacterial lipoprotein and synthetic lipopeptide analogues. *Immunobiology* 177, 158–170.

Holt, S.C. (1978) Anatomy and chemistry of spirochetes. *Microbiological Reviews* 42, 114–160.

Hossain, H., Wellensiek, H., Geyer, R. and Lochnit, G. (2001) Structural analysis of glycolipids from *Borrelia burgdorferi*. *Biochimie* 83, 683–692.

Hovind Hougen, K. (1984) Ultrastructure of spirochetes isolated from *Ixodes ricinus* and *Ixodes dammini*. *Yale Journal of Biology and Medicine* 57, 543–548.

Howe, T.R., LaQuier, F.W. and Barbour, A.G. (1986) Organization of genes encoding two outer membrane proteins of the Lyme disease agent *Borrelia burgdorferi* within a single transcriptional unit. *Infection and Immunity* 54, 207–212.

Hu, C.M., Humair, P.F., Wallich, R. and Gern, L. (1997) *Apodemus* sp. rodents, reservoir hosts for *Borrelia afzelii* in an endemic area in Switzerland. *Zentralblatt für Bakteriologie* 285, 558–564.

Hubálek, Z. and Halouzka, J. (1997) Distribution of *Borrelia burgdorferi* sensu lato genomic groups in Europe, a review. *European Journal of Epidemiology* 13, 951–957.

Hudson, B.J., Stewart, M., Lennox, V.A., Fukunaga, M., Yabuki, M., Macorison, H. and Kitchener-Smith, J. (1998) Culture-positive Lyme borreliosis. *Medical Journal of Australia* 168, 500–502.

Hyde, F.W. and Johnson, R.C. (1984) Genetic relationship of Lyme disease spirochetes to *Borrelia, Treponema*, and *Leptospira* spp. *Journal of Clinical Microbiology* 20, 151–154.

Indest, K.J. and Philipp, M.T. (2000) DNA-binding proteins possibly involved in regulation of the post-logarithmic-phase expression of lipoprotein P35 in

Borrelia burgdorferi. *Journal of Bacteriology* 182, 522–525.
Indest, K.J., Ramamoorthy, R., Sole, M., Gilmore, R.D., Johnson, B.J. and Philipp, M.T. (1997) Cell-density-dependent expression of *Borrelia burgdorferi* lipoproteins *in vitro*. *Infection and Immunity* 65, 1165–1171.
Jauris-Heipke, S., Liegl, G., Preac-Mursic, V., Rossler, D., Schwab, E., Soutschek, E., Will, G. and Wilske, B. (1995) Molecular analysis of genes encoding outer surface protein C (OspC) of *Borrelia burgdorferi* sensu lato: relationship to ospA genotype and evidence of lateral gene exchange of ospC. *Journal of Clinical Microbiology* 33, 1860–1866.
Jauris-Heipke, S., Rossle, B., Wanner, G., Habermann, C., Rossler, D., Fingerle, V., Lehnert, G., Lobentanzer, R., Pradel, I., Hillenbrand, B., Schulte-Spechtel, U. and Wilske, B. (1999) Osp17, a novel immunodominant outer surface protein of *Borrelia afzelii*: recombinant expression in *Escherichia coli* and its use as a diagnostic antigen for serodiagnosis of Lyme borreliosis. *Medical Microbiology and Immunology* 187, 213–219.
Johnson, R.C., Schmid, G.P., Hyde, F.W., Steigerwalt, A.G. and Brenner, D.J. (1984) *Borrelia burgdorferi* sp. nov.: etiologic agent of Lyme disease. *International Journal of Systematic Bacteriology* 34, 496–497.
Jones, J.D., Bourell, K.W., Norgard, M.V. and Radolf, J.D. (1995) Membrane topology of *Borrelia burgdorferi* and *Treponema pallidum* lipoproteins. *Infection and Immunity* 63, 2424–2434.
Jonsson, M. and Bergström, S. (1995) Transcriptional and translational regulation of the expression of the major outer surface proteins in Lyme disease *Borrelia* strains. *Microbiology* 141, 1321–1329.
Jonsson, M., Noppa, L., Barbour, A.G. and Bergström, S. (1992) Heterogeneity of outer membrane proteins in *Borrelia burgdorferi*: comparison of *osp* operons of three isolates of different geographic origins. *Infection and Immunity* 60, 1845–1853.
Jwang, B., Dewing, P., Fikrig, E. and Flavell, R.A. (1995) The hook protein of *Borrelia burgdorferi*, encoded by the *flgE* gene, is serologically recognized in Lyme disease. *Clinical and Diagnostic Laboratory Immunology* 2, 609–615.
Katona, L.I., Beck, G. and Habicht, G.S. (1992) Purification and immunological characterization of a major low-molecular-weight lipoprotein from *Borrelia burgdorferi*. *Infection and Immunity* 60, 4995–5003.
Kawabata, H., Masuzawa, T. and Yanagihara, Y. (1993) Genomic analysis of *Borrelia japonica* sp. nov. isolated from *Ixodes ovatus* in Japan. *Microbiology and Immunology* 37, 843–848.
Kimsey, R.B. and Spielman, A. (1990) Motility of Lyme disease spirochetes in fluids as viscous as the extracellular matrix. *Journal of Infectious Diseases* 162, 1205–1208.
Kinashi, H., Shimaji, M. and Sakai, A. (1987) Giant linear plasmids in *Streptomyces* which code for antibiotic biosynthesis genes. *Nature* 328, 454–456.
Kitten, T. and Barbour, A.G. (1990) Juxtaposition of expressed variable antigen genes with a conserved telomere in the bacterium *Borrelia hermsii*. *Proceedings of the National Academy of Sciences, USA* 87, 6077–6081.
Knight, S.W., Kimmel, B.J., Eggers, C.H. and Samuels, D.S. (2000) Disruption of the *Borrelia burgdorferi gac* gene, encoding the naturally synthesized GyrA C-terminal domain. *Journal of Bacteriology* 182(7), 2048–2051.
Koopman, M.B., Baats, E., van Vorstenbosch, C.J., van der Zeijst, B.A. and Kusters, J.G. (1992) The periplasmic flagella of *Serpulina* (*Treponema*) *hyodysenteriae* are composed of two sheath proteins and three core proteins. *Journal of General Microbiology* 138, 2697–2706.

Kornacki, J.A. and Oliver, D.B. (1998) Lyme disease-causing *Borrelia* species encode multiple lipoproteins homologous to peptide-binding proteins of ABC-type transporters. *Infection and Immunity* 66, 4115–4122.

Kryuchechnikov, V.N., Korenberg, E.I., Scherbakov, S.V., Kovalevsky Iu, V. and Levin, M.L. (1988) Identification of *Borrelia* isolated in the USSR from *Ixodes persulcatus* (Schulze) ticks. *Journal of Microbiology, Epidemiology and Immunobiology* 12, 41–44.

Kurtenbach, K., Peacey, M., Rijpkema, S.G., Hoodless, A.N., Nuttall, P.A. and Randolph, S.E. (1998) Differential transmission of the genospecies of *Borrelia burgdorferi* sensu lato by game birds and small rodents in England. *Applied Environmental Microbiology* 64, 1169–1174.

Lahdenne, P., Porcella, S.F., Hagman, K.E., Akins, D.R., Popova, T.G., Cox, D.L., Katona, L.I., Radolf, J.D. and Norgard, M.V. (1997) Molecular characterization of a 6.6-kilodalton *Borrelia burgdorferi* outer membrane-associated lipoprotein (lp6.6) which appears to be downregulated during mammalian infection. *Infection and Immunity* 65, 412–421.

Lam, T.T., Nguyen, T.P., Fikrig, E. and Flavell, R.A. (1994a) A chromosomal *Borrelia burgdorferi* gene encodes a 22-kilodalton lipoprotein, P22, that is serologically recognized in Lyme disease. *Journal of Clinical Microbiology* 32, 876–883.

Lam, T.T., Nguyen, T.P., Montgomery, R.R., Kantor, F.S., Fikrig, E. and Flavell, R.A. (1994b) Outer surface proteins E and F of *Borrelia burgdorferi*, the agent of Lyme disease. *Infection and Immunity* 62, 290–298.

Le Fleche, A., Postic, D., Girardet, K., Peter, O. and Baranton, G. (1997) Characterization of *Borrelia lusitaniae* sp. nov. by 16S ribosomal DNA sequence analysis. *International Journal of Systematic Bacteriology* 47, 921–925.

Levy, S.A., Lissman, B.A. and Ficke, C.M. (1993) Performance of a *Borrelia burgdorferi* bacterin in borreliosis-endemic areas. *Journal of the American Veterinary Medical Association* 202, 1834–1838.

Lex, A., Wiesmuller, K.H., Jung, G. and Bessler, W.G. (1986) A synthetic analogue of *Escherichia coli* lipoprotein, tripalmitoyl pentapeptide, constitutes a potent immune adjuvant. *Journal of Immunology* 137, 2676–2681.

Li, H., Dunn, J.J., Luft, B.J. and Lawson, C.L. (1997) Crystal structure of Lyme disease antigen outer surface protein A complexed with an Fab. *Proceedings of the National Academy of Sciences, USA* 94, 3584–3589.

Li, M., Masuzawa, T., Takada, N., Ishiguro, F., Fujita, H., Iwaki, A., Wang, H., Wang, J., Kawabata, M. and Yanagihara, Y. (1998) Lyme disease *Borrelia* species in northeastern China resemble those isolated from far eastern Russia and Japan. *Applied Environmental Microbiology* 64, 2705–2709.

Li, Z., Dumas, F., Dubreuil, D. and Jacques, M. (1993) A species-specific periplasmic flagellar protein of *Serpulina* (*Treponema*) *hyodysenteriae*. *Journal of Bacteriology* 175, 8000–8007.

Liang, F.T. and Philipp, M.T. (1999) Analysis of antibody response to invariable regions of VlsE, the variable surface antigen of *Borrelia burgdorferi*. *Infection and Immunity* 67, 6702–6706.

Lin, B., Short, S.A., Eskildsen, M., Klempner, M.S. and Hu, L.T. (2001) Functional testing of putative oligopeptide permease (Opp) proteins of *Borrelia burgdorferi*: a complementation model in opp(–) *Escherichia coli*. *Biochimica et Biophysica Acta* 1499, 222–231.

Lin, Y.S., Kieser, H.M., Hopwood, D.A. and Chen, C.W. (1993) The chromosomal DNA of *Streptomyces lividans* 66 is linear. *Molecular Microbiology* 10, 923–933.

Livey, I., Gibbs, C.P., Schuster, R. and Dorner, F. (1995) Evidence for lateral transfer and recombination in OspC variation in Lyme disease *Borrelia*. *Molecular Microbiology* 18, 257–269.

Luft, B.J., Jiang, W., Munoz, P., Dattwyler, R.J. and Gorevic, P.D. (1989) Biochemical and immunological characterization of the surface proteins of *Borrelia burgdorferi*. *Infection and Immunity* 57, 3637–3645.

Magnarelli, L.A., Anderson, J.F. and Johnson, R.C. (1987) Cross-reactivity in serological tests for Lyme disease and other spirochetal infections. *Journal of Infectious Diseases* 156, 183–188.

Marconi, R.T. and Garon, C.F. (1992a) Phylogenetic analysis of the genus *Borrelia*: a comparison of North American and European isolates of *Borrelia burgdorferi*. *Journal of Bacteriology* 174, 241–244.

Marconi, R.T. and Garon, C.F. (1992b) Identification of a third genomic group of *Borrelia burgdorferi* through signature nucleotide analysis and 16S rRNA sequence determination. *Journal of General Microbiology* 138, 533–536.

Marconi, R.T., Samuels, D.S., Landry, R.K. and Garon, C.F. (1994) Analysis of the distribution and molecular heterogeneity of the *ospD* gene among the Lyme disease spirochetes: evidence for lateral gene exchange. *Journal of Bacteriology* 176, 4572–4582.

Marconi, R.T., Liveris, D. and Schwartz, I. (1995) Identification of novel insertion elements, restriction fragment length polymorphism patterns, and discontinuous 23S rRNA in Lyme disease spirochetes: phylogenetic analyses of rRNA genes and their intergenic spacers in *Borrelia japonica* sp. nov. and genomic group 21038 (*Borrelia andersonii* sp. nov.) isolates. *Journal of Clinical Microbiology* 33, 2427–2434.

Margolis, N. and Rosa, P.A. (1993) Regulation of expression of major outer surface proteins in *Borrelia burgdorferi*. *Infection and Immunity* 61, 2207–2210.

Marti Ras, N., Postic, D., Foretz, M. and Baranton, G. (1997) *Borrelia burgdorferi* sensu stricto, a bacterial species 'made in the U.S.A.'? *International Journal of Systematic Bacteriology* 47, 1112–1117.

Masuzawa, T., Komikado, T., Iwaki, A., Suzuki, H., Kaneda, K. and Yanagihara, Y. (1996) Characterization of *Borrelia* sp. isolated from *Ixodes tanuki*, *I. turdus*, and *I. columnae* in Japan by restriction fragment length polymorphism of rrf (5S)–rrl (23S) intergenic spacer amplicons. *FEMS Microbiology Letters* 142, 77–83.

Masuzawa, T., Takada, N., Kudeken, M., Fukui, T., Yano, Y., Ishiguro, F., Kawamura, Y., Imai, Y. and Ezaki, T. (2001) *Borrelia sinica* sp. nov., a Lyme disease-related *Borrelia* species isolated in China. *International Journal of Systematic and Evolutionary Microbiology* 51, 1817–1824.

Meinhardt, F., Schaffrath, R. and Larsen, M. (1997) Microbial linear plasmids. *Applied Microbiology and Biotechnology* 47, 329–336.

de Michelis, S., Sewell, H.S., Collares-Pereira, M., Santos-Reis, M., Schouls, L.M., Benes, V., Holmes, E.C. and Kurtenbach, K. (2000) Genetic diversity of *Borrelia burgdorferi* sensu lato in ticks from mainland Portugal. *Journal of Clinical Microbiology* 38, 2128–2133.

van Mierlo, P., Jacob, W. and Dockx, P. (1993) Erythema chronicum migrans: an electron-microscopic study. *Dermatology* 186, 306–310.

Miller, J.C., Bono, J.L., Babb, K., El-Hage, N., Casjens, S. and Stevenson, B. (2000) A second allele of eppA in *Borrelia burgdorferi* strain B31 is located on the previously undetected circular plasmid cp9-2. *Journal of Bacteriology* 182, 6254–6258.

Motaleb, M.A., Corum, L., Bono, J.L., Elias, A.F., Rosa, P., Samuels, D.S. and Charon, N.W. (2000) *Borrelia burgdorferi* periplasmic flagella have both skeletal and motility functions. *Proceedings of the National Academy of Sciences, USA* 97, 10899–10904.

Neubert, U., Schaller, M., Januschke, E., Stolz, W. and Schmieger, H. (1993) Bacteriophages induced by ciprofloxacin in a *Borrelia burgdorferi* skin isolate. *Zentralblatt für Bakteriologie* 279, 307–315.

Noppa, L., Östberg, Y., Lavrinovicha, M. and Bergström, S. (2001) P13, an integral membrane protein of *Borrelia burgdorferi*, is C-terminally processed and contains surface-exposed domains. *Infection and Immunity* 69, 3323–3334.

Norris, S.J. (1993) Polypeptides of *Treponema pallidum*: progress toward understanding their structural, functional, and immunologic roles. *Treponema pallidum* Polypeptide Research Group [published erratum appears in *Microbiological Reviews* (1994) 58, 291]. *Microbiological Reviews* 57, 750–779.

Norris, S.J., Charon, N.W., Cook, R.G., Fuentes, M.D. and Limberger, R.J. (1988) Antigenic relatedness and N-terminal sequence homology define two classes of periplasmic flagellar proteins of *Treponema pallidum* subsp. *pallidum* and *Treponema phagedenis*. *Journal of Bacteriology* 170, 4072–4082.

Norris, S.J., Carter, C.J., Howell, J.K. and Barbour, A.G. (1992) Low-passage-associated proteins of *Borrelia burgdorferi* B31: characterization and molecular cloning of OspD, a surface-exposed, plasmid-encoded lipoprotein. *Infection and Immunity* 60, 4662–4672.

Norris, S.J., Howell, J.K., Garza, S.A., Ferdows, M.S. and Barbour, A.G. (1995) High- and low-infectivity phenotypes of clonal populations of *in vitro*-cultured *Borrelia burgdorferi*. *Infection and Immunity* 63, 2206–2212.

Nosek, J., Tomaska, L., Fukuhara, H., Suyama, Y. and Kovac, L. (1998) Linear mitochondrial genomes: 30 years down the line. *Trends in Genetics* 14, 184–188.

Ohnishi, J., Piesman, J. and de Silva, A.M. (2001) Antigenic and genetic heterogeneity of *Borrelia burgdorferi* populations transmitted by ticks. *Proceedings of the National Academy of Sciences, USA* 98, 670–675.

Olsen, B., Jaenson, T.G., Noppa, L., Bunikis, J. and Bergström, S. (1993) A Lyme borreliosis cycle in seabirds and *Ixodes uriae* ticks. *Nature* 362, 340–342.

Olsen, B., Duffy, D.C., Jaenson, T.G., Gylfe, Å., Bonnedahl, J. and Bergström, S. (1995) Transhemispheric exchange of Lyme disease spirochetes by seabirds. *Journal of Clinical Microbiology* 33, 3270–3274.

Olsen, G.J., Woese, C.R. and Overbeek, R. (1994) The winds of (evolutionary) change: breathing new life into microbiology. *Journal of Bacteriology* 176, 1–6.

Pal, U., de Silva, A.M., Montgomery, R.R., Fish, D., Anguita, J., Anderson, J.F., Lobet, Y. and Fikrig, E. (2000) Attachment of *Borrelia burgdorferi* within *Ixodes scapularis* mediated by outer surface protein A. *Journal of Clinical Investigation* 106, 561–569.

Payne, J.W. and Gilvarg, C. (1968) Size restriction on peptide utilization in *Escherichia coli*. *Journal of Biological Chemistry* 243, 6291–6299.

Persing, D.H., Rutledge, B.J., Rys, P.N., Podzorski, D.S., Mitchell, P.D., Reed, K.D., Liu, B., Fikrig, E. and Malawista, S.E. (1994) Target imbalance: disparity of *Borrelia burgdorferi* genetic material in synovial fluid from Lyme arthritis patients. *Journal of Infectious Diseases* 169, 668–672.

Peters, D.J. and Benach, J.L. (1997) *Borrelia burgdorferi* adherence and injury to undifferentiated and differentiated neural cells *in vitro*. *Journal of Infectious Diseases* 176, 470–477.

Picken, R.N., Cheng, Y., Strle, F. and Picken, M.M. (1996) Patient isolates of *Borrelia burgdorferi* sensu lato with genotypic and phenotypic similarities of strain 25015. *Journal of Infectious Diseases* 174, 1112–1115.

Plasterk, R.H., Simon, M.I. and Barbour, A.G. (1985) Transposition of structural genes to an expression sequence on a linear plasmid causes antigenic variation in the bacterium *Borrelia hermsii*. *Nature* 318, 257–263.

Porcella, S.F., Popova, T.G., Akins, D.R., Li, M., Radolf, J.D. and Norgard, M.V. (1996) *Borrelia burgdorferi* supercoiled plasmids encode multicopy tandem open reading frames and a lipoprotein gene family. *Journal of Bacteriology* 178, 3293–3307.

Porcella, S.F., Fitzpatrick, C.A. and Bono, J.L. (2000) Expression and immunological analysis of the plasmid-borne mlp genes of *Borrelia burgdorferi* strain B31. *Infection and Immunity* 68, 4992–5001.

Postic, D., Assous, M.V., Grimont, P.A. and Baranton, G. (1994) Diversity of *Borrelia burgdorferi* sensu lato evidenced by restriction fragment length polymorphism of *rrf* (5S)–*rrl* (23S) intergenic spacer amplicons. *International Journal of Systematic Bacteriology* 44, 743–752.

Postic, D., Ras, N.M., Lane, R.S., Hendson, M. and Baranton, G. (1998) Expanded diversity among Californian *Borrelia* isolates and description of *Borrelia bissettii* sp. nov. (formerly *Borrelia* group DN127). *Journal of Clinical Microbiology* 36, 3497–3504.

Probert, W.S. and Johnson, B.J. (1998) Identification of a 47 kDa fibronectin-binding protein expressed by *Borrelia burgdorferi* isolate B31. *Molecular Microbiology* 30, 1003–1015.

Probert, W.S., Kim, J.H., Hook, M. and Johnson, B.J. (2001) Mapping the ligand-binding region of *Borrelia burgdorferi* fibronectin-binding protein BBK32. *Infection and Immunity* 69, 4129–4133.

Pugsley, A.P. (1993) The complete general secretory pathway in Gram-negative bacteria. *Microbiological Reviews* 57, 50–108.

Purser, J.E. and Norris, S.J. (2000) Correlation between plasmid content and infectivity in *Borrelia burgdorferi*. *Proceedings of the National Academy of Sciences, USA* 97, 13865–13870.

Radolf, J.D. (1994) Role of outer membrane architecture in immune evasion by *Treponema pallidum* and *Borrelia burgdorferi*. *Trends in Microbiology* 2, 307–311.

Radolf, J.D., Bourell, K.W., Akins, D.R., Brusca, J.S. and Norgard, M.V. (1994) Analysis of *Borrelia burgdorferi* membrane architecture by freeze–fracture electron microscopy. *Journal of Bacteriology* 176, 21–31.

Radolf, J.D., Goldberg, M.S., Bourell, K., Baker, S.I., Jones, J.D. and Norgard, M.V. (1995a) Characterization of outer membranes isolated from *Borrelia burgdorferi*, the Lyme disease spirochete. *Infection and Immunity* 63, 2154–2163.

Radolf, J.D., Arndt, L.L., Akins, D.R., Curetty, L.L., Levi, M.E., Shen, Y., Davis, L.S. and Norgard, M.V. (1995b) *Treponema pallidum* and *Borrelia burgdorferi* lipoproteins and synthetic lipopeptides activate monocytes/macrophages. *Journal of Immunology* 154, 2866–2877.

Restrepo, B.I. and Barbour, A.G. (1994) Antigen diversity in the bacterium *B. hermsii* through 'somatic' mutations in rearranged *vmp* genes. *Cell* 78, 867–876.

Restrepo, B.I., Carter, C.J. and Barbour, A.G. (1994) Activation of a *vmp* pseudogene in *Borrelia hermsii*: an alternate mechanism of antigenic variation during relapsing fever. *Molecular Microbiology* 13, 287–299.

Rosa, P.A., Schwan, T. and Hogan, D. (1992) Recombination between genes encoding major outer surface proteins A and B of *Borrelia burgdorferi*. *Molecular Microbiology* 6, 3031–3040.

Rosey, E.L., Kennedy, M.J., Petrella, D.K., Ulrich, R.G. and Yancey, R.J. Jr (1995) Inactivation of *Serpulina hyodysenteriae flaA1* and *flaB1* periplasmic flagellar genes by electroporation-mediated allelic exchange. *Journal of Bacteriology* 177, 5959–5970.

Rybchin, V.N. and Svarchevsky, A.N. (1999) The plasmid prophage N15: a linear DNA with covalently closed ends. *Molecular Microbiology* 33, 895–903.

Sadziene, A., Thomas, D.D., Bundoc,

V.G., Holt, S.C. and Barbour, A.G. (1991) A flagella-less mutant of *Borrelia burgdorferi*. Structural, molecular, and *in vitro* functional characterization. *Journal of Clinical Investigation* 88, 82–92.

Sadziene, A., Wilske, B., Ferdows, M.S. and Barbour, A.G. (1993a) The cryptic *ospC* gene of *Borrelia burgdorferi* B31 is located on a circular plasmid. *Infection and Immunity* 61, 2192–2195.

Sadziene, A., Barbour, A.G., Rosa, P.A. and Thomas, D.D. (1993b) An OspB mutant of *Borrelia burgdorferi* has reduced invasiveness *in vitro* and reduced infectivity *in vivo*. *Infection and Immunity* 61, 3590–3596.

Sadziene, A., Thomas, D.D. and Barbour, A.G. (1995) *Borrelia burgdorferi* mutant lacking Osp: biological and immunological characterization. *Infection and Immunity* 63, 1573–1580.

Sadziene, A., Thompson, P.A. and Barbour, A.G. (1996) A flagella-less mutant of *Borrelia burgdorferi* as a live attenuated vaccine in the murine model of Lyme disease. *Journal of Infectious Diseases* 173, 1184–1193.

Samuels, D.S., Mach, K.E. and Garon, C.F. (1994a) Genetic transformation of the Lyme disease agent *Borrelia burgdorferi* with coumarin-resistant *gyrB*. *Journal of Bacteriology* 176, 6045–6049.

Samuels, D.S., Marconi, R.T., Huang, W.M. and Garon, C.F. (1994b) *gyrB* mutations in coumermycin A1-resistant *Borrelia burgdorferi*. *Journal of Bacteriology* 176, 3072–3075.

Sankaran, K., Gupta, S.D. and Wu, H.C. (1995) Modification of bacterial lipoproteins. *Methods in Enzymology* 250, 683–697.

Sartakova, M., Dobrikova, E. and Cabello, F.C. (2000) Development of an extrachromosomal cloning vector system for use in *Borrelia burgdorferi*. *Proceedings of the National Academy of Sciences, USA* 97, 4850–4855.

Schmid, G.P., Steigerwalt, A.G., Johnson, S.E., Barbour, A.G., Steere, A.C., Robinson, I.M. and Brenner, D.J. (1984) DNA characterization of the spirochete that causes Lyme disease. *Journal of Clinical Microbiology* 20, 155–158.

Schoenfeld, R., Araneo, B., Ma, Y., Yang, L.M. and Weis, J.J. (1992) Demonstration of a B-lymphocyte mitogen produced by the Lyme disease pathogen, *Borrelia burgdorferi*. *Infection and Immunity* 60, 455–464.

Schwan, T.G. and Burgdorfer, W. (1987) Antigenic changes of *Borrelia burgdorferi* as a result of *in vitro* cultivation. *Journal of Infectious Diseases* 156, 852–853.

Schwan, T.G. and Piesman, J. (2000) Temporal changes in outer surface proteins A and C of the Lyme disease-associated spirochete, *Borrelia burgdorferi*, during the chain of infection in ticks and mice. *Journal of Clinical Microbiology* 38, 382–388.

Schwan, T.G., Burgdorfer, W. and Garon, C.F. (1988) Changes in infectivity and plasmid profile of the Lyme disease spirochete, *Borrelia burgdorferi*, as a result of *in vitro* cultivation. *Infection and Immunity* 56, 1831–1836.

Schwan, T.G., Piesman, J., Golde, W.T., Dolan, M.C. and Rosa, P.A. (1995) Induction of an outer surface protein on *Borrelia burgdorferi* during tick feeding. *Proceedings of the National Academy of Sciences, USA* 92, 2909–2913.

Seshu, J. and Skare, J.T. (2000) The many faces of *Borrelia burgdorferi*. *Journal of Molecular Microbiology and Biotechnology* 2, 463–472.

Shi, W., Yang, Z., Geng, Y., Wolinsky, L.E. and Lovett, M.A. (1998) Chemotaxis in *Borrelia burgdorferi*. *Journal of Bacteriology* 180, 231–235.

Shoberg, R.J. and Thomas, D.D. (1993) Specific adherence of *Borrelia burgdorferi* extracellular vesicles to human endothelial cells in culture. *Infection and Immunity* 61, 3892–3900.

Shoberg, R.J. and Thomas, D.D. (1995) *Borrelia burgdorferi* vesicle production occurs via a mechanism independent of immunoglobulin M involvement. *Infection and Immunity* 63, 4857–4861.

Sigal, L.H., Zahradnik, J.M., Lavin, P., Patella, S.J., Bryant, G., Haselby, R., Hilton, E., Kunkel, M., Adler-Klein, D., Doherty, T., Evans, J. and Malawista, S.E. (1998) A vaccine consisting of recombinant *Borrelia burgdorferi* outer-surface protein A to prevent Lyme disease. Recombinant Outer-Surface Protein A Lyme Disease Vaccine Study Consortium. *New England Journal of Medicine* 339, 216–222.

de Silva, A.M., Telford, S.R., 3rd, Brunet, L.R., Barthold, S.W. and Fikrig, E. (1996) *Borrelia burgdorferi* OspA is an arthropod-specific transmission-blocking Lyme disease vaccine. *Journal of Experimental Medicine* 183, 271–275.

de Silva, A.M., Zeidner, N.S., Zhang, Y., Dolan, M.C., Piesman, J. and Fikrig, E. (1999) Influence of outer surface protein A antibody on *Borrelia burgdorferi* within feeding ticks. *Infection and Immunity* 67, 30–35.

Simpson, W.J., Garon, C.F. and Schwan, T.G. (1990) Analysis of supercoiled circular plasmids in infectious and non-infectious *Borrelia burgdorferi*. *Microbial Pathogenicity* 8, 109–118.

Simpson, W.J., Schrumpf, M.E., Hayes, S.F. and Schwan, T.G. (1991a) Molecular and immunological analysis of a polymorphic periplasmic protein of *Borrelia burgdorferi*. *Journal of Clinical Microbiology* 29, 1940–1948.

Simpson, W.J., Burgdorfer, W., Schrumpf, M.E., Karstens, R.H. and Schwan, T.G. (1991b) Antibody to a 39-kilodalton *Borrelia burgdorferi* antigen (P39) as a marker for infection in experimentally and naturally inoculated animals. *Journal of Clinical Microbiology* 29, 236–243.

Skare, J.T., Shang, E.S., Foley, D.M., Blanco, D.R., Champion, C.I., Mirzabekov, T., Sokolov, Y., Kagan, B.L., Miller, J.N. and Lovett, M.A. (1995) Virulent strain associated outer membrane proteins of *Borrelia burgdorferi*. *Journal of Clinical Investigation* 96, 2380–2392.

Skare, J.T., Champion, C.I., Mirzabekov, T.A., Shang, E.S., Blanco, D.R., Erdjument-Bromage, H., Tempst, P., Kagan, B.L., Miller, J.N. and Lovett, M.A. (1996) Porin activity of the native and recombinant outer membrane protein Oms28 of *Borrelia burgdorferi*. *Journal of Bacteriology* 178, 4909–4918.

Skare, J.T., Mirzabekov, T.A., Shang, E.S., Blanco, D.R., Erdjument-Bromage, H., Bunikis, J., Bergström, S., Tempst, P., Kagan, B.L., Miller, J.N. and Lovett, M.A. (1997) The Oms66 (p66) protein is a *Borrelia burgdorferi* porin. *Infection and Immunity* 65, 3654–3661.

de Souza, M.S., Fikrig, E., Smith, A.L., Flavell, R.A. and Barthold, S.W. (1992) Nonspecific proliferative responses of murine lymphocytes to *Borrelia burgdorferi* antigens. *Journal of Infectious Diseases* 165, 471–478.

Steere, A.C., Sikand, V.K., Meurice, F., Parenti, D.L., Fikrig, E., Schoen, R.T., Nowakowski, J., Schmid, C.H., Laukamp, S., Buscarino, C. and Krause, D.S. (1998) Vaccination against Lyme disease with recombinant *Borrelia burgdorferi* outer-surface lipoprotein A with adjuvant. *New England Journal of Medicine* 339, 209–215.

Steere, A.C., Gross, D., Meyer, A.L. and Huber, B.T. (2001) Autoimmune mechanisms in antibiotic treatment-resistant Lyme arthritis. *Journal of Autoimmunity* 16, 263–268.

Stevenson, B., Tilly, K. and Rosa, P.A. (1996) A family of genes located on four separate 32-kilobase circular plasmids in *Borrelia burgdorferi* B31. *Journal of Bacteriology* 178, 3508–3516.

Stevenson, B., Bono, J.L., Schwan, T.G. and Rosa, P. (1998) *Borrelia burgdorferi* Erp proteins are immunogenic in mammals infected by tick bite, and their synthesis is inducible in cultured bacteria. *Infection and Immunity* 66, 2648–2654.

Stevenson, B., Zuckert, W.R. and Akins, D.R. (2000a) Repetition, conservation, and variation: the multiple cp32 plasmids of *Borrelia* species. *Journal of Molecular Microbiology and Biotechnology* 2, 411–422.

Stevenson, B., Porcella, S.F., Oie, K.L., Fitzpatrick, C.A., Raffel, S.J., Lubke, L., Schrumpf, M.E. and Schwan, T.G. (2000b) The relapsing fever spirochete *Borrelia hermsii* contains multiple, antigen-encoding circular plasmids that are homologous to the cp32 plasmids of Lyme disease spirochetes. *Infection and Immunity* 68, 3900–3908.

Stewart, P.E., Thalken, R., Bono, J.L. and Rosa, P. (2001) Isolation of a circular plasmid region sufficient for autonomous replication and transformation of infectious *Borrelia burgdorferi*. *Molecular Microbiology* 39, 714–721.

Strle, F., Picken, R.N., Cheng, Y., Cimperman, J., Maraspin, V., Lotric-Furlan, S., Ruzic-Sabljic, E. and Picken, M.M. (1997) Clinical findings for patients with Lyme borreliosis caused by *Borrelia burgdorferi* sensu lato with genotypic and phenotypic similarities to strain 25015. *Clinical Infectious Disease* 25, 273–280.

Suk, K., Das, S., Sun, W., Jwang, B., Barthold, S.W., Flavell, R.A. and Fikrig, E. (1995) *Borrelia burgdorferi* genes selectively expressed in the infected host. *Proceedings of the National Academy of Sciences, USA* 92, 4269–4273.

Sung, S.Y., McDowell, J.V., Carlyon, J.A. and Marconi, R.T. (2000) Mutation and recombination in the upstream homology box-flanked *ospE*-related genes of the Lyme disease spirochetes result in the development of new antigenic variants during infection. *Infection and Immunity* 68, 1319–1327.

Swancutt, M.A., Riley, B.S., Radolf, J.D. and Norgard, M.V. (1989) Molecular characterization of the pathogen-specific, 34-kilodalton membrane immunogen of *Treponema pallidum*. *Infection and Immunity* 57, 3314–3323.

Takayama, K., Rothenberg, R.J. and Barbour, A.G. (1987) Absence of lipopolysaccharide in the Lyme disease spirochete, *Borrelia burgdorferi*. *Infection and Immunity* 55, 2311–2313.

Theisen, M., Borre, M., Mathiesen, M.J., Mikkelsen, B., Lebech, A.M. and Hansen, K. (1995) Evolution of the *Borrelia burgdorferi* outer surface protein OspC. *Journal of Bacteriology* 177, 3036–3044.

Thomas, D.D. and Comstock, L.E. (1989) Interaction of Lyme disease spirochetes with cultured eucaryotic cells. *Infection and Immunity* 57, 1324–1336.

Tilly, K., Casjens, S., Stevenson, B., Bono, J.L., Samuels, D.S., Hogan, D. and Rosa, P. (1997) The *Borrelia burgdorferi* circular plasmid cp26: conservation of plasmid structure and targeted inactivation of the ospC gene. *Molecular Microbiology* 25(2), 361–373.

Tilly, K., Lubke, L. and Rosa, P. (1998) Characterization of circular plasmid dimers in *Borrelia burgdorferi*. *Journal of Bacteriology* 180(21), 5676–5681.

Tilly, K., Elias, A.F., Bono, J.L., Stewart, P. and Rosa, P. (2000) DNA exchange and insertional inactivation in spirochetes. *Journal of Molecular Microbiology and Biotechnology* 2, 433–442.

Tilly, K., Elias, A.F., Errett, J. Fischer, E. Iyer, R., Schwartz, I., Bono, J.L. and Rosa, P. (2001) Genetics and regulation of chitobiose utilization in *Borrelia burgdorferi*. *Journal of Bacteriology* 183(19), 5544–5553.

Tomb, J.F., White, O., Kerlavage, A.R., Clayton, R.A., Sutton, G.G., Fleischmann, R.D., Ketchum, K.A., Klenk, H.P., Gill, S., Dougherty, B.A., Nelson, K., Quackenbush, J., Zhou, L., Kirkness, E.F., Peterson, S., Loftus, B., Richardson, D., Dodson, R., Khalak, H.G., Glodek, A., McKenney, K.,

Fitzegerald, L.M., Lee, N., Adams, M.D., Hickey, E.K., Berg, D.E., Gocayne, J.D., Utterback, T.R., Peterson, J.D., Kelley, J.M., Cotton, M.D., Weidman, J.M., Fujii, C., Bowman, C., Watthey, L., Wallin, E., Hayes, W.S., Borodovsky, M., Karp, P.D., Smith, H.O., Fraser, C.M. and Venter, J.C. (1997) The complete genome sequence of the gastric pathogen *Helicobacter pylori*. *Nature* 388, 539–547.

Trollmo, C., Meyer, A.L., Steere, A.C., Hafler, D.A. and Huber, B.T. (2001) Molecular mimicry in Lyme arthritis demonstrated at the single cell level: LFA-1 alpha L is a partial agonist for outer surface protein A-reactive T cells. *Journal of Immunology* 166, 5286–5291.

Trueba, G.A., Bolin, C.A. and Zuerner, R.L. (1992) Characterization of the periplasmic flagellum proteins of *Leptospira interrogans*. *Journal of Bacteriology* 174, 4761–4768.

Trueba, G.A., Old, I.G., Saint Girons, I. and Johnson, R.C. (1997) A *cheA* and *cheW* operon in *Borrelia burgdorferi*. *Research in Microbiology* 148, 191–200.

Valsangiacomo, C., Balmelli, T. and Piffaretti, J.C. (1997) A phylogenetic analysis of *Borrelia burgdorferi* sensu lato based on sequence information from the *hbb* gene, coding for a histone-like protein. *International Journal of Systematic Bacteriology* 47, 1–10.

Volff, J.N. and Altenbuchner, J. (2000) A new beginning with new ends: linearisation of circular chromosomes during bacterial evolution. *FEMS Microbiology Letters* 186, 143–150.

Walker, E.M., Borenstein, L.A., Blanco, D.R., Miller, J.N. and Lovett, M.A. (1991) Analysis of outer membrane ultrastructure of pathogenic *Treponema* and *Borrelia* species by freeze–fracture electron microscopy. *Journal of Bacteriology* 173, 5585–5588.

Wallich, R., Moter, S.E., Simon, M.M., Ebnet, K., Heiberger, A. and Kramer, M.D. (1990) The *Borrelia burgdorferi* flagellum-associated 41-kilodalton antigen (flagellin): molecular cloning, expression, and amplification of the gene. *Infection and Immunity* 58, 1711–1719.

Wallich, R., Simon, M.M., Hofmann, H., Moter, S.E., Schaible, U.E. and Kramer, M.D. (1993) Molecular and immunological characterization of a novel polymorphic lipoprotein of *Borrelia burgdorferi*. *Infection and Immunity* 61, 4158–4166.

Wang, G., van Dam, A.P., Le Fleche, A., Postic, D., Peter, O., Baranton, G., de Boer, R., Spanjaard, L. and Dankert, J. (1997) Genetic and phenotypic analysis of *Borrelia valaisiana* sp. nov. (*Borrelia* genomic groups VS116 and M19). *International Journal of Systematic Bacteriology* 47, 926–932.

Wang, G., van Dam, A.P. and Dankert, J. (1999a) Phenotypic and genetic characterization of a novel *Borrelia burgdorferi* sensu lato isolate from a patient with Lyme borreliosis. *Journal of Clinical Microbiology* 37, 3025–3028.

Wang, G., van Dam, A.P. and Dankert, J. (1999b) Evidence for frequent OspC gene transfer between *Borrelia valaisiana* sp. nov. and other Lyme disease spirochetes. *FEMS Microbiology Letters* 177, 289–296.

Wang, G., van Dam, A.P., Schwartz, I. and Dankert, J. (1999c) Molecular typing of *Borrelia burgdorferi* sensu lato: taxonomic, epidemiological, and clinical implications. *Clinical Microbiology Reviews* 12, 633–653.

Wang, G., van Dam, A.P. and Dankert, J. (2000) Two distinct *ospA* genes among *Borrelia valaisiana* strains. *Research in Microbiology* 151, 325–331.

Wang, I.N., Dykhuizen, D.E., Qiu, W., Dunn, J.J., Bosler, E.M. and Luft, B.J. (1999) Genetic diversity of ospC in a local population of *Borrelia burgdorferi* sensu stricto. *Genetics* 151, 15–30.

Wayne, L.G., Brenner, D.J., Colwell, R.R., Grimont, P.A.D., Kandler, O., Krichevsky, M.I., Moore, L.H., Moore, W.E.C., Murray, R.G.E., Stackebrandt, E., Starr, M.P. and Truper, H.G. (1987) Report of the Ad Hoc Committee on reconciliation of approaches to bacterial systematics. *International Journal of Systematic Bacteriology* 37, 463–464.

Wheeler, C.M., Garcia Monco, J.C., Benach, J.L., Golightly, M.G., Habicht, G.S. and Steere, A.C. (1993) Nonprotein antigens of *Borrelia burgdorferi*. *Journal of Infectious Diseases* 167, 665–674.

Wilske, B., Preac-Mursic, V., Gobel, U.B., Graf, B., Jauris, S., Soutschek, E., Schwab, E. and Zumstein, G. (1993) An OspA serotyping system for *Borrelia burgdorferi* based on reactivity with monoclonal antibodies and OspA sequence analysis. *Journal of Clinical Microbiology* 31, 340–350.

Wilske, B., Jauris-Heipke, S., Lobentanzer, R., Pradel, I., Preac-Mursic, V., Rössler, D., Soutschek, E. and Johnson, R.C. (1995) Phenotypic analysis of outer surface protein C (OspC) of *Borrelia burgdorferi* sensu lato by monoclonal antibodies: relationship to genospecies and OspA serotype. *Journal of Clinical Microbiology* 33, 103–109.

Wilske, B., Busch, U., Fingerle, V., Jauris-Heipke, S., Preac Mursic, V., Rössler, D. and Will, G. (1996) Immunological and molecular variability of OspA and OspC. Implications for *Borrelia* vaccine development. *Infection* 24, 208–212.

Wlodarzyk, M. and Nowicka, B. (1988) Preliminary evidence for the linear nature of *Thiobacillus versutus* pTAV2 plasmid. *FEMS Microbiology Letters* 55, 125–128.

Woese, C.R. (1987) Bacterial evolution. *Microbiological Reviews* 51, 221–271.

Woese, C.R., Olsen, G.J., Ibba, M. and Soll, D. (2000) Aminoacyl-tRNA synthetases, the genetic code, and the evolutionary process. *Microbiology and Molecular Biology Reviews* 64, 202–236.

Xu, Y., Kodner, C., Coleman, L. and Johnson, R.C. (1996) Correlation of plasmids with infectivity of *Borrelia burgdorferi* sensu stricto type strain B31. *Infection and Immunity* 64, 3870–3876.

Yang, X., Popova, T.G., Hagman, K.E., Wikel, S.K., Schoeler, G.B., Caimano, M.J., Radolf, J.D. and Norgard, M.V. (1999) Identification, characterization, and expression of three new members of the *Borrelia burgdorferi* Mlp (2.9) lipoprotein gene family. *Infection and Immunity* 67, 6008–6018.

Yang, X., Goldberg, M.S., Popova, T.G., Schoeler, G.B., Wikel, S.K., Hagman, K.E. and Norgard, M.V. (2000) Interdependence of environmental factors influencing reciprocal patterns of gene expression in virulent *Borrelia burgdorferi*. *Molecular Microbiology* 37, 1470–1479.

Zhang, J.R. and Norris, S.J. (1998a) Kinetics and *in vivo* induction of genetic variation of *vlsE* in *Borrelia burgdorferi*. *Infection and Immunity* 66, 3689–3697.

Zhang, J.R. and Norris, S.J. (1998b) Genetic variation of the *Borrelia burgdorferi* gene *vlsE* involves cassette-specific, segmental gene conversion. *Infection and Immunity* 66, 3698–3704.

Zhang, J.R., Hardham, J.M., Barbour, A.G. and Norris, S.J. (1997) Antigenic variation in Lyme disease *Borreliae* by promiscuous recombination of VMP-like sequence cassettes. *Cell* 89, 275–285.

Zhang, Q.Y., DeRyckere, D., Lauer, P. and Koomey, M. (1992) Gene conversion in *Neisseria gonorrhoeae*: evidence for its role in pilus antigenic variation. *Proceedings of the National Academy of Sciences, USA* 89, 5366–5370.

Zuckert, W.R. and Meyer, J. (1996) Circular and linear plasmids of Lyme disease spirochetes have extensive homology: characterization of a repeated DNA element. *Journal of Bacteriology* 178, 2287–2298.

Vectors of *Borrelia burgdorferi* sensu lato

4

Lars Eisen and Robert S. Lane

Department of Environmental Science, Policy and Management, University of California, Berkeley, CA 94720, USA

Introduction

Lyme borreliosis (LB) is the most common arthropod-borne human disease in temperate regions of the northern hemisphere, with more than 16,800 cases occurring in the USA alone in 1998 (CDC, 2000). The primary causative agents of LB include three of the 11 genospecies currently comprising *Borrelia burgdorferi* sensu lato (s.l.), i.e. *B. burgdorferi* sensu stricto (s.s.), *Borrelia afzelii* and *Borrelia garinii*. These spirochaetal bacteria are maintained in enzootic spirochaete–tick vector–vertebrate cycles and incidental human exposure occurs when enzootic tick vectors bridge the gap from such maintenance cycles by biting humans. The primary bridging vectors of LB spirochaetes to humans include four tick species of the *Ixodes ricinus* complex (Acari: Ixodidae: *Ixodes*: subgenus *Ixodes*; Filippova, 1999; Keirans *et al.*, 1999), i.e. the castor bean tick, *I. ricinus*, and the taiga tick, *Ixodes persulcatus*, in Eurasia and the western black-legged tick, *Ixodes pacificus*, and the black-legged tick, *Ixodes scapularis*, in North America (Fig. 4.1; Table 4.1). The above-mentioned tick species also include the primary vectors of the disease agents causing tick-borne encephalitis, human babesiosis and human granulocytic ehrlichiosis (see Table 4.1).

The main goals of this chapter are to summarize the results of experimental studies of vector competence for *B. burgdorferi* s.l., give a general description of tick–*B. burgdorferi* s.l. relationships, provide an outline of the biology of the primary bridging vectors of LB spirochaetes and examine how differences in the ecology of these ticks affect their vector potential. Our literature review, concluded by August 2001, is restricted to the seminal reference(s) and examples of recent studies.

Table 4.1. Human disease agents transmitted by the primary bridging vectors of Lyme borreliosis spirochaetes.

Ixodes vector[a]	Geographical distribution	Pathogen						
		Bb ss[b]	B afz[b]	B gar[b]	TBE-virus[c]	Bab div[d]	Bab mic[d]	Ana pha[e]
I. persulcatus	Far-eastern Europe, Asia		X[f]	X	X			
I. pacificus	Western North America	X						X
I. scapularis	Eastern North America	X			X[g]		X	X
I. ricinus	Europe, far-western Asia, far-northern Africa	X	X	X	X	X		X

[a] I. scapularis includes records for I. dammini (Oliver et al., 1993a).
[b] Bb ss, Borrelia burgdorferi sensu stricto; B afz, B. afzelii; B gar, B. garinii; causative agents of Lyme borreliosis.
[c] TBE-virus, tick-borne encephalitis subgroup viruses; causative agents of TBE in Europe (European or western TBE), Asia (Far-eastern TBE or Russian spring–summer encephalitis), and North America (Powassan virus). See reviews by Korenberg (1994) and Nuttall and Labuda (1994).
[d] Bab div, Babesia divergens; Bab mic, Babesia microti; primary causative agents of human babesiosis. See reviews by Spielman et al. (1985) and Kjemtrup and Conrad (2000).
[e] Ana pha Anaplasma (Ehrlichia) phagocytophila (Dulmer et al., 2001); causative agent of human granulocytic ehrlichiosis. See reviews by Woldehiwet (1983) and Dumler and Bakken (1998).
[f] Vector competence indicated by presence of B. afzelii in I. persulcatus and rodents (e.g. Postic et al., 1997; Alekseev et al., 2001); experimental confirmation of vector status lacking.
[g] I. scapularis is a confirmed vector of Powassan virus (which includes a genotype called deer tick virus; Telford et al., 1997) (Costero and Grayson, 1996), but Ixodes cookei may be the primary bridging vector to humans (Artsob, 1988).

Vector Competence for *B. burgdorferi* s.l.

A literature review revealed that natural infections with *B. burgdorferi* s.l. have been recorded from at least 25 species of *Ixodes* ticks, some 15 other ixodid tick species in six genera (*Amblyomma, Boophilus, Dermacentor, Haemaphysalis, Hyalomma, Rhipicephalus*) and two argasid tick species representing two genera (*Argas, Ornithodoros*). Some *Ixodes* ticks appear to be associated with one or a few *B. burgdorferi* s.l. genospecies, whereas *I. ricinus* harbours at least five genospecies (reviewed by Baranton *et al.*, 1998; Wang *et al.*, 1999).

In some studies, *B. burgdorferi* s.l. were observed in partially or fully fed ticks removed from hosts, which demonstrates only that the ticks ingested infected body fluids. In others, *B. burgdorferi* s.l. were detected in questing ticks, indicating that spirochaetes ingested by larvae or nymphs survived the moult and were passed to the resulting nymphs or adults (transstadial passage). However, neither finding demonstrates that the ticks are vectors for *B. burgdorferi* s.l. As used in this chapter, *vector competence* for *B. burgdorferi* s.l. means that a tick species can acquire spirochaetes when feeding on an infective host, pass them transstadially and, subsequently, transmit the spirochaetes to a susceptible host while feeding. However, a tick species may be a vector for *B. burgdorferi* s.l. but only make a minor contribution to enzootic spirochaete–tick vector–vertebrate cycles (referred to as enzootic cycles) or as a bridging vector to humans. Therefore, *vector potential* (taking into account any factors that increase or decrease vector efficiency; (see Kahl *et al.*, Chapter 2) will be used to assess the importance of tick species/populations as enzootic vectors or bridging vectors to humans.

Vector competence for *B. burgdorferi* s.l. has been experimentally confirmed for 12 tick species. These include six of the 14 members of the *I. ricinus* complex (*I. affinis, I. jellisoni, I. pacificus, I. persulcatus, I. ricinus, I. scapularis*) and six other *Ixodes* spp. (*I. angustus, I. dentatus, I. hexagonus, I. minor, I. muris, I. spinipalpis*) (Table 4.2). Published vector competence studies have included only four *B. burgdorferi* s.l. genospecies (*B. burgdorferi* s.s., *B. afzelii, B. bissettii, B. garinii*) (see Table 4.2). In contrast to most other *Borrelia* spp. spirochaetes, several *B. burgdorferi* s.l. genospecies appear to have a low vector specificity; vectors for *B. burgdorferi* s.s., for example, include at least nine *Ixodes* spp. (Sonenshine, 1993; Table 4.2). In addition to *B. burgdorferi* s.l., *I. scapularis* was recently demonstrated to be naturally infected with and a vector of a hitherto unknown relapsing fever group spirochaete closely related to *Borrelia miyamotoi* (Scoles *et al.*, 2001), a spirochaete previously detected in *I. persulcatus* from Japan (Fukunaga *et al.*, 1995).

I. pacificus, I. persulcatus, I. ricinus and *I. scapularis* are the primary bridging vectors of LB spirochaetes to humans in temperate regions of the northern hemisphere (Fig. 4.1, Table 4.1). There is also evidence indicating that *I. angustus*, a tick species occurring in moist, cool habitats in North America and eastern Asia, acts as a secondary bridging vector (Damrow *et al.*, 1989). The majority of the remaining confirmed vectors feed primarily on rodents and/or lagomorphs (i.e. *I. dentatus, I. jellisoni, I. muris* and *I. spinipalpis* in North America and *I. minor* in North and South America), whereas *I. affinis* infests a wide variety of mammals in North and South America, and *I. hexagonus* is found on various medium-sized mammals in Europe and north-western Africa (Jaenson *et al.*, 1994; Durden and Keirans, 1996). Their host preferences render these tick species unlikely to act as bridging vectors to humans but some of them probably have a significant impact on enzootic spirochaete maintenance. In addition to the experimentally confirmed vectors, the presence of spirochaetes in ticks and their primary hosts suggest vector competence for *B. burgdorferi* s.l. of several other *Ixodes* spp. These include the cosmopolitan seabird tick *I. uriae* (Olsen *et al.*, 1995), the Eurasian rodent tick *I. trianguliceps* (Postic *et al.*, 1997) and the Asian ticks *I. columnae, I. granulatus, I. nipponensis, I. tanuki* and *I. turdus* (e.g. Fukunaga *et al.*, 1996; Masuzawa *et al.*, 1999).

Seven tick species evaluated for vector competence appear unable to transmit *B. burgdorferi* s.l. (*Amblyomma americanum, Dermacentor andersoni, D. occidentalis, D. variabilis, Ixodes cookei, I. holocyclus, I. ovatus*; Table 4.3). In most cases these ticks

Table 4.2. Tick species experimentally confirmed as vectors of *Borrelia burgdorferi* sensu lato.

Ixodes tick species[a]	Borrelia burgdorferi sensu lato	Genospecies and isolate used[b] Borrelia burgdorferi sensu stricto	Borrelia afzelii	Borrelia bissettii	Borrelia garinii	Selected references[c]
I. affinis		SI-1				1
I. angustus		CA4				2
I. dentatus	Wild[d]					3
I. hexagonus		B31				4
I. jellisoni	CA404, 409, 445					5
I. minor		SH-2-82				1
I. muris		B31, SH-2-82				1, 6
I. pacificus		JD1, CA4		SCW-30		7–8, 24
I. persulcatus				CA589,592		9–10
I. ricinus	Wild	Wild, ZS7	Wild, PGau		JEM6	11–14
I. scapularis	Wild, NC-2, MI-129	B31, JD1, SI-1, SH-2-82		MI-6, SCW-30	Wild[e], VS286	1, 13, 15–21
I. spinipalpis	Wild	B31	Pgau	Wild, CA589, 592	VS286	21–24

[a] *I. scapularis* includes records for *I. dammini* (Oliver et al., 1993a) and *I. spinipalpis* for *I. neotomae* (Norris et al., 1997).
[b] In some cases the genospecies was not given in the original study, but could be elucidated by later genetic studies of the isolate used. Isolates not characterized genetically, to our knowledge, are listed as *B. burgdorferi* sensu lato.
[c] Studies where transmission of spirochaetes was accomplished by partially fed ticks, infected by artificial feeding or partial blood meals on infected hosts, are excluded; (1) Oliver, 1996; J.H. Oliver Jr, Georgia Southern University, 2001, personal communication; (2) Peavey et al., 2000; (3) Telford and Spielman, 1989a; (4) Gern et al., 1991; (5) Lane et al., 1999; (6) Dolan et al., 2000; (7) Lane et al., 1994; (8) Peavey and Lane, 1995; (9) Nakao and Sato, 1996; (10) Sato and Nakao, 1997; (11) Burgdorfer et al., 1983; (12) Gern and Rais, 1996; (13) Dolan et al., 1998; (14) Hu et al., 2001; (15) Burgdorfer et al., 1982; (16) Piesman and Sinsky, 1988; (17) Ryder et al., 1992; (18) Oliver et al., 1993b; (19) Sanders and Oliver, 1995; (20) Piesman and Happ, 1997; (21) Dolan et al., 1997; (22) Brown and Lane, 1992; (23) Burkot et al., 2000; (24) L. Eisen and R.S. Lane, unpublished data.
[d] Spirochaetes originating from field-captured ticks or mammals, no isolate designation given.
[e] Including *B. garinii* OspA serotype 4 (Hu et al., 2001).

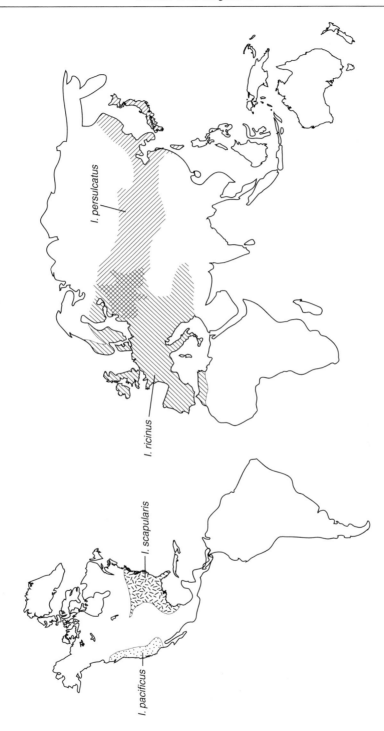

Fig. 4.1. The geographical distributions of the primary vectors of Lyme borreliosis spirochaetes. (Designed by B. Kaye.)

Table 4.3. Tick species apparently unable to serve as vectors of *Borrelia burgdorferi* sensu lato.

Tick species	Genospecies and isolate used[a]				Isolates demonstrated to be:			References[b]
	Borrelia burgdorferi sensu lato	*Borrelia burgdorferi* sensu stricto	*Borrelia bissettii*	*Borrelia garinii*	Acquired	Passed transstadially	Transmitted	
Amblyomma americanum	NC-2	JD1, SI-1, SH-2-82			+/−[c]	+/−[c]	−	1–7
Dermacentor andersonii	Wild	B31			+	−	−	8
Dermacentor occidentalis	CA13, 79, 85	CA4, 5, 11			+	−	−[d]	9–11
Dermacentor variabilis	NC-2, MI-119, 129	JD1, SI-1	MI-6		+/−[e]	+/−[e]	−[e]	1–3, 6–7
Ixodes cookei	LI-231	SH-2-82	MI-6		+	+/−[f]	−	4, 12
Ixodes holocyclus		JD1			+	−	NT[g]	13
Ixodes ovatus				JEM4, 5, 6	+	−	NT[g]	14

[a] Examples of isolates used. In some cases the genospecies was not given in the original study but could be elucidated from later genetic studies of the isolate used. Isolates not characterized genetically, to our knowledge, are listed as *B. burgdorferi* sensu lato.
[b] (1) Piesman and Sinsky, 1988; (2) Mather and Mather, 1990; (3) Mukolwe *et al.*, 1992; (4) Ryder *et al.*, 1992; (5) Oliver *et al.*, 1993b; (6) Sanders and Oliver, 1995; (7) Piesman and Happ, 1997; (8) Dolan *et al.*, 1997; (9) Brown and Lane, 1992; (10) Lane *et al.*, 1994; (11) Li and Lane, 1996; (12) Barker *et al.*, 1993; (13) Piesman and Stone, 1991; (14) Nakao and Miyamoto, 1994.
[c] Acquisition of spirochaetes during feeding was observed in three of five studies examining newly engorged larvae. Transstadial transmission was recorded by Ryder *et al.* (1992) (one of 60 examined nymphs infected with SH-2-82), but absent in four other studies.
[d] Examined with artificially infected ticks; transmission of *B. burgdorferi* sensu stricto (CA4) from infected adult ticks to rabbits was observed for *I. pacificus* but not for *D. occidentalis* (Li and Lane, 1996).
[e] Acquisition of spirochaetes during feeding was observed in three of four studies examining newly engorged larvae. Transstadial transmission was recorded by Piesman and Happ (1997) for isolates NC-2 (one infected nymph) and MI-119 (two infected nymphs), but absent in three other studies. Spirochaetes were not isolated from any host exposed to *D. variabilis* in three studies, although one of six exposed rabbits reportedly seroconverted in one of these studies (Mukolwe *et al.*, 1992).
[f] Isolate LI-231 was passed transstadially to 4–16% of nymphs fed as larvae on various hosts (Barker *et al.*, 1993), whereas no transstadial passage was recorded for isolate SH-2-82 (Ryder *et al.*, 1992).
[g] NT, not tested.

acquired *B. burgdorferi* s.l. when feeding on infected hosts but transstadial passage was rare or absent, and there was no evidence of spirochaete transmission during feeding. *Haemaphysalis leporispalustris* also appears to be a poor potential vector of *B. burgdorferi* s.l.; spirochaetes were absent from 24 unfed nymphs fed as larvae on wild-caught rabbits, whereas approximately 80% of unfed *I. dentatus* nymphs fed as larvae on the same individual hosts were infected (Telford and Spielman, 1989b). Until vector competence has been demonstrated for *Amblyomma, Argas, Boophilus, Dermacentor, Haemaphysalis, Hyalomma, Ornithodoros* or *Rhipicephalus* tick species, they should not be assumed to serve as bridging vectors for *B. burgdorferi* s.l. to humans, or to play any role in the enzootic maintenance of these spirochaetes. Also, the apparent vector incompetence of *I. cookei* for *B. burgdorferi* s.s. and *I. ovatus* for *B. garinii* (Table 4.3) underscores the need for experimental verification of *Ixodes* species suspected to be vectors for *B. burgdorferi* s.l.

In addition to ticks, *B. burgdorferi* s.l. have been detected in numerous insect species, including mosquitoes, tabanid flies and fleas (e.g. Magnarelli and Anderson, 1988; Hubálek *et al.*, 1998). There are some published accounts of LB following insect bites but no evidence that insects can serve as vectors for LB spirochaetes has been presented. Although insects may prove incompetent to serve as vectors for *B. burgdorferi* s.l., the possibility of occasional mechanical transmission should not be discounted. For example, people could be exposed to *B. burgdorferi* s.l. by crushing an infected insect and thereby introducing spirochaetes into the bite site or through broken skin.

Problems in the design of experimental vector-competence studies are discussed by Kahl *et al.* (Chapter 2) and were previously addressed by Lane (1994) and Randolph and Nuttall (1994). Main problem areas include usage of culture-adapted spirochaete strains, transmission of the disease agent by tick-bite versus needle inoculation, choice of tick host and disparity in the geographical origin of the ticks and the spirochaetal isolates used.

Tick–*Borrelia* Relationships

Recent studies have provided new insights into the dynamics at the tick–*B. burgdorferi* s.l. interface, as well as possible explanations for vector incompetence. Spirochaetes ingested during a blood meal multiply rapidly within the midgut of *I. scapularis* larvae during the first 2 weeks post-repletion but their numbers drop dramatically during the subsequent pre-moult period (Piesman *et al.*, 1990). In moulted unfed *Ixodes* spp. ticks the spirochaetes are generally restricted to the midgut; during a blood meal the spirochaetes multiply in the gut before they escape through the gut wall into the haemocoel and migrate to and invade the salivary glands (e.g. Ribeiro *et al.*, 1987; Gern *et al.*, 1990; de Silva and Fikrig, 1995). *B. burgdorferi* s.s. and *B. afzelii* downregulate expression of outer surface protein A (OspA) and upregulate OspC during tick feeding (e.g. Schwan *et al.*, 1995; de Silva *et al.*, 1996; Leuba-Garcia *et al.*, 1998). This change in surface protein profile may constitute the mechanism enabling the spirochaetes to escape the tick midgut and/or invade the salivary glands; OspA mediates binding of spirochaete–spirochaete and spirochaete–*I. scapularis* tick gut (Pal *et al.*, 2000), and conditions that limit upregulation of OspC during the blood meal result in reduced spirochaete migration from the midgut to the salivary glands (de Silva *et al.*, 1999; Gilmore and Piesman, 2000).

B. burgdorferi s.s. can appear in the salivary glands of *I. scapularis* nymphs within 36 h of attachment but spirochaete numbers apparently increase dramatically after approximately 54 h (de Silva and Fikrig, 1995; Ohnishi *et al.*, 2001). The time elapsed from tick attachment to spirochaete transmission may differ between vector species, between *B. burgdorferi* s.l. genospecies, and even between strains of the same genospecies. *I. ricinus* nymphs transmitted *B. burgdorferi* s.l. to approximately 50% of gerbil hosts within 17 h of attachment and all hosts developed viable infections after 48 h of nymphal feeding (Kahl *et al.*, 1998). In contrast, only 11% of deer mice (*Peromyscus maniculatus*) fed on for 48 h by *B. burgdorferi* s.s.-infected *I. pacificus* nymphs developed viable infections, and approximately 78 h of nymphal attachment was required to transmit spirochaetes to 50% of the mice (Peavey and Lane, 1995).

Early studies showed that *I. scapularis* nymphs transmitted *B. burgdorferi* s.s. strain JD1 to 7–14% of rodent hosts within 24–36 h of attachment, and 36–100% of the hosts developed viable infections after 48 h of nymphal feeding (e.g. Piesman *et al.*, 1987; Shih and Spielman, 1993). Intriguingly, viable infections in white mice exposed to another *B. burgdorferi* s.s. strain (B31) occurred only for mice fed on by infected *I. scapularis* nymphs for at least 53 h, after which period spirochaete numbers increased dramatically in nymphal salivary glands and the majority of spirochaetes in the salivary glands failed to produce either OspA or OspC (Ohnishi *et al.*, 2001). Finally, a recent comparative study recorded similar transmission rates of *B. burgdorferi* s.s. strains JD1 and B31 by *I. scapularis* nymphs, with 0–12% of white mice developing viable infections after 48 h of nymphal feeding and 56–71% after 72 h (des Vignes *et al.*, 2001).

In addition to the antigenic changes *B. burgdorferi* s.l. undergoes during the blood meal, the ability of these spirochaetes to infect the tick host successfully may be related to immunosuppressive factors present in tick saliva (e.g. Ribeiro and Mather, 1998; Valenzuela *et al.*, 2000). For example, *I. ricinus* salivary gland extract inhibited killing of *B. afzelii* by mouse macrophages (Kuthejlová *et al.*, 2001). Similarly, salivary components facilitated transmission of tick-borne encephalitis virus from tick to host (saliva-activated transmission) (Nuttall, 1999).

The factor(s) ultimately determining vector competence are still unknown. Under experimental conditions, *B. burgdorferi* s.s. is readily detected in North American *Dermacentor* species during the first 2 weeks after detachment from an infective host but is usually absent after 3 weeks (e.g. Piesman and Sinsky, 1988; Dolan *et al.*, 1997; Piesman and Happ, 1997). This phenomenon may be related to tick immune responses to *B. burgdorferi* s.l.; Johns *et al.* (2001) recorded strong borreliacidal activity in the haemolymph of *D. variabilis*, but not in that of *I. scapularis*. However, spirochaetal infection prevalences as high as 5.4% for *A. americanum* and 11.3% for *Dermacentor reticulatus* have been recorded in questing ticks (Schulze *et al.*, 1986; Kahl *et al.*, 1992). This suggests that, under natural conditions, transstadial passage of *B. burgdorferi* s.l. in these ticks may be more common than indicated in laboratory studies (see Table 4.3). Vector incompetence could ultimately result from an inability of spirochaetes to escape the midgut during the blood meal, as suggested for *D. occidentalis* (Li and Lane, 1996), or spirochaete mortality caused by exposure to the haemocoel during the migration to the salivary glands (Johns *et al.*, 2001).

Biology of the Primary LB Vectors

As for all ixodid ticks, *I. pacificus*, *I. persulcatus*, *I. ricinus* and *I. scapularis* undergo four developmental stages: egg, larva, nymph and adult (Figs 4.2 and 4.3; Balashov, 1972; Oliver, 1989; Sonenshine, 1991). A single large blood meal is ingested in the larval, nymphal and adult female stages. After each blood meal three-host ticks such as *Ixodes* spp. detach from the host before they start moulting to the next stage or, in the case of adult females, laying eggs (Balashov, 1972). As in other *Ixodes* spp., adult male *I. ricinus* and *I. persulcatus* are considered facultative blood feeders and usually do not require a blood meal to fertilize females (Balashov, 1972). However, males may repeatedly take small blood meals and can remain on a host for weeks to months in search of females (Oliver, 1989). Mating can take place either in the vegetation or on the host while the female tick is feeding (e.g. Gray, 1987; Yuval and Spielman, 1990a). The mated female then lays a single egg batch; the average number of eggs produced typically exceeds 850 for *I. pacificus* and *I. scapularis* fed on laboratory rabbits, and 1000 for *I. ricinus* fed on sheep (Gray, 1981; Oliver *et al.*, 1993a; Schoeler and Lane, 1993). The length of the developmental cycle depends on temperature conditions, seasonal timing of questing and host availability. *I. scapularis* ticks typically complete their developmental cycle in 2–3 years and *I. pacificus*, *I. persulcatus* and *I. ricinus* in 3 years, but ranges of 2–4 years for *I. scapularis* and 2–5/6 years for *I. persulcatus* and *I. ricinus* have been reported (e.g. Balashov, 1972; Yuval and Spielman, 1990b; Gray, 1991; Lindsay *et al.*, 1998; Padgett and Lane, 2001).

These ticks use an ambush strategy to contact hosts and rarely move more than a few metres while questing (e.g. Balashov, 1972; Gray, 1985; Loye and Lane, 1988; Falco and Fish, 1991). Hosts are detected by vibrations caused by animal movements, odours, body heat and shadows (Balashov, 1972; Sonenshine, 1993). The subadult stages feed on a wide range of hosts, including lizards, birds and mammals ranging in size from minuscule insectivores to large ungulates, whereas the adults are restricted to feeding on medium-sized to large mammals (Fig. 4.2; Anastos, 1957; Furman and Loomis, 1984; Jaenson *et al.*, 1994; Keirans *et al.*, 1996). The feeding process consists of an initial phase of slow feeding, characterized by ingestion of small amounts of blood but intensive cuticular growth in the tick midgut, followed by a phase of rapid ingestion of blood during the last 12–36 h of the blood meal; during the blood meal weight increases range from approximately 10–20-fold for larvae to more than 100-fold for females

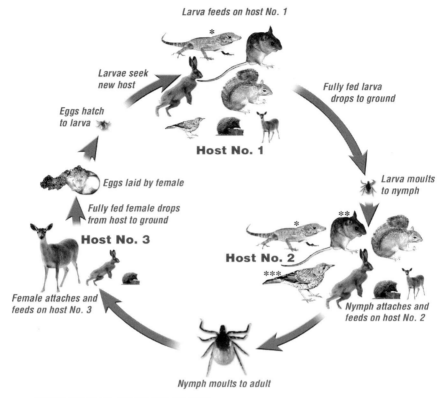

* In the western and southeastern United States
** In the northeastern United States
*** In parts of Europe

Fig. 4.2. Generalized developmental cycle for *Ixodes pacificus, I. persulcatus, I. ricinus* and *I. scapularis*. Tick hosts include various rodents (exemplified by a mouse and a squirrel), lizards, birds, insectivores (hedgehog) and medium-sized (hare) and large (deer) mammals. The importance of these hosts for each life stage is indicated by the scale of depiction. (Designed by B. Kaye.)

(Fig. 4.3; Balashov, 1972). Larval *I. ricinus* and *I. persulcatus*, for example, complete their feeding in 2–5 days, whereas the nymphs take 2–7 days, and the female ticks 6–11 days (Balashov, 1972). The feeding duration and success rate vary with choice of host, attachment site, host immune response and, in the case of exothermic lizard hosts, temperature (e.g. Balashov, 1972; Davidar *et al.*, 1989; James and Oliver, 1990; Sonenshine, 1991; Lane, 1994). The long feeding period of ixodid ticks makes them vulnerable to host responses and, consequently, their saliva contains an array of components that serve as anticoagulants and mediate host inflammatory and immune responses (reviewed by Wikel, 1996; Bowman *et al.*, 1997; Gillespie *et al.*, 2000).

Fig. 4.3. Life stages of *Ixodes ricinus*. Left to right: larva, nymph, adult female, adult male; top row, unfed stages; bottom row, fed stages (female potentially fed; males do not engorge). (Photo by Gunilla Olsson, modified by Bernard Kaye.)

The density of questing ticks is related to climatic conditions and the abundance of vertebrate tick hosts. Local abundances of *I. ricinus* and *I. scapularis* ticks are typically positively correlated with those of deer (e.g. Gray *et al.*, 1992; Wilson, 1998), which serve as important hosts for the adult tick stages (e.g. Tälleklint and Jaenson, 1994). Also, the density of *I. scapularis* nymphs in the northeastern United States was positively correlated with that of white-footed mice (*Peromyscus leucopus*) in the previous year (Ostfeld *et al.*, 2001). However, less than 5% of a 2–3 year developmental cycle is spent in direct contact with hosts. Off-host survival is usually positively correlated with relative humidities surpassing 80% (MacLeod, 1934; Balashov, 1972; Stafford, 1994), when atmospheric water vapour can be actively taken up by the tick through the secretion and then ingestion of hygroscopic fluids (Lees, 1946; Kahl and Knülle, 1988). Active questing is also typically interrupted by periods when the ticks seek out humid microhabitats to regain lost body water (Lees and Milne, 1951; Loye and Lane, 1988). Because they are sensitive to desiccation, questing ticks tend to survive for longer periods of time in sheltered woodlands compared with more open, exposed habitats (e.g. Daniel *et al.*, 1977; Bertrand and Wilson, 1996; Peavey and Lane, 1996; Lindsay *et al.*, 1998), although in areas with wet and cool climates, such as the British Isles, tick survival can be similar in sheltered and open habitats (Gray, 1981).

Seasonal patterns of questing activity can differ between tick species and with climatic conditions for the same tick species. For example, peak questing activity of the adult stage usually occurs during spring and early summer for *I. ricinus* and *I. persulcatus*, autumn for *I. scapularis* and winter for *I. pacificus* (e.g. Ai *et al.*, 1990; Lane, 1990; Gray, 1991; Sonenshine, 1993; Korenberg, 1994). Depending, in part, on climatic conditions, the seasonal questing activity of *I. ricinus* can be continuous from spring to autumn, bimodal with peaks in spring and late summer or early autumn, or, at the southern limit of its geographical distribution, restricted to the winter months (Gray, 1991). Such seasonal patterns of questing activity are regulated primarily by the ticks entering long periods of diapause at various stages in their developmental cycles. This enables them to avoid questing during periods of the year with inclement climatic conditions, such as during the winter at high latitudes and during the hot summer months in southern Europe and the western USA. Diapause phenomena have been studied extensively for *I. ricinus* and *I. persulcatus* (reviewed by Belozerov, 1982; Gray, 1991), whereas fewer studies have focused on *I. scapularis* (Yuval and Spielman, 1990b; Daniels *et al.*, 1996; Lindsay *et al.*, 1998) and *I. pacificus* (Peavey and Lane, 1996; Padgett and Lane, 2001).

On a large spatial scale, unfavourable climatic conditions limit the geographical distributions of the primary bridging vectors of LB spirochaetes. For example, these ticks cannot establish populations in areas with very short warm seasons; *I. ricinus* ticks are virtually absent in far northern Scandinavia and at high altitudes in central Europe (Daniel, 1993; Tälleklint and Jaenson, 1998). Likewise, *I. pacificus* ticks are limited to especially moist habitats in arid regions of the western USA, such as along rivers or at altitudes where snow melt provides sufficient moisture (Dennis *et al.*, 1998).

Vector Potential for *B. burgdorferi* s.l. in Relation to Tick Ecology

Tick vector potential for *B. burgdorferi* s.l. is influenced by factors intrinsic to the tick, e.g. tick questing behaviour, diapause phenomena and host preference, as well as phenomena extrinsic to the tick, such as climatic conditions, host abundance and behaviour, host immune response, host susceptibility to spirochaetes and genetic variation in infectivity of spirochaetes (Lane, 1994). Here we will give some examples of how the above-mentioned factors affect tick vector potential.

Seasonal patterns of questing activity

Differences in the seasonality of questing by subadult ticks may influence vector potential for *B. burgdorferi* s.l. Peak infestation on rodents by *I. scapularis* nymphs in the northeastern USA and *I. persulcatus* nymphs in China precedes that of the larval stages by at least 1 month (e.g. Wilson and Spielman, 1985; Ai *et al.*, 1991), a time interval sufficient for white-footed mice infected with *B. burgdorferi* s.l. by feeding nymphs to have become infective to feeding ticks at the time of larval peak infestation (Donahue *et al.*, 1987). These inverted patterns of seasonal activity of subadult ticks probably serve to intensify the enzootic transmission of *B. burgdorferi* s.l. (Spielman *et al.*, 1985). No such clear temporal separation occurs for the peak abundances of larval and nymphal *I. pacificus* infesting lizards in California (Lane and Loye, 1989; Eisen *et al.*, 2001) or for *I. ricinus* on rodents in Europe (e.g. Matuschka *et al.*, 1990; Randolph *et al.*, 1999), although peak activity by *I. ricinus* nymphs may precede that of the larvae in exposed habitats (Gray, 1991). Intriguingly, *I. ricinus* larvae can become infected with *B. burgdorferi* s.l. when feeding in close proximity to infected nymphs on uninfected hosts (co-feeding transmission; Gern and Rais, 1996). Although this phenomenon may increase the efficiency of enzootic transmission of *B. burgdorferi* s.l. in situations where peak larval and nymphal infestation coincide (Ogden *et al.*, 1997), further studies are needed to assess the potential importance of co-feeding transmission in various enzootic cycles.

Role of different life stages as bridging vectors to humans

The relative importance of different tick stages as bridging vectors of LB spirochaetes depends on how commonly they attach to humans, how frequently they are infected with spirochaetes and how likely they are to be detected and removed before spirochaetes are transmitted. The larval stage is of little importance as a bridging vector because larvae are not aggressive human biters and the prevalence of infection with *B. burgdorferi* s.l. is usually low in unfed larvae (reviewed by Lane, 1994; Mather and Ginsberg, 1994; Hubálek and Halouzka, 1998).

The nymphal stage is considered the primary bridging vector for *I. scapularis* in the eastern USA and *I. ricinus* in Europe. In the eastern USA, most LB cases occur from June to August (Sonenshine, 1993; Falco *et al.*, 1999; CDC, 2000), a time period with peak activity of questing *I. scapularis* nymphs but low activity of adult ticks (e.g. Sonenshine, 1993; Stafford and Magnarelli, 1993; Falco *et al.*, 1999). Detected human bites by *I. scapularis* nymphs and adult females apparently are equally likely to result in LB (Sood *et al.*, 1997). However, nymphal ticks typically feed on humans for a longer period of time than female ticks before being discovered (e.g. Yeh *et al.*, 1995; Falco *et al.*, 1996), and nymphs probably account for the vast majority of the up to 85% of bites that can go undetected in people developing erythema migrans (e.g. Walker, 1998). In Europe, the seasonal distribution of LB cases cannot be used to elucidate which tick stage is the primary bridging vector because the seasonal activity periods of nymphs and adults are similar (Gray, 1991). However, three European studies showed that nymphs accounted for as much as 54, 70 and 82%, respectively, of all detected human bites by *I. ricinus* ticks, as compared with 43, 22 and 3% for female ticks (Maiwald *et al.*, 1998; Liebisch and Liebisch, 1999; Robertson *et al.*, 2000).

In contrast, *I. persulcatus* females are the primary bridging vector of LB spirochaetes to humans in far-eastern Europe and Asia because nymphs only rarely infest humans (Ai *et al.*, 1990; Korenberg, 1994). Notably, the questing behaviour of the adult stage is more aggressive in *I. persulcatus* than in *I. ricinus* (Uspensky, 1993). Also, generalized infections (i.e. infections not restricted to the tick gut) were more common for *B. afzelii* or *B. garinii* in *I. persulcatus* adults (12.9% of infected ticks with generalized infections) than for *B. burgdorferi* s.s. in *I. scapularis* adults (2.4%) (Korenberg and Moskvitina, 1996). This finding suggests that *I. persulcatus* females may be more likely to transmit spirochaetes shortly after attachment.

The relative importance of nymphal versus female ticks is harder to discern in western North America. The seasonal distribution of human LB cases in relation to the peak activity periods of *I. pacificus* nymphs (spring) and adults (late autumn to winter) indicated that the nymphal stage was the primary bridging vector in one rural area in northern California (Clover and Lane, 1995). However, there is probably considerable spatial variation in the importance of *I. pacificus* nymphs versus females as bridging vectors: humans are exposed to nymphs, but rarely to adults, in dense woodland with a ground cover dominated by leaf-litter, whereas adults, but very few nymphs, are collected by drag sampling in woodlands with emergent vegetation or open grasslands (e.g. Clover and Lane, 1995).

Complexity of enzootic *B. burgdorferi* s.l.–tick vector–vertebrate cycles

Ironically, the most intensively studied enzootic maintenance cycle for *B. burgdorferi* s.l., occurring in the northeastern and north-central USA, may be the least complicated. In this region, the enzootic cycle appears to be dominated by a single genospecies (*B. burgdorferi* s.s.), a single tick vector (*I. scapularis*) and one primary vertebrate spirochaete reservoir (the white-footed mouse), although other vertebrate hosts for *I. scapularis* do contribute to the enzootic cycle and *I. dentatus* and lagomorphs maintain a *Borrelia andersonii* transmission cycle (see Piesman, Chapter 9). In contrast, at least two genospecies of *B. burgdorferi* s.l. (*B. burgdorferi* s.s., *B. bissettii*) are frequently isolated from ticks and/or rodents in the western and southeastern USA. The enzootic cycles in these regions also include multiple tick vectors of *B. burgdorferi* s.l. (*I. angustus*, *I. pacificus* and *I. spinipalpis* in the west and *I. affinis*, *I. minor* and *I. scapularis* in the southeast), and important vertebrate spirochaete reservoirs include woodrats/cotton rats and *Peromyscus* mice (see Piesman, Chapter 9). Finally, in the western USA *I. pacificus* subadults feed abundantly on lizards having a borreliacidal factor in their blood (e.g. Lane and Quistad, 1998), a phenomenon that probably serves to decrease the intensity of enzootic transmission of *B. burgdorferi* s.l. (zooprophylaxis; Spielman *et al.*, 1985).

In Europe, the enzootic cycle(s) maintained primarily by *I. ricinus* involves at least five different genospecies of *B. burgdorferi* s.l. (*B. burgdorferi* s.s., *B. afzelii*, *B. garinii*, *B. lusitaniae*, *B. valaisiana*; see Gern and Huwait, Chapter 6). The roles of various vertebrates in the enzootic maintenance cycles of these spirochaetes need further study, although it appears that *B. garinii* and *B. valaisiana* may be associated primarily with birds and *B. afzelii* and *B. burgdorferi* s.s. with small mammals (see Kurtenbach *et al.*, Chapter 5). In Asia, *I. persulcatus* and its vertebrate hosts maintain enzootic cycles of *B. afzelii* and *B. garinii*, and data on tick–spirochaete associations indicate the presence of several additional enzootic cycles including *I. ovatus* and *B. japonica*, *I. tanuki* and *B. tanukii*, *I. turdus* and *B. turdi*, and *I. columnae*, *I. granulatus* and *I. nipponensis* and *B. valaisiana* in eastern Asia (see Chapters 7 and 8). Notably, the geographical distributions of *I. ricinus* and *I. persulcatus* overlap in far-eastern Europe (see Chapter 6), which allows direct comparison of the roles of these species in enzootic transmission cycles and as bridging vectors to humans (Korenberg *et al.*, 2001).

The role of transovarially infected larvae in the maintenance of enzootic cycles is controversial (Randolph and Craine, 1995). Transovarially infected larvae may be more efficient enzootic vectors than nymphs in situations where extremely few nymphs feed on rodent reservoirs for *B. burgdorferi* s.l. and other enzootic tick vectors are rare or absent (e.g. Kurtenbach *et al.*, 1995; Randolph and Craine, 1995). Spirochaete transmission by transovarially infected larvae was demonstrated by exposing white mice to groups of 30–50 *I. ricinus* larvae (Stanek *et al.*, 1986), whereas a single *B. burgdorferi* s.l.-infected *I. ricinus*, *I. pacificus* or *I. scapularis* nymph can infect a susceptible rodent (Stanek *et al.*, 1986; Donahue *et al.*, 1987; Peavey and Lane, 1995). Further experimental studies are needed to evaluate the potential of transovarially infected larvae, as compared with transstadially infected nymphs, to transmit different *B. burgdorferi* s.l. genospecies to natural tick hosts.

Summary

Vector competence for *B. burgdorferi* s.l. indicates that a tick species can acquire spirochaetes when feeding on an infective host, pass them transstadially and, subsequently, transmit the spirochaetes to a susceptible host while feeding. Experimental studies are required to demonstrate vector competence; observations of natural infections in ticks collected from hosts or vegetation do not provide definitive evidence of vector competence. Vector competence for *B. burgdorferi* s.l. has been confirmed for 12 species of ticks, including six of the 14 members of the *I. ricinus* complex (*I. affinis*, *I. jellisoni*, *I. pacificus*, *I. persulcatus*, *I. ricinus*, *I. scapularis*) and six other *Ixodes* species (*I. angustus*, *I. dentatus*, *I. hexagonus*, *I. minor*, *I. muris*, *I. spinipalpis*). In contrast, seven tick species evaluated for vector competence appear unable to transmit *B. burgdorferi* s.l. (*A. americanum*, *D. andersoni*, *D. occidentalis*, *D. variabilis*, *I. cookei*, *I. holocyclus*, *I. ovatus*). The potential of tick species, or their different life stages, to act as enzootic vectors or bridging vectors to humans is influenced by numerous factors, such as seasonality of questing activity, questing behaviour, host range and time elapsed from tick attachment to spirochaete transmission.

Acknowledgements

We are indebted to Dr J.H. Oliver Jr, Georgia Southern University, for generously sharing unpublished data on vector competence studies. This work was supported by a grant from the National Institutes of Health (AI-22501) to R.S. Lane.

References

Ai, C., Hu, R., Hyland, K.E., Wen, Y., Zhang, Y., Qiu, Q., Li, D., Liu, X., Shi, Z., Zhao, J. and Cheng, D. (1990) Epidemiological and aetiological evidence for transmission of Lyme disease by adult *Ixodes persulcatus* in an endemic area in China. *International Journal of Epidemiology* 19, 1061–1065.

Ai, C., Qiu, G., Shi, Z., Wu, X., Liu, X. and Zhao, J. (1991) Host associations and seasonal abundance of immature *Ixodes persulcatus* (Acari: Ixodidae) in a Lyme-disease-endemic area in China. *Experimental and Applied Acarology* 12, 251–257.

Alekseev, A.N., Dubinina, H.V., van de Pol, I. and Schouls, L.M. (2001) Identification of *Ehrlichia* spp. and *Borrelia burgdorferi* in *Ixodes* ticks in the Baltic Regions of Russia. *Journal of Clinical Microbiology* 39, 2237–2242.

Anastos, G. (1957) *The Ticks, or Ixodides, of the U.S.S.R. – a Review of the Literature.* United States Department of Health, Education, and Welfare, Public Health Service Publication No. 458, 397 pp.

Artsob, H. (1988) Powassan encephalitis. In: Monath, T.P. (ed.) *The Arboviruses: Epidemiology and Ecology*, Vol. IV. CRC Press, Boca Raton, Florida, pp. 29–49.

Balashov, Y.S. (1972) Bloodsucking ticks (Ixodoidea) – vectors of diseases of man and animals. *Miscellaneous Publications of the Entomological Society of America* 8, 163–376.

Baranton, G., Marti Ras, N. and Postic, D. (1998) Molecular epidemiology of the aetiological agents of Lyme borreliosis. *Wiener Klinische Wochenschrift* 110, 850–855.

Barker, I.K., Lindsay, L.R., Campbell, G.D., Surgeoner, G.A. and McEwen, S.A. (1993) The groundhog tick *Ixodes cookei* (Acari: Ixodidae): a poor potential vector of Lyme borreliosis. *Journal of Wildlife Diseases* 29, 416–422.

Belozerov, V.N. (1982) Diapause and biological rhythms in ticks. In: Obenchain, F.D. and Galun, R. (eds) *Physiology of Ticks*. Pergamon Press, Oxford, UK, pp. 469–500.

Bertrand, M.R. and Wilson, M.L. (1996) Microclimate-dependent survival of unfed adult *Ixodes scapularis* (Acari: Ixodidae) in nature: life cycle and study design implications. *Journal of Medical Entomology* 33, 619–627.

Bowman, A.S., Coons, L.B., Needham, G.R. and Sauer, J.R. (1997) Tick saliva: recent advances and implications for vector competence. *Medical and Veterinary Entomology* 11, 277–285.

Brown, R.N. and Lane, R.S. (1992) Lyme disease in California: a novel enzootic transmission cycle of *Borrelia burgdorferi*. *Science* 256, 1439–1442.

Burgdorfer, W., Barbour, A.G., Hayes, S.F., Benach, J.L., Grunwaldt, E. and Davis, J.P. (1982) Lyme disease – a tickborne spirochetosis? *Science* 216, 1317–1319.

Burgdorfer, W., Barbour, A.G., Hayes, S.F., Péter, O. and Aeschlimann, A. (1983) Erythema chronicum migrans – a tickborne spirochetosis. *Acta Tropica* 40, 79–83.

Burkot, T.R., Schneider, B.S., Pieniazek, N.J., Happ, C.M., Rutherford, J.S., Slemenda, S.B., Hoffmeister, E., Maupin, G.O. and Zeidner, N.S. (2000) *Babesia microti* and *Borrelia bissettii* transmission by *Ixodes spinipalpis* ticks among prairie voles, *Microtus ochrogaster*, in Colorado. *Parasitology* 121, 595–599.

CDC (2000) Surveillance for Lyme disease – United States, 1992–1998. *Morbidity and Mortality Weekly Report* 49 (SS03), 1–11.

Clover, J.R. and Lane, R.S. (1995) Evidence implicating nymphal *Ixodes pacificus* (Acari: Ixodidae) in the epidemiology of Lyme disease in California. *American Journal of Tropical Medicine and Hygiene* 53, 237–240.

Costero, A. and Grayson, M.A. (1996) Experimental transmission of Powassan virus (Flaviviridae) by *Ixodes scapularis* ticks (Acari: Ixodidae). *American Journal of Tropical Medicine and Hygiene* 55, 536–546.

Damrow, T., Freedman, H., Lane, R.S. and Preston, K.L. (1989) Is *Ixodes (Ixodiopsis) angustus* a vector of Lyme disease in Washington State? *The Western Journal of Medicine* 150, 580–582.

Daniel, M. (1993) Influence of the microclimate on the vertical distribution of the tick *Ixodes ricinus* (L.) in Central Europe. *Acarologia* 34, 105–113.

Daniel, M., Cerny, V., Dusbábek, F., Honzáková, E. and Olejnicek, J. (1977) Influence of microclimate on the life cycle of the common tick *Ixodes ricinus* (L.) in an open area in comparison with forest habitats. *Folia Parasitologica* 24, 149–160.

Daniels, T.J., Falco, R.C., Curran, K.L. and Fish, D. (1996) Timing of *Ixodes scapularis* (Acari: Ixodidae) oviposition and larval activity in southern New York. *Journal of Medical Entomology* 33, 140–147.

Davidar, P., Wilson, M. and Ribeiro, J.M.C. (1989) Differential distribution of immature *Ixodes dammini* (Acari: Ixodidae) on rodent hosts. *Journal of Parasitology* 75, 898–904.

Dennis, D.T., Nekomoto, T.S., Victor, J.C., Paul, W.S. and Piesman, J. (1998) Reported distribution of *Ixodes scapularis* and *Ixodes pacificus* (Acari: Ixodidae) in the United States. *Journal of Medical Entomology* 35, 629–638.

Dolan, M.C., Maupin, G.O., Panella, N.A., Golde, W.T. and Piesman, J. (1997) Vector competence of *Ixodes scapularis*, *I. spinipalpis*, and *Dermacentor andersoni* (Acari: Ixodidae) in transmitting *Borrelia burgdorferi*, the etiologic agent of Lyme disease. *Journal of Medical Entomology* 34, 128–135.

Dolan, M.C., Piesman, J., Mbow, M.L., Maupin, G.O., Péter, O., Brossard, M. and Golde, W.T. (1998) Vector competence of *Ixodes scapularis* and *Ixodes ricinus* (Acari: Ixodidae) for three genospecies of *Borrelia burgdorferi*. *Journal of Medical Entomology* 35, 465–470.

Dolan, M.C., Lacombe, E.H. and Piesman, J. (2000) Vector competence of *Ixodes muris* (Acari: Ixodidae) for *Borrelia burgdorferi*. *Journal of Medical Entomology* 37, 766–768.

Donahue, J.G., Piesman, J. and Spielman, A. (1987) Reservoir competence of white-footed mice for Lyme disease spirochetes. *American Journal of Tropical Medicine and Hygiene* 36, 92–96.

Dumler, J.S. and Bakken, J.S. (1998) Human ehrlichiosis: newly recognized infections transmitted by ticks. *Annual Review of Medicine* 49, 201–213.

Dumler, J.S., Barbet, A.F., Bekker, C.P.J., Dasch, G.A., Palmer, G.H., Ray, S.C., Rikihisa, Y. and Rurangirwa, F.R. (2001) Reorganization of genera in the families *Rickettsiaceae* and *Anaplasmataceae* in the order *Rickettsiales*: unification of some species of *Ehrlichia* with *Anaplasma*, *Cowdria* with *Ehrlichia* and *Ehrlichia* with *Neorickettsia*, descriptions of six new species combinations and designation of *Ehrlichia equi* and 'HGE-agent' as subjective synonyms of *Ehrlichia phagocytophila*. *International Journal of Systematic and Evolutionary Microbiology* 51, 2145–2165.

Durden, L.A. and Keirans, J.E. (1996) *Nymphs of the Genus Ixodes (Acari: Ixodidae) of the United States: Taxonomy, Identification Key, Distribution, Hosts, and Medical/Veterinary Importance*. The Entomological Society of America, Lanham, Maryland, 95 pp.

Eisen, R.J., Eisen, L. and Lane, R.S. (2001) Prevalence and abundance of *Ixodes pacificus* immatures (Acari: Ixodidae) infesting western fence lizards (*Sceloporus occidentalis*) in northern California: temporal trends and environmental correlates. *Journal of Parasitology* 87, 1301–1307.

Falco, R.C. and Fish, D. (1991) Horizontal movement of adult *Ixodes dammini* (Acari: Ixodidae) attracted to CO_2-baited traps. *Journal of Medical Entomology* 28, 726–729.

Falco, R.C., Fish, D. and Piesman, J. (1996) Duration of tick bites in a Lyme disease-endemic area. *American Journal of Epidemiology* 143, 187–192.

Falco, R.C., McKenna, D.F., Daniels, T.J., Nadelman, R.B., Nowakowski, J., Fish, D. and Wormser, G.P. (1999) Temporal relation between *Ixodes scapularis* abundance and risk for Lyme disease associated with erythema migrans. *American Journal of Epidemiology* 149, 771–776.

Filippova, N.A. (1999) Systematic relationships of the *Ixodes ricinus* complex in the palearctic faunal region. In: Needham, G.R., Mitchell, R., Horn, D.J and Welbourn, W.C. (eds) *Acarology IX*: Vol. 2, *Symposia*. Ohio Biological Survey, Columbus, Ohio, pp. 355–361.

Fukunaga, M., Takahashi, Y., Tsuruta, Y., Matsushita, O., Ralph, D., McClelland, M. and Nakao, M. (1995) Genetic and phenotypic analysis of *Borrelia miyamotoi* sp. nov., isolated from the ixodid tick *Ixodes persulcatus*, the vector for Lyme disease in Japan. *International Journal of Systematic Bacteriology* 45, 804–810.

Fukunaga, M., Hamase, A., Okada, K., Inoue, H., Tsuruta, Y., Miyamoto, K. and Nakao, M. (1996) Characterization of spirochetes isolated from ticks (*Ixodes tanuki*, *Ixodes turdus*, and *Ixodes columnae*) and comparison of the sequences with those of *Borrelia burgdorferi* sensu lato strains. *Applied and Environmental Microbiology* 62, 2338–2344.

Furman, D.P. and Loomis, E.C. (1984) The ticks of California (Acari: Ixodida). *Bulletin of the California Insect Survey* 25, 1–239.

Gern, L. and Rais, O. (1996) Efficient transmission of *Borrelia burgdorferi* between cofeeding *Ixodes ricinus* ticks (Acari: Ixodidae). *Journal of Medical Entomology* 33, 189–192.

Gern, L., Zhu, Z. and Aeschlimann, A. (1990) Development of *Borrelia burgdorferi* in *Ixodes ricinus* females during blood feeding. *Annales de Parasitologie Humaine et Comparée* 65, 89–93.

Gern, L., Toutoungi, L.N., Hu, C.M. and Aeschlimann, A. (1991) *Ixodes* (*Pholeoixodes*) *hexagonus*, an efficient vector of *Borrelia burgdorferi* in the laboratory. *Medical and Veterinary Entomology* 5, 431–435.

Gillespie, R.D., Mbow, M.L. and Titus, R.G. (2000) The immunomodulatory factors of bloodfeeding arthropod saliva. *Parasite Immunology* 22, 319–331.

Gilmore, R.D. Jr and Piesman, J. (2000) Inhibition of *Borrelia burgdorferi* migration from the midgut to the salivary glands following feeding by ticks on OspC-immunized mice. *Infection and Immunity* 68, 411–414.

Gray, J.S. (1981) The fecundity of *Ixodes ricinus* (L.) (Acarina: Ixodidae) and the mortality of its developmental stages under field conditions. *Bulletin of Entomological Research* 71, 533–542.

Gray, J.S. (1985) A carbon dioxide trap for prolonged sampling of *Ixodes ricinus* L. populations. *Experimental and Applied Acarology* 1, 35–44.

Gray, J.S. (1987) Mating and behavioural diapause in *Ixodes ricinus* L. *Experimental and Applied Acarology* 3, 61–71.

Gray, J.S. (1991) The development and seasonal activity of the tick *Ixodes ricinus*: a vector of Lyme borreliosis. *Review of Medical and Veterinary Entomology* 79, 323–333.

Gray, J.S., Kahl, O., Janetzki, C. and Stein, J. (1992) Studies on the ecology of Lyme disease in a deer forest in County Galway, Ireland. *Journal of Medical Entomology* 29, 915–920.

Hu, C.M., Wilske, B., Fingerle, V., Lobet, Y. and Gern, L. (2001) Transmission of *Borrelia garinii* OspA serotype 4 to BALB/c mice by *Ixodes ricinus* ticks collected in the field. *Journal of Clinical Microbiology* 39, 1169–1171.

Hubálek, Z. and Halouzka, J. (1998) Prevalence rates of *Borrelia burgdorferi* sensu lato in host-seeking *Ixodes ricinus* ticks in Europe. *Parasitology Research* 84, 167–172.

Hubálek, Z., Halouzka, J. and Juricová, Z. (1998) Investigation of haematophagous arthropods for borreliae – summarized data, 1988–1996. *Folia Parasitologica* 45, 67–72.

Jaenson, T.G.T., Tälleklint, L., Lundqvist, L., Olsen, B., Chirico, J. and Mejlon, H.A. (1994) Geographical distribution, host associations, and vector roles of ticks (Acari: Ixodidae, Argasidae) in Sweden. *Journal of Medical Entomology* 31, 240–256.

James, A.M. and Oliver, J.H. Jr (1990) Feeding and host preference of immature *Ixodes dammini*, *I. scapularis*, and *I. pacificus* (Acari: Ixodidae). *Journal of Medical Entomology* 27, 324–330.

Johns, R., Ohnishi, J., Broadwater, A., Sonenshine, D.E., de Silva, A.M. and Hynes, W.L. (2001) Contrasts in tick innate immune responses to *Borrelia burgdorferi* challenge: immunotolerance in *Ixodes scapularis* versus immunocompetence in *Dermacentor variabilis* (Acari: Ixodidae). *Journal of Medical Entomology* 38, 99–107.

Kahl, O. and Knülle, W. (1988) Water vapour uptake from subsaturated atmospheres by engorged immature ixodid ticks. *Experimental and Applied Acarology* 4, 73–83.

Kahl, O., Janetzki, C., Gray, J.S., Stein, J. and Bauch, R.J. (1992) Tick infection rates with *Borrelia*: *Ixodes ricinus* versus *Haemaphysalis concinna* and *Dermacentor reticulatus* in two locations in eastern Germany. *Medical and Veterinary Entomology* 6, 363–366.

Kahl, O., Janetzki-Mittman, C., Gray, J.S., Jonas, R., Stein, J. and de Boer, R. (1998) Risk of infection with *Borrelia burgdorferi* sensu lato for a host in relation to the duration of nymphal *Ixodes ricinus* feeding and the method of tick removal. *Zentralblatt für Bakteriologie* 287, 41–52.

Keirans, J.E., Hutcheson, H.J., Durden, L.A. and Klompen, J.S.H. (1996) *Ixodes* (*Ixodes*) *scapularis* (Acari: Ixodidae): redescription of all active stages, distribution, hosts, geographical variation, and medical and veterinary importance. *Journal of Medical Entomology* 33, 297–318.

Keirans, J.E., Needham, G.R. and Oliver, J.H. Jr (1999) The *Ixodes ricinus* complex worldwide: diagnosis of the species in the complex, hosts and distribution. In: Needham, G.R., Mitchell, R., Horn, D.J and Welbourn, W.C. (eds) *Acarology IX*: Vol. 2, *Symposia*. Ohio Biological Survey, Columbus, Ohio, pp. 341–347.

Kjemtrup, A.M. and Conrad, P.A. (2000) Human babesiosis: an emerging tick-borne disease. *International Journal for Parasitology* 30, 1323–1337.

Korenberg, E.I. (1994) Comparative ecology and epidemiology of Lyme disease and tick-borne encephalitis in the former Soviet Union. *Parasitology Today* 10, 157–160.

Korenberg, E.I. and Moskvitina, G.G. (1996) Interrelationships between different *Borrelia* genospecies and their principal vectors. *Journal of Vector Ecology* 21, 178–185.

Korenberg, E.I., Kovalevskii, Y.V., Levin, M.L. and Shchyogoleva, T.V. (2001) The prevalence of *Borrelia burgdorferi* sensu lato in *Ixodes persulcatus* and *I. ricinus* ticks in the zone of their sympatry. *Folia Parasitologica* 48, 63–68.

Kurtenbach, K., Kampen, H., Dizij, A., Arndt, S., Seitz, H.M., Schaible, U.E. and Simon, M.M. (1995) Infestation of rodents with larval *Ixodes ricinus* (Acari: Ixodidae) is an important factor in the transmission cycle of *Borrelia burgdorferi* s.l. in German woodlands. *Journal of Medical Entomology* 32, 807–817.

Kuthejlová, M., Kopecky, J., Stepanová, G. and Macela, A. (2001) Tick salivary gland extract inhibits killing of *Borrelia afzelii* spirochetes by mouse macrophages. *Infection and Immunity* 69, 575–578.

Lane, R.S. (1990) Seasonal activity of two human-biting ticks. *California Agriculture* 44, 23–25.

Lane, R.S. (1994) Competence of ticks as vectors of microbial agents with an emphasis on *Borrelia burgdorferi*. In: Sonenshine, D.E. and Mather, T.N. (eds) *Ecological Dynamics of Tick-borne Zoonoses*. Oxford University Press, Oxford, UK, pp. 45–67.

Lane, R.S. and Loye, J.E. (1989) Lyme disease in California: interrelationship of *Ixodes pacificus* (Acari: Ixodidae), the western fence lizard (*Sceloporus occidentalis*), and *Borrelia burgdorferi*. *Journal of Medical Entomology* 26, 272–278.

Lane, R.S. and Quistad, G.B. (1998) Borreliacidal factor in the blood of the western fence lizard (*Sceloporus occidentalis*). *Journal of Parasitology* 84, 29–34.

Lane, R.S., Brown, R.N., Piesman, J. and Peavey, C.A. (1994) Vector competence of *Ixodes pacificus* and *Dermacentor occidentalis* (Acari: Ixodidae) for various isolates of Lyme disease spirochetes. *Journal of Medical Entomology* 31, 417–424.

Lane, R.S., Peavey, C.A., Padgett, K.A. and Hendson, M. (1999) Life history of *Ixodes (Ixodes) jellisoni* (Acari: Ixodidae) and its vector competence for *Borrelia burgdorferi* sensu lato. *Journal of Medical Entomology* 36, 329–340.

Lees, A.D. (1946) The water balance in *Ixodes ricinus* L. and certain other species of ticks. *Parasitology* 37, 1–20.

Lees, A.D. and Milne, A. (1951) The seasonal and diurnal activities of individual sheep ticks (*Ixodes ricinus* L.). *Parasitology* 41, 189–208.

Leuba-Garcia, S., Martinez, R. and Gern, L. (1998) Expression of outer surface proteins A and C of *Borrelia afzelii* in *Ixodes ricinus* ticks and in the skin of mice. *Zentralblatt für Bakteriologie* 287, 475–484.

Li, X. and Lane, R.S. (1996) Vector competence of ixodid ticks (Acari) for *Borrelia burgdorferi* as determined with a capillary-feeding technique. *Journal of Spirochetal and Tick-borne Diseases* 3, 116–123.

Liebisch, A. and Liebisch, G. (1999) Hard ticks (Ixodidae) biting humans in Germany and their infection with *Borrelia burgdorferi*. In: Needham, G.R., Mitchell, R., Horn, D.J. and Welbourn, W.C. (eds) *Acarology IX: Vol. 2, Symposia*. Ohio Biological Survey, Columbus, Ohio, pp. 465–468.

Lindsay, L.R., Barker, I.K., Surgeoner, G.A., McEwen, S.A., Gillespie, T.J. and Addison, E.M. (1998) Survival and development of the different life stages of *Ixodes scapularis* (Acari: Ixodidae) held within four habitats on Long Point, Ontario, Canada. *Journal of Medical Entomology* 35, 189–199.

Loye, J.E. and Lane, R.S. (1988) Questing behavior of *Ixodes pacificus* (Acari: Ixodidae) in relation to meteorological and seasonal factors. *Journal of Medical Entomology* 25, 391–398.

MacLeod, J. (1934) *Ixodes ricinus* in relation to its physical environment: the influence of climate on development. *Parasitology* 26, 282–305.

Magnarelli, L.A. and Anderson, J.F. (1988) Ticks and biting insects infected with the etiologic agent of Lyme disease, *Borrelia burgdorferi*. *Journal of Clinical Microbiology* 26, 1482–1486.

Maiwald, M., Oehme, R., March, O., Petney, T.N., Kimmig, P., Naser, K., Zappe, H.A., Hassler, D. and von Knebel Doeberitz, M. (1998) Transmission risk of *Borrelia burgdorferi* sensu lato from *Ixodes ricinus* ticks to humans in southwest Germany. *Epidemiology and Infection* 121, 103–108.

Masuzawa, T., Fukui, T., Miyake, M., Oh, H.-B., Cho, M.-K., Chang, W.-H., Imai, Y. and Yanagihara, Y. (1999) Determination of members of a *Borrelia afzelii*-related group isolated from *Ixodes nipponensis* in Korea as *Borrelia valaisiana*. *International Journal of Systematic Bacteriology* 49, 1409–1415.

Mather, T.N. and Ginsberg, H.S. (1994) Vector–host–pathogen relationships: transmission dynamics of tick-borne infections. In: Sonenshine, D.E. and Mather, T.N. (eds) *Ecological Dynamics of Tick-borne Zoonoses*. Oxford University Press, Oxford, UK, pp. 68–90.

Mather, T.N. and Mather, M.E. (1990) Intrinsic competence of three ixodid ticks (Acari) as vectors of the Lyme disease spirochete. *Journal of Medical Entomology* 27, 646–650.

Matuschka, F.-R., Lange, R., Spielman, A., Richter, D. and Fischer, P. (1990) Subadult *Ixodes ricinus* (Acari: Ixodidae) on rodents in Berlin, West Germany. *Journal of Medical Entomology* 27, 385–390.

Mukolwe, S.W., Kocan, A.A., Barker, R.W., Kocan, K.M. and Murphy, G.L. (1992) Attempted transmission of *Borrelia burgdorferi* (Spirochaetales: Spirochaetaceae) (JD1 strain) by *Ixodes scapularis* (Acari: Ixodidae), *Dermacentor variabilis*, and *Amblyomma americanum*. *Journal of Medical Entomology* 29, 673–677.

Nakao, M. and Miyamoto, K. (1994) Susceptibility of *Ixodes persulcatus* and *I. ovatus* (Acari: Ixodidae) to Lyme disease spirochetes isolated from humans in Japan. *Journal of Medical Entomology* 31, 467–473.

Nakao, M. and Sato, Y. (1996) Refeeding activity of immature ticks of *Ixodes persulcatus* and transmission of Lyme disease spirochete by partially fed larvae. *Journal of Parasitology* 84, 669–672.

Norris, D.E., Klompen, J.S.H., Keirans, J.E., Lane, R.S., Piesman, J. and Black, W.C. IV (1997) Taxonomic status of *Ixodes neotomae* and *I. spinipalpis* (Acari: Ixodidae) based on mitochondrial DNA evidence. *Journal of Medical Entomology* 34, 696–703.

Nuttall, P.A. (1999) Pathogen–tick–host interactions: *Borrelia burgdorferi* and TBE virus. *Zentralblatt für Bakteriologie* 289, 492–505.

Nuttall, P.A. and Labuda, M. (1994) Tick-borne encephalitis subgroup. In: Sonenshine, D.E. and Mather, T.N. (eds) *Ecological Dynamics of Tick-borne Zoonoses*. Oxford University Press, Oxford, pp. 351–391.

Ogden, N.H., Nuttall, P.A. and Randolph, S.E. (1997) Natural Lyme disease cycles maintained via sheep by cofeeding ticks. *Parasitology* 115, 591–599.

Ohnishi, J., Piesman, J. and de Silva, A.M. (2001) Antigenic and genetic heterogeneity of *Borrelia burgdorferi* populations transmitted by ticks. *Proceedings of the National Academy of Sciences, USA* 98, 670–675.

Oliver, J.H. Jr (1989) Biology and systematics of ticks (Acari: Ixodida). *Annual Review of Ecology and Systematics* 20, 397–430.

Oliver, J.H. Jr (1996) Lyme borreliosis in the southern United States: a review. *Journal of Parasitology* 82, 926–935.

Oliver, J.H. Jr, Owsley, M.R., Hutcheson, H.J., James, A.M., Chen, C., Irby, W.S., Dotson, E.M. and McLain, D.K. (1993a) Conspecificity of the ticks *Ixodes scapularis* and *I. dammini* (Acari: Ixodidae). *Journal of Medical Entomology* 30, 54–63.

Oliver, J.H. Jr, Chandler, F.W. Jr, Luttrell, M.P., James, A.M., Stallknecht, D.E., McGuire, B.S., Hutcheson, H.J., Cummins, G.A. and Lane, R.S. (1993b) Isolation and transmission of the Lyme disease spirochete from the southeastern United States. *Proceedings of the National Academy of Sciences, USA* 90, 7371–7375.

Olsen, B., Duffy, D.C., Jaenson, T.G.T., Gylfe, Å., Bonnedahl, J. and Bergström, S. (1995) Transhemispheric exchange of Lyme disease spirochetes by seabirds. *Journal of Clinical Microbiology* 33, 3270–3274.

Ostfeld, R.S., Schauber, E.M., Canham, C.D., Keesing, F., Jones, C.G. and Wolff, J.O. (2001) Effects of acorn production and mouse abundance on abundance and *Borrelia burgdorferi* infection prevalence of nymphal *Ixodes scapularis* ticks. *Vector Borne and Zoonotic Diseases* 1, 55–63.

Padgett, K.A. and Lane, R.S. (2001) Life cycle of *Ixodes pacificus* (Acari: Ixodidae): timing of developmental processes under field and laboratory conditions. *Journal of Medical Entomology* 38, 684–693.

Pal, U., de Silva, A.M., Montgomery, R.R., Fish, D., Anguita, J., Anderson, J.F., Lobet, Y. and Fikrig, E. (2000) Attachment of *Borrelia burgdorferi* within *Ixodes scapularis* mediated by outer surface protein A. *Journal of Clinical Investigation* 106, 561–569.

Peavey, C.A. and Lane, R.S. (1995) Transmission of *Borrelia burgdorferi* by *Ixodes pacificus* nymphs and reservoir competence of deer mice (*Peromyscus maniculatus*) infected by tick-bite. *Journal of Parasitology* 81, 175–178.

Peavey, C.A. and Lane, R.S. (1996) Field and laboratory studies on the timing of oviposition and hatching of the western black-legged tick, *Ixodes pacificus* (Acari: Ixodidae). *Experimental and Applied Acarology* 20, 695–711.

Peavey, C.A., Lane, R.S. and Damrow, T. (2000) Vector competence of *Ixodes angustus* (Acari: Ixodidae) for *Borrelia burgdorferi* sensu stricto. *Experimental and Applied Acarology* 24, 77–84.

Piesman, J. and Happ, C.M. (1997) Ability of the Lyme disease spirochete *Borrelia burgdorferi* to infect rodents and three species of human-biting ticks (Blacklegged tick, American dog tick, Lone star tick) (Acari: Ixodidae). *Journal of Medical Entomology* 34, 451–456.

Piesman, J. and Sinsky, R.J. (1988) Ability of *Ixodes scapularis*, *Dermacentor variabilis*, and *Amblyomma americanum* (Acari: Ixodidae) to acquire, maintain, and transmit Lyme disease spirochetes (*Borrelia burgdorferi*). *Journal of Medical Entomology* 25, 336–339.

Piesman, J. and Stone, B.F. (1991) Vector competence of the Australian paralysis tick, *Ixodes holocyclus*, for the Lyme disease spirochete *Borrelia burgdorferi*. *International Journal for Parasitology* 21, 109–111.

Piesman, J., Mather, T.N., Sinsky, R.J. and Spielman, A. (1987) Duration of tick attachment and *Borrelia burgdorferi* transmission. *Journal of Clinical Microbiology* 25, 557–558.

Piesman, J., Oliver, J.R. and Sinsky, R.J. (1990) Growth kinetics of the Lyme disease spirochete (*Borrelia burgdorferi*) in vector ticks (*Ixodes dammini*). *American Journal of Tropical Medicine and Hygiene* 42, 352–357.

Postic, D., Korenberg, E., Gorelova, N., Kovalevski, Y.V., Bellenger, E. and

Baranton, G. (1997) *Borrelia burgdorferi* sensu lato in Russia and neighbouring countries: high incidence of mixed isolates. *Research in Microbiology* 148, 691–702.

Randolph, S.E. and Craine, N.G. (1995) General framework for comparative quantitative studies on transmission of tick-borne diseases using Lyme borreliosis in Europe as an example. *Journal of Medical Entomology* 32, 765–777.

Randolph, S.E. and Nuttall, P.A. (1994) Nearly right or precisely wrong? Natural versus laboratory studies of vector-borne diseases. *Parasitology Today* 10, 458–462.

Randolph, S.E., Miklisová, D., Lysy, J., Rogers, D.J. and Labuda, M. (1999) Incidence from coincidence: patterns of tick infestations on rodents facilitate transmission of tick-borne encephalitis virus. *Parasitology* 118, 177–186.

Ribeiro, J.M.C. and Mather, T.N. (1998) *Ixodes scapularis*: salivary kininase activity is a metallo dipeptidyl carboxypeptidase. *Experimental Parasitology* 89, 213–221.

Ribeiro, J.M.C., Mather, T.N., Piesman, J. and Spielman, A. (1987) Dissemination and salivary delivery of Lyme disease spirochetes in vector ticks (Acari: Ixodidae). *Journal of Medical Entomology* 24, 201–205.

Robertson, J.N., Gray, J.S. and Stewart, P. (2000) Tick bite and Lyme borreliosis risk at a recreational site in England. *European Journal of Epidemiology* 16, 647–652.

Ryder, J.W., Pinger, R.R. and Glancy, T. (1992) Inability of *Ixodes cookei* and *Amblyomma americanum* nymphs (Acari: Ixodidae) to transmit *Borrelia burgdorferi*. *Journal of Medical Entomology* 29, 525–530.

Sanders, F.H. Jr and Oliver, J.H. Jr (1995) Evaluation of *Ixodes scapularis*, *Amblyomma americanum*, and *Dermacentor variabilis* (Acari: Ixodidae) from Georgia as vectors of a Florida strain of the Lyme disease spirochete, *Borrelia burgdorferi*. *Journal of Medical Entomology* 32, 402–406.

Sato, Y. and Nakao, M. (1997) Transmission of the Lyme disease spirochete, *Borrelia garinii*, between infected and uninfected immature *Ixodes persulcatus* during cofeeding on mice. *Journal of Parasitology* 83, 547–550.

Schoeler, G.B. and Lane, R.S. (1993) Efficiency of transovarial transmission of the Lyme disease spirochete, *Borrelia burgdorferi*, in the western blacklegged tick, *Ixodes pacificus* (Acari: Ixodidae). *Journal of Medical Entomology* 30, 80–86.

Schulze, T.L., Lakat, M.F., Parkin, W.E., Shisler, J.K., Charette, D.J. and Bosler, E.M. (1986) Comparison of rates of infection by the Lyme disease spirochete in selected populations of *Ixodes dammini* and *Amblyomma americanum* (Acari: Ixodidae). *Zentralblatt für Bakteriologie, Mikrobiologie, und Hygiene Series A* 263, 72–78.

Schwan, T.G., Piesman, J., Golde, W.T., Dolan, M.C. and Rosa, P.A. (1995) Induction of an outer surface protein on *Borrelia burgdorferi* during tick feeding. *Proceedings of the National Academy of Sciences, USA* 92, 2909–2913.

Scoles, G.A., Papero, M., Beati, L. and Fish, D. (2001) A relapsing fever group spirochete transmitted by *Ixodes scapularis* ticks. *Vector Borne and Zoonotic Diseases* 1, 21–34.

Shih, C.-M. and Spielman, A. (1993) Accelerated transmission of Lyme disease spirochetes by partially fed vector ticks. *Journal of Clinical Microbiology* 31, 2878–2881.

de Silva, A.M. and Fikrig, E. (1995) Growth and migration of *Borrelia burgdorferi* in *Ixodes* ticks during blood feeding. *American Journal of Tropical Medicine and Hygiene* 53, 397–404.

de Silva, A.M., Telford, S.R. III, Brunet, L.R., Barthold, S.W. and Fikrig, E. (1996) *Borrelia burgdorferi* OspA is an arthropod-specific transmission-blocking Lyme disease vaccine. *Journal of Experimental Medicine* 183, 271–275.

de Silva, A.M., Zeidner, N.S., Zhang, Y., Dolan, M.C., Piesman, J. and Fikrig, E. (1999) Influence of outer surface protein A antibody on *Borrelia burgdorferi* within feeding ticks. *Infection and Immunity* 67, 30–35.

Sonenshine, D.E. (1991) *Biology of Ticks*, Vol. 1. Oxford University Press, Oxford, UK, 447 pp.

Sonenshine, D.E. (1993) *Biology of Ticks*, Vol. 2. Oxford University Press, Oxford, UK, 465 pp.

Sood, S.K., Salzman, M.B., Johnson, B.J.B., Happ, C.M., Feig, K., Carmody, L., Rubin, L.G., Hilton, E. and Piesman, J. (1997) Duration of tick attachment as a predictor of the risk of Lyme disease in an area in which Lyme disease is endemic. *Journal of Infectious Diseases* 175, 996–999.

Spielman, A., Wilson, M.L., Levine, J.F. and Piesman, J. (1985) Ecology of *Ixodes dammini*-borne human babesiosis and Lyme disease. *Annual Review of Entomology* 30, 439–460.

Stafford, K.C. III (1994) Survival of immature *Ixodes scapularis* (Acari: Ixodidae) at different relative humidities. *Journal of Medical Entomology* 31, 310–314.

Stafford, K.C. III and Magnarelli, L.A. (1993) Spatial and temporal patterns of *Ixodes scapularis* (Acari: Ixodidae) in southeastern Connecticut. *Journal of Medical Entomology* 30, 762–771.

Stanek, G., Burger, I., Hirschl, A., Wewalka, G. and Radda, A. (1986) *Borrelia* transfer by ticks during their life cycle. *Zentralblatt für Bakteriologie, Mikrobiologie, und Hygiene Series A* 263, 29–33.

Tälleklint, L. and Jaenson, T.G.T. (1994) Transmission of *Borrelia burgdorferi* s.l. from mammal reservoirs to the primary vector of Lyme borreliosis, *Ixodes ricinus* (Acari: Ixodidae), in Sweden. *Journal of Medical Entomology* 31, 880–886.

Tälleklint, L. and Jaenson, T.G.T. (1998) Increasing geographical distribution and density of *Ixodes ricinus* (Acari: Ixodidae) in central and northern Sweden. *Journal of Medical Entomology* 35, 521–526.

Telford, S.R. III and Spielman, A. (1989a) Competence of a rabbit-feeding *Ixodes* (Acari: Ixodidae) as a vector of the Lyme disease spirochete. *Journal of Medical Entomology* 26, 118–121.

Telford, S.R. III and Spielman, A. (1989b) Enzootic transmission of the agent of Lyme disease in rabbits. *American Journal of Tropical Medicine and Hygiene* 41, 482–490.

Telford, S.R. III, Armstrong, P.M., Katavolos, P., Foppa, I., Olmeda Garcia, A.S., Wilson, M.L. and Spielman, A. (1997) A new tick-borne encephalitis-like virus infecting New England deer ticks, *Ixodes dammini*. *Emerging Infectious Diseases* 3, 165–170.

Uspensky, I. (1993) Ability of successful attack in two species of ixodid ticks (Acari: Ixodidae) as a manifestation of their aggressiveness. *Experimental and Applied Acarology* 17, 673–683.

Valenzuela, J.G., Charlab, R., Mather, T.N. and Ribeiro, J.M.C. (2000) Purification, cloning, and expression of a novel salivary anticomplement protein from the tick, *Ixodes scapularis*. *Journal of Biological Chemistry* 275, 18717–18723.

des Vignes, F., Piesman, J., Heffernan, R., Schulze, T.L., Stafford, K.C. III and Fish, D. (2001) Effect of tick removal on transmission of *Borrelia burgdorferi* and *Ehrlichia phagocytophila* by *Ixodes scapularis* nymphs. *Journal of Infectious Diseases* 183, 773–778.

Walker, D.H. (1998) Tick-transmitted infectious diseases in the United States. *Annual Review of Public Health* 19, 237–269.

Wang, G., van Dam, A.P., Schwartz, I. and Dankert, J. (1999) Molecular typing of *Borrelia burgdorferi* sensu lato: taxonomic, epidemiological, and clinical implications. *Clinical Microbiology Reviews* 12, 633–653.

Wikel, S.K. (1996) Host immunity to ticks. *Annual Review of Entomology* 41, 1–22.

Wilson, M.L. (1998) Distribution and abundance of *Ixodes scapularis* (Acari: Ixodidae) in North America: ecological processes and spatial analysis. *Journal of Medical Entomology* 35, 446–457.

Wilson, M.L. and Spielman, A. (1985) Seasonal activity of immature *Ixodes dammini* (Acari: Ixodidae). *Journal of Medical Entomology* 22, 408–414.

Woldehiwet, Z. (1983) Tick-borne fever: a review. *Veterinary Research Communications* 6, 163–175.

Yeh, M.-T., Bak, J.M., Hu, R., Nicholson, M.C., Kelly, C. and Mather, T.N. (1995) Determining the duration of *Ixodes scapularis* (Acari: Ixodidae) attachment to tick-bite victims. *Journal of Medical Entomology* 32, 853–858.

Yuval, B. and Spielman, A. (1990a) Sperm precedence in the deer tick *Ixodes dammini*. *Physiological Entomology* 15, 123–128.

Yuval, B. and Spielman, A. (1990b) Duration and regulation of the developmental cycle of *Ixodes dammini* (Acari: Ixodidae). *Journal of Medical Entomology* 27, 196–201.`

5 Borrelia burgdorferi sensu lato in the Vertebrate Host

Klaus Kurtenbach,[1] Stefanie M. Schäfer,[1,2] Simona de Michelis,[1] Susanne Etti[1,2] and Henna-Sisko Sewell[1,2]

[1]*Department of Infectious Disease Epidemiology, Imperial College of Science, Technology and Medicine, St Mary's Campus, Norfolk Place, London W2 1PG, UK;* [2]*Department of Zoology, University of Oxford, Oxford, UK*

Introduction

Borrelia burgdorferi sensu lato (s.l.) is a bacterial species complex, at present comprising 11 delineated and named genospecies (Postic *et al.*, 1994, 1998; Le Fleche *et al.*, 1997; Masuzawa *et al.*, 2001). In addition, several as yet unnamed genomic groups contribute to the increasingly recognized diversity of *B. burgdorferi* s.l. Like all the other members of the genus *Borrelia*, the life style of *B. burgdorferi* s.l. is obligate parasitic, with no free-living stages known. *B. burgdorferi* s.l. is maintained in nature by complex zoonotic transmission cycles involving a large variety of mammalian and avian hosts and hard ticks of the genus *Ixodes* as vectors. Humans are a dead-end host. Larvae, nymphs and female ticks each feed once on a host for several consecutive days. Once infected by an infectious vector tick, reservoir hosts develop a prolonged or even persistent spirochaete infection, which can efficiently be passed on to uninfected ticks (Piesman *et al.*, 1987, 1991; Piesman, 1988, 1993a). Natural hosts do not appear to develop clinical disease (Donahue *et al.*, 1987; Gern *et al.*, 1994; Kurtenbach *et al.*, 1994, 1995); however, minor disease symptoms may escape detection as they are possibly confounded by other factors such as increased predation. Rodents are considered to be the principal reservoir hosts (Donahue *et al.*, 1987), but the role avian hosts play in the ecology of Lyme borreliosis (LB) is increasingly gaining recognition (Humair *et al.*, 1998; Kurtenbach *et al.*, 1998a,b; Gylfe *et al.*, 2000; Richter *et al.*, 2000). A variety of laboratory animals are known to develop disease (Schaible *et al.*, 1989, 1994; Barthold, 1995; Barthold *et al.*, 1999).

In this chapter the major principles governing transmission, dissemination, tissue invasion, survival and pathogenesis of *B. burgdorferi* s.l. in the vertebrate host are outlined. In addition, special emphasis is placed on the diversity of these processes, as differences in genetic background, physiology, immunology, behaviour and ecology of vertebrate hosts and *Borrelia* spp. genotypes are important elements in the eco-epidemiology and pathogenesis of LB.

Current Concepts of Transmission, Dissemination, Persistence and Pathogenesis of *B. burgdorferi* s.l. in the Vertebrate Host

Transmission of *B. burgdorferi* s.l. from the tick to the host

Transmission of *B. burgdorferi* s.l. from the tick to the host is not a mechanical event, but rather a complex biological process that has only recently been recognized (Fikrig *et al.*, 1992; de Silva and Fikrig, 1997; de Silva *et al.*, 1999; Ohnishi *et al.*, 2001). One of the major features that has emerged is that the life cycle of *B. burgdorferi* s.l. is highly adapted and is analogous to transmission cycles of eukaryotic vector-borne parasites, for example *Plasmodium* spp. (de Silva and Fikrig, 1997).

Most members of the *Ixodes ricinus/persulcatus* species complex are known to be vector-competent for *B. burgdorferi* s.l. According to the current model of transmission, *B. burgdorferi* s.l. is normally confined to the midgut of unfed infected ticks (Lebet and Gern, 1994; de Silva and Fikrig, 1997; de Silva *et al.*, 1999; Ohnishi *et al.*, 2001). It is now well documented that *B. burgdorferi* s.l. selectively expresses outer surface proteins (Osps) in the midgut of the unfed tick that are not expressed in the vertebrate host and vice versa (Fikrig *et al.*, 1992; Schwan *et al.*, 1995; de Silva *et al.*, 1999; Fingerle *et al.*, 2000). The most abundant and best studied of the more than 150 (putative) Osps is OspA, followed by OspB, C, D, E, F, pG, VlsE, p66 and a few others (Wallich *et al.*, 1995; Fraser *et al.*, 1997; Zhang *et al.*, 1997; Casjens *et al.*, 2000; Nordstrand *et al.*, 2000). OspA has been shown to act as an adhesin that binds to midgut epithelial cells of ixodid ticks, suggesting an important function of this lipoprotein in tick–spirochaete interactions (Pal *et al.*, 2000).

The first step in the transmission of *B. burgdorferi* s.l. is the attachment of the infected tick to the host. Most ticks of the genus *Ixodes* feed for 2–7 consecutive days depending on the developmental stage. The uptake of blood triggers *B. burgdorferi* s.l. to multiply in the midgut of the tick (Piesman *et al.*, 1990). A fraction of the bacterial population present in the tick penetrates the midgut barrier and invades the haemocoel. This process requires the use of the host-derived plasminogen activation system, a mechanism whereby the spirochaetes acquire extracellular proteolytic activity from the host (Fuchs *et al.*, 1994, 1996; Coleman *et al.*, 1995, 1997; Coleman and Benach, 1999). A number of spirochaetal polypeptides, including OspA, have been identified as receptors of plasminogen. During the blood meal of the tick, *B. burgdorferi* s.l. starts to undergo antigenic phase variation at a variety of loci. In particular, an increasing fraction of the spirochaete population in the tick expresses OspC (Schwan *et* al., 1995; de Silva and Fikrig, 1997; de Silva et al., 1999; Fingerle et al., 2000). It has been proposed that only the subpopulation of spirochaetes that express OspC escapes the midgut lumen. This proposition is supported by findings that antibodies to OspC block the migration of spirochaetes from the midgut to the haemocoel (Gilmore and Piesman, 2000). From the salivary glands, *B. burgdorferi* s.l. is delivered into the feeding lesion that the tick has created.

Dissemination in the host

Once inoculated into a host, *B. burgdorferi* s.l. must disseminate in order to survive in the long term (basic reproduction number, $R_0 > 1$) (Anderson and May, 1991). As a first step in dissemination through the vertebrate host, *B. burgdorferi* s.l. adheres to host tissue by a variety of binding mechanisms. Most notably, spirochaetes bind to platelets via integrin $\alpha_{IIb}\beta_3$ (Coburn et al., 1998) and to endothelial cells via integrins $\alpha_v\beta_1$ and $\alpha_5\beta_1$ (Boeggemeyer et al., 1994; Schaible et al., 1994; Sellati et al., 1996). *B. burgdorferi* s.l. also adheres to glycosaminoglycans. The recognition of glycosaminoglycans seems to vary amongst different *B. burgdorferi* s.l. strains, perhaps explaining cell-type specific binding and tissue tropism (Parveen et al., 1999). These *in vitro* observations are consistent with *in vivo* experimental, clinical and population genetic studies on *B. burgdorferi* s.l. that have provided evidence that the bacteria are differentially invasive (Seinost et al., 1999). Invasive strains disseminate from the feeding lesions and infiltrate other host tissues, whereas non-invasive strains remain localized to the feeding lesions and may never infect the host systemically. Dissemination from the site of inoculation into adjacent sites of the dermis is a relatively slow process, even in vertebrate hosts that are highly reservoir-competent for a given *B. burgdorferi* s.l. strain. For example, rodents require a period of 2–4 weeks to develop a systemic infection with *B. burgdorferi* sensu stricto (s.s.) (Piesman, 1988; Shih et al., 1992). The kinetics of dissemination suggests that haematogenous spread of *B. burgdorferi* s.l. is less important than extravascular migration. This is further corroborated by the finding that *B. burgdorferi* s.l. binds to decorin, a collagen-associated extracellular matrix proteoglycan of the host tissue, mediated by two adhesins of *B. burgdorferi* s.l., the decorin-binding proteins, DbpA and DbpB, which are expressed within the vertebrate host (Guo et al., 1995). Furthermore, *B. burgdorferi* s.l. interacts with fibroblasts and fibronectin as well as with fibrocytes *in vitro* (Klempner et al., 1993; Grab et al., 1998, 1999). The interaction with fibrocytes may be a mechanism of immune evasion, thereby facilitating persistence in the infected host. *B. burgdorferi* s.l. has been shown to be phagocytosed by professional phagocytes via a mechanism termed 'coiling phagocytosis' (Rittig et al., 1998). It is, however, possible that macrophages are involved in dissemination, because spirochaetes occasionally survive in macrophages (Montgomery and Malavista, 1996).

Although it can be assumed that invasiveness is governed by a large and as yet unknown variety of molecular interactions, resistance to the complement system is a major prerequisite for dissemination. The complement system of vertebrates is one of the major effector components of defence against microbes (Sim and Dodds, 1997). As part of the innate immune system it can rapidly respond to microorganisms, long before specific antibodies are generated. The complement system is composed of more than 30 proteins, which interact with each other when the system is activated. Microorganisms can activate complement through the classical or the alternative pathway. Both pathways result in the activation of C3 and the binding of C3b to activator surfaces, followed by an amplification loop and formation of the membrane attack complex (C5b–C9). If not repaired, this eventually leads to cell damage and lysis. Discrimination between 'self' and 'non-self' is achieved by preferential binding of regulatory proteins that control the amplification loop at different stages. On 'self-surfaces' factor H has a higher affinity to C3b than factor B. C3b that binds factor H is catabolized by factor I into iC3b. Degradation of C3b prevents the generation of the alternative pathway C3 convertase, C3bBb, which drives the amplification loop (Sim and Dodds, 1997).

Microbial pathogens that are invasive have developed mechanisms of resistance to complement in their evolution. A variety of different mechanisms to escape the complement-mediated attack have been described (Lachmann, 1986; Sim and Dodds, 1997). Recently, it has been discovered that complement-resistant *B. burgdorferi* s.l. strains bind the two host-derived complement control proteins: factor H-like protein 1/reconectin (FHL-1) and factor H (Kraiczy et al., 2001a,b). Five proteins of *B. burgdorferi* s.l. have so far been found to interact with complement control proteins, collectively termed CRASPs (complement regulatory acquiring surface proteins) (Kraiczy et al., 2001b). Amongst the CRASPs, the outer surface lipoprotein OspE has been identified as a specific ligand for factor H, with the binding site localized to the C-terminal short consensus repeat domains 15–20 of factor H, leaving the N-terminal regulatory domains free to carry out their inhibitory activities (Hellwage et al., 2001). At the time of writing, no sequence data for the other CRASPs are available. The best-studied Osp of *B. burgdorferi* s.l., OspA, does not interact with complement (Hellwage et al., 2001).

Binding of host-derived complement control proteins by *B. burgdorferi* s.l. is an example of exploitation of host molecules for the deflection of immunity (Finlay and Cossart, 1997). The exploitation of host-derived complement inhibitors is thus an efficient mechanism to avoid lysis and, possibly, complement-mediated enhancement of adaptive immunity.

As already mentioned, *B. burgdorferi* s.l. binds to the plasminogen system of the host. The plasminogen system plays a significant role in fibrinolysis (Fuchs et al., 1994; Coleman et al., 1995; Coleman and Benach, 1999). In the presence of specific activators, plasminogen that is bound to bacterial surfaces is readily converted into plasmin (Fuchs et al., 1996). Plasmin is an aggressive proteolytic enzyme with a broad substrate specificity, e.g. it degrades fibrin, fibronectin and, thus, extracellular matrices. *In vitro*, such bound plasmin substantially enhances the migration of *B. burgdorferi* s.s. through endothelial monolayers (Coleman et al., 1995). The plasmin(ogen) system enhances the dissemination of *B. burgdorferi* s.l. through both skin and organ invasion, allowing the bacteria to avoid fibrin-based immobilization (Coleman *et al.*, 1997). Plasmin is also effectively used by other invasive bacteria and by tumour cells to disseminate in tissues (Coleman and Benach, 1999). It remains to be elucidated whether binding to plasmin(ogen) is variable depending on the combination of *B. burgdorferi* s.l. genotype and vertebrate species.

Persistence in the host

Once a *B. burgdorferi* s.l. strain has disseminated from the site of inoculation, infections tend to persist in most animals and in humans lifelong if untreated (Gern *et al.*, 1994; Kurtenbach *et al.*, 1994; Zhong *et al.*, 1997). The lack of self-limitation of *B. burgdorferi* s.l. infections is remarkable as *B. burgdorferi* s.l. induces strong cellular and humoral immune responses that target a variety of Osps expressed in the host (Roehrig *et al.*, 1992; Golde *et al.*, 1993; Schaible *et al.*, 1993).

Both the quality and the quantity of antibodies produced in response to *B. burgdorferi* s.l. are dependent on the mode of infection. A large number of studies have been conducted that have analysed this phenomenon using laboratory animals as well as natural vertebrate hosts (Roehrig *et al.*, 1992; Gern *et al.*, 1993; Kurtenbach *et al.*, 1994). It was found that infections with *B. burgdorferi* s.l. induced by tick bites are normally characterized by a lack of antibodies to OspA and B. However, strong and consistent humoral responses, independent of the genetic background of the vertebrate host, are mounted to other lipoproteins such as OspC, DbpA and DbpB, flagellin, BmpA (formerly known as p39) and VlsE. In contrast, intradermal syringe-induced infections using high numbers of cultured spirochaetes ($> 10^4$) generated antibodies to OspA and B in the hosts (Schaible *et al.*, 1993; Kurtenbach *et al.*, 1994). Inoculation with high numbers of killed spirochaetes was found to induce a multivalent antibody response similar to that induced by viable spirochaetes, except for antibodies to BmpA. The presence of antibodies to BmpA, therefore, distinguishes between active and inactive infections in the vertebrate host (Golde *et al.*, 1993). Intradermal infections with low numbers of cultured viable spirochaetes, however, induce antibody profiles with similar specificities to those seen after tick-induced infections. For this reason it has been suggested that the quality of the antibody response depends on the dose of antigen injected (Gern *et al.*, 1993; Schaible *et al.*, 1993). Subsequent studies revealed the mechanism underlying this observation: antigenic phase variation ('switch on–off') of *B. burgdorferi* s.l. Thus, the organisms delivered by the tick are antigenically distinct from cultured forms (Schwan *et al.*, 1995; de Silva and Fikrig, 1997; de Silva *et al.*, 1999; Ohnishi *et al.*, 2001). The stages of *B. burgdorferi* s.l. delivered by the tick appear to be pre-adapted to the host environment and do not, or only at low abundance, express OspA and B. OspA and B are thus 'concealed' antigens. Consequently, these lipoproteins are only occasionally, if at all, recognized by the immune system of naturally infected hosts. The finding that midgut-specific stages of *B. burgdorferi* s.l. selectively express OspA provides the rationale behind the recently approved transmission-blocking vaccine (LYMErix, SmithKline Beecham) that targets OspA in the feeding tick. Antibodies taken up by the feeding tick kill the pathogens before they are delivered into the hosts through the salivary glands (Fikrig *et al.*, 1992).

Irrespective of the mode of infection and the antibody profile, *B. burgdorferi* s.l. infections, once successfully introduced, persist in the host. Several concepts and hypotheses have been proposed to explain the persistence of *B. burgdorferi* s.l. in the presence of specific immunity. One hypothesis is that *B. burgdorferi* s.l. resides in immunoprivileged sites of the host, thereby escaping effective clearance by the immune system (Nordstrand *et al.*, 2000). Recently, an alternative working model of transmission has been proposed, suggesting that *B. burgdorferi* s.l. populations avoid sterile immunity via a novel pattern of antigenic phase variation (Ohnishi *et al.*, 2001). According to this model, the tick delivers an antigenically heterogeneous population of *B. burgdorferi* s.l. into the host of which only the subpopulation that is depleted of OspA and OspC disseminates. Thus, OspC-specific immune responses as seen in natural infections would be a by-product that do not target the infectious forms of *B. burgdorferi* s.l. Alternatively, naturally induced OspC-specific antibodies may target *B. burgdorferi* s.l. *in situ*, but may be unable to lyse the bacteria in the host (Zhong *et al.*, 1997, 1999).

Phylogenetic and experimental studies suggest that *ospC* and other genes that encode *in vivo*-expressed antigens recombine genetically (Jauris-Heipke *et al.*, 1995; Zhang *et al.*, 1997; Stevenson *et al.*, 1998; Wang *et al.*, 1999). For example, the gene-tree topologies of *ospC* are disconcordant with the gene trees based on the *ospA* or the *fla* gene (Figs 5.1–5.3), suggesting horizontal gene transfer between different strains of *B. burgdorferi* s.l. at the *ospC* locus. Genetic recombination is supported by a major study analysing the genetic (in-) stability of the *ospC* gene in experimental chains of tick–host infections using *Ixodes scapularis* and mice (Ryan *et al.*, 1998). In contrast, in other studies analysing genetic stability of needle infections in mice, recombination was not observed (Hodzic *et al.*, 2000). It is, therefore, possible that genetic recombination at the *ospC* locus requires the natural transmission cycle involving the invertebrate vector. For the *vlsE* locus, recombination has been observed both in the feeding tick and in the infected host (Zhang *et al.*, 1997). Various genetic mechanisms have been proposed for the *vlsE* system, in particular intragenic recombination and complex DNA rearrangements similar to those seen in relapsing fever spirochaetes (Zhang *et al.*, 1997; Casjens *et al.*, 2000). Altogether, genetic recombination of *B. burgdorferi* s.l. genes is an attractive explanation for persistence in the host, as recombination may lead to the emergence of new genotypes that evade effective immunity.

In relapsing fever spirochaetes such as *Borrelia hermsii*, recombination generates new variable membrane protein (*vmp*) genotypes in the course of infection. New genotypes that are not yet recognized by the host are positively selected, which leads to a temporarily high spirochaetaemia that is later cleared by strong

ospA

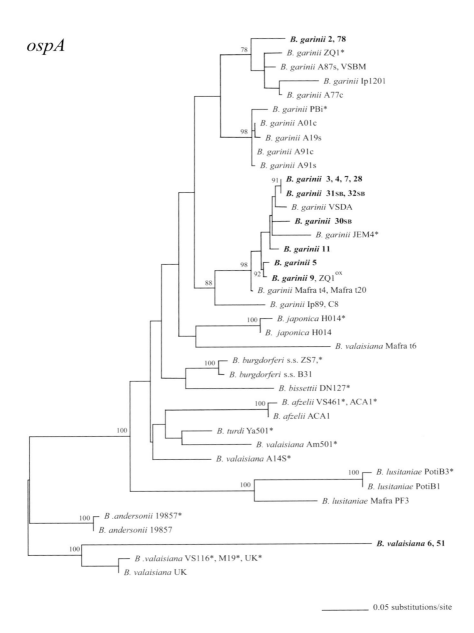

Fig. 5.1. Midpoint rooted maximum likelihood tree based on *ospA* sequences derived from *Ixodes uriae* ticks (shown in bold), various *Borrelia burgdorferi* s.l. isolates, and sequences downloaded from GenBank (indicated with an asterisk *). Bootstrap values under 70 are not shown. The bar indicates nucleotide substitutions per site.

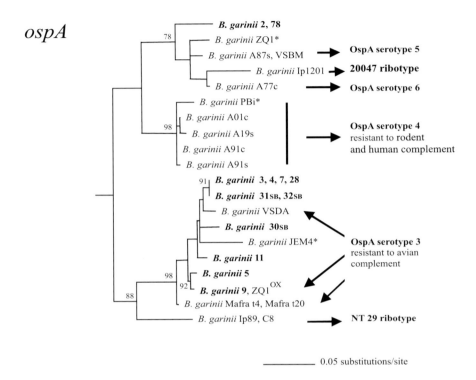

Fig. 5.2. As Fig. 5.1 but showing only *B. garinii* sequences and their correlation with OspA serotypes and ribotypes.

and specific humoral immunity, followed by the rise of another new variant. Such an extreme form of frequency-dependent selection mediated by strong adaptive immunity does not seem to operate in *B. burgdorferi* s.l. infections (Qui et al., 1997). Here, the spirochaetal load in the vertebrate host seems to be very low at any time of infection, as indicated by results based on PCR, culturing and histology. The low number of *B. burgdorferi* s.l. in the host indicates that, despite persistence and the failure to clear the infection, spirochaetemia is under some, probably multifactorial, control involving innate and adaptive components of immunity. This is corroborated by a recent study using a migratory bird species as model, showing that *B. burgdorferi* s.l. infections can be reactivated in birds during experimentally induced migratory restlessness, whereas spirochaetes could not be recovered by culturing from sedentary birds (Gylfe et al., 2000). Upon experimental induction of physiological restlessness, spirochaetes could be reisolated successfully from the birds. The findings suggest that control mechanisms in the host may break down under certain conditions. At present, the mechanisms of reactivation of *B. burgdorferi* s.l. infection in the migratory bird system are unclear.

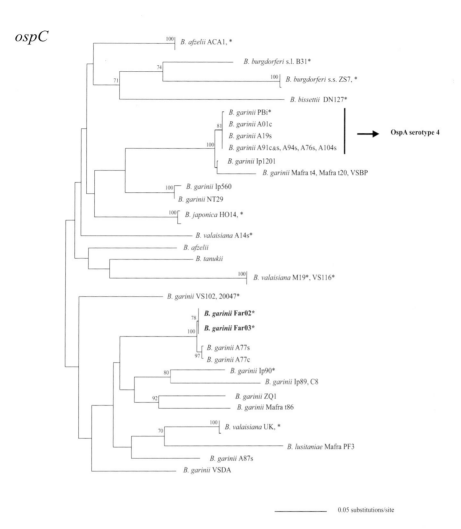

Fig. 5.3. Midpoint rooted maximum likelihood tree based on *ospC* sequences derived from a pelagic habitat (shown in bold), various terrestrial *Borrelia burgdorferi* s.l. isolates, and sequences downloaded from GenBank (indicated with an asterisk). Bootstrap values below 70 are not shown. The bar indicates nucleotide substitutions per site. This gene tree is disconcordant with the *ospA* tree, indicating genetic recombination. The Mafra strains t4 and t20, which cluster together with OspA serotype 3 strains at the *ospA* locus, display close genetic similarity with hyperinvasive OspA serotype 4 strains at the *ospC* locus. As the *ospC* locus is a marker for invasiveness, the finding suggests that these bird-associated strains from Portugal (as indicated by the *ospA* sequences) resemble OspA serotype 4 strains in terms of pathology.

The number of *B. burgdorferi* s.l. cells residing in the host varies with the organ or tissue, suggesting a limited degree of tissue tropism that differs amongst the genotypes of *B. burgdorferi* s.l. (van Dam *et al.*, 1993; Pennington *et al.*, 1997). For example, in humans *Borrelia afzelii* seems to be associated with skin, *Borrelia garinii* with nervous tissue and *B. burgdorferi* s.s. with joints. As the genome project of *B. burgdorferi* s.l. has found evidence for the presence of chemotaxis genes, the mechanisms underlying differential tissue tropism may, apart from differential binding patterns to host cells, also be related to allelic variation of chemotaxis genes (Casjens *et al.*, 2000).

In conclusion, while innate and adaptive components of natural immunity fail to clear *B. burgdorferi* s.l. from the host, host immunity substantially suppresses spirochaetemia.

Transmission from the host to the tick

As in transmission from the vector to the host, acquisition of spirochaetes by a tick occurs during the blood meal. Ticks may acquire spirochaetes from a variety of vertebrate host species. Murine hosts are the best studied, and a wealth of information indicates that mice remain infectious lifelong, irrespective of their genetic background (Donahue *et al.*, 1987; Gern *et al.*, 1994; Kurtenbach *et al.*, 1994; Hofmeister *et al.*, 1999a,b; Humair *et al.*, 1999). In contrast to tick-to-host transmission, which requires time, *B. burgdorferi* s.l. is normally more rapidly acquired by a tick from an infected host (Piesman, 1991). However, in order for *B. burgdorferi* s.l. to be transmitted transstadially, the blood meal must be completed, because below a certain threshold partially fed ticks do not moult (Dizij and Kurtenbach, 1995).

Not all ticks acquire a spirochaete infection from an infected host during any one blood meal. Depending on the system studied, the probability of spirochaete acquisition is highly variable (Donahue *et al.*, 1987; Piesman, 1988, 1993b; Gern *et al.*, 1994; Kurtenbach *et al.*, 1994, 1995; Humair *et al.*, 1999). Transmissibility from the host to the tick is unlikely to be simply correlated with spirochaete burden, as the numbers of spirochaetes residing in the skin of the vertebrate host always seem to be low. An increasing body of evidence indicates that transmissibility of *B. burgdorferi* s.l. from the host to the tick is determined by innate and, to some degree, naturally acquired immunity (Kurtenbach *et al.*, 1995, 1998c; Qui *et al.*, 1997; Hu *et al.*, 2001). In addition, for a variety of animal models there is evidence that transmission is influenced by immunological processes induced by the tick (Dizij and Kurtenbach, 1995).

Experimental studies, furthermore, show that the efficiency with which *B. burgdorferi* s.l. is transmitted from infected hosts to feeding ticks is dependent on the mode of spirochaete inoculation (Gern *et al.*, 1994; Kurtenbach *et al.*, 1994). For example, studies in a variety of natural and laboratory mouse models reported much higher parameter values of host-to-tick transmission coefficients after natural (i.e. tick-borne) infections than after infections induced by needle inoculation with high numbers of cultured spirochaetes. It has been demonstrated that transmission is inversely correlated with the amount of OspA-specific antibodies defined by the monoclonal antibody LA-2 (Fikrig *et al.*, 1992; Kurtenbach *et al.*, 1994, 1997).

Once a tick has completed its blood meal, *B. burgdorferi* s.l. is, in permissive systems, retained throughout the moult and transmitted to the next developmental stage. This stage, usually the nymph, can then act as a vector, thus completing the *B. burgdorferi* s.l. transmission cycle.

Unlike the kinetics of protein expression and migration patterns of *B. burgdorferi* s.l. in the feeding vector tick, antigenic phase variation and possible migration during acquisition of spirochaetes from hosts by ticks are not well studied.

Pathogenesis of Lyme borreliosis

LB primarily affects humans and some domestic animals, whereas natural reservoir hosts do not appear to develop disease. However, it is possible that some natural hosts do develop mild disease manifestations that readily escape detection. It is difficult to assess disease in natural hosts as they are normally outbred and therefore polymorphic. Very few natural host species have so far been analysed for pathogenesis in experimental studies, and in the field it is even more difficult to detect disease caused by *B. burgdorferi* s.l. The main reason for this difficulty is that natural hosts are normally carriers of a variety of other endoparasites (such as other bacteria, protozoa or helminths) and ectoparasites (such as lice, fleas and ticks) that may confound rare and/or mild symptoms of disease due to *B. burgdorferi* s.l.

The genome of one strain of *B. burgdorferi* s.l. has recently been sequenced fully, now also including most of the linear and circular plasmids (Fraser *et al.*, 1997; Casjens *et al.*, 2000). It was hoped that the molecular basis of pathogenicity of *B. burgdorferi* s.l. would be identified in the course of this genome project. Most interestingly, the project failed to identify any virulence factors or pathogenicity islands in *B. burgdorferi* s.l. Furthermore, apart from the lack of any defined virulence factors, more than 40% of the open reading frames located on the chromosome and more than 80% of the open reading frames located on the plasmids did not show any similarity with sequences found anywhere else in nature. Therefore, putative proteins inferred from the genome cannot yet be assigned to any biological function. The unique genome and obscure biology render *B. burgdorferi* s.l. one of the least understood and most interesting pathogens of humans and animals.

Why do humans and some domestic and laboratory animals develop LB? The answer to this question can only come from clinical studies, from experimental studies using defined animal models and from functional genomics. Many laboratory animal disease models have already been analysed in detail since the 1980s. The advantage of such models in pathogenesis research is that they are genetically and microbiologically defined. This allows more precise distinction between host-derived factors and spirochaete-associated factors in the onset of disease.

The most important laboratory animal species and strains that have been investigated include hamsters (Schmitz et al., 1989), gerbils (Preac-Mursic et al., 1990), guinea pigs (Sonnensyn et al., 1993), laboratory mice (Schaible et al., 1989; Barthold et al., 1990; Bockenstedt et al., 1997), rats (Barthold, 1995), rabbits (Barthold et al., 1988), dogs (Appel et al., 1993) and monkeys (Philipp et al., 1993). Amongst these animal models, inbred and recombinant strains of mice have been particularly valuable in providing insight into the pathogenesis of spirochaete infection. Therefore, in this section emphasis will be placed on laboratory mouse models of LB.

Some strains of mice, like humans, develop multisystemic disease, mainly arthritis and cardiovascular symptoms (Schaible et al., 1989; Museteanu et al., 1991). One of the major findings using mouse models was that disease seems to be mediated by spirochaete-induced inflammatory responses that are often out of proportion to the spirochaetal load. This suggests that *B. burgdorferi* s.l. is able to trigger amplifying inflammatory processes that cause the disease manifestations (Nordstrand et al., 2000). *In vitro* studies showed that *B. burgdorferi* s.l. is able to activate macrophages and to induce oxygen radicals and NO (Modelell et al., 1994) as well as a variety of proinflammatory cytokines like interleukin (IL)-1β, IL-6 and tumour necrosis factor-α (Ma and Weiss, 1993; Radolf et al., 1995). Furthermore, the induction of a large variety of chemokines, such as IL-8 (Ebnet et al., 1996a,b; Burns et al., 1997), which activate various types of white blood cell, suggests a crucial role in pathogenesis (Sprenger et al., 1997). Although no virulence factors could be inferred from the genome itself, it is likely that the lipoproteins and glycoproteins associated with the outer surface of *B. burgdorferi* s.l. (many of which are encoded by genes located on plasmids) trigger the disease-causing processes (Radolf et al., 1995). For example, preparations of these lipoproteins directly stimulate macrophages and B cells and provide co-stimulatory signals for T cells (Simon et al., 1995; Knigge et al., 1996). This suggests that pathological processes are driven by interactions between the outer surface lipo(glyco)proteins of the spirochaetes and infiltrating and tissue-specific cells (Simon et al., 1993; Nordstrand et al., 2000).

In vivo studies using immunocompetent (Barthold et al., 1990) and immunocompromised strains of mice (Schaible et al., 1989) have provided evidence that pathogenesis of *B. burgdorferi* s.l. is linked to the genetic background of the animal model and to the genotype of *B. burgdorferi* s.l. (Pennington et al., 1997). These studies have also established that the adaptive immune response is a critical element in pathogenesis, being either ameliorative or pathogenic. A variety of disease-susceptible and resistant mouse models have been developed and investigated. Outbred and inbred mice are all susceptible to *B. burgdorferi* s.l. infection. However, susceptibility to disease differs between inbred strains of mice representing different haplotypes of the major histocompatibility complex (MHC). This reinforces the notion that infection and disease have to be dissociated in LB (Thomas et al., 2001). The role that acquired immunity plays in pathogenesis is underlined by the fact that the various H-2 haplotypes of mice differ in their cellular immune response to *B. burgdorferi* s.l. For example, the immunocompetent mouse strains BALB/cByJ (H-2^d) and C57BL/6J (H-2^b) have been found to be relatively disease-resistant to *B. burgdorferi* s.s., whereas C3H/HeJ (H-2^k) or SWR/J (H-2^q) mice were found to develop severe disease symptoms after infection with the same pathogenic borrelial strains (Simon *et al.*, 1993; Barthold, 1995). For this reason the C3H mouse strain has become one of the most widely used disease-susceptible mouse strains in LB pathogenesis research (Oteo *et al.*, 1998). Of all the MHC haplotypes analysed, H-2^d strains of inbred mice seem to be the most disease-resistant. The conclusion that acquired immunity controls disease is further supported by the finding that some of the disease-resistant MHC haplotypes, in particular H-2^d strains such as BALB/C, are susceptible to disease when infected at a very young age, i.e. at an age at which they are still immunocompromised (Barthold *et al.*, 1990).

On the other hand, studies using mice with severe combined immunodeficiency (scid) syndromes have shown that disease is also controlled by genes that regulate innate susceptibility. Scid mice lack any functional B or T cells and are therefore not able to produce immunoglobulins. For example, comparison of the immunocompromised strains C3H.scid, C.B.-17.scid, RAG-1D/D or RAG-2 D/D showed that the first two strains do develop severe arthritis, in contrast with the RAG strains, which do not (Schaible *et al.*, 1989; Simon *et al.*, 1993). The relative importance of acquired versus innate immunity to control or mediate disease remains a matter of debate.

Apart from the genetic background of the host, there seem to be intrinsic differences amongst the different genospecies of *B. burgdorferi* s.l. in disease induction. In humans, for example, of the 11 genospecies delineated so far, only *B. afzelii*, *B. garinii* and *B. burgdorferi* s.s. have unambiguously been associated with disease (van Dam *et al.*, 1993). However, some genospecies previously considered to be apathogenic, such as *Borrelia lusitaniae*, have recently been found to cause disease in C3H mice (N. Zeidner, Fort Collins, USA, personal communication). On the other hand, *B. burgdorferi* s.l. isolates originally pathogenic may lose entire plasmids after multiple *in vitro* passages and this has been associated with the loss of pathogenic potential of the bacteria (Schwan *et al.*, 1988; Siebers *et al.*, 1999; Casjens *et al.*, 2000).

Pathogenesis research in LB has passed into the genomic era and it is likely that functional genomics will shed more light on disease-inducing molecules of *B. burgdorferi* s.l. in the near future (Fraser *et al.*, 1997; Nordstrand *et al.*, 2000).

Diversity of Tick–Host–Spirochaete Interactions

The processes governing transmission, persistence and pathogenesis of *B. burgdorferi* s.l. are surprisingly diverse. We are only beginning to understand the extent and impact of this diversity. First of all, the bacterium displays considerable genetic and phenotypic diversity. Secondly, the range of vertebrate species found to be susceptible to infection and also transmission-competent for one or more strains of *B. burgdorferi* s.l. is increasing. Of equal interest are those vertebrate species that are not transmission-competent because they may reduce the basic reproduction number (R_0) of *B. burgdorferi* s.l. Thirdly, the tick–host relationships differ substantially from each other depending on the species involved, with the acquisition of resistance to ticks on one side of the biological range and enhanced tolerance to repeated tick bites on the other (Dizij and Kurtenbach, 1995). These diverse interactions determine the life cycle and ecology of *B. burgdorferi* s.l.

Diversity of *B. burgdorferi* s.l.

Since 1992, *B. burgdorferi* s.l. is no longer considered to be a single bacterial species, but rather a diverse species complex now comprising 11 named genospecies: *B. afzelii*, *B. andersonii*, *B. bissettii*, *B. burgdorferi* s.s., *B. garinii*, *B. japonica*, *B. lusitaniae*, *B. sinica*, *B. tanukii*, *B. turdi* and *B. valaisiana*. Some of the genospecies of *B. burgdorferi* s.l. display considerable intraspecific diversity. *B. garinii* appears to be the most polymorphic genospecies (Marti-Ras *et al.*, 1997; Masuzawa *et al.*, 1997; Postic *et al.*, 1998).

The phylogenetic and evolutionary relationships amongst the different strains of *B. burgdorferi* s.l. are not yet fully resolved (Marti-Ras *et al.*, 1997). The degree of clonality is one of the controversial issues discussed (Dykhuizen *et al.*, 1993; Wang *et al.*, 1999; Dykhuizen and Baranton, 2001). Horizontal gene transfer between different spirochaetal strains, and other mechanisms of recombination, may disrupt the clonal frame of *B. burgdorferi* s.l. (Stevenson *et al.*, 1998; Sung *et al.*, 1998; Wang *et al.*, 1999). However, the rates of genetic recombination of *B. burgdorferi* s.l. loci are not known.

The increasingly recognized genetic and phenotypic diversity of the bacteria has an important impact on the diversity of tick–host–spirochaete interactions discussed in this chapter, and provides a framework for eco-epidemiological studies.

Diversity of vertebrate hosts

Globally, more than 100 natural vertebrate host species have been identified as playing a role in the ecology of *B. burgdorferi* s.l. (Anderson, 1991; Oliver, 1996; Gern *et al.*, 1998). As the physiology, immunology and behavioural ecology of these vertebrates are highly diverse, each animal species (or subpopulation thereof) may contribute differently to the biology of LB.

The most extensively studied natural hosts are small mammals of the genera *Neotoma* and *Peromyscus* in North America (Mather *et al.*, 1989; Brunet *et al.*, 1995; Oliver, 1996) and *Apodemus* and *Clethrionomys* in Eurasia (Kurtenbach *et al.*, 1994, 1995; Humair *et al.*, 1999). Rodents substantially contribute to the basic reproduction number (R_0) of *B. burgdorferi* s.l. in nature by virtue of their high capacity to infect larval ticks. Perhaps for this reason, rodents have been considered to be the principal reservoir hosts of *B. burgdorferi* s.l. In recent years, however, it has become apparent that some bird species are also reservoir hosts for *B. burgdorferi* s.l. (Olsen *et al.*, 1996; Piesman *et al.*, 1996; Humair *et al.*, 1998; Kurtenbach *et al.*, 1998a; Rand *et al.*, 1998; Richter *et al.*, 2000). Furthermore, avian hosts, particularly migratory birds, contribute to the continental and even transhemispheric dispersal of *B. burgdorferi* s.l. (Olsen *et al.*, 1993, 1995a,b).

Many natural vertebrate hosts are unable to infect ticks with *B. burgdorferi* s.l. and are therefore considered to be reservoir-incompetent. The most prominent example is deer (apparently most, if not all, species) (Telford *et al.*, 1988; Matuschka *et al.*, 1993; Gern *et al.*, 1998; Nelson *et al.*, 2000). For this reason, it has been concluded that deer play a zooprophylactic role in the ecology of LB (Matuschka *et al.*, 1993). However, reservoir incompetence does not necessarily imply zooprophylaxis. Deer often feed large numbers of immature and adult ticks and boost the tick population. As the population density of ticks is an important variable in the equation of the basic reproduction number R_0 (Randolph, 1998), mathematical modelling is required to determine whether deer reduce or increase R_0 of LB spirochaetes.

Host specificity of *B. burgdorferi* s.l.

One of the most important findings in LB research in recent years is that the different genospecies or strains of *B. burgdorferi* s.l. are maintained in nature by different spectra of vertebrate host species. For example, the woodmouse, *Apodemus sylvaticus*, is frequently infected with *B. afzelii* in central parts of Europe (Humair *et al.*, 1995, 1999; Kurtenbach *et al.*, 1998b, 2001). In central parts of North America, woodrats (*Neotoma* spp.) are infected predominantly with *B. bissettii* (Maupin *et al.*, 1994; Schneider *et al.*, 2000). In New England, USA, mice of the genus *Peromyscus* are mainly infected with *B. burgdorferi* s.s. (Donahue *et al.*, 1987; Mather *et al.*, 1989). In Japan, rodents are predominantly carriers of *B. japonica* (Nakao *et al.*, 1994a,b; Masuzawa *et al.*, 1997). Seabirds analysed on islands in the North Sea are infected with *B. garinii*, but not with *B. afzelii* (Olsen *et al.*, 1993, 1995a).

It is evident that a variety of extrinsic ecological factors (such as climate, vegetation or geology) determine the relative role a vertebrate host population plays in the maintenance and geographical distribution of a particular strain of *B. burgdorferi* s.l. (Kurtenbach *et al.*, 1995; Randolph and Craine, 1995; Kirstein *et al.*, 1997; Randolph, 1998; de Michelis *et al.*, 2000). It has, however, become apparent that intrinsic factors also play important roles in the biology of *B. burgdorferi* s.l. A growing body of experimental and ecological evidence indicates that different genotypes of *B. burgdorferi* s.l. differ in their transmissibility between hosts and ticks. This suggests a strong element of host-specificity that is independent of extrinsic ecological factors (Kurtenbach *et al.*, 1998b; Gylfe *et al.*, 1999; Humair *et al.*, 1999; Richter *et al.*, 2000; Hu *et al.*, 2001). Thus, reservoir competence is variable depending on the combination of spirochaetal strain and vertebrate host species. The overall pattern that has emerged is that small mammals are reservoir-competent for a different set of *B. burgdorferi* s.l. strains from that of avian hosts. This specific transmission pattern suggests that many genotypes or strains of *B. burgdorferi* s.l. are adapted to either mammalian or avian hosts but not to both. In contrast, *B. burgdorferi* s.s. appears to be less specialized and often seems to be infectious for both avian and mammalian hosts. Based on the findings on differential transmission patterns, *B. burgdorferi* s.l. strains can be divided into three groups: (i) strains that are adapted to small mammals; (ii) strains that are adapted to birds; and (iii) strains that are not specialized (Fig. 5.4).

The findings on differential transmissibility and host-specificity of *B. burgdorferi* s.l. require a cautious use of the terms 'reservoir competence', 'reservoir incompetence' and 'zooprophylaxis' because one vertebrate species may be reservoir-competent for one strain of *B. burgdorferi* s.l. but reservoir-incompetent for another.

The association of spirochaetal genotypes with particular vertebrates has an important implication for the dispersal of *B. burgdorferi* s.l. Unlike small mammals, which have a migration rate of only 200–300 m per generation, many birds have high rates of migration, a parameter in the population biology and ecology of *B. burgdorferi* s.l. that has only recently been addressed. High rates of migration of *B. garinii* genotypes have been found by various research groups (Olsen *et al.*, 1995a; Gylfe *et al.*, 1999). For example, the authors of this chapter have isolated strains of *B. garinii* with almost identical *ospA* and flagellin sequences

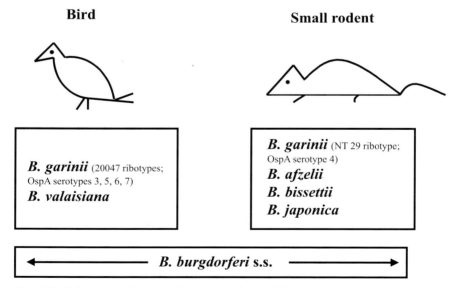

Fig. 5.4. Schematic diagram of host-specificity of *B. burgdorferi* s.l. Avian hosts are reservoir-competent for a different set of strains from that of small rodents.

from questing ticks in Portugal and from seabird ticks (*Ixodes uriae*) collected on islands in the North Sea (unpublished observations; figs 1–3).

Altogether, it is now established that *B. burgdorferi* s.l. is a multi-host bacterial species complex that is ecologically differentiated (Fig. 5.4). Ecological differentiation, however, is not always consistent with genospecies classification, as a genospecies may comprise divergent 'ecotypes'. Thus, whilst the current genospecies concept has certain benefits, it obviously lacks precise predictive power in ecological studies of LB.

Because of independent transmission, the values for R_0 of *B. burgdorferi* s.l. need to be estimated mathematically for each genospecies. It is likely that the sum of the species-specific values for the basic reproduction number is lower than previously calculated for the species complex as a single unit (Randolph and Craine, 1995; Randolph, 1998).

The complement system as a determinant of host-association specificity

It is now accepted that the complement system is a major determinant of host-specificity (i.e. niche adaptation) of *B. burgdorferi* s.l. (Kurtenbach *et al.*, 1998c; Kuo *et al.*, 2000; Nelson *et al.*, 2000). Therefore, this aspect of the biology of *B. burgdorferi* s.l. deserves special attention.

Cultured strains of *B. burgdorferi* s.l. are highly heterogeneous with respect to complement-mediated lysis. Brade and co-workers first reported on differences in serum resistance amongst strains of *B. burgdorferi* s.l. (Brade *et al.*, 1992; Breitner-Ruddock *et al.*, 1997). The studies on resistance/sensitivity patterns to complement have since been extended to a large number of *B. burgdorferi* s.l. strains and different vertebrate species including avian, rodent and ruminant hosts (van Dam *et al.*, 1997; Kurtenbach *et al.*, 1998b) (Tables 5.1 and 5.2).

A clear pattern of resistance/sensitivity by *B. burgdorferi* s.l. to complement has emerged that correlates with patterns of traits that are biologically significant, for example OspA serotype or transmissibility. For example, *B. afzelii*, a genospecies that is not transmitted by avian hosts, is lysed rapidly *in vitro* by avian complement. Conversely, *B. valaisiana* and the subtypes of *B. garinii* that are transmitted by avian species (20047 or OspA serotypes 3, 5, 6, 7) are resistant to avian complement. Invasive strains of *B. garinii* that belong to the OspA serotype 4 group of isolates from cerebrospinal fluid of patients and those that are related to the Asian ribotype NT29 are relatively resistant to human and rodent complement, but sensitive to avian complement (Kurtenbach *et al.*, 1998c; K. Kurtenbach *et al.*, unpublished observations).

Various studies have attempted to identify the mechanisms that confer resistance to complement in the *Borrelia* system. All *B. burgdorferi* s.l. strains invariably bind C3. It is now established that resistant and sensitive strains differ with respect to deposition of C5b–C9, the membrane attack complex. As already described, resistance to complement is mediated by species-specific binding of two complement control proteins: factor H-like protein 1/reconectin and factor H. Species-restricted binding of complement control proteins is mediated by

Table 5.1. Resistance of *Borrelia burgdorferi* sensu lato to host complement.

	B. burgdorferi s.s	B. afzelii	B. japonica	B. lusitaniae	B. valaisiana	B. garinii 20047	B. garinii NT29
Vole	R	R	R	R	R	S	R
Hamster	R	R	R	R	R	S	R
Squirrel	R	R	R	R	R	S	R
Yellow-necked mouse	R	R	R	R	R	S	R
Woodmouse	R	R	R	R	R	S	R
Pheasant	P	S	S	S	R	S	S
Guillemot	P	S	S	S	R	S	S
Blackbird	P	S	S	S	R	S	S
Lizard (*Lacerta*)	P	S	S	S	R	S	S
Sheep	P	S	S	S	S	S	S
Horse	P	S	S	S	S	S	S
Pig	P	S	S	S	S	S	S
Cow	P	S	S	S	S	S	S
Deer	P	S	S	S	S	S	S

R = Fully resistant to host complement.
P = Partially resistant to host complement.
S = Not resistant (i.e. sensitive) to complement.

B. burgdorferi s.l. CRASPs (Hellwage *et al.*, 2001; Kraiczy *et al.*, 2001a,b). Since OspE, a member of a large, plasmid-borne multigene family (see Bergström *et al.*, Chapter 3) is one of the CRASPs, it is possible that the other CRASPs are also members of this family.

A model of transmission

The studies on resistance/sensitivity patterns were all performed *in vitro* using cultured strains. In terms of protein expression, cultured forms resemble the stages of *B. burgdorferi* s.l. found in the midgut of the tick (de Silva and Fikrig, 1997), suggesting that the CRASPs that serve as specific ligands for host-derived complement control proteins are also expressed in the midgut of the tick. Although only data on OspE exist (Hellwage *et al.*, 2001), this suggestion is strongly corroborated by the precise match of complement-resistance patterns found *in vitro* with patterns of transmission (Kurtenbach *et al.*, 1998b,c).

Here a working model is proposed that depicts how complement may critically operate in the transmission cycle of *B. burgdorferi* s.l. Infected ticks that contain *B. burgdorferi* s.l. in the midgut take up host complement in large quantities during the blood meal along with other host-derived molecules such as plasminogen. It

Table 5.2. Bacteriolysis of *B. burgdorferi* sensu lato by different concentrations of untreated sera after 2, 4 and 22 h of incubation.[a]

Animal species/spirochaete	Serum conc. (%)	% bacteriolysis at incubation time (h)		
		2	4	22
Mouse/*B. garinii* (ZQ 1)	10	5	24	61
	25	42	100	100
	50	>95	100	100
Pheasant/*B. afzelii* (ACA 1)	10	0	11	31
	25	11	19	87
	50	17	28	100
Sheep/*B. burgdorferi* s.s. (ZS 7)	10	5	13	21
	25	9	12	43
	50	14	18	84
Deer/*B. garinii* (ZQ 1)	10	72	78	100
	25	100	100	100
	50	100	100	100

[a]For each combination of animal species and spirochaetal strain, the per cent mortality of spirochaetes is given for one representative untreated serum sample (from Kurtenbach *et al.*, 1998c).

has been shown, at least for *I. ricinus*, that complement is active in the feeding tick (Papatheodorou and Brossard, 1987) despite the possible uptake of salivary anti-complement proteins (Valenzuela *et al.*, 2000). As seen *in vitro*, the model predicts that ingested complement interacts with the midgut stages of *B. burgdorferi* s.l. Resistant strains, for example *B. afzelii* encountering rodent complement during the blood meal of a tick on a mouse, specifically bind rodent factor H, conferring resistance to rodent complement. Consequently *B. afzelii* survives the blood meal, escapes the midgut lumen and invades the salivary glands from where the bacteria are delivered to the host (Fig. 5.5). Complement-resistant strains are also transmitted transstadially through the moult to the next developmental stage of the tick.

In contrast, sensitive strains, for example *B. afzelii* that encounter deer or avian complement, activate the complement cascade, which kills the bacteria in the midgut before they are transmitted to the host (Figs 5.5 and 5.6). In addition, elimination of complement-sensitive spirochaetes from the midgut prevents transstadial transmission of the bacteria to the next developmental stage of the tick. The model proposed here suggests that the spirochaetes are selectively vulnerable to complement in the midgut of the feeding tick. The differential lysis or survival of the bacteria within the midgut renders the tick's midgut a bottleneck in terms of transmission of *B. burgdorferi* s.l. Selective elimination/retention of *B. burgdorferi* s.l. in the midgut of the tick may manifest itself in the replacement of *B. burgdorferi* s.l. strains during development from the nymph to the adult tick, as each developmental stage may feed on a different host. For example, a

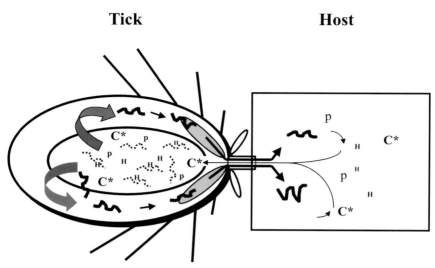

Fig. 5.5. Schematic figure depicting the proposed model of transmission of *B. burgdorferi* s.l. through the tick into the host. Plasminogen (P), complement (C*) and complement control proteins (H) are taken up by the feeding tick. In permissive systems, *B. burgdorferi* s.l. binds complement control proteins in the midgut (broken line) and is protected against complement-mediated lysis. Dissemination through the tick into the salivary glands takes at least 48 h and is facilitated by binding of plasminogen/plasmin. Spirochaetes that have undergone antigenic phase variation (i.e. downregulation of OspA; solid line) are eventually delivered to the host. In non-permissive systems, spirochaetes do not bind complement control proteins, resulting in the formation of the membrane attack complex and lysis of bacteria in the midgut before they are transmitted to the host. In systemically infected ticks, spirochaetes are already present in the salivary glands at the beginning of the blood meal. In such ticks, complement-mediated lysis of spirochaetes in the midgut may not prevent rapid transmission to the host.

questing nymph infected with *B. afzelii* that feeds on a bird that is infected with *B. garinii* (e.g. 20047 or OspA serotype 6) is likely to lose the *B. afzelii* infection but acquire *B. garinii*. In fact, in a study from Europe mixed infections of *B. afzelii* and *B. garinii* in individual questing adult ticks were found significantly less frequently than expected under neutralist assumptions (Kurtenbach *et al.*, 2001).

However, the working model presented here needs to be reconciled with a conflicting observation. Wild rodents have occasionally been found to be carriers of 20047-like variants of *B. garinii* and OspA serotype 6 strains (Kurtenbach *et al.*, 1998b; K. Kurtenbach *et al.*, unpublished observations). As these strains are highly sensitive to rodent complement *in vitro*, one would expect that vector ticks carrying such variants would lose their infection. This raises the important question: how do hosts whose complement system lyses *B. burgdorferi* s.l. *in vitro* become occasionally infected with such strains in the first place? According to

1)   Tick * ⟶ Host * ⟶ Tick *
     Tick * ↺

2)   Tick * — // ⟶ Host — // ⟶ Tick
     Tick ↺̸

3)   Tick ^syst. ⟶ Host * — // ⟶ Tick
     Tick ↺

Fig. 5.6. Schematic diagram of possible transmission routes and blockades of *B. burgdorferi* s.l. from infected vector ticks to hosts and recipient ticks. An asterisk means the tick or host is infected. (1) Permissive system with horizontal and vertical transmission through the cycle. Example: *B. afzelii*-infected tick feeding on a rodent. (2) Non-permissive system. No horizontal transmission from the infected tick (midgut infection) to the host and no vertical transmission to the next developmental stage of the tick (i.e. elimination of midgut infection in vector). Example: tick infected with *B. afzelii* feeding on an avian host or deer. (3) Semi-permissive system. Horizontal transmission to the host by systemically infected vector ticks and possible vertical transmission through the moult, but no further transmission to recipient ticks. Example: ticks systemically infected with *B. garinii* (e.g. OspA serotypes 5, 6) may infect rodents. However, rodents carrying such strains do not infect feeding ticks.

the model of transmission presented here, *B. burgdorferi* s.l. should be killed in the midgut of the tick, blocking any further development of the bacteria in the tick, thereby preventing transmission to that host. The answer may come from systemically infected ticks (Fig. 5.6). Systemically infected ticks may carry forms of *B. burgdorferi* s.l. in the salivary glands that are already pre-adapted to the host environment. We propose that *B. burgdorferi* s.l. delivered by systemically infected ticks may escape complement-mediated destruction. It remains to be seen whether resistance to complement of *B. burgdorferi* s.l. in the host is mediated by the CRASPs identified in cultured strains or by other mechanisms. Here it is hypothesized that resistance to complement of spirochaetes within the host is not species- or strain-restricted and that it is conferred by structural protection (e.g. glycosylation) and not by exploitation of host-derived complement control proteins. Functional genomics will provide further insight into the complex interactions of *B. burgdorferi* s.l. and the complement system.

The discovery that genotypes of *B. burgdorferi* s.l. are eliminated from the feeding tick by a host-derived factor adds a novel parameter to models of the life cycle of *B. burgdorferi* s.l. Until recently, most models of transmission assumed that the coefficients of transstadial transmission of *B. burgdorferi* s.l. (β_{T-T}) were close to 1 (Randolph and Craine, 1995; Randolph, 1998). It was assumed that transstadial transmission is highly and invariably efficient, irrespective of the source of the blood meal or the spirochaetal genotype. In Fig. 5.6, the three possible transmission pathways of *B. burgdorferi* s.l. are summarized schematically. The observation that sensitive genotypes are eliminated by host complement taken up by the tick during the blood meal is reminiscent of the elimination of spirochaetes by OspA-specific antibodies.

The model of transmission presented here does not deny the importance of other factors that may determine transmission of *B. burgdorferi* s.l. through ticks and hosts. In fact, it is highly probable that a large number of different and as yet unknown specific molecular interactions have evolved that define a permissive condition for transmission. However, specific binding of *B. burgdorferi* s.l. proteins to host-derived complement control proteins in the tick is a prerequisite of successful transmission. In other words, failure of *B. burgdorferi* s.l. to bind complement control proteins results in a blockade of transmission. Resistance/sensitivity patterns of *B. burgdorferi* s.l to complement can be used as an *in vitro* tool to predict or rule out reservoir competence of hosts for *B. burgdorferi* strains whose ecology is unknown, for example European strains of *B. bissettii* and *B. lusitaniae*. The prediction that OspA serotype 4 strains of *B. garinii* are transmissible through rodents to ticks has already been validated (van Dam *et al.*, 1997; Marconi *et al.*, 1999; Hu *et al.*, 2001).

In summary, it has now been established that interaction with complement is a critical element in the life cycle of *B. burgdorferi* s.l. and therefore in the global ecology of LB (Kurtenbach *et al.*, 1998b; Lane and Quistad, 1998; Kuo *et al.*, 2000; Nelson *et al.*, 2000).

Concluding Remarks

B. burgdorferi s.l. is a remarkable and unique bacterium. The majority of genes encoding proteins of the outer surface coat do not match anything in the tree of life. The bacterium causes disease, and yet no virulence factors have been found. Very few bacterial cells reside in the vertebrate host, and yet they may trigger pathological reactions that are entirely out of proportion. *B. burgdorferi* s.l. strongly stimulates the adaptive immune system, and yet it persists in the vertebrate host, often lifelong. It performs a sophisticated form of deflection of the innate immune system for its own benefit. Diversity is created by 'localized sex', perhaps in the tick. Its ecological niche is defined by the micro-environment at the interface of tick and host, rather than solely in the host. It is, in evolutionary terms, 'old' and certainly not an emerging bacterium, and yet was detected only in the 1980s. Adaptive radiation of *B. burgdorferi* s.l. has enabled this species complex to occupy a large variety of ecological niches. Amongst the genospecies, speciation and niche adaptation are most obvious for *B. garinii*. This genospecies displays pronounced subspecies structure with clades that are adapted to avian-tick transmission cycles and others that are adapted to rodent-tick transmission cycles. It is possible that *ospC* alleles that mark hyper-invasiveness are horizontally transferred into bird-associated spirochaetal strains. This, along with dramatically changing patterns of avian migration (Berthold, 1998), could pose a threat to public health.

Important biological questions remain to be addressed in LB research. Is *B. burgdorferi* s.l. originally a tick-associated commensal or a microparasite of mammalian, avian or even reptilian species? When in natural history did *B. burgdorferi* s.l. switch to its present strictly vector-borne lifestyle, relying on the alternating passage between vertebrate host and invertebrate vector? Functional genomics, molecular phylogeny, population genetics coupled with ecology and experimental transmission studies are likely to answer such questions. In some ways, the genome project of *B. burgdorferi* s.l. has closed doors, but in many ways it has opened new doors. The post-genomic era ahead is likely to become a productive period in LB research.

Acknowledgements

We would like to thank Markus Simon, Peter Kraiczy, Volker Brade, Margarida Collares-Pereira, Margarida Santos-Reis, Rainer Wallich, Lise Gern, Patricia Nuttall, Sarah Randolph, Edward Holmes, Olaf Kahl, Nick Ogden, Joseph Piesman, Amy Fagerberg, Michael Donaghy, Milan Labuda, Klara Hanincová and Jeremy Gray for constructive advice. Some of the work presented was supported by the Boehringer Ingelheim Fonds, Stuttgart, the German Academic Exchange Service (DAAD), Bonn, Germany, the Natural Environment Research Council, Swindon, UK, and the Wellcome Trust, London, UK.

References

Anderson, J. (1991) Epizootiology of Lyme borreliosis. *Scandinavian Journal of Infectious Diseases Suppl.* 77, 23–34.

Anderson, R.M. and May, R.M. (1991) *Infectious Diseases of Humans, Dynamics and Control*. Oxford University Press, Oxford, UK, 575 pp.

Appel, M.J.G., Allan, S. and Jacobson, R.H. (1993) Experimental Lyme disease in dogs produces arthritis and persistent infection. *Journal of Infectious Diseases* 167, 651–664.

Barthold, S.W. (1995) Animal models for Lyme disease. *Laboratory Investigation* 72, 127.

Barthold, S.W., Moody, K.D., Trewilliger, G.A., Duray, P.H., Jacoby, R.O. and Steere, A.C. (1988) Experimental Lyme arthritis in rats infected with *Borrelia burgdorferi*. *Journal of Infectious Diseases* 157, 842–846.

Barthold, S.W., Beck, D.S., Hansen, G.M., Terwilliger, G.A. and Moody, K.D. (1990) Lyme borreliosis in selected strains and ages of laboratory mice. *Journal of Infectious Diseases* 162, 133–138.

Berthold, P. (1998) Bird migration. Genetic programs with high adaptability. *Zoology – Analysis of Complex Systems* 101, 235–245.

Bockenstedt, L.K., Hidzic, E., Feng, S., Bourrel, K.W., de Silva, A., Montgomery, R.R., Fikrig, E., Radolf, J.D. and Barthold, S.W. (1997) *Borrelia burgdorferi* strain-specific OspC-mediated immunity in mice. *Infection and Immunity* 65, 4661–4670.

Boeggemeyer, E., Stehle, T., Schaible, U.E., Hahne, M., Vestweber, D. and Simon, M.M. (1994) *Borrelia burgdorferi* upregulates the adhesion molecules E-selectin, P-selectin, ICAM-1 and VCAM-1 on mouse endothelioma cells *in vitro*. *Cell Adhesion Communication* 2, 145–157.

Brade, V., Kleber, I. and Acker, G. (1992) Differences of two *Borrelia burgdorferi* strains in complement activation and serum resistance. *Immunobiology* 185, 453–465.

Breitner-Ruddock, S., Wuerzner, R., Schulze, J. and Brade, V. (1997) Heterogeneity in the complement-dependent bacteriolysis within the species of *Borrelia burgdorferi*. *Medical Microbiology and Immunology* 185, 253–260.

Brunet, L.R., Selitto, C, Spielman, A. and Telford, S.R. III (1995) Antibody response of the mouse reservoir to *Borrelia burgdorferi* in nature. *Infection and Immunity* 63, 3030–3036.

Burns, M.J., Sellati, T.J., Teng, E.I. and Furie, M.B. (1997) Production of interleukin-8 (IL-8) by cultured endothelial cells in response to *Borrelia burgdorferi* occurs independently of secretion of IL-1 and tumor necrosis factor alpha and is required for subsequent transendothelial migration of neutrophils. *Infection and Immunity* 65, 1217–1222.

Casjens, S., Palmer, N., van Vugt, R., Huang, W.M., Stevenson, B., Rosa, P., Lathigra, R., Sutton, G., Peterson, J., Dobson, R.J., Haft, D., Hickey, E., Gwinn, M., White, O. and Fraser, C.M. (2000) A bacterial genome in flux: the twelve linear and nine circular extrachromosomal DNAs in an infectious isolate of the Lyme disease spirochete *Borrelia burgdorferi*. *Molecular Microbiology* 35, 490–516.

Coburn, J., Magoun, L., Bodary, S.C. and Leong, J.M. (1998) Integrins $\alpha_v\beta_3$ and $\alpha_5\beta_1$ mediate attachment of Lyme disease spirochaetes to human cells. *Infection and Immunity* 66, 1946–1952.

Coleman, J.L. and Benach, J.L. (1999) Use of the plasminogen activation system by micro-organisms. *Journal of Laboratory Clinical Medicine* 134, 567–574.

Coleman, J.L., Sellati, T.J., Testa, J.E., Kew, R.R., Furie, M.B. and Benach, J.L. (1995) *Borrelia burgdorferi* binds plasminogen, resulting in enhanced penetration of endothelial monolayers. *Infection and Immunity* 63, 2478–2484.

Coleman, J.L., Gebbia, J.A., Piesman, J., Degen, J.L., Bugge, T.H. and Benach, J.L. (1997) Plasminogen is required for efficient dissemination of *B. burgdorferi* in ticks and for enhancement of spirochetemia in mice. *Cell* 89, 1111–1119.

van Dam, A.P., Kuiper, H., Vos, K., Widjojokusumo, A., de Jongh, B.M., Spanjaard, L., Ramselar, A.C., Kramer, M.D. and Dankert, J. (1993) Different genospecies of *Borrelia burgdorferi* are associated with distinct clinical manifestations of Lyme borreliosis. *Clinical Infectious Diseases* 17, 708–717.

van Dam, A.P, Oei, A., Jaspars, R., Fijen, C., Wilske, B., Spanjaard, L. and Dankert, J. (1997) Complement-mediated serum activity among spirochetes that cause Lyme disease. *Infection and Immunity* 65, 1228–1236.

Dizij, A. and Kurtenbach, K. (1995) *Clethrionomys glareolus*, but not *Apodemus flavicollis*, acquires resistance to *Ixodes ricinus* (Acari: Ixodidae), the main European vector of *Borrelia burgdorferi*. *Parasite Immunology* 17, 177–183.

Donahue, J.G., Piesman, J. and Spielman, A. (1987) Reservoir competence of white-footed mice for lyme-disease spirochetes. *American Journal of Tropical Medicine and Hygiene* 36, 92–96.

Dykhuizen, D.E. and Baranton, G. (2001) The implications of low rate of horizontal transfer in borrelia. *Trends in Microbiology* 9, 344–350.

Dykhuizen, D.E., Polin, D.S., Dunn, J.J., Wilske, B., Preac-Mursic, V., Dattwyler, R.J. and Luft, B. (1993) *Borrelia burgdorferi* is clonal: implications for taxonomy and vaccine development. *Proceedings of the National Academy of Sciences, USA* 90, 10163–10167.

Ebnet, K., Brown, K.D., Siebenlist, U.K., Simon, M.M. and Shaw, S. (1996a) *Borrelia burgdorferi* activates nuclear factor-kappa B and is a potent inducer of chemokine and adhesion molecule gene expression in endothelial cells and fibroblasts. *Journal of Immunology* 158, 3285–3292.

Ebnet, K., Simon, M.M. and Shaw, S. (1996b) Regulation of chemokine gene expression in human endothelial cells by proinflammatory cytokines and *Borrelia burgdorferi*. *Annals of the New York Academy of Sciences* 797, 107–117.

Fikrig, E., Telford, S.R. III, Barthold, S.W., Kantor, F.S., Spielman, A. and Flavell, R.A. (1992) Elimination of *Borrelia burgdorferi* from vector ticks feeding on OspA-immunized mice. *Proceedings of the National Academy of Sciences, USA* 89, 5418–5421.

Fingerle, V., Laux, H., Munderloh, U.G., Schulte-Spechtel, U. and Wilske, B. (2000) Differential expression of outer surface proteins A and C by individual *Borrelia burgdorferi* in different genospecies. *Medical Microbiology and Immunology* 189, 59–66.

Finlay, B.B. and Cossart, P. (1997) Exploitation of mammalian host cell functions by bacterial pathogens. *Science* 276, 718–725.

Fraser, C.M., Casjens, S., Huang, W.M., Sutton, G.G., Clayton, R., Lathigra, R., White, O., Ketchum, K.A., Dodson, R., Hickey, E.K., Gwinn, M., Dougherty, B., Tomb, J.F., Fleischmann, R.D., Richardson, D., Peterson, J., Kerlavage, A.R., Quackenbush, J., Salzberg, S., Hanson, M., van Vugt, R., Palmer, N., Adams, M.D., Gocayne, J., Weidman, J., Utterback, T., Watthey, L., McDonald, L., Artiach, P., Bowman, C., Garland, S., Fujii, C., Cotton, M.D., Horst, K., Roberts, K., Hatch, B., Smith, H.O. and Venter, J.C. (1997) Genomic sequence of a Lyme disease spirochaete, *Borrelia burgdorferi*. *Nature* 390, 580–586.

Fuchs, H., Wallich, R., Simon, M.M. and Kramer, M.D. (1994) The outer surface protein A of the spirochete *Borrelia burgdorferi* is a plasmin(ogen) receptor. *Proceedings of the National Academy of Sciences, USA* 91, 12594–12598.

Fuchs, H., Simon, M.M., Wallich, R., Bechtel, M. and Kramer, M.D. (1996) *Borrelia burgdorferi* induces secretion of pro-urokinase-type plasminogen activator by human monocytes. *Infection and Immunity* 64, 4307–4312.

Gern, L., Schaible, U.E. and Simon, M.M. (1993) Mode of inoculation of the Lyme disease agent *Borrelia burgdorferi* influences infection and immune responses in inbred strains of mice. *Journal of Infectious Diseases* 167, 971–975.

Gern, L., Siegenthaler, M.C., Hu, M., Leuba-Garcia, S., Humair, P.F. and Moret, J. (1994) *Borrelia burgdorferi* in rodents (*Apodemus flavicollis* and *A. sylvaticus*): duration and enhancement of infectivity for *Ixodes ricinus* ticks. *European Journal of Epidemiology* 10, 75–80.

Gern, L., Estrada-Pena, A., Frandsen, F., Gray, J.S., Jaenson, T.G.T., Jongejan, F., Kahl, O., Korenberg, E., Mehl, R. and Nuttall, P.A. (1998) European reservoir hosts of *Borrelia burgdorferi* sensu lato. *Zentralblatt für Bakteriologie* 287, 196–204.

Gilmore, R.D. and Piesman, J. (2000) Inhibition of *Borrelia burgdorferi* migration from the midgut to the salivary glands following feeding by ticks on OspC-immunized mice. *Infection and Immunity* 68, 411–414.

Golde, W.T., Burkot, T.R., Sviat, S., Keen, M.G., Mayer, L.W., Johnson, B.J.B. and Piesman. J. (1993) The major histocompatibility complex-restricted response of recombinant inbred strains of mice to natural tick transmission of *Borrelia burgdorferi*. *Journal of Experimental Medicine* 177, 9–17.

Grab, D.J., Givens, C. and Kennedy, R. (1998) Fibronectin-binding activity in *Borrelia burgdorferi*. *Biochimica et Biophysica Acta* 1407, 135–145.

Grab, D.J., Lanners, H.-N., Martin, L.N., Chesney, J., Cai, C., Adkisson, H.D. and Bucala, R. (1999) Interaction of *Borrelia burgdorferi* with peripheral blood fibrocytes, antigen-presenting cells with the potential for connective tissue targeting. *Molecular Medicine* 5, 46–54.

Guo, B.P., Norris, S.J., Rosenberg, L.C. and Hook, M. (1995) Adherence of *Borrelia burgdorferi* to the proteoglycan decorin. *Infection and Immunity* 63, 3467–3472.

Gylfe, Å., Olsen, B., Strasevicius, D., Ras, N.M., Weihe, P., Noppa, L., Ostberg, Y., Baranton, G. and Bergström, S. (1999) Isolation of Lyme disease *Borrelia* from puffins (*Fratercula arctica*) and seabird ticks (*Ixodes uriae*) on the Faroe Islands. *Journal of Clinical Microbiology* 37, 890–896.

Gylfe, Å., Bergström, S., Lunstrom, J. and Olsen, B. (2000) Epidemiology – reactivation of *Borrelia* infection in birds. *Nature* 403, 724–725.

Hellwage, J., Meri, T., Heikkilae, T., Alitalo, A., Panelius, J., Lahdenne, P., Seppaelae, I.J.T. and Meri, S. (2001) The complement regulator factor H binds to the surface protein OspE of *Borrelia burgdorferi*. *Journal of Biological Chemistry* 276, 8427–8435.

Hodzic, E., Febg, S. and Barthold, S.W. (2000) Stability of *Borrelia burgdorferi* outer surface protein C under immune selection pressure. *Journal of Infectious Diseases* 181, 750–753.

Hofmeister, E.K., Ellis, B.A., Glass, G.E. and Childs, J.E. (1999a) Longitudinal study of infection with *Borrelia burgdorferi* in a population of *Peromyscus leucopus* at a Lyme disease-enzootic site in Maryland. *American Journal of Tropical Medicine and Hygiene* 60, 598–609.

Hofmeister, E.K., Glass, G.E., Childs, J.E. and Persing, D.H. (1999b) Population dynamics of a naturally occurring heterogeneous mixture of *Borrelia burgdorferi* clones. *Infection and Immunity* 67, 5709–5716.

Hu, C.M., Wilske, B., Fingerle, V., Lobet, Y. and Gern, L. (2001) Transmission of *Borrelia garinii* OspA serotype 4 to BALB/c mice by *Ixodes ricinus* ticks collected in the field. *Journal of Clinical Microbiology* 39, 1169–1171.

Humair, P.F., Peter, O., Wallich, R. and Gern, L. (1995) Strain variation of Lyme disease spirochetes isolated from *Ixodes ricinus* ticks and rodents collected in two endemic areas in Switzerland. *Journal of Medical Entomology* 32, 433–438.

Humair, P.F., Postic, D., Wallich, R. and Gern, L. (1998) An avian reservoir (*Turdus merula*) of the Lyme borreliosis spirochetes. *Zentralblatt für Bakteriologie* 287, 521–538.

Humair, P.F., Rais, O. and Gern, L. (1999) Transmission of *Borrelia afzelii* from *Apodemus* mice and *Clethrionomys* voles to *Ixodes ricinus* ticks, differential transmission pattern and overwintering maintenance. *Parasitology* 118, 33–42.

Jauris-Heipke, S., Liegl, G., Preac-Mursic, V., Roessler, D., Schwab, E., Soutschek, E., Will, G. and Wilske, B. (1995) Molecular analysis of genes encoding outer surface protein C (OspC) of *Borrelia burgdorferi* sensu lato: relationship to *ospA* genotype and evidence of lateral gene exchange of *ospC*. *Journal of Clinical Microbiology* 33, 1860–1866.

Kirstein, F., Rijpkema, S., Molkenboer, M. and Gray, J.S. (1997) Local variations in the distribution of *Borrelia burgdorferi* sensu lato genomospecies in *Ixodes ricinus* ticks. *Applied and Environmental Microbiology* 3, 1102–1106.

Klempner, M.S., Noring, R. and Rogers, R.A. (1993) Invasion of human skin fibroblasts by the Lyme disease spirochaete, *Borrelia burgdorferi*. *Journal of Infectious Diseases* 167, 1074–1081.

Knigge, H., Simon, M.M., Meuer, S.C., Kramer, M.D. and Wallich, R. (1996) The outer surface lipoprotein OspA of *Borrelia burgdorferi* provides co-stimulatory signals to normal human peripheral CD4+ and CD8+ T lymphocytes. *European Journal of Immunology* 26, 2299–2303.

Kraiczy, P., Skerka, C., Kirschfink, M., Brade, V. and Zipfel, P.F. (2001a) Immune evasion of *Borrelia burgdorferi* by acquisition of human complement regulators FHL-1/reconectin and factor H. *European Journal of Immunology* 31, 1674–1684.

Kraiczy, P., Skerka, C., Kirschfink, M., Zipfel, P.F. and Brade, V. (2001b) Mechanism of complement resistance of pathogenic *Borrelia burgdorferi* isolates. *International Immunopharmacology* 1, 393–401.

Kuo, M.M., Lane, R.S. and Giclas, P.C. (2000) A comparative study of mammalian and reptilian alternative pathway of complement-mediated killing of the Lyme disease spirochete (*Borrelia burgdorferi*). *Journal of Parasitology* 86, 1223–1228.

Kurtenbach, K., Dizij, A., Seitz, H.-M., Margos, G., Moter, S.E., Kramer, M.D., Wallich, R., Schaible, U.E. and Simon, M.M. (1994) Differential immune responses to *Borrelia burgdorferi* in European wild rodent species influence spirochete transmission to *Ixodes ricinus* L. (Acari: Ixodidae). *Infection and Immunity* 62, 5344–5352.

Kurtenbach, K., Kampen, H., Dizij, A., Arndt, S., Seitz, H.-M., Schaible, U.E. and Simon, M.M. (1995) Infestation of rodents with larval *Ixodes ricinus* L. (Acari: Ixodidae) is an important factor in the transmission cycle of *Borrelia burgdorferi* s.l. in German woodlands. *Journal of Medical Entomology* 32, 807–817.

Kurtenbach, K., Dizij, A., Voet, P., Hauser, P. and Simon, M.M. (1997) Vaccination of natural reservoir hosts with recombinant lipidated OspA induces a transmission-blocking immunity against Lyme disease spirochaetes associated with high levels of LA-2 equivalent antibodies. *Vaccine* 15, 1670–1674.

Kurtenbach, K., Carey, D., Hoodless, A., Nuttall, P.A. and Randolph, S.E. (1998a) Competence of pheasants as reservoirs for Lyme disease spirochetes. *Journal of Medical Entomology* 35, 77–81.

Kurtenbach, K., Peacey, M., Rijpkema, S.G.T., Hoodless, A.N., Nuttall, P.A. and Randolph, S.E. (1998b) Differential transmission of the genospecies of *Borrelia burgdorferi* sensu lato by game birds and small rodents in England. *Applied and Environmental Microbiology* 64, 1169–1174.

Kurtenbach, K., Sewell, H.-S., Ogden, N.H., Randolph, S.E. and Nuttall, P.A. (1998c) Serum complement sensitivity as a key factor in Lyme disease ecology. *Infection and Immunity* 66, 1248–1251.

Kurtenbach, K., de Michelis, S., Sewell, H.-S., Etti, S., Schäfer, M., Hails, R., Collares-Pereira, M., Hanincová, K.,

Labuda, M., Bormane, A. and Donaghy, M. (2001) Distinct combinations of *Borrelia burgdorferi* sensu lato genospecies found in individual questing ticks from Europe. *Applied and Environmental Microbiology* 67, 4926–4929.

Lachmann, P.J. (1986) A common form of killing. *Nature* 321, 560.

Lane, R.S. and Quistad, G.B. (1998) Borreliacidal factor in the blood of the western fence lizard (*Sceloporus occidentalis*). *Journal of Parasitology* 84, 29–34.

Lebet, N. and Gern, L. (1994) Histological examination of *Borrelia burgdorferi* infections in unfed *Ixodes ricinus* nymphs. *Experimental and Applied Acarology* 18, 177–183.

Le Fleche, A., Postic, D., Girardet, K., Peter, O. and Baranton, G. (1997) Characterization of *Borrelia lusitaniae* sp. nov. by 16S ribosomal DNA sequence analysis. *International Journal of Systematic Bacteriology* 47, 921–925.

Ma, Y. and Weiss, J.J. (1993) *Borrelia burgdorferi* outer surface lipoproteins OspA and OspB possess B-cell mitogenic and cytokine-stimulatory properties. *Infection and Immunity* 61, 3843–3853.

Marconi, R.T., Hohenberger, S., Jauris-Heipke, S., Schulte-Spechtel, U., Lavoie, C.P., Rößler, D. and Wilske, B. (1999) Genetic analysis of *Borrelia garinii* OspA serotype 4 strains associated with neuroborreliosis: evidence for genetic homogeneity. *Journal Clinical Microbiology* 37, 3965–3970.

Marti-Ras, N., Postic, D., Foretz, M. and Baranton, G. (1997) *Borrelia burgdorferi* sensu stricto, a bacterial species 'made in the U.S.A.'? *International Journal of Systematic Bacteriology* 47, 1112–1117.

Masuzawa, T., Iwaki, A., Sato, Y., Miyamoto, K., Korenberg, E.I. and Yanagihara, Y. (1997) Genetic diversity of *Borrelia burgdorferi* sensu lato isolated in far eastern Russia. *Microbiology and Immunology* 41, 595–600.

Masuzawa, T., Takada, N., Kudeken, M., Fukui, T., Yano, Y., Ishiguro, F., Kawamura, Y., Imai, Y. and Ezaki, T. (2001) *Borrelia sinica* sp. nov., a Lyme disease-related *Borrelia* species isolated in China. *International Journal of Systematic and Evolutionary Microbiology* 51, 1817–1824.

Mather, T.N., Wilson, M.L., Moore, S.I., Ribeiro, J.M. and Spielman, A. (1989) Comparing the relative potential of rodents as reservoirs of the Lyme disease spirochete (*Borrelia burgdorferi*). *American Journal of Epidemiology* 130, 143–150.

Matuschka, F.R., Heiler, M., Eiffert, H., Fischer, P., Lotter, H. and Spielman, A. (1993) Diversionary role of hoofed game in the transmission of Lyme disease spirochetes. *American Journal of Tropical Medicine and Hygiene* 48, 693–699.

Maupin, G.O., Gage, K.L., Piesman, J., Montenieri, S.L., Sviat, L., Vander-Zanden, C.M., Happ, M., Dolan, M. and Johnson, B.J.B. (1994) Discovery of an enzootic cycle of *Borrelia burgdorferi* in *Neotoma mexicana* and *Ixodes spinipalpis* from northern Colorado, an area where Lyme disease is nonendemic. *Journal of Infectious Diseases* 170, 636–643.

de Michelis, S., Sewell, H.-S., Collares-Pereira, M., Santos-Reis, M., Schouls, L.M., Benes, V., Holmes, E.C. and Kurtenbach, K. (2000) Genetic diversity of *Borrelia burgdorferi* sensu lato in ticks from mainland Portugal. *Journal of Clinical Microbiology* 38, 2128–2133.

Modolell, M., Schaible, U.E., Rittig, M. and Simon, M.M. (1994) Killing of *Borrelia burgdorferi* by macrophages is dependent on oxygen radicals and nitric oxide and can be enhanced by antibodies to outer surface proteins of the spirochete. *Immunology Letters* 40, 139–146.

Montgomery, R.R. and Malavista, S.E. (1996) Entry of *Borrelia burgdorferi* into macrophages is end-on and leads to degradation in lysosomes. *Infection and Immunity* 64, 2867–2972.

Museteanu, C., Schaible, U.E., Stehle, T., Kramer, M.D. and Simon, M.M. (1991) Myositis in mice inoculated with *Borrelia burgdorferi*. *American Journal of Pathology* 139, 1267–1271.

Nakao, M., Miyamoto, K. and Fukunaga, M. (1994a) Lyme disease spirochetes in Japan: enzootic transmission cycles in birds, rodents, and *Ixodes persulcatus* ticks. *Journal of Infectious Diseases* 170, 878–882.

Nakao, M., Miyamoto, K. and Fukunaga, M. (1994b) *Borrelia japonica* in nature: genotypic identification of spirochetes isolated from Japanese small mammals. *Microbiology and Immunology* 38, 805–808.

Nelson, D.R., Rooney, S., Miller, N.J. and Mather, T.N. (2000) Complement-mediated killing of *Borrelia burgdorferi* by non-immune sera from sika deer. *Journal of Parasitology* 86, 1232–1238.

Nordstrand, A., Barbour, A.G. and Bergström, S. (2000) *Borrelia* pathogenesis research in the post-genomic and post-vaccine era. *Current Opinion in Microbiology* 3, 86–92.

Ohnishi, J., Piesman, J. and de Silva, A. (2001) Antigenic and genetic heterogeneity of *Borrelia burgdorferi* populations transmitted by ticks. *Proceedings of the National Academy of Sciences, USA* 98, 670–675.

Oliver, J.H. (1996) Lyme borreliosis in the southern United States: a review. *Journal of Parasitology* 82, 926–935.

Olsen, B., Jaenson, T.G.T., Noppa, L., Bunikis, J. and Bergström, S.A. (1993) A Lyme borreliosis cycle in seabirds and *Ixodes uriae* ticks. *Nature* 362, 340–342.

Olsen, B., Duffy, D.C., Jaenson, T.G.T., Gylfe, A., Bonnedahl, J. and Bergström, S.A. (1995a) Transhemispheric exchange of Lyme disease spirochaetes by seabirds. *Journal of Clinical Microbiology* 33, 3270–3274.

Olsen, B., Jaenson, T.G.T. and Bergström, S.A. (1995b) Prevalence of *Borrelia burgdorferi* sensu lato-infected ticks on migrating birds. *Applied and Environmental Microbiology* 61, 3082–3087.

Olsen, B., Gylfe, A. and Bergström, S.A. (1996) Canary finches (*Serinus canaria*) as an avian infection model for Lyme borreliosis. *Microbial Pathogenesis* 20, 319–324.

Oteo, J.A., Backenson, P.B., Vitucia, M.M., Garcia Monco, J.C., Rodriguez, L., Escudero, R. and Anda, P. (1998) Use of C3H/He Lyme disease mouse model for the recovery of a Spanish isolate of *Borrelia garinii* from erythema migrans lesions. *Research in Microbiology* 149, 39–46.

Pal, U., de Silva, A.M., Montgomery, R.R., Fish, D., Anguita, J., Anderson, J.F., Lobet, Y. and Fikrig, E. (2000) Attachment of *Borrelia burgdorferi* within *Ixodes scapularis* mediated by outer surface protein A. *Journal of Clinical Investigation* 106, 561–569.

Papatheodorou, V. and Brossard, M. (1987) C-3 levels in the sera of rabbits infested and reinfested with *Ixodes ricinus* L. and in midguts of fed ticks. *Experimental and Applied Acarology* 3, 53–59.

Parveen, N., Robbins, D. and Leong, J.M. (1999) Strain variation in glycosaminoglycan recognition influences cell-type specific binding by Lyme disease spirochaetes. *Infection and Immunity* 67, 1743–1749.

Pennington, P.M., Allred, C.D., West, C.S., Alvarez, R. and Barbour, A.G. (1997) Arthritis severity and spirochete burden are determined by serotype in the *Borrelia turicatae*–mouse model of Lyme disease. *Infection and Immunity* 65, 285–292.

Philipp, M.T., Aydintug, M.K., Bohm, R.P. Jr, Cogswell, F.B., Dennis, V.A., Lanners, H.N., Lowrie, R.C. Jr, Roberts, E.D., Conway, M.D., Karacorlu, M. et al. (1993) Early and early-disseminated phases of Lyme disease in the rhesus monkey: a model for infections in humans. *Infection and Immunity* 61, 3047–3059.

Piesman, J. (1988) Intensity and duration of *Borrelia burgdorferi* and *Babesia microti* infectivity in rodent hosts. *International Journal of Parasitology* 18, 687–689.

Piesman, J. (1991) Experimental acquisition of the Lyme disease spirochete, *Borrelia burgdorferi*, by larval *Ixodes dammini* (Acari: Ixodidae) during partial blood meals. *Journal of Medical Entomology* 28, 259–262.

Piesman, J. (1993a) Dynamics of *Borrelia burgdorferi* transmission by nymphal *Ixodes dammini* ticks. *Journal of Infectious Diseases* 167, 1082–1085.

Piesman, J. (1993b) Standard system for infecting ticks (Acari: Ixodidae) with the Lyme disease spirochete, *Borrelia burgdorferi*. *Journal of Medical Entomology* 30, 199–203.

Piesman, J., Mather, T.N., Sinsky, R.J. and Spielman, A. (1987) Duration of tick attachment and *Borrelia burgdorferi* transmission. *Journal of Clinical Microbiology* 25, 557–558.

Piesman, J., Oliver, J.O. and Sinsky, R.J. (1990) Growth kinetics of the Lyme disease spirochete (*Borrelia burgdorferi*) in vector ticks (*Ixodes dammini*). *American Journal of Tropical Medicine and Hygiene* 42, 352–357.

Piesman, J., Maupin, G.O., Campos, E.G. and Happ, C.M. (1991) Duration of adult female *Ixodes dammini* attachment and transmission of *Borrelia burgdorferi*, with description of a needle aspiration isolation method. *Journal of Infectious Diseases* 163, 895–897.

Piesman, J., Dolan, M.C., Schriefer, M.E. and Burkot, T.R. (1996) Ability of experimentally infected chickens to infect ticks with the Lyme disease spirochete, *Borrelia burgdorferi*. *American Journal of Tropical Medicine and Hygiene* 54, 294–298.

Postic, D., Assous, M., Grimont, P.A.D. and Baranton, G. (1994) Diversity of *Borrelia burgdorferi* sensu lato evidenced by restriction fragment length polymorphism of *rrf* (5S)–*rrl* (23S) intergenic spacer amplicons. *International Journal of Systematic Bacteriology* 44, 743–752.

Postic, D., Marti-Ras, N., Lane, R.S., Hendson, M. and Baranton, G. (1998) Expanded diversity among Californian *Borrelia* isolates and description of *Borrelia bissettii* sp. nov. (formerly *Borrelia* group DN127). *Journal of Clinical Microbiology* 36, 3497–3504.

Preac-Mursic, V., Patsouris, E., Wilske, B., Reinhardt, S., Gos, B. and Mehraein, P. (1990) Persistence of *Borrelia burgdorferi* and histopathological alterations in experimentally infected animals: comparison with histopathological findings in human Lyme disease. *Infection* 18, 332–341.

Qui, W.-G., Bosler, E., Campbell, J.R., Ugine, G.D., Wang, L.-N., Luft, B.J. and Dykhuizen, D.E. (1997) A population genetic study of *Borrelia burgdorferi* sensu stricto from eastern Long Island, New York, suggested frequency-dependent selection, gene flow and host adaptation. *Hereditas* 127, 203–216.

Radolf, J.D., Arndt, L.L., Akins, D.R., Curetty, L.L., Levi, M.E., Davis, L.S. and Norgard, M.V. (1995) *Treponema pallidum* and *Borrelia burgdorferi* lipoproteins and synthetic lipoproteins activate monocytes/macrophages. *Journal of Immunology* 154, 2866–2877.

Rand, P.W., Lacombe, E.H., Smith, R.P. and Ficker, J. (1998) Participation of birds (Aves) in the emergence of Lyme disease in Southern Maine. *Journal of Medical Entomology* 35, 270–276.

Randolph, S.E. (1998) Ticks are not insects: consequences of contrasting vector biology for transmission potential. *Parasitology Today* 14, 186–192.

Randolph, S.E. and Craine, N.G.A. (1995) A general framework for comparative quantitative studies on the transmission of tick-borne diseases using Lyme borreliosis in Europe as an example. *Journal of Medical Entomology* 32, 765–777.

Richter, D., Spielman, A., Komar, N. and Matuschka, F.R. (2000) Competence of American robins as reservoir hosts for Lyme disease spirochetes. *Emerging Infectious Diseases* 6, 133–138.

Rittig, M.G., Jagoda, J.C., Wilske, B., Murgia, R., Cinco, M., Repp, R., Burmester, G.R. and Krause, R. (1998) Coiling phagocytosis discriminates between different spirochetes and is enhanced by phorbol myristate acetate and granulocyte–macrophage colony-stimulating factor. *Infection and Immunity* 66, 627–635.

Roehrig, J.T., Piesman, J., Hunt, A.R., Keen, M.G., Happ, C.M. and Johnson, B.J.B. (1992) The hamster immune response to tick-transmitted *Borrelia burgdorferi* differs from the response to needle-inoculated, cultured organisms. *Journal of Immunology* 149, 3648–3653.

Ryan, J.R., Levine, J.F., Apperson, C.S., Lubke, L., Wirtz, R.A., Spears, P.A. and Orndorff, P.E. (1998) An experimental chain of infection reveals that distinct *Borrelia burgdorferi* populations are selected in arthropod and mammalian hosts. *Molecular Microbiology* 30, 365–379.

Schaible, U.E., Kramer, M.D., Museteanu, C., Zimmer, G., Mossmann, H. and Simon, M.M. (1989) The severe combined immunodeficiency (scid) mouse. A laboratory model for the analysis of Lyme arthritis and carditis. *Journal of Experimental Medicine* 170, 1427–1432.

Schaible, U.E., Gern, L., Wallich, R., Kramer, M.D., Prester, M. and Simon, M.M. (1993) Distinct patterns of protective antibodies are generated against *Borrelia burgdorferi* in mice experimentally inoculated with high and low doses of antigen. *Immunology Letters* 36, 219–226.

Schaible, U.E., Vestweber, D., Butcher, E.G., Stehle, T. and Simon, M.M. (1994) Expression of endothelial cell adhesion molecules in joints and heart during *Borrelia burgdorferi* infection of mice. *Cell Adhesion Communication* 2, 465–479.

Schmitz, J.L., Hejke, A., England, D.M., Callister, S.M. and Schell, R.F. (1989) Induction of Lyme arthritis in LSH hamsters. *American Journal of Pathology* 134, 1113–1123.

Schneider, B.S., Zeidner, N.S., Burkot, T.R., Maupin, G.O. and Piesman, J. (2000) *Borrelia* isolates in Northern Colorado identified as *Borrelia bissettii*. *Journal of Clinical Microbiology* 38, 3103–3105.

Schwan, T.G., Burgdorfer, W. and Garon, C.F. (1988) Changes in infectivity and plasmid profile of the Lyme disease spirochete, *Borrelia burgdorferi*, as a result of *in vitro* cultivation. *Infection and Immunity* 56, 1831–1836.

Schwan, T.G., Piesman, J., Golde, W.T., Dolan, M.C. and Rosa, P.A. (1995) Induction of an outer surface protein on *Borrelia burgdorferi* during tick feeding. *Proceedings of the National Academy of Sciences, USA* 92, 2909–2913.

Seinost, G., Dykhuizen, D.E., Datwyler, R.J., Golde, W.T., Dunn, J.J., Wang, L.-N., Wormser, G.P., Schriefer, M.E. and Luft, B.J. (1999) Four clones of *Borrelia burgdorferi* sensu stricto cause invasive infections in humans. *Infection and Immunity* 67, 3518–3524.

Sellati, T.J., Abrescia, L.D., Radolf, J.D. and Furie, M.B. (1996) Outer surface lipoproteins of *Borrelia burgdorferi* activate vascular endothelium *in vitro*. *Infection and Immunity* 64, 3170–3180.

Shih, C.M., Pollack, R.J., Telford, S.R. III and Spielman, A. (1992) Delayed dissemination of Lyme disease spirochetes from the site of deposition in the skin of mice. *Journal of Infectious Diseases* 166, 827–831.

Siebers, A., Zhong, W., Wallich, R. and Simon, M.M. (1999) Loss of pathogenic potential after cloning of the low-passage *Borrelia burgdorferi* ZS7 tick isolate, a cautionary note. *Medical Microbiology and Immunology* 188, 125–130.

de Silva, A.M. and Fikrig, E. (1997) *Borrelia burgdorferi* genes selectively expressed in ticks and mammals. *Parasitology Today* 13, 267–270.

de Silva, A.M., Zeidner, N.S., Zhang, Y., Dolan, M.C., Piesman, J. and Fikrig, E. (1999) Influence of outer surface protein

A antibody on *Borrelia burgdorferi* within feeding ticks. *Infection and Immunity* 67, 30–35.

Sim, R.B. and Dodds, A.W. (1997) The complement system: an introduction. In: Dodds, A.W. and Sim, R.B. (eds) *Complement: a Practical Approach*. IRL Press, Oxford, pp. 1–18.

Simon, M.M., Kramer, M.D., Wallich, R. and Schaible, U.E. (1993) Lyme arthritis: pathogenic principles emerging from studies in man and mouse. In: Panaysi, G.S. (ed.) *Immunology of Connective Tissue Diseases*. Kluwer Academic Publishers, Dordrecht, The Netherlands, pp. 205–229.

Simon, M.M., Nerz, G., Kramer, M.D., Hurtenbach, U., Schaible, U.E. and Wallich, R. (1995) The outer surface lipoprotein A of *Borrelia burgdorferi* provides direct and indirect augmenting/co-stimulatory signals for the activation of $CD4^+$ and $CD8^+$ T cells. *Immunology Letters* 45, 137–142.

Sonnensyn, S.W., Manivel, J.C., Johnson, R.C. and Goodman, J.L. (1993) A guinea pig model for Lyme disease. *Infection and Immunity* 61, 4777–4784.

Sprenger, H., Krause, A., Kaufmann, A., Priem, S., Fabian, D., Burmester, G.R., Gemsa, D. and Rittig, M.G. (1997) *Borrelia burgdorferi* induces chemokines in human monocytes. *Infection and Immunity* 65, 4384–4388.

Stevenson, B., Casjens, S. and Rosa, P. (1998) Evidence of past recombination events among genes encoding the Erp antigens of *Borrelia burgdorferi*. *Microbiology* 144, 1869–1879.

Sung, S.Y., Lavoie, C.P., Carlyon, J.A. and Marconi, R.T. (1998) Genetic divergence and evolutionary instability of *ospE*-related members of the upstream homology box gene family in *Borrelia burgdorferi* sensu lato complex isolates. *Infection and Immunity* 66, 4656–4668.

Telford, S.R. III, Mather, T.N., Moore, S.I., Wilson, M.L. and Spielman, A. (1988) Incompetence of deer as reservoirs of the Lyme disease spirochete. *American Journal Tropical Medicine and Hygiene* 39, 105–109.

Thomas, V., Anguita, J., Samanta, S., Rosa, P.A., Stewart, P., Barthold, S.W. and Fikrig, E. (2001) Dissociation of infectivity and pathogenicity in *Borrelia burgdorferi*. *Infection and Immunity* 69, 3507–3509.

Valenzuela, J.G., Charlab, R., Mather, T.N. and Ribeiro, J.M.C. (2000) Purification, cloning, and expression of a novel salivary anticomplement protein from the tick, *Ixodes scapularis*. *Journal of Biological Chemistry* 23, 18717–18723.

Wallich, R., Brenner, C., Kramer, M.D. and Simon, M.M. (1995) Molecular cloning and immunological characterization of a novel linear-plasmid-encoded gene, pG, of *Borrelia burgdorferi* expressed only *in vivo*. *Infection and Immunity* 63, 3327–3335.

Wang, G., van Dam, A.P. and Dankert, J. (1999) Evidence for frequent OspC gene transfer between *Borrelia valaisiana* sp. nov. and other Lyme disease spirochetes. *FEMS Microbiology Letters* 177, 289–296.

Zhang, J.R., Hardham, J.M., Barbour, A.G. and Norris, S.J. (1997) Antigenic variation in Lyme disease borreliae by promiscuous recombination of VMP-like sequence cassettes. *Cell* 89, 275–285.

Zhong, W., Stehle, T., Museteanu, C., Siebers, A., Gern, L., Kramer, M., Wallich, R. and Simon, M.M. (1997) Therapeutic passive vaccination against chronic Lyme disease in mice. *Proceedings of the National Academy of Sciences, USA* 94, 12533–12538.

Zhong, W., Gern, L., Stehle, T., Museteanu, C., Kramer, M., Wallich, R. and Simon, M.M. (1999) Resolution of experimental and tick-borne *Borrelia burgdorferi* infection in mice by passive, but not active immunization using recombinant OspC. *European Journal of Immunology* 29, 946–957.

Ecology of *Borrelia burgdorferi* sensu lato in Europe

Lise Gern and Pierre-François Humair

Institute of Zoology, University of Neuchâtel, Emile Argand 11, 2007 Neuchâtel, Switzerland

Introduction

Borrelia burgdorferi, the aetiologic agent of the tick-borne zoonosis Lyme borreliosis (LB), circulates between ticks and vertebrate hosts. However, as in the case of most other vector-borne zoonoses, the presence of the pathogen in nature was revealed through clinical manifestations in humans.

It took almost 100 years after the first description, in Europe, of clinical manifestations of the disease now known as LB for biologists to discover borreliae in *Ixodes scapularis* ticks from northeastern USA (Burgdorfer *et al.*, 1982). Thus, the pathogen and the vector of this disease were discovered simultaneously. Similar spirochaetes were observed in Europe shortly thereafter in a closely related tick species, *Ixodes ricinus* (Burgdorfer *et al.*, 1983). Finally, the spirochaetes were recognized as a new species, *B. burgdorferi* (Johnson *et al.*, 1984), and as the causative agent of the new clinical entity, Lyme disease or LB. It is now known that LB is the most prevalent tick-borne disease in the northern hemisphere.

As in other vector-borne diseases, understanding the ecological relationships that exist between pathogens, vectors and wildlife hosts in LB is essential to the understanding of the epidemiology of the disease in human populations. In addition to the diversity of vector and host species in the large geographical area occupied by *B. burgdorferi* worldwide, a great diversity of *Borrelia* species within the complex *B. burgdorferi* sensu lato (s.l.) has been discovered recently.

The transmission and maintenance cycles of *B. burgdorferi* s.l. in enzootic areas show a complicated picture with regional and local peculiarities, and our present view of the ecology of these spirochaetes in Europe is like a collection of temporal and local snapshots, the whole picture being still fragmentary. This chapter is a review of the current knowledge of *B. burgdorferi* s.l. ecology in Europe, but because of the relatively few publications from eastern Europe it focuses particularly on western and central regions. The maintenance of *B. burgdorferi* s.l. requires the interaction of three groups of organisms: the pathogens, vectors and hosts. These three essential components of LB will be described first, before attention is focused on their interrelationships.

The Spirochaetes

With the increasing number of isolates of *B. burgdorferi* obtained from various geographical and biological sources, it became obvious first in Europe and Asia and then also in North America that the spirochaete exhibited phenotypic and genotypic heterogeneity. This resulted in the description of 11 genomic groups or genospecies gathered under the name *Borrelia burgdorferi* s.l.: *B. burgdorferi* sensu stricto (s.s.) (Johnson *et al.*, 1984), *Borrelia garinii* (Baranton *et al.*, 1992), *Borrelia afzelii* (Canica *et al.*, 1993), *Borrelia japonica* (Kawabata *et al.*, 1993), *Borrelia andersonii* (Marconi *et al.*, 1995), *Borrelia tanukii* and *Borrelia turdi* (Fukunaga *et al.*, 1996), *Borrelia valaisiana* (Wang *et al.*, 1997), *Borrelia lusitaniae* (Le Fleche *et al.*, 1997), *Borrelia bissettii* (Postic *et al.*, 1998) and *Borrelia sinica* (Masuzawa *et al.*, 2001). In Europe, five species (*B. burgdorferi* s.s., *B. garinii*, *B. afzelii*, *B. valaisiana* and *B. lusitaniae*) have been isolated from ticks and *B. burgdorferi* s.l. has been reported from 26 countries from Italy to Iceland, and from Portugal to Russia (Hubálek and Halouzka, 1997). Two additional genospecies have been reported on the European continent from patient tissues: *B. bissettii*, a species present in North America, has been obtained from patients in Slovenia (Picken *et al.*, 1996; Strle *et al.*, 1997), and a novel *B. burgdorferi* s.l. genospecies has been cultured from an erythema migrans biopsy of a patient who contracted the disease in The Netherlands (Wang *et al.*, 1999). The European vector ticks and natural hosts of these two genospecies have not been determined as yet.

In addition to the classification of *B. burgdorferi* into 11 different species, an intraspecific heterogeneity has been documented with respect to outer surface protein (Osp) A, which is expressed by most *B. burgdorferi* s.l. spirochaetes (Will *et al.*, 1995). Wilske *et al.* (1993) developed a comprehensive serotyping system of *B. burgdorferi* s.l. by using immunoblotting and a variety of monoclonal antibodies against OspA. A strong correlation exists between OspA phenotypes and genospecific grouping of Lyme disease borreliae. This OspA serotyping was confirmed by sequence analysis of the *osp* genes (Jauris-Heipke *et al.*, 1995; Will *et al.*, 1995). These studies revealed that *B. garinii* was phenotypically the most heterogeneous genospecies with five different serotypes (serotypes 3–7).

Vectors and Carriers

In Europe, *B. burgdorferi* s.l. is mainly transmitted by *I. ricinus*. Two additional tick species, *Ixodes hexagonus* and *Ixodes uriae*, are also vectors and maintain

B. burgdorferi s.l. in secondary transmission cycles without necessarily involving *I. ricinus*.

The sheep tick or castor bean tick, *I. ricinus*, the principal vector of *B. burgdorferi* s.l., is the most common tick species in Europe. *I. ricinus* occurs from the Atlantic coast to as far east as 50–60° longitude in central Asia and from Iceland to as far south as the Atlas Mountains of North Africa (Hillyard, 1996). The eastern distribution of *I. ricinus* overlaps the western distribution of the related *Ixodes persulcatus* (Fig. 6.1).

The habitat of *I. ricinus* is situated in the leaf litter and the low strata vegetation of temperate deciduous woodlands and mixed forests. In areas with high rainfall, *I. ricinus* also occurs in high densities in coniferous forests and in open areas such as grassland (Gray *et al.*, 1998). *I. ricinus* has particular requirements regarding the humidity of its habitat and a relative humidity (RH) value of > 80% is important for the activity and survival of free-living *I. ricinus*. A low

Fig. 6.1. Distribution of *Ixodes ricinus* and *Ixodes persulcatus*, the major vectors of *Borrelia burgdorferi* s.l. in Europe. Note that, in southern areas and in Iceland and northern Scandinavia *I. ricinus* populations are sparse and scattered.

saturation deficit, which is calculated by using values of RH and temperature, can also be closely correlated with optimal conditions for *I. ricinus* (Randolph and Storey, 1999; Perret *et al.*, 2000).

Although studied for more than 50 years, knowledge of the biology of *I. ricinus* remains incomplete, partly because of the geographical and genetic heterogeneity of this tick species and also because of the diverse habitats it occupies. *I. ricinus* is an exophilic (or non-nidicolous) tick and the questing tick waits passively on the vegetation for passing hosts. A wide range of vertebrates such as reptiles, birds, small-, medium- and large-sized mammals serve as hosts for *I. ricinus* (Aeschlimann, 1972). Small mammals, ground-foraging birds and reptiles are common hosts for *I. ricinus* immatures but not for adults. Medium- to large-sized mammals are parasitized by *I. ricinus* adults (females) and immatures. Males occasionally take a blood meal but do not engorge. Immature stages and adult females can attach to humans. The life cycle of *I. ricinus* lasts 2–6 years, typically 3 years, since each stage takes about a year to develop to the next instar (Gray, 1991). The duration of the life cycle can vary from one habitat to another and also regionally, and can be affected by microclimatic factors and host density (Gray, 1991). *I. ricinus* ticks are active from spring to autumn, generally from February to November and seasonal activity of the three active stages usually shows a bimodal pattern with an activity peak in spring and another one in autumn. Tick activity in spring is usually greater than in autumn, except for larvae, which show the oppposite situation in some areas. Activity peaks can be indistinct and a pseudo-bimodal pattern may occur, as observed in Switzerland (Gigon, 1985; Perret *et al.*, 2000).

Besides the classical vector, *I. ricinus*, two other tick species, *I. hexagonus* and *I. uriae*, contribute to the circulation of *B. burgdorferi* s.l. in Europe. The hedgehog tick, *I. hexagonus*, is widespread in Europe and is also present in North Africa (Arthur, 1968). The presence of *B. burgdorferi* s.l. in *I. hexagonus* was first discovered in tick samples collected from hedgehogs (Liebisch *et al.*, 1989). The vector competence of *I. hexagonus* for *B. burgdorferi* s.l. has been demonstrated under both laboratory conditions (Gern *et al.*, 1991; Toutoungi and Gern, 1993) and field conditions (Gern *et al.*, 1997). In contrast to *I. ricinus*, *I. hexagonus* is an endophilic (or nidicolous) tick species and its habitat is the nest, burrow or cave of its vertebrate host. The host range of *I. hexagonus* is therefore more restricted than that of *I. ricinus* and it feeds primarily on carnivores such as foxes and mustelids, and on hedgehogs, but also, less frequently, on rodents, hares and rabbits (Arthur, 1953; Toutoungi *et al.*, 1991). *I. hexagonus* has occasionally been collected from birds (*Pica pica*, *Falco tinnunculus*) and deer (*Capreolus capreolus*) (Toutoungi *et al.*, 1991; Hubbard *et al.*, 1998). Domestic animals such as cats, dogs, horses, goats and cows have also been found to be infested (Arthur, 1968; Toutoungi *et al.*, 1991; Bernasconi *et al.*, 1997). *I. hexagonus* also bites humans but apparently less frequently than *I. ricinus* (Arthur, 1953; Liebisch and Liebisch, 1996; Liebisch *et al.*, 1998a). The occurrence of *I. hexagonus* in the urban environment is due to the presence of suitable hosts such as hedgehogs, cats and dogs in gardens and public parks (Gern *et al.*, 1991, 1997).

The seabird tick, *I. uriae*, has been found to be infected by *B. garinii* (Olsen et al., 1993; Nuttall et al., 1994; Hubbard et al., 1998) and is the third ixodid species implicated in the circulation of *B. burgdorferi* s.l. in Europe. *I. uriae* infests more than 50 seabird species living in high latitude areas in both hemispheres (Olsen, 1995). In Europe, *I. uriae* is present on the coasts of France, Great Britain, Ireland, Denmark, Norway, Sweden and Iceland (Olsen et al., 1995a; Hillyard, 1996). Apparently, *I. uriae* can infest other avian hosts such as passerines resting in marine habitats (Arthur, 1968). Mammals such as seals, river otters and humans can also occasionally be parasitized by *I. uriae* (Arthur, 1968; Eley, 1977; Mehl and Traavik, 1983; Hubbard et al., 1998). People engaged in ornithological activity, such as bird-ringing, or in exploitation of colonial seabirds, as in the Faeroe Islands, are likely to be exposed to *I. uriae* bites (Olsen et al., 1995a; Hubbard et al., 1998; Gylfe et al., 1999). The vector competence of *I. uriae* for *B. burgdorferi* s.l. has never been demonstrated under laboratory conditions, but its involvement as a vector of borrelial spirochaetes in transmission cycles in seabird colonies has been shown (Olsen et al., 1993, 1995a).

The presence of *B. burgdorferi* s.l. in vector arthropods results from blood meals taken from infectious vertebrate hosts. In competent vectors, borrelial infection persists in the tick throughout the moulting process to the next instar. This phenomenon, called transstadial transmission, or more appropriately transstadial maintenance, is fundamental to the long-term survival of *B. burgdorferi* s.l. in nature and its efficiency is one of the factors determining vector competence. Transstadial maintenance has been demonstrated in two European tick vectors, *I. ricinus* and *I. hexagonus* (Monin et al., 1989; Gern et al., 1991; Toutoungi and Gern, 1993; Bellet-Edimo, 1997), and apparently occurs also in *I. uriae*, since unfed *I. uriae* have been found to be infected (Hubbard et al., 1998) and since a cycle has been described between seabirds and *I. uriae* (Olsen et al., 1993).

B. burgdorferi s.l. is ingested during the tick blood meal, colonizes the tick midgut and usually remains there after the moult. In some cases, spirochaetes may migrate to other tick organs, causing generalized or systemic infections in unfed *I. ricinus* ticks (Lebet and Gern, 1994; Leuba-Garcia et al., 1994). In female ticks, *B. burgdorferi* s.l. may be present in ovaries and in the developing embryos (eggs). This mode of infection of the larval progeny through the ovarian tissue of the mother tick is called transovarial or vertical transmission. Experimental transovarial transmission of *B. burgdorferi* s.s. (isolate Sta14, now identified as NE14) in *I. ricinus* was successfully demonstrated (Monin et al., 1989). Studies on the efficiency of transovarial transmission showed that this phenomenon rarely occurs in *I. ricinus*, as only about 1% of field-collected females transmitted *B. burgdorferi* s.l. to *I. ricinus* larvae (Bellet-Edimo, 1997). However, when transovarial transmission does occur, the prevalence of infection in *I. ricinus* eggs or subsequent larvae is high (43–100%) (Burgdorfer et al., 1983; Bellet-Edimo, 1997). Transovarial transmission of *B. burgdorferi* s.s. (isolate B31) also occurs in *I. hexagonus* and has been shown experimentally (Gern et al., 1991). In contrast to *I. ricinus*, many more *I. hexagonus* females can transmit *B. burgdorferi* s.l. to their progeny, but infection rates within batches of infected progenies are lower than in *I. ricinus*. This discrepancy might be related to tick biological and behavioural features: in the case of the endophilic *I. hexagonus*, a high prevalence of infection in larvae is less important than in the case of the exophilic *I. ricinus*, because of closer contacts in a confined environment between endophilic ticks and hosts. However, the prevalence of inherited infection in field-derived *I. hexagonus* larvae remains unknown. At present, no data are available concerning the existence of transovarial transmission of *B. burgdorferi* s.l. in *I. uriae*.

B. burgdorferi s.l. is reportedly present in all *I. ricinus* populations examined so far in Europe (Hubálek and Halouzka, 1998). Consequently, the infection risk for humans seems to be present in any *I. ricinus* habitat. The infection prevalence of questing *I. ricinus* ticks in Europe averages 1.9% for larvae (range 0–11%), 10.8% for nymphs (range 2–43%) and 17.4% for adults (range 3–75%) (Hubálek and Halouzka, 1998; de Michelis *et al.*, 2000) and varies geographically and also according to the methods used for the detection of *B. burgdorferi* s.l. in unfed *I. ricinus* ticks. Cultivation in BSK medium seems to be the least sensitive method compared with microscopy (dark-field, phase-contrast or immunofluorescence) or PCR (Hubálek and Halouzka, 1998; Gern *et al.*, 1999).

The prevalence of *B. burgdorferi* s.l. in unfed *I. hexagonus* is unknown. Concerning *I. uriae*, the prevalence of *B. garinii* was reported by Olsen (1995), who detected borreliae in 1/100 unengorged ticks, and by Hubbard *et al.* (1998), who detected infection in 13/13 field-collected engorged or semi-engorged museum specimens. Given the biology of *I. hexagonus* and *I. uriae*, inspection of host nests is needed for the collection of additional data for unfed free-living ticks of these two species.

As already mentioned, five genospecies of *B. burgdorferi* s.l. (*B. burgdorferi* s.s., *B. garinii*, *B. afzelii*, *B. valaisiana* and *B. lusitaniae*) are present in European ixodid vectors. The three pathogenic species (*B. burgdorferi* s.s., *B. garinii* and *B. afzelii*) have been frequently isolated from or detected in *Ixodes* spp. ticks, whereas *B. valaisiana* and *B. lusitaniae* have rarely been reported. Since the discovery of *B. burgdorferi* in *I. ricinus* ticks, valuable data concerning the occurrence of *B. burgdorferi* s.l. genospecies have been collected throughout Europe. However, it is difficult to draw a precise map of the distribution and frequency of the different *B. burgdorferi* s.l. genospecies on the European continent until more data have been collected. Current knowledge shows that *B. garinii* is the most frequent genospecies found in ticks, followed by *B. afzelii* and *B. burgdorferi* s.s. (Hubálek and Halouzka, 1997; Saint Girons *et al.*, 1998). *B. garinii* and *B. afzelii* occur throughout the European continent from the Atlantic coasts to the Ural mountains. In contrast, the frequency of *B. burgdorferi* s.s. seems to decrease from west to east (Saint Girons *et al.*, 1998) and appears to be absent in *I. persulcatus*. *B. valaisiana* and *B. lusitaniae* appear to be far less common and have been reported from certain European countries only. *B. valaisiana* occurs in Switzerland (Péter and Bretz, 1992; Péter *et al.*, 1995; Humair *et al.*, 1998), The Netherlands (Rijpkema *et al.*, 1995), Germany (Liebisch *et al.*, 1998a), Croatia (Rijpkema *et al.*, 1996), Great Britain (Cutler *et al.*, 1989; Kurtenbach *et al.*, 1998b), Ireland (Kirstein *et al.*, 1997), Italy (Cinco *et al.*, 1998) and Slovakia (Gern *et al.*, 1999).

B. lusitaniae has been reported in Portugal (Nuncio *et al.*, 1993; de Michelis *et al.*, 2000), Spain (Escudero *et al.*, 2000), Slovakia (Gern *et al.*, 1999), Belorussia, the Czech Republic (Le Fleche *et al.*, 1997), Moldavia, Ukraine (Postic *et al.*, 1997) and interestingly in North Africa (Zhioua *et al.*, 1999; Younsi *et al.*, 2001). It is true that *B. valaisiana* and *B. lusitaniae* have been described only recently but this cannot explain their lower frequency compared with pathogenic *Borrelia* spp. Very few isolates of *B. valaisiana* and *B. lusitaniae* have been obtained to date, and cultivation in the usual BSK medium seems to be problematic for the isolation of these two genomic groups. However, PCR-based studies have demonstrated a high prevalence of *B. valaisiana* in unfed *I. ricinus* ticks in Ireland (Kirstein *et al.*, 1997) and of *B. lusitaniae* in ticks in Portugal (de Michelis *et al.*, 2000). The question remains whether the DNA detected reflects the presence of viable borreliae. Further studies are needed to determine if *B. valaisiana* and *B. lusitaniae* are indeed patchily distributed in Europe or if they have been relatively undetected previously.

Infections by multiple *B. burgdorferi* s.l. genospecies have been observed in ticks in many parts of Europe, including The Netherlands (Rijpkema *et al.*, 1995), Croatia (Rijpkema *et al.*, 1996), Switzerland (Leuba-Garcia *et al.*, 1994), France (Pichon *et al.*, 1995), Belgium (Misonne *et al.*, 1998), Estonia, Kirghizia, Moldavia, Russia and Ukraine (Postic *et al.*, 1997), Ireland (Kirstein *et al.*, 1997), Italy (Cinco *et al.*, 1998) and Germany (Liebisch *et al.*, 1998a; Hu *et al.*, 2001). Different combinations of mixed infection with two or three genospecies have been detected. *B. burgdorferi* s.s. and *B. lusitaniae* are the two *Borrelia* spp. least involved in mixed infections. In most cases, mixed infections in ticks have been detected using PCR methods. Detection of mixed infections using cultivation might be difficult because one genospecies may overgrow another.

Certain tick or insect species found to be infected, but without evidence of vector competence, are considered to be carriers (Table 6.1). Transstadial transmission appears to occur in some tick carrier species since unfed ticks have been found infected by borreliae. It should be noted that spirochaetes in these insect and tick carriers have not always been clearly characterized as *B. burgdorferi* s.l.

Vertebrate Hosts

B. burgdorferi s.l. is maintained in natural cycles involving not only vectors but also wild vertebrates. The vertebrate hosts that are required to maintain tick populations may also act as reservoir hosts for tick-transmitted pathogens. Reservoir hosts as defined by Kahl *et al.* (Chapter 2) have mainly been identified in Europe by xenodiagnosis and by comparison of infection rates in questing ticks with those in ticks removed from hosts.

These reservoir hosts are among the large spectrum of different host species of the three competent tick vectors, *I. ricinus*, *I. hexagonus* and *I. uriae*. *I. ricinus* feeds on the largest variety of hosts on the European continent, infesting more than 300 different species including mammals, birds and reptiles (Anderson,

Table 6.1. Arthropod carrier species of *Borrelia burgdorferi* sensu lato in Europe.

Species	Countries	References
ACARI		
Ixodidae		
Ixodes trianguliceps	France	Doby *et al.* (1990)
	Russia	Postic *et al.* (1997)
	UK	Hubbard *et al.* (1998)
Ixodes acuminatus	France	Doby *et al.* (1990)
Ixodes canisuga	France	Doby *et al.* (1991)
	Germany	Liebisch *et al.* (1998b)
	Spain	Estrada-Peña *et al.* (1995)
Ixodes frontalis	France	Doby *et al.* (1995)
	Spain	Estrada-Peña *et al.* (1995)
Dermacentor reticulatus	Germany	Kahl *et al.* (1992)
	France	Doby *et al.* (1994)
	UK	Hubbard *et al.* (1998)
	Germany	Liebisch *et al.* (1998b)
Haemaphysalis punctata	Spain	Marquez and Constan (1990)
	UK	Nuttall *et al.* (1994)
	Sweden	Tälleklint (1996)
	UK	Hubbard *et al.* (1998)
Haemaphysalis inermis	France	Macaigne and Perez-Eid (1991)
Haemaphysalis concinna	France	Doby *et al.* (1994)
	Czech Republic	Hubálek *et al.* (1998a)
Haemaphysalis marginatum	Portugal	de Michelis *et al.* (2000)
Rhipicephalus sanguineus	UK	Hubbard *et al.* (1998)
Argasidae		
Argas reflexus	Italy	Stanek and Simeoni (1989)
Argas vespertilionis	UK	Hubbard *et al.* (1998)
INSECTA		
Siphonaptera (fleas)		
Ctenophtalmus baeticus arvernus	France	Doby *et al.* (1990)
Ctenophtalmus agyrtes	Czech Republic	Hubálek *et al.* (1998a)
Ctenophtalmus solutus	Slovakia	cited in Hubálek *et al.* (1998a)
Megabothris turbidus	France	Doby *et al.* (1990)
	Slovakia	cited in Hubálek *et al.* (1998a)
Spilopsyllus cuniculi	France	Doby *et al.* (1991)
Anoplura (sucking lice)		
Haematopinus suis	France	Doby *et al.* (1994)

Table 6.1. Continued

Species	Countries	References
Diptera		
Aedes vexans	Czech Republic	Halouzka (1993)
	Czech Republic	Halouzka *et al.* (1998)
	Czech Republic	Hubálek *et al.* (1998a)
Aedes cantans	Czech Republic	Halouzka *et al.* (1998)
	Czech Republic	Hubálek *et al.* (1998a)
Aedes sticticus	Czech Republic	Halouzka *et al.* (1998)
	Czech Republic	Hubálek *et al.* (1998a)
Culex pipiens pipiens	Czech Republic	Halouzka *et al.* (1998)
	Czech Republic	Hubálek *et al.* (1998a)
Culex pipiens molestus	Czech Republic	Halouzka (1993)
	Czech Republic	Halouzka *et al.* (1998)
	Czech Republic	Halouzka *et al.* (1999)
	Czech Republic	Hubálek *et al.* (1998a)
Lipoptena cervi	France	Doby *et al.* (1994)

1991). Consequently, each of these hosts could potentially be a reservoir of *B. burgdorferi* s.l. and therefore the reservoir status of each host species must be individually evaluated.

Of about 300 vertebrate species that are hosts for *I. ricinus*, only a small number have been examined so far for their borrelial infectivity to ticks. Small mammals are certainly the vertebrate group that has been the most extensively investigated up to now, mainly because they can be easily captured, handled and maintained in the laboratory. Several species of mice, voles, rats and shrews have been shown to be competent reservoirs of *B. burgdorferi* s.l. in Europe (Gern *et al.*, 1998). In particular, strong evidence for reservoir competence has been obtained in many European countries for the mice, *Apodemus flavicollis*, *Apodemus sylvaticus*, *Apodemus agrarius* and for the vole, *Clethrionomys glareolus*. These rodents transmit *B. burgdorferi* s.l. to a large number of *I. ricinus* larvae (Aeschlimann *et al.*, 1986; Matuschka *et al.*, 1992; de Boer *et al.*, 1993; Humair *et al.*, 1993a, 1999; Gern *et al.*, 1994a; Kurtenbach *et al.*, 1994a, 1995, 1998b; Tälleklint and Jaenson, 1994; Randolph and Craine, 1995; Hu *et al.*, 1997; Richter *et al.*, 1999).

More limited information has been obtained for other small- and medium-sized mammals. An additional species of vole, *Microtus agrestis*, has been described as a reservoir in Sweden by Tälleklint and Jaenson (1994), and black rats (*Rattus rattus*) and Norway rats (*Rattus norvegicus*) contribute to the transmission of *B. burgdorferi* s.l. in urbanized environments in continental Europe (Matuschka *et al.*, 1996, 1997). In an urban park in Germany, for example, *R. norvegicus* rats were less abundant than *A. flavicollis*, but showed a higher infestation by *I. ricinus* subadults and a greater infectivity for ticks (Matuschka *et al.*, 1996). Black rats and Norway rats also perpetuate LB spirochaetes in Madeira, a subtropical island where native terrestrial mammals are absent (Matuschka *et al.*, 1994b).

Edible dormice (*Glis glis*) (Matuschka *et al.*, 1994a) and garden dormice (*Eliomys quercinus*) (Matuschka *et al.*, 1999) have been shown to be reservoir hosts in Germany and in France, respectively. Edible and garden dormice were more heavily infested by larval and nymphal *I. ricinus* than were mice and voles in the studied habitats. This suggests that in certain habitats where dormice are abundant, these rodent species could be the primary reservoirs (Matuschka *et al.*, 1994b, 1999). Other rodents like grey squirrels (*Sciurus carolinensis*) in England (Craine *et al.*, 1997) and red squirrels (*Sciurus vulgaris*) in Switzerland (Humair and Gern, 1998) also contribute to the amplification of the infection in vector ticks. Grey squirrels, which were introduced into Great Britain from North America, were shown to be more important hosts than small rodents for *I. ricinus* larvae and nymphs (Craine *et al.*, 1995). Moreover, the abundance of grey squirrels makes this species an important source of infection for ticks wherever it is present. Similarly, in continental Europe red squirrels have also been found to be heavily infested with ticks (Matuschka *et al.*, 1996; Humair and Gern, 1998) and feeding ticks showed a high prevalence of infection (69%) (Humair and Gern, 1998). Although red squirrel density has not been evaluated, it is evident that they are potent reservoir hosts and that they contribute to the maintenance of *B. burgdorferi* s.l. in enzootic areas (Humair and Gern, 1998).

Among insectivores, the reservoir competence of three species of shrews (*Neomys fodiens*, *Sorex minutus* and *Sorex araneus*) has been demonstrated in Sweden (Tälleklint and Jaenson, 1994). Evidence of reservoir competence of hedgehogs has been obtained from Ireland (Gray *et al.*, 1994), Germany (Liebisch *et al.*, 1996) and Switzerland (Gern *et al.*, 1997). In Switzerland, an enzootic transmission cycle has been described in an urban environment involving hedgehogs and *I. hexagonus* in the absence of *I. ricinus* (Gern *et al.*, 1997).

Lagomorphs such as the brown hare (*Lepus europaeus*) and the varying hare (*Lepus timidus*) also contribute to the maintenance of *B. burgdorferi* s.l. in nature, as observed in Sweden in habitats where hares coexist with small mammals (Tälleklint and Jaenson, 1993, 1994), and on islands where hares are the only terrestrial mammal species permanently present (Jaenson and Tälleklint, 1996). Although the European rabbit (*Oryctolagus cuniculus*) can apparently function as a reservoir host for *B. afzelii* (Matuschka *et al.*, 2000), the reservoir capacity of this species seems limited since only one out of seven rabbits (14%) was infective to ticks.

Among large-sized mammals, two studies in Germany (Kahl and Geue, 1998; Liebisch *et al.*, 1998b) have implicated red foxes as reservoirs, but these animals did not appear to be very potent reservoirs since the borreliae were apparently poorly transmitted to ticks.

Birds in Europe were initially identified as non-reservoirs (Matuschka and Spielman, 1992). They were described not only as incompetent reservoirs, but also as having a zooprophylactic role. Zooprophylactic hosts are considered as dead-end hosts for *B. burgdorferi* s.l. because they eliminate borrelial infection in feeding ticks, which interrupts the maintenance cycle of the spirochaete. Although the involvement of birds has remained controversial for a long time, their role as reservoir hosts is now clearly established. Infected ticks have been collected from various bird species in Sweden (Olsen *et al.*, 1993, 1995a,b), Switzerland (Humair *et al.*, 1993b, 1998), England (Craine *et al.*, 1997; Kurtenbach *et al.*, 1998b) and the Czech Republic (Hubálek *et al.*, 1996). Ground-foraging birds such as thrushes (*Turdus* spp.), blackbirds (*Turdus merula*), robins (*Erithacus rubecula*) and pheasants (*Phasianus colchicus*) were particularly involved, as well as colonial seabirds.

B. *burgdorferi* s.l. was isolated from *I. uriae* ticks fed on seabirds (Olsen *et al.*, 1993, 1995a; Gylfe *et al.*, 1999), from *I. ricinus* ticks fed on blackbirds (Hubálek *et al.*, 1996; Humair *et al.*, 1998), from the skin of blackbirds (*T. merula*) (Humair *et al.*, 1998) and from the blood of puffins (*Fratercula artica*) (Gylfe *et al.*, 1999), suggesting that live spirochaetes can survive in birds and in bird-feeding ticks. Strong evidence of reservoir competence was eventually obtained for two bird species: blackbird (*T. merula*) (Humair *et al.*, 1998) and pheasant (*P. colchicus*) (Kurtenbach *et al.*, 1998a) were investigated by tick xenodiagnosis, which clearly demonstrated the efficient transmission of borrelial infection from birds to ticks. B. *burgdorferi* s.l. also circulates between seabirds such as razorbills (*Alca torda*) and puffins (*F. artica*) and the seabird tick *I. uriae* in isolated environments (Olsen *et al.*, 1993, 1995a; Gylfe *et al.*, 1999). Borrelia-infected *I. uriae* ticks have been observed in seabird colonies in the northern and southern hemispheres (Olsen *et al.*, 1995a). The similarity of DNA sequences of B. *burgdorferi* s.l. from *I. uriae* of various locations strongly suggests that seabirds transfer borrelial spirochaetes between seabird colonies (Olsen *et al.*, 1995a). Reactivation of latent borrelial infection has been observed with passerines experimentally submitted to stressful conditions that simulate migration (progressive decrease of photoperiod) (Gylfe *et al.*, 2000). This suggests that migrating birds could spread B. *burgdorferi* s.l. spirochaetes to ticks encountered along the migration route.

The determination of the reservoir competence of a host species is important but should also include quantification of the contribution of a particular host species in a particular habitat. The immune status of the host may have an impact on the circulation of B. *burgdorferi* s.l. in nature by affecting the ability of ticks to acquire infection from reservoir hosts. *Apodemus* spp. mice and *Clethrionomys* spp. voles are a good illustration of this phenomenon. These host species seem to have developed different responses towards tick infestation and borrelial infection, and this influences their respective reservoir importance. Kurtenbach *et al.* (1994b) showed in a laboratory study that *C. glareolus* voles express a low immunity to spirochaetes and as a consequence develop high levels of infection and show high transmission rates to ticks. In contrast, *Apodemus* spp. mice control spirochaete infection more effectively by a specific immune response that maintains their borrelial infection at a low level and also at a low transmission rate to ticks (Kurtenbach *et al.*, 1994b). In parallel, Dizij and Kurtenbach (1995) demonstrated experimentally that voles, but not mice, acquire resistance to *I. ricinus* ticks with consecutive infestations. Interestingly, some of the phenomena demonstrated by these laboratory results have been confirmed under natural conditions. Spirochaetes are more easily isolated from ear biopsies of *Clethrionomys* spp. voles than of *Apodemus* spp. mice (Petney *et al.*, 1996; Humair *et al.*, 1999). This high level of *B. burgdorferi* infection in voles also allows a higher transmission rate of spirochaetes from voles to the ticks feeding on them (Humair *et al.*, 1999). However, voles progressively acquire resistance to ticks after several tick infestations. Ticks that feed on these resistant voles suffer a retarded development due to a reduction of their engorgement weights as well as an increased mortality since they do not moult successfully. The low moulting success of ticks fed on voles lessens the relative contribution of voles in endemic areas, despite their high infection rate and their high infectivity for ticks (Humair *et al.*, 1999).

Reservoir potential (as defined by Mather *et al.*, 1989) or coefficient of transmission (Randolph and Craine, 1995) is an attempt to quantify the contribution as a reservoir of a particular host species in a particular habit. The reservoir potential is influenced by three main factors: (i) host infectivity to ticks; (ii) degree of tick–host contact (mean infestation rate by larval ticks); and (iii) host density in the habitat (Mather *et al.*, 1989). The higher the values of these three parameters, the higher the reservoir potential of the host and the more important its contribution as a reservoir in the studied habitat. The evaluation of reservoir potential is not an easy task since it requires the assessment of various factors (Tälleklint and Jaenson, 1994), but this multifaceted approach should be encouraged in the future.

The determination of a vertebrate species as a reservoir is obviously important in understanding the ecology of LB. The identification of a host as a non-reservoir or as a zooprophylactic host is also important since the circulation of *B. burgdorferi* s.l. in nature is impaired by such hosts. The identification of a non-reservoir is in fact much more difficult than that of a reservoir host, since repeated investigations are required. Until now, only a few vertebrate species have been described as non-reservoirs or zooprophylactic hosts in Europe.

The lizard (*Lacerta viridis*) was reported to have a zooprophylactic effect on the island of Madeira since spirochaetes were destroyed in the midgut of ticks feeding on this host (Matuschka *et al.*, 1994b). Several other vertebrates such as blackbirds (Matuschka and Spielman, 1992) and some ungulates (roe deer, *C. capreolus*; red deer, *Cervus elaphus*; sika deer, *Cervus nippon* (cited by Gern *et al.*, 1998); moose, *Alces alces*; fallow deer, *Dama dama*; cattle, *Bos taurus*; sheep, *Ovis aries*) (Gray *et al.*, 1992, 1995; Jaenson and Tälleklint, 1992; Tälleklint and Jaenson, 1994) have been reported as zooprophylactic or non-reservoir hosts. However, further studies have shown that some of these host species are in fact reservoir-competent for some genospecies of *B. burgdorferi*, for example blackbirds (Humair *et al.*, 1998), and others, for example sheep (Ogden *et al.*, 1997) and sika deer (Kimura *et al.*, 1995), may also contribute to the circulation of *B. burgdorferi* s.l. in nature.

Various factors may lead to the conclusion that a host is non-reservoir or is zooprophylactic. For example, most methods for the determination of the reservoir competence of a host are based on detection of systemic infections. However, co-feeding transmission, in which uninfected engorging ticks may acquire borrelia infections from localized sites where infected ticks feed simultaneously (Gern and Rais, 1996; Randolph *et al.*, 1996), has been reported in sheep in Britain (Ogden *et al.*, 1997) and sika deer in Japan (Kimura *et al.*, 1995). Even if some ungulates do not sustain systemic borrelia infection, those in which co-feeding transmission occurs can be considered as amplifying hosts. Quantification of co-feeding transmission in ungulates, particularly in deer, and its importance in maintaining disease foci remain to be evaluated.

Another factor complicating the determination of reservoir competence of a vertebrate host is the possibility of host specificity of different *Borrelia* genospecies. This aspect has been neglected for a long time and even now knowledge of the relationships between the different *B. burgdorferi* s.l. species and their hosts is poorly documented. Indications of this phenomenon in Europe were first provided by the research of Humair *et al.* (1995) in Switzerland, who showed that borrelial isolates from 14 ear biopsies taken from ten *C. glareolus* voles and one *A. flavicollis* mouse (three isolates were obtained from two recaptured voles) captured in two endemic areas belonged exclusively to *B. afzelii*, although *B. burgdorferi* s.s. and *B. garinii* were present in questing ticks in the same areas. Interestingly, the first borrelial isolate obtained from a rodent in Sweden (Hovmark *et al.*, 1988) was characterized as *B. afzelii* (Postic *et al.*, 1994). More recently, *B. afzelii* was once again isolated from ear biopsies of mice and voles collected in a third study site in Switzerland, suggesting that a specific association occurs between *B. afzelii* and small rodents (Humair *et al.*, 1999). The use of tick xenodiagnosis on mice and voles showed that *B. afzelii* is clearly the dominant genospecies transmitted from small rodents to *I. ricinus* ticks in two enzootic areas in Switzerland (Hu *et al.*, 1997; Humair *et al.*, 1999). Out of 124 positive xenodiagnostic ticks, 120 ticks (97%) were infected by *B. afzelii*, two ticks contained *B burgdorferi* s.s. (one with a co-infection with *B. afzelii*) and two ticks contained uncharacterized spirochaetes. The isolation of *B. afzelii* from xenodiagnostic ticks demonstrates the viability of spirochaetes discovered in rodent-feeding ticks, which is a crucial point in the transmission of *B. burgdorferi* s.l. from rodents. All these results strongly suggest that an association exists between *Apodemus* spp. mice, *Clethrionomys* spp. voles and *B. afzelii*, since this genospecies is almost exclusively transmitted to *I. ricinus* larvae from these animals. Dolan *et al.* (1998) obtained similar results with *I. ricinus* ticks and laboratory mice in a Centres for Disease Control and Prevention (CDC) laboratory, where *B. afzelii* was isolated in all ear biopsy cultures from outbred laboratory mice that had been infested with *I. ricinus* ticks collected from the field near Neuchâtel (Switzerland). Additionally, xenodiagnostic *I. ricinus* larvae fed on outbred mice inoculated with other strains of *B. burgdorferi* s.l. (*B. afzelii* strain PGau.C3, *B. garinii* strains VS286 and VSBP, *B. burgdorferi* s.s. strains B-31 and B-31.D1) acquired all three *Borrelia* genospecies. However, the prevalence of *B. afzelii* infection in ticks was particularly high (90%) compared with low infection prevalences (3–5%) with *B. garinii* and *B. burgdorferi* s.s. Further, only *B. afzelii* survived the transstadial maintenance to nymphs (Dolan *et al.*, 1998). Additional observations were obtained by Hubálek *et al.* (1998a), who isolated *B. afzelii* from two *A. flavicollis*, one *C. glareolus* and two fleas (*Ctenophthalmus agyrtes*) fed on *C. glareolus* in the Czech Republic.

In parallel with the observations made by Humair *et al.* (1995) on *B. afzelii* and small rodents, Olsen *et al.* (1995b) described the predominant presence of *B. garinii* in ticks feeding on migrating birds collected in Sweden and Denmark: *B. garinii* DNA was found in 70% of *I. ricinus* larvae (23/33), *B. burgdorferi* s.s. DNA in 20% (7/33) and *B. afzelii* DNA in 9% (3/33). *B. garinii* was detected in all *I. ricinus* larvae ($n = 11$) collected from autumn-migrating birds and was also most prevalent in *I. ricinus* larvae collected from spring-migrating birds. In addition to migrating passerines, the existence of an association between birds and *B. garinii* was clearly illustrated by Olsen *et al.* (1995a) in the case of seabirds and the vector *I. uriae*. In the isolated maintenance cycle involving seabirds and the seabird tick (Olsen *et al.*, 1993), *B. garinii* was isolated and characterized from *I. uriae* ticks and more recently from seabird tissues (Olsen *et al.*, 1995a; Gylfe *et al.*, 1999).

In pheasants, more than 50% of *I. ricinus* ticks were infected by borreliae, particularly *B. garinii* (27%) and *B. valaisiana* (16%) (Kurtenbach *et al.*, 1998b). Hubálek *et al.* (1996) in the Czech Republic obtained a *B. garinii* isolate from a nymphal *I. ricinus* fed on *T. merula*. Shortly afterwards, *B. garinii* and *B. valaisiana* infections (seven isolates and four DNA detections) were revealed in five *T. merula* and one *Turdus philomelos* (Humair *et al.*, 1998). No infection with *Borrelia* genospecies other than *B. garinii* and *B. valaisiana* has been observed in European *Turdus* spp.

Characterization of borreliae from xenodiagnostic ticks fed on blackbirds in Switzerland demonstrated that *B. valaisiana* was most prevalent in these ticks (30%) when compared with *B. garinii* (3%) and *B. afzelii* (3%). Co-infections with *B. valaisiana* and *B. garinii* or *B. afzelii* were observed in 4% of the ticks. Interestingly, only *B. valaisiana* and *B. garinii* were isolated successfully from these xenodiagnostic ticks, whereas the presence of *B. afzelii* was revealed by DNA detection only (Humair *et al.*, 1998). This shows that blackbirds transmitted viable *B. valaisiana* and *B. garinii* to ticks feeding on them whereas the viability of *B. afzelii* spirochaetes in bird-feeding ticks remains uncertain.

More recently, Gylfe *et al.* (2000) demonstrated that *B. garinii* latent infection can be reactivated in experimentally infected redwing thrushes (*Turdus iliacus*) placed under stressful conditions simulating migration and causing migratory restlessness. This suggests that *Turdus* spp. passerines can sustain *B. garinii* infection for a long time and that migrating birds placed under conditions of autumnal migration transmit *B. garinii* to ticks. This corroborates the previous observation of *B. garinii* in all examined *I. ricinus* ticks from autumnal migrant birds (Olsen *et al.*, 1995b). Birds have a high body temperature, and Hubálek *et al.* (1998b) reported that the optimal growth temperature of *B. garinii* is higher than that of the other two pathogenic species, *B. afzelii* and *B. burgdorferi* s.s. Overall, these data strongly suggest that an association exists between birds and *B. garinii* and also *B. valaisiana*, a more recently described genospecies. Interestingly, an association between *B. garinii* and birds has been observed in Japan (Nakao *et al.*, 1994; Miyamoto *et al.*, 1997; Ishiguro *et al.*, 2000). However, in areas outside the geographical distribution of *B. garinii*, such as North America, *B. burgdorferi* s.s. is the *Borrelia* sp. circulating between birds and ticks (Richter *et al.*, 2000).

In addition to the associations of small rodents with *B. afzelii*, and birds with *B. garinii* and *B. valaisiana*, another association has been described in Switzerland and in the UK. As previously mentioned, grey and red squirrels (*S. carolinensis* and *S. vulgaris*) are reservoirs for *B. burgdorferi* s.l. In Switzerland, *B. burgdorferi* s.s. and *B. afzelii* were the only two genospecies isolated from skin samples of red squirrels (*S. vulgaris*) and were the two most prevalent genospecies found in *I. ricinus* ticks engorging on red squirrels, representing 96% of the isolates (Humair and Gern, 1998). The high frequency of *B. burgdorferi* s.s. and *B. afzelii* in the skin of squirrels as well as in squirrel-feeding *I. ricinus* suggests that both genospecies are preferentially transmitted from squirrels to ticks. Similar findings were obtained by Craine *et al.* (1997) in the UK: *B. afzelii* was identified in *I. ricinus* nymphs fed on a naturally infected grey squirrel and another grey squirrel sustained an experimental infection with *B. burgdorferi* s.s.

However, other borrelia–host associations that may depend on the local ecosystems have been described. In some parts of Europe, in addition to *B. afzelii*, *B. garinii* was also observed to be associated with rodents. Thus, data from Austria, Germany and Russia showed that different genospecies can be isolated from or DNA detected (by PCR) in the urinary bladder, heart and spleen of small mammals or in xenodiagnostic ticks fed on rats and yellow-necked mice (Khanakah *et al.*, 1994; Gorelova *et al.*, 1995; Richter *et al.*, 1999). Differences in biological and ecological factors occurring in some parts of Europe, such as the presence of different tick and rodent species or of different *B. garinii* subtypes, could explain the observed discrepancy. For example, *B. garinii* serotype 4, which has only been described in the cerebrospinal fluid of patients in Germany, The Netherlands, Denmark and Slovenia (Wilske *et al.*, 1993, 1996; van Dam *et al.*, 1997), is apparently rare in nature and has so far only been found in *I. ricinus* ticks collected in Germany (Eiffert *et al.*, 1995; Hu *et al.*, 2001). The difficulty in obtaining infected ticks from laboratory mice infected with *B. garinii* (usually associated with birds) has been documented previously (Gern *et al.*, 1994b; Dolan *et al.*, 1998). However, it was shown recently that laboratory mice infected with the serotype 4 of *B. garinii* easily transmitted this serotype to ticks (Hu *et al.*, 2001), which could be explained by the resistance of this serotype to complement (van Dam *et al.*, 1997). Therefore, one explanation for the observation of *B. garinii* in rodents in Austria, Germany and Russia may be the presence of *B. garinii* serotype 4 or related serotypes in these areas. In Japan, a similar association has been described between rodents and *B. garinii* ribotype IV, since this particular ribotype was predominant in rodents (*Apodemus speciosus* and *Clethrionomys rutilus*) and rodent-feeding *I. persulcatus* ticks (Nakao *et al.*, 1994). These observations show how diverse the ecology of *B. burgdorferi* s.l. can be.

Although some of the reservoirs of *B. afzelii*, *B. burgdorferi* s.s., *B. garinii* and *B. valaisiana* have been documented, those of *B. lusitaniae* still remain to be identified. De Michelis *et al.* (2000) suggested that avian and rodent species might be reservoir hosts for *B. lusitaniae* because DNA of this genospecies has been detected in adults of *Hyalomma marginatum*, a tick species that mainly feeds on birds and rodents as subadults (Hillyard, 1996). Other borrelia types identified in European patients by Picken *et al.* (1996) and Wang *et al.* (1999) have not been described in either ticks or wild animals.

Specific associations between reservoirs and *Borrelia* spp. spirochaetes imply that in certain cases *B. burgdorferi* s.l. genospecies are also transmitted to inadequate hosts. The specific association between birds and *B. garinii/B. valaisiana* in Europe may explain why the reservoir role of birds has remained controversial for a long time and why they were first described as zooprophylactic hosts. The first European studies on birds were performed before the description of different genospecies within the *B. burgdorferi* s.l. species complex. The existence of specific associations explains why blackbirds (*T. merula*) were first described as incompetent reservoirs and even as zooprophylactic hosts for LB spirochaetes by Matuschka and Spielman (1992). In this experiment, blackbirds were exposed to infected *I. ricinus* nymphs that acquired infection after feeding as larvae on black-striped mice (*A. agrarius*). Thus, considering the association of small rodents and *B. afzelii*, it is most probable that the *I. ricinus* nymphs used to infect blackbirds carried mouse-adapted *Borrelia* genospecies. Consequently, blackbirds were effectively inadequate hosts for the *Borrelia* spp. to which they had been exposed, because they are incompetent reservoirs and zooprophylactic hosts for rodent-adapted genospecies.

The existence of specific spirochaetal transmission patterns, of zooprophylactic and of non-reservoir hosts can be explained by the specific borreliacidal effect of host complement (Kurtenbach *et al.*, 1998c). This effect may prevent infection of the vertebrate host by some *Borrelia* species and may result in loss of spirochaetes by ticks that feed on inadequate hosts. The pattern of serum sensitivity of different *Borrelia* genospecies matches the known reservoir status of many vertebrate species for *B. burgdorferi* s.l. For example, *B. afzelii* was resistant to rodent sera whereas a strain of *B. garinii* was readily lysed, and the absence of systemic infection in deer correlates with the borreliacidal activity of deer sera against all the genospecies tested (Kurtenbach *et al.*, 1998c).

Conclusions

The ecological picture of LB is particularly complex because of the existence of diversity within the three main components of this zoonosis: pathogens, vectors and reservoirs. At present in Europe, five *B. burgdorferi* s.l. genospecies, three vector species and about 35 reservoir host species are linked in a web of relationships between each other. By focusing on this web, it can be seen that the links are organized, since each element of the web has privileged relationships with some others. For example, the seabird tick, *I. uriae*, is only infected by *B. garinii*, whereas *I. ricinus* may be infected by any of the five genospecies present in Europe. Similarly, *I. uriae* exclusively infests seabirds, whereas *I. ricinus* infests a large range of mammals and birds. It is obvious from this example that *B. garinii* is associated with seabirds, whereas similar associations between hosts of *I. ricinus* and *B. burgdorferi* s.l. genospecies are less obvious.

Although *I. ricinus* infection by *B. burgdorferi* s.l. genospecies and host infestation by ticks are generally well documented, the relationships between *I. ricinus* hosts and borreliae needs clarification in many respects. Additionally, the web varies geographically as well as temporally.

A huge amount of knowledge has been collected on the ecology of *B. burgdorferi* s.l. in Europe but further studies are needed for a better understanding of the situation. Emphasis has to be put on the distribution of the various *Borrelia* species and subtypes throughout Europe, the identification of additional reservoir hosts, their relationships with the various *Borrelia* species and the immune response of hosts towards ticks and borreliae. Because of regional and local pecularities, these studies should be performed in as many habitats as possible. The collection of these local snapshots will help towards a better picture of the ecology of *B. burgdorferi* s.l. in Europe.

References

Aeschlimann, A. (1972) *Ixodes ricinus*, Linné, 1758 (Ixodoidea: Ixodidae). Essai préliminaire de synthèse sur la biologie de cette espèce en Suisse. *Acta Tropica* 29, 321–340.

Aeschlimann, A., Chamot, E., Gigon, F., Jeanneret, J.P., Kesseler, D. and Walther, C. (1986) *B. burgdorferi* in Switzerland. *Zentralblatt für Bakteriologie, Mikrobiologie und Hygiene A* 263, 450–458.

Anderson, J.F. (1991) Epizootiology of Lyme borreliosis. *Scandinavian Journal of Infectious Diseases Suppl.* 77, 23–34.

Arthur D.R. (1953) The host relationships of *Ixodes hexagonus* Leach in Britain. *Parasitology* 43, 227–238.

Arthur, D.R. (1968) *British Ticks*. Butterworths, London, 213 pp.

Baranton, G., Postic, D., Saint Girons, I., Boerlin, P., Piffaretti, J.C., Assous, M. and Grimont, P.A.D. (1992) Delineation of *Borrelia burgdorferi* sensu stricto, *Borrelia garinii* sp. nov., and group VS461 associated with Lyme borreliosis. *International Journal of Systematic Bacteriology* 42, 378–383.

Bellet-Edimo, O.R. (1997) Importance de la transmission transstadiale et de la transmission transovarienne du spirochète *Borrelia burgdorferi* (Spirochaetales: Spirochaetaceae) chez la tique *Ixodes ricinus* (Acari: Ixodidae) dans l'épidémiologie de la borréliose de Lyme. PhD thesis, University of Neuchâtel, Neuchâtel, Switzerland.

Bernasconi, M.V., Valsangiacomo, C., Balmelli, T., Péter, O. and Piffaretti, J.C. (1997) Tick zoonoses in the southern part of Switzerland (Canton Ticino): occurrence of *Borrelia burgdorferi* sensu lato and *Rickettsia* sp. *European Journal of Epidemiology* 13, 209–215.

de Boer, R., Hovius, K.E., Nohlmans, M.K.E. and Gray, J.S. (1993) The woodmouse (*Apodemus sylvaticus*) as a reservoir of tick-transmitted spirochetes (*Borrelia burgdorferi*) in the Netherlands. *Zentralblatt für Bakteriologie* 279, 404–416.

Burgdorfer, W., Barbour, A.G., Hayes, S.F., Benach, J.L., Grunwaldt, E. and Davies, J.P. (1982) Lyme disease – a tick-borne spirochetosis? *Science* 220, 321–322.

Burgdorfer, W., Barbour, A.G., Hayes, S.F., Péter, O. and Aeschlimann, A. (1983) Erythema chronicum migrans – a tick-borne spirochetosis. *Acta Tropica* 40, 79–83.

Canica, M.M., Nato, F., du Merle, L., Mazie, J.C., Baranton, G. and Postic,

D. (1993) Monoclonal antibodies for identification of *Borrelia afzelii* sp. nov. associated with late cutaneous manifestations of Lyme borreliosis. *Scandinavian Journal of Infectious Diseases* 25, 441–448.

Cinco, M., Padovan, D., Murgia, R., Poldini, L., Frusteri, L., van de Pol, I., Verbeek-De Kruif, N., Rijpkema, S.G. and Marioli, M. (1998) Rate of infection of *Ixodes ricinus* ticks with *Borrelia burgdorferi* sensu stricto, *Borrelia garinii*, *Borrelia afzelii* and group VS116 in an endemic focus of Lyme disease in Italy. *European Journal of Clinical Microbiology and Infectious Diseases* 17, 90–94.

Craine, N.G., Randolph, S.E. and Nuttall P.A. (1995) Seasonal variation in the role of grey squirrels as hosts of *Ixodes ricinus*, the tick vector of the Lyme disease spirochaete, in a British woodland. *Folia Parasitologica* 42, 73–80.

Craine, N.G., Nuttall, P.A., Marriott, A.C. and Randolph, S.E. (1997) Role of grey squirrels and pheasants in the transmission of *Borrelia burgdorferi* sensu lato, the Lyme disease spirochaete, in the UK. *Folia Parasitologica* 44, 155–160.

Cutler, S.J., Williams, N.G. and Wright, D.J.M. (1989) *Borrelia* species from an endemic area for Lyme disease in the United Kingdom. In: *4th European Congress of Clinical Microbiology*, Nice, France, Abstract 908.

van Dam, A.P., Oei, A., Jaspars, R., Fijen, C., Wilske, B., Spanjaard, L. and Dankert, J. (1997) Complement-mediated serum sensitivity among spirochetes that cause Lyme disease. *Infection and Immunity* 65, 1228–1236.

Dizij, A. and Kurtenbach, K. (1995) *Clethrionomys glareolus*, but not *Apodemus flavicollis*, acquires resistance to *Ixodes ricinus* L., the main European vector of *Borrelia burgdorferi*. *Parasite Immunology* 17, 177–183.

Doby, J.M., Bigaignon, G., Launay, H., Costil, C. and Lorvellec, O. (1990) Présence de *Borrelia burgdorferi*, agent de spirochétoses à tiques, chez *Ixodes* (*Exopalpiger*) *trianguliceps* Birula, 1895 et *Ixodes* (*Ixodes*) *acuminatus* Neumann, 1901 (Acariens Ixodidae) et chez *Ctenophtalmus baeticus arvernus* Jordan, 1931 et *Megabothris turbidus* Rothschild, 1909 (Insectes Siphonaptera), ectoparasites de micromammifères des forêts dans l'Ouest de la France. *Bulletin de la Société Française de Parasitologie* 8, 311–322.

Doby, J.M., Bigaignon, G., Aubert, M. and Imbert, G. (1991) Ectoparasites du renard et borréliose de Lyme. Recherche de *Borrelia burgdorferi* chez les tiques Ixodidae et insectes Siphonaptera. *Bulletin de la Société Française de Parasitologie* 9, 279–288.

Doby, J.M., Bigaignon, G., Degeilh, B. and Guigen, C. (1994) Ectoparasites des grands mammifères sauvages, cervidés et suidés, et borréliose de Lyme. Recherche de *Borrelia burgdorferi* chez plus de 1400 tiques, poux, pupipares et puces. *Revue Médicale et Vétérinaire* 145, 743–748.

Doby, J.M., Degeilh, B., Cayouette, S. and Guigen, C. (1995) Présence de *Borrelia burgdorferi* sensu lato chez *Ixodes* (*Trichotoixodes*) *pari* Leach, 1815 (Acari: Ixodidae), tique strictement spécifique des oiseaux. *Bulletin de la Société de Pathologie Exotique* 88, 185–186.

Dolan, M.C., Piesman, J., Mbow, M.L., Maupin, G.O., Péter, O., Brossard, M. and Golde, W.T. (1998) Vector competence of *Ixodes scapularis* and *Ixodes ricinus* (Acari: Ixodidae) for three genospecies of *Borrelia burgdorferi*. *Journal of Medical Entomology* 35, 465–470.

Eiffert, H., Ohlenbusch, A., Christen, H.J., Thomssen, R., Spielman, A. and Matuschka, F.R. (1995) Nondifferentiation between Lyme disease spirochetes from vector ticks and human cerebrospinal fluid. *Journal of Infectious Diseases* 171, 476–479.

Eley, T.J. (1977) *Ixodes uriae* (Acari, Ixodidae) from a river otter. *Journal of Medical Entomology* 13, 506.

Escudero, R., Barral, M., Pérez, A.,

Vitutia, M.M., Garcia-Perez, A.L., Jimenez, S., Sellek, R.E. and Anda, P. (2000) Molecular and pathogenic characterization of *Borrelia burgdorferi* sensu lato isolates from Spain. *Journal of Clinical Microbiology* 38, 4026–4033.

Estrada-Peña, A., Oteo, J.A., Estrada-Peña, R., Gortázar, C., Osácar, J.J., Moreno, J.A. and Castellá, J. (1995) *Borrelia burgdorferi* sensu lato in ticks (Acari: Ixodidae) from two different foci in Spain. *Experimental and Applied Acarology* 19, 173–180.

Fukunaga, M., Hamase, A., Okada, K. and Nakao, M. (1996) *Borrelia tanuki* sp. nov. and *Borrelia turdae* sp. nov. found from ixodid ticks in Japan: rapid species identification by 16S rRNA. *Microbiology and Immunology* 40, 877–881.

Gern, L. and Rais, O. (1996) Efficient transmission of *Borrelia burgdorferi* between cofeeding *Ixodes ricinus* ticks (Acari: Ixodidae). *Journal of Medical Entomology* 33, 189–192.

Gern, L., Toutoungi, L.N., Hu, C.M. and Aeschlimann, A. (1991) *Ixodes (Pholeoixodes) hexagonus*, an efficient vector of *Borrelia burgdorferi* in the laboratory. *Medical and Veterinary Entomology* 5, 431–435.

Gern, L., Siegenthaler, M., Hu, C.M., Leuba-Garcia, S., Humair, P.F. and Moret, J. (1994a) *Borrelia burgdorferi* in rodents (*Apodemus flavicollis* and *A. sylvaticus*): duration and enhancement of infectivity for *Ixodes ricinus* ticks. *European Journal of Epidemiology* 10, 75–80.

Gern, L., Rais, O., Capiau, C., Hauser, P., Lobet, Y., Simoen, E., Voet, P. and Petre, J. (1994b) Immunization of mice by recombinant OspA preparations and protection against *Borrelia burgdorferi* infection induced by *Ixodes ricinus* tick bites. *Immunology Letters* 39, 249–258.

Gern, L., Rouvinez, E., Toutoungi, L.N. and Godfroid, E. (1997) Transmission cycles of *Borrelia burgdorferi* sensu lato involving *Ixodes ricinus* and/or *I. hexagonus* ticks and the European hedgehog, *Erinaceus europaeus*, in suburban and urban areas in Switzerland. *Folia Parasitologica* 44, 309–314.

Gern, L., Estrada-Peña, A., Frandsen, F., Gray, J.S., Jaenson, T.G.T., Jongejan, F., Kahl, O., Korenberg, E., Mehl, R. and Nuttall, P.A. (1998) European reservoir hosts of *Borrelia burgdorferi* sensu lato. *Zentralblatt für Bakteriologie* 287, 196–204.

Gern, L., Hu, C.M., Kocianova, E., Vyrostekova, V. and Rehacek, J. (1999) Genetic diversity of *Borrelia burgdorferi* sensu lato isolates obtained from *Ixodes ricinus* ticks collected in Slovakia. *European Journal of Epidemiology* 15, 665–669.

Gigon, F. (1985) Biologie d'*Ixodes ricinus* L. sur le Plateau Suisse – une contribution à l'écologie de ce vecteur. PhD thesis, University of Neuchâtel, Neuchâtel, Switzerland.

Gorelova, N.B., Korenberg, E.I., Kovalevskii, Y.V. and Shcherbakov, S.V. (1995) Small mammals as reservoir hosts for borrelia in Russia. *Zentralblatt für Bakteriologie* 282, 315–322.

Gray, J.S. (1991) The development and seasonal activity of the tick *Ixodes ricinus*: a vector of Lyme borreliosis. *Review of Medical and Veterinary Entomology* 79, 323–333.

Gray, J.S., Kahl, O., Janetzki, C. and Stein, J. (1992) Studies on the ecology of Lyme disease in a deer forest in County Galway, Ireland. *Journal of Medical Entomology* 29, 915–920.

Gray, J.S., Kahl, O., Janetzki-Mittman, C., Stein, J. and Guy, E. (1994) Acquisition of *Borrelia burgdorferi* by *Ixodes ricinus* ticks fed on the European hedgehog, *Erinaceus europaeus* L. *Experimental and Applied Acarology* 18, 485–491.

Gray, J.S., Kahl, O., Janetzki, C., Stein, J. and Guy, E. (1995) The spatial distribution of *Borrelia burgdorferi*-infected *Ixodes ricinus* in the Connemara region of Co. Galway, Ireland. *Experimental and Applied Acarology* 18, 485–491.

Gray, J.S., Kahl, O., Robertson, J.N.,

Daniel, M., Estrada-Pena, A., Gettinby, G., Jaenson, T.G.T., Jaenson, P., Jongejan, F., Korenberg, E., Kurtenbach, K. and Zeman, P. (1998) Lyme borreliosis habitat assessment. *Zentralblatt für Bakteriologie und Hygiene* 287, 211–228.

Gylfe, Å., Olsen, B., Strasevicius, D., Ras, N.M., Weihe, P., Noppa, L., Ostberg, Y., Baranton, G. and Bergström, S. (1999) Isolation of Lyme disease *Borrelia* from puffins (*Fratercula arctica*) and seabird ticks (*Ixodes uriae*) on the Faeroe Islands. *Journal of Clinical Microbiology* 37, 890–896.

Gylfe, Å., Bergström, S., Lunström, J. and Olsen, B. (2000) Reactivation of *Borrelia* infection in birds. *Nature* 403, 724–725.

Halouzka, J. (1993) Borreliae in *Aedes vexans* and hibernating *Culex pipiens molestus* mosquitoes. *Biológia* 48, 123–124.

Halouzka, J., Postic, D. and Hubálek, Z. (1998) Isolation of the spirochaete *Borrelia afzelii* from the mosquito *Aedes vexans* in the Czech Republic. *Medical and Veterinary Entomology* 12, 103–105.

Halouzka, J., Wilske, B., Stünzer, D., Sanogo, Y.O. and Hubálek, Z. (1999) Isolation of *Borrelia afzelii* from overwintering *Culex pipiens* biotype *molestus* mosquitoes. *Infection* 27, 275–277.

Hillyard, P.D. (1996) *Ticks of North-west Europe*, Vol. 52. Edited by Barnes, R.S.K. and Crothers, J.H. Field Studies Council, Shrewsbury, UK, 178 pp.

Hovmark, A., Jaenson, T.G.T., Åsbrink, E., Forsman, A. and Jansson, E. (1988) First isolations of *Borrelia burgdorferi* from rodents collected in Northern Europe. *Acta Pathologica, Microbiologica et Immunologica Scandinavica Section B* 96, 917–920.

Hu, C.M., Humair, P.F., Wallich, R. and Gern, L. (1997) *Apodemus* sp. rodents, reservoir hosts for *Borrelia afzelii* in an endemic area in Switzerland. *Zentralblatt für Bakteriologie* 285, 558–564.

Hu, C.M., Wilske, B., Fingerle, V., Lobet, Y. and Gern, L. (2001) Transmission of *Borrelia garinii* OspA serotype 4 to BALB/c mice by *Ixodes ricinus* ticks collected in the field. *Journal of Clinical Microbiology* 39, 1169–1171.

Hubálek, Z. and Halouzka, J. (1997) Distribution of *Borrelia burgdorferi* sensu lato genomic groups in Europe, a review. *European Journal of Epidemiology* 13, 951–957.

Hubálek, Z. and Halouzka, J. (1998) Prevalence rates of *Borrelia burgdorferi* sensu lato in host-seeking *Ixodes ricinus* ticks in Europe. *Parasitology Research* 84, 167–172.

Hubálek, Z., Anderson, J.F., Halouzka, J. and Hájek, V. (1996) Borreliae in immature *Ixodes ricinus* (Acari: Ixodidae) ticks parasitizing birds in the Czech Republic. *Journal of Medical Entomology* 33, 766–771.

Hubálek, Z., Halouzka, J. and Juricová, Z. (1998a) Investigation of haematophagous arthropods for borreliae – summarized data, 1988–1996. *Folia Parasitologica* 45, 67–72.

Hubálek, Z., Halouzka, J. and Heroldova, M. (1998b) Growth temperature ranges of *Borrelia burgdorferi* sensu lato strains. *Journal of Medical Microbiology* 47, 929–932.

Hubbard, M.J., Baker, A.S. and Cann, K.J. (1998) Distribution of *Borrelia burgdorferi* s.l. spirochaete DNA in British ticks (Argasidae and Ixodidae) since the 19th century, assessed by PCR. *Medical and Veterinary Entomology* 12, 89–97.

Humair, P.F. and Gern, L. (1998) Relationship between *Borrelia burgdorferi* sensu lato species, red squirrels (*Sciurus vulgaris*) and *Ixodes ricinus* in enzootic areas in Switzerland. *Acta Tropica* 69, 213–227.

Humair, P.F., Turrian, N., Aeschlimann, A. and Gern, L. (1993a) *Borrelia burgdorferi* in a focus of Lyme borreliosis: epizootiologic contribution of small mammals. *Folia Parasitologica* 40, 65–70.

Humair, P.F., Turrian, N., Aeschlimann, A. and Gern, L. (1993b) *Ixodes ricinus* immatures on birds in a focus of

Lyme borreliosis. *Folia Parasitologica* 40, 237–242.

Humair, P.F., Péter, O., Wallich, R. and Gern, L. (1995) Strain variation of Lyme disease spirochetes isolated from *Ixodes ricinus* ticks and rodents collected in two endemic areas in Switzerland. *Journal of Medical Entomology* 32, 433–438.

Humair, P.F., Postic, D., Wallich, R. and Gern, L. (1998) An avian reservoir (*Turdus merula*) of the Lyme borreliosis spirochetes. *Zentralblatt für Bakteriologie* 287, 521–538.

Humair, P.F., Rais, O. and Gern, L. (1999) Transmission of *Borrelia afzelii* from *Apodemus* mice and *Clethrionomys* voles to *Ixodes ricinus* ticks: differential transmission pattern and overwintering maintenance. *Parasitology* 118, 33–42.

Ishiguro, F., Takada, N., Masuzawa, T. and Fukui, T. (2000) Prevalence of Lyme disease *Borrelia* spp. in ticks from migratory birds on the Japanese mainland. *Applied and Environmental Microbiology* 66, 982–986.

Jaenson, T.G.T. and Tälleklint, L. (1992) Incompetence of roe deer as reservoirs of the Lyme borreliosis spirochete. *Journal of Medical Entomology* 29, 813–817.

Jaenson, T.G.T. and Tälleklint, L. (1996) Lyme borreliosis spirochetes in *Ixodes ricinus* (Acari: Ixodidae) and the varying hare on isolated islands in the Baltic sea. *Journal of Medical Entomology* 33, 339–343.

Jauris-Heipke, S., Liegl, G., Preac-Mursic, V., Rössler, D., Schwab, E., Soutschek, E., Will, G. and Wilske, B. (1995) Molecular analysis of genes encoding outer surface protein C (OspC) of *Borrelia burgdorferi* sensu lato: relationship to *ospA* genotype and evidence of lateral gene exchange of *ospC*. *Journal of Clinical Microbiology* 33, 1860–1866.

Johnson, R.C., Schmid, G.P., Hyde, F.W., Steigerwalt, A.G. and Brenner, D.J. (1984) *Borrelia burgdorferi* sp. nov.: etiologic agent of Lyme disease. *International Journal of Systematic Bacteriology* 34, 496–497.

Kahl, O. and Geue, L. (1998) Laboratory study on the possible role of the European fox, *Vulpes vulpes* as a potential reservoir of *Borrelia burgdorferi* s.l. In: Coons, L. and Rothschild, M. (eds) *2nd International Conference on Tick-Borne Pathogens at the Host–Vector Interface: a Global Perspective, S.I.* The Conference, Kruger National Park, South Africa, p. 29.

Kahl, O., Janetzki, C., Gray, J.S., Stein, J. and Bauch, R.J. (1992) Tick infection rates with *Borrelia*: *Ixodes ricinus* versus *Haemaphysalis concinna* and *Dermacentor reticulatus* in two locations in eastern Germany. *Medical and Veterinary Entomology* 6, 363–366.

Kawabata, H., Masuzawa, T. and Yanagihara, Y. (1993) Genomic analysis of *Borrelia japonica* sp. nov. isolated from *Ixodes ovatus* in Japan. *Microbiology and Immunology* 37, 843–848.

Khanakah, G., Kmety, E., Radda, A. and Stanek, G. (1994) Micromammals as reservoir of *Borrelia burgdorferi* in Austria. In: Cevenini, R., Sambri, V. and La Placa, M. (eds) *Abstract Book of the VI Conference on Lyme Borreliosis*. Bologna, Italy, Abstract P077W.

Kimura, K., Isogai, E., Isogai, H., Kamewaka, Y., Nishikawa, T., Ishii, N. and Fujii, N. (1995) Detection of Lyme disease spirochetes in the skin of naturally infected wild sika deer (*Cervus nippon yeosoensis*) by PCR. *Applied and Environmental Microbiology* 61, 1641–1642.

Kirstein, F., Rijpkema, S., Molkenboer, M. and Gray, J.S. (1997) The distribution and prevalence of *B. burgdorferi* genomospecies in *Ixodes ricinus* ticks in Ireland. *European Journal of Epidemiology* 13, 67–72.

Kurtenbach, K., Dizij, A., Kampen, H., Maier, W.A., Seitz, H.M., Schaal, K.P., Moter, S.E., Kramer, M.D., Wallich,

R., Schaible, U.E. and Simon, M.M. (1994a) The ecology of *Borrelia burgdorferi* in an endemic focus of Lyme borreliosis near Bonn. *Applied Parasitology* 35, 149–150.

Kurtenbach, K., Dizij, A., Seitz, H.M., Margos, G., Moter, S.E., Kramer, M.D., Wallich, R., Schaible, U.E. and Simon, M.M. (1994b) Differential immune responses to *Borrelia burgdorferi* in European wild rodent species influence spirochete transmission to *Ixodes ricinus* L. (Acari: Ixodidae). *Infection and Immunity* 62, 5344–5352.

Kurtenbach, K., Kampen, H., Dizij, A., Arndt, S., Seitz, H.M., Schaible, U.E. and Simon, M.M. (1995) Infestation of rodents with larval *Ixodes ricinus* (Acari: Ixodidae) is an important factor in the transmission cycle of *Borrelia burgdorferi* s.l. in German woodlands. *Journal of Medical Entomology* 32, 807–817.

Kurtenbach, K., Carey, D., Hoodless, A.N., Nuttall, P.A. and Randolph, S.E. (1998a) Competence of pheasants as reservoirs for Lyme disease spirochetes. *Journal of Medical Entomology* 35, 77–81.

Kurtenbach, K., Peacey, M., Rijpkema, S.G.T., Hoodless, A.N., Nuttall, P.A. and Randolph, S.E. (1998b) Differential transmission of the genospecies of *Borrelia burgdorferi* sensu lato by game birds and small rodents in England. *Applied and Environmental Microbiology* 64, 1169–1174.

Kurtenbach, K., Sewell, H.S., Ogden, N.H., Randolph, S.E. and Nuttall, P.A. (1998c) Serum complement sensitivity as a key factor in Lyme disease ecology. *Infection and Immunity* 66, 1248–1251.

Lebet, N. and Gern, L. (1994) Histological examination of *Borrelia burgdorferi* infections in unfed *Ixodes ricinus* nymphs. *Experimental and Applied Acarology* 18, 177–183.

Le Fleche, A., Postic, D., Girardet, K., Péter, O. and Baranton, G. (1997) Characterization of *Borrelia lusitaniae* sp. nov. by 16S ribosomal DNA sequence analysis. *International Journal of Systematic Bacteriology* 47, 921–925.

Leuba-Garcia, S., Kramer, M.D., Wallich, R. and Gern, L. (1994) Characterization of *Borrelia burgdorferi* isolated from different organs of *Ixodes ricinus* ticks collected in nature. *Zentralblatt für Bakteriologie* 280, 468–475.

Liebisch, A. and Liebisch, G. (1996) Hard ticks (Ixodidae) biting humans in Germany and their infection with *Borrelia burgdorferi*. In: Mitchell, R., Horn, D.J., Needham, G.N. and Welbourn, W.C. (eds) *Proceedings of Acarology IX*. Ohio Biological Survey, Columbus, pp. 465–467.

Liebisch, A., Olbrich, S., Brand, A., Liebisch, G. and Mourettou-Kunitz, M. (1989) Natürliche Infektionen der Zeckenart *Ixodes hexagonus* mit Borrelien (*Borrelia burgdorferi*). *Tierärztliche Umschau* 44, 809–810.

Liebisch, G., Finkbeiner-Weber, B. and Liebisch, A. (1996) The infection with *Borrelia burgdorferi* s.l. in the European hedgehog (*Erinaceus europaeus*) and its ticks. *Parasitologia* 38, 385.

Liebisch, G., Sihns, B. and Bautsch, W. (1998a) Detection and typing of *Borrelia burgdorferi* sensu lato in *Ixodes ricinus* ticks attached to human skin by PCR. *Journal of Clinical Microbiology* 36, 3355–3358.

Liebisch, G., Dimpfl, B., Finkbeiner-Weber, B., Liebisch, A. and Frosch, M. (1998b) The red fox (*Vulpes vulpes*) a reservoir competent host for *Borrelia burgdorferi sensu lato*. In: Coons, L. and Rothschild, M. (eds) *2nd International Conference on Tick-Borne Pathogens at the Host-Vector Interface: a Global Perspective*, S.I. The Conference, Kruger National Park, South Africa, p. 238.

Macaigne, F. and Perez-Eid, C. (1991) Présence de *Borrelia* affines de *B. burgdorferi* chez *Haemaphysalis* (*Alloceraea*) *inermis* Birula, 1895 (Acarina, Ixodoidea), dans le Sud-ouest de la France. *Annales de Parasitologie Humaine et Comparée* 66, 269–271.

Marconi, R.T., Liveris, D. and Schwartz, I. (1995) Identification of novel insertion elements, restriction fragment length polymorphism patterns, and discontinuous 23S rRNA in Lyme disease spirochetes: phylogenetic analyses of rRNA genes and their intergenic spacers in *Borrelia japonica* sp.nov. and genomic group 21038 (*Borrelia andersonii* sp. nov.) isolates. *Journal of Clinical Microbiology* 33, 2427–2434.

Marquez, F.J. and Constan, M.C. (1990) Infection d'*Ixodes ricinus* (L., 1758) et *Haemaphysalis punctata* Canestrini et Fanzago, 1877 (*Acarina, Ixodidae*) par *Borrelia burgdorferi* dans le nord de la Péninsule Ibérique (Pays Basque espagnol et Navarre). *Bulletin de la Société Française de Parasitologie* 8, 323–330.

Masuzawa, T., Takada, N., Kudeken, M., Fukui, T., Yano, Y., Ishiguro, F., Kawamura, Y., Imai, Y. and Ezaki, T. (2001) *Borrelia sinica* sp. nov., a Lyme disease-related *Borrelia* species isolated in China. *International Journal of Systematic and Evolutionary Microbiology* 51, 1817–1824.

Mather, T.N., Wilson, M.L., Moore, S.I., Ribeiro, J.M.C. and Spielman, A. (1989) Comparing the relative potential of rodents as reservoirs of the Lyme disease spirochete (*Borrelia burgdorferi*). *American Journal of Epidemiology* 130, 143–150.

Matuschka, F.R. and Spielman, A. (1992) Loss of Lyme disease spirochetes from *Ixodes ricinus* ticks feeding on European blackbirds. *Experimental Parasitology* 74, 151–158.

Matuschka, F.R., Fischer, P., Heiler, M., Richter, D. and Spielman, A. (1992) Capacity of European animals as reservoir hosts for the Lyme disease spirochete. *Journal of Infectious Diseases* 165, 479–483.

Matuschka, F.R., Eiffert, H., Ohlenbusch, A. and Spielman, A. (1994a) Amplifying role of edible dormice in Lyme disease transmission in Central Europe. *Journal of Infectious Diseases* 170, 122–127.

Matuschka, F.R., Eiffert, H., Ohlenbusch, A., Richter, D., Schein, E. and Spielman, A. (1994b) Transmission of the agent of Lyme disease on a subtropical island. *Tropical Medicine and Parasitology* 45, 39–44.

Matuschka, F.R., Endepols, S., Richter, D., Ohlenbusch, A., Eiffert, H. and Spielman, A. (1996) Risk of urban Lyme disease enhanced by the presence of rats. *Journal of Infectious Diseases* 174, 1108–1111.

Matuschka, F.R., Endepols, S., Richter, D. and Spielman, A. (1997) Competence of urban rats as reservoir hosts for Lyme disease spirochetes. *Journal of Medical Entomology* 34, 489–493.

Matuschka, F.R., Allgöwer, R., Spielman, A. and Richter, D. (1999) Characteristics of garden dormice that contribute to their capacity as reservoirs for Lyme disease spirochetes. *Applied and Environmental Microbiology* 65, 707–711.

Matuschka, F.R., Schinkel, T.W., Klug, B., Spielman, A. and Richter, D. (2000) Relative incompetence of European rabbits for Lyme disease spirochaetes. *Parasitology* 121, 297–302.

Mehl, R. and Traavik, T. (1983) The tick *Ixodes uriae* (Acari, Ixodidae) in seabird colonies in Norway. *Fauna Norvegica Serie B* 30, 94–107.

de Michelis, S., Sewell, H.S., Collares-Pereira, M., Santos-Reis, M., Schouls, L.M., Benes, V., Holmes, E.C. and Kurtenbach, K. (2000) Genetic diversity of *Borrelia burgdorferi* sensu lato in ticks from mainland Portugal. *Journal of Clinical Microbiology* 38, 2128–2133.

Misonne, M.C., van Impe, G. and Hoet, P.P. (1998) Genetic heterogeneity of *Borrelia burgdorferi* sensu lato in *Ixodes ricinus* ticks collected in Belgium. *Journal of Clinical Microbiology* 36, 3352–3354.

Miyamoto, K., Sato, Y., Okada, K., Fukunaga, M. and Sato, F. (1997) Competence of a migratory bird, red-bellied thrush (*Turdus chrysolaus*), as an avian reservoir for the Lyme disease spirochetes in Japan. *Acta Tropica* 65, 43–51.

Monin, R., Gern, L. and Aeschlimann, A. (1989) A study of the different modes of transmission of *Borrelia burgdorferi* by *Ixodes ricinus*. *Zentralblatt für Bakteriologie, Suppl.* 18, 14–20.

Nakao, M., Miyamoto, K. and Fukunaga, M. (1994) Lyme disease spirochetes in Japan: enzootic transmission cycles in birds, rodents, and *Ixodes persulcatus* ticks. *Journal of Infectious Diseases* 170, 878–882.

Nuncio, M.S., Péter, O., Alves, M.J., Bacellar, F. and Filipe, A.R. (1993) Isolamento e caracterizaçao de borrélias de *Ixodes ricinus* L. em Portugal. *Revista Portuguesa Doenças Infecciosas* 16, 175–179.

Nuttall, P., Randolph, S., Carey, D., Craine, N., Livesley, A. and Gern, L. (1994) The ecology of Lyme borreliosis in the UK. In: Axford, J.S. and Rees, D.H.E. (eds) *Lyme Borreliosis*. Plenum Press, New York, pp. 125–129.

Ogden, N.H., Nuttall, P.A. and Randolph, S.E. (1997) Natural Lyme disease cycles maintained via sheep by cofeeding ticks. *Parasitology* 115, 591–599.

Olsen, B. (1995) Birds and *Borrelia*. PhD thesis, Umeå University, Umeå, Sweden.

Olsen, B., Jaenson, T.G.T., Noppa, L., Bunikis, J. and Bergström, S. (1993) A Lyme borreliosis cycle in seabirds and *Ixodes uriae* ticks. *Nature* 362, 340–342.

Olsen, B., Duffy, D.C., Jaenson, T.G.T., Gylfe, Å., Bonnedahl, J. and Bergström, S. (1995a) Transhemispheric exchange of Lyme disease spirochetes by seabirds. *Journal of Clinical Microbiology* 33, 3270–3274.

Olsen, B., Jaenson, T.G.T. and Bergström, S. (1995b) Prevalence of *Borrelia burgdorferi* sensu lato-infected ticks on migrating birds. *Applied and Environmental Microbiology* 61, 3082–3087.

Perret, J.L., Guigoz, E., Rais, O. and Gern, L. (2000) Influence of saturation deficit and temperature on *Ixodes ricinus* tick questing activity in a Lyme borreliosis-endemic area (Switzerland). *Parasitology Research* 86, 554–557.

Péter, O. and Bretz, A.G. (1992) Polymorphism of outer surface proteins of *Borrelia burgdorferi* as a tool for the classification. *Zentralblatt für Bakteriologie* 277, 28–33.

Péter, O., Bretz, A.G. and Bee, D. (1995) Occurrence of different genospecies of *Borrelia burgdorferi* sensu lato in ixodid ticks of Valais, Switzerland. *European Journal of Epidemiology* 11, 463–467.

Petney, T.N., Hassler, D., Bröckner, M. and Maiwald, M. (1996) Comparison of urinary bladder and ear biopsy samples for determining prevalence of *Borrelia burgdorferi* in rodents in central Europe. *Journal of Clinical Microbiology* 34, 1310–1312.

Pichon, B., Godfroid, E., Hoyois, B., Bollen, A., Rodhain, F. and Perez, C. (1995) Simultaneous infection of *Ixodes ricinus* by two *Borrelia burgdorferi* sensu lato species: possible implications for clinical manifestations. *Emerging Infectious Diseases* 1, 89–90.

Picken, R.N., Cheng, Y., Strle, F. and Picken, M.M. (1996) Patient isolates of *Borrelia burgdorferi* sensu lato with genotypic and phenotypic similarities to strain 25015. *Journal of Infectious Diseases* 174, 1112–1115.

Postic, D., Assous, M.V., Grimont, P.A.D. and Baranton, G. (1994) Diversity of *Borrelia burgdorferi* sensu lato evidenced by restriction fragment length polymorphism of *rrf* (5S)–*rrl* (23S) intergenic spacer amplicons. *International Journal of Systematic Bacteriology* 44, 743–752.

Postic, D., Korenberg, E., Gorelova, N., Kovalevski, Y.V., Bellenger, E. and Baranton G. (1997) *Borrelia burgdorferi* sensu lato in Russia and neighbouring countries: high incidence of mixed isolates. *Research in Microbiology* 148, 691–702.

Postic, D., Marti Ras, N., Lane, R.S., Hendson, M. and Baranton, G. (1998) Expanded diversity among Californian *Borrelia* isolates and description of

Borrelia bissettii sp.nov. (formerly Borrelia group 127). *Journal of Clinical Microbiology* 36, 3497–3504.

Randolph, S.E. and Craine, N.G. (1995) General framework for comparative quantitative studies on transmission of tick-borne diseases using Lyme borreliosis in Europe as an example. *Journal of Medical Entomology* 32, 765–777.

Randolph, S. and Storey, K. (1999) Impact of microclimate on immature tick–rodent host interactions (*Acari: Ixodidae*): implications for parasite transmission. *Journal of Medical Entomology* 36, 741–748.

Randolph, S.E., Gern, L. and Nuttall, P.A. (1996) Co-feeding ticks: epidemiological significance for tick-borne pathogen transmission. *Parasitology Today* 12, 472–479.

Richter, D., Endepols, S., Ohlenbusch, A., Eiffert, H., Spielman, A and Matuschka, F.R. (1999) Genospecies diversity of Lyme disease spirochetes in rodent reservoirs. *Emerging Infectious Diseases* 5, 291–296.

Richter, D., Spielman, A., Komar, N. and Matuschka, F.R. (2000) Competence of American robins as reservoir hosts for Lyme disease spirochetes. *Emerging Infectious Diseases* 6, 133–138.

Rijpkema, S.G.T., Molkenboer, M.J.C.H., Schouls, L.M., Jongejan, F. and Schellekens, J.F.P. (1995) Simultaneous detection and genotyping of three genomic groups of *Borrelia burgdorferi* sensu lato in Dutch *Ixodes ricinus* ticks by characterization of the amplified intergenic spacer region between 5S and 23S rRNA genes. *Journal of Clinical Microbiology* 33, 3091–3095.

Rijpkema, S., Golubic, D., Molkenboer, M., Verbeek-De Kruif, N. and Schellekens, J. (1996) Identification of four genomic groups of *Borrelia burgdorferi* sensu lato in *Ixodes ricinus* ticks collected in a Lyme borreliosis endemic region of northern Croatia. *Experimental and Applied Acarology* 20, 23–30.

Saint Girons, I., Gern, L., Gray, J.S., Guy, E.C., Korenberg, E., Nuttall, P.A., Rijpkema, S.G.T., Schönberg, A., Stanek, G. and Postic, D. (1998) Identification of *Borrelia burgdorferi* sensu lato species in Europe. *Zentralblatt für Bakteriologie* 287, 190–195.

Stanek, G. and Simeoni, J. (1989) Are pigeons' ticks transmitters of *Borrelia burgdorferi* to humans? *Zentralblatt für Bakteriologie Suppl.* 18, 42–43.

Strle, F., Picken, R.N., Cheng, Y., Cimperman, J., Maraspin, V., Lotric-Furlan, S., Ruzic-Sabljic, E. and Picken, M.M. (1997) Clinical findings for patients with Lyme borreliosis caused by *Borrelia burgdorferi* sensu lato with genotypic and phenotypic similarities to strain 25015. *Clinical Infectious Diseases* 25, 273–280.

Tälleklint, L. (1996) Lyme borreliosis spirochetes in *Ixodes ricinus* and *Haemaphysalis punctata* ticks (Acari: Ixodidae) on three islands in the Baltic Sea. *Experimental and Applied Acarology* 20, 467–476.

Tälleklint, L. and Jaenson, T.G.T. (1993) Maintenance by hares of European *Borrelia burgdorferi* in ecosystems without rodents. *Journal of Medical Entomology* 30, 273–276.

Tälleklint, L. and Jaenson, T.G.T. (1994) Transmission of *Borrelia burgdorferi* s.l. from mammal reservoirs to the primary vector of Lyme borreliosis, *Ixodes ricinus* (Acari: Ixodidae), in Sweden. *Journal of Medical Entomology* 31, 880–886.

Toutoungi, L. and Gern, L. (1993) Ability of transovarially and subsequent transstadially infected *Ixodes hexagonus* ticks to maintain and transmit *Borrelia burgdorferi* in the laboratory. *Experimental and Applied Acarology* 17, 581–586.

Toutoungi, L.N., Gern, L., Aeschlimann, A. and Debrot, S. (1991) A propos du genre *Pholeoixodes*, parasite des carnivores en Suisse. *Acarologia* XXXII, 311–328.

Wang, G., van Dam, A.P., Le Fleche, A., Postic, D., Péter, O., Baranton, G., de

Boer, R., Spanjaard, L. and Dankert, J. (1997) Genetic and phenotypic analysis of *Borrelia valaisiana* sp. nov. (*Borrelia* genomic groups VS116 and M19). *International Journal of Systematic Bacteriology* 47, 926–932.

Wang, G.Q., van Dam, A.P. and Dankert, J. (1999) Phenotypic and genetic characterization of a novel *Borrelia burgdorferi* sensu lato isolate from a patient with Lyme borreliosis. *Journal of Clinical Microbiology* 37, 3025–3028.

Will, G., Jauris-Heipke, S., Schwab, E., Busch, U., Rössler, D., Soutschek, E., Wilske, B. and Preac-Mursic, V. (1995) Sequence analysis of *ospA* genes shows homogeneity within *Borrelia burgdorferi* sensu stricto and *Borrelia afzelii* strains but reveals major subgroups within the *Borrelia garinii* species. *Medical Microbiology and Immunology* 184, 73–80.

Wilske, B., Preac-Mursic, V., Göbel, U.B., Graf, B., Jauris, S., Soutschek, E., Schwab, E. and Zumstein, G. (1993) An OspA serotyping system for *Borrelia burgdorferi* based on reactivity with monoclonal antibodies and OspA sequence analysis. *Journal of Clinical Microbiology* 31, 340–350.

Wilske, B., Busch, U., Eiffert, H., Fingerle, V., Pfister, H.W., Rössler, D. and Preac-Mursic, V. (1996) Diversity of OspA and OspC among cerebrospinal fluid isolates of *Borrelia burgdorferi* sensu lato from patients with neuroborreliosis in Germany. *Medical Microbiology and Immunology* 184, 195–201.

Younsi, H., Postic, D., Baranton, G. and Bouattour, A. (2001) High prevalence of *Borrelia lusitaniae* in *Ixodes ricinus* ticks in Tunisia. *European Journal of Epidemiology* 17, 53–56.

Zhioua, E., Bouattour, A., Hu, C.M., Gharbi, M., Aeschlimann, A., Ginsberg, H. and Gern, L. (1999) Infections of *Ixodes ricinus* (Acari: Ixodidae) by *Borrelia burgdorferi* sensu lato in North Africa (Tunisia). *Journal of Medical Entomology* 36, 216–218.

7
Ecology of *Borrelia burgdorferi* sensu lato in Russia

Edward I. Korenberg, Natalya B. Gorelova, and Yurii V. Kovalevskii

Gamaleya Research Institute for Epidemiology and Microbiology, Russian Academy of Medical Sciences, 18 Gamaleya Street, Moscow, 123098 Russia

Introduction

Purposeful investigations of borrelioses transmitted by hard ticks began in Russia in 1984, immediately after the corresponding spirochaetes were found in European *Ixodes ricinus* ticks (Burgdorfer *et al.*, 1983) and the aetiological agent of Lyme disease was defined as a new species, *Borrelia burgdorferi* sp. nov. (Johnson *et al.*, 1984). However, Russian dermatologists had described clinical manifestations of the disease under different names – atrophia cutis maculata, atrophia cutis idiopathica progressiva acquisita, idiopathic acquired skin atrophy, or autonomous skin atrophy – as early as the turn of the 19th and 20th centuries (Nikol'skii, 1896; Mescherskii, 1898; Pisemskii, 1902; Pospelov, 1905, 1906). Nearly 25 years before the causes of the disease were revealed, Lisovskaya (1958) characterized in detail the clinical picture of chronic atrophic acarodermatitis in her dissertation. Moreover, since the 1940s, neuropathologists have devoted special attention to the diseases that occurred after tick bites in the regions where tick-borne encephalitis (TBE) was endemic and which were accompanied by extensive 'acaroerythema' or 'annular acaroerythema' (as it was called at that time) and various neurological disturbances, and where serological tests for TBE were negative. Such cases were recorded in the northwestern, central and eastern areas of European Russia, in the Urals, West Siberia and the Russian Far East. In the 1960s, several specialists in clinical medicine concluded that these were aetiologically independent diseases unrelated to TBE. The detailed descriptions of their general clinical picture and of migratory erythema observed after tick bites provide evidence so that many of these cases can be diagnosed retrospectively as Lyme disease (LD) or Lyme borreliosis (LB). Both terms actually refer to the whole group of aetiologically independent diseases. Their pathogens

© 2002 CAB *International. Lyme Borreliosis: Biology, Epidemiology and Control* (eds J. Gray, O. Kahl, R.S. Lane and G. Stanek)

have become associated ecologically with certain ticks of the genus *Ixodes*. It was proposed to name this group ixodid tick-borne borrelioses (ITBB) in contrast to argasid tick-borne borrelioses (ATBB) associated with soft ticks (Korenberg, 1993, 1995, 1996; Korenberg and Kryuchechnikov, 1996).

In Russia, the cases of ITBB were first identified serologically in 1985 (Korenberg *et al.*, 1986). Patients with these infections have been recorded in 53 administrative territories of the Russian Federation, from the Baltic Sea to the Far East and southern Sakhalin. Thus, much of the worldwide range of *B. burgdorferi* s.l. is on Russian territory (Korenberg *et al.*, 1987b; Korenberg, 1994, 1995). As diagnostics improve, ITBB morbidity is steadily approaching the predicted level of 10,000–12,000 fresh cases per year, and in some years this level may be even higher (Korenberg, 1995). Thus, the total numbers of ITBB cases in Russia in 1998 and 1999 were 8606 and 8470, or 5.9 and 5.8 cases per 100,000 people, respectively. These infections are now amongst the foremost causes of morbidity within the group of zoonoses with natural focality, with the greatest number of cases being recorded in the Urals, West Siberia and the Volga–Vyatka region.

Studies on natural ITBB foci in Russia began in the northwest and the far east (Korenberg *et al.*, 1987a, 1988, 1989, 1991a,b). Those results provided the basis for the following principal conclusions: (i) the tick *Ixodes persulcatus* Schultze is the main vector of borreliae in the greater part of the ITBB range; (ii) the most active foci are found in broad-leaved, mixed and southern taiga forests; (iii) the tick *I. ricinus* (L.) is a borrelia vector in the European part of Russia, but is less efficient than *I. persulcatus*; and (iv) the distribution of ITBB and the basic features of their epidemiology in Russia are similar to those of TBE. These conclusions were fully confirmed in subsequent studies (Korenberg, 1994; Naumov, 1999). These two tick species were the source of the first *B. burgdorferi* s.l. isolates obtained and serotyped in Russia (Kryuchechnikov *et al.*, 1988, 1990, 1993). Some of them (Ip 3, Ip 21, Ip 89, Ip 90, Ir 210 and others) were used by specialists from several countries in studies on the genetic structure, taxonomy and phylogeny of *Borrelia* spp. (Korenberg and Kryuchechnikov, 1996), including characterization of their genogroups and original description of individual genospecies (for example, Baranton *et al.*, 1992; Belfaiza *et al.*, 1993; Fukunaga *et al.*, 1993; Assous *et al.*, 1994; Postic *et al.*, 1994; Liveris *et al.*, 1995). The data accumulated in studies on the main aspects of *B. burgdorferi* s.l. ecology in Russia are presented below.

Vectors and *B. burgdorferi* Genospecies Present

Isolates of borreliae were obtained from three tick species: *I. persulcatus*, *I. ricinus*, and *Ixodes (Exopalpiger) trianguliceps*. However, attempts to isolate borreliae from *Dermacentor reticulatus* ticks collected in northwestern Russia and from *Haemaphysalis concinna* and *Haemaphysalis japonica* ticks from the far east of Russia were not successful (Korenberg *et al.*, 1997). Information concerning borrelial infection in

D. reticulatus, Dermacentor marginatus, Dermacentor nuttalii, H. concinna and *H. japonica* ticks in West Siberia and the Far East is based on the results of dark-field microscopy only (Ivanov *et al.*, 1996; Rudakova and Matuschenko, 1996) and has not been confirmed by more objective microbiological methods.

The taiga tick *I. persulcatus* and the forest tick *I. ricinus* are the main vectors of borreliae in natural foci and the source of these pathogens for man. The taiga tick is the vector in the southern forest zone of Eurasia, the vast area extending from the Far East to the Russian northwest (beyond this zone, its distribution is limited mainly to some Asian countries), and the forest tick is the vector in similar forests of Europe. In much of European Russia, where these ticks are sympatric (Figs 7.1 and 7.2), natural foci with the two vector species exist, as in the case of TBE (Korenberg, 1994; Filippova, 1999). Indices of borrelial prevalence in adult *I. persulcatus* and *I. ricinus* ticks in different Russian regions, as determined by dark-field microscopy, are shown in Table 7.1. Prevalence is high everywhere, but it appears that in general slightly higher indices are characteristic of the Ural, West Siberian and, probably, the Far-East regions. The prevalence of borreliae in *I. ricinus* ticks is markedly lower than in *I. persulcatus*, both in individual regions, for example Leningrad and Moscow oblasts (a large Russian administrative area), and in general. In this connection it should be noted that the presence and proportion of one vector in an ecosystem does not have any significant effect on the

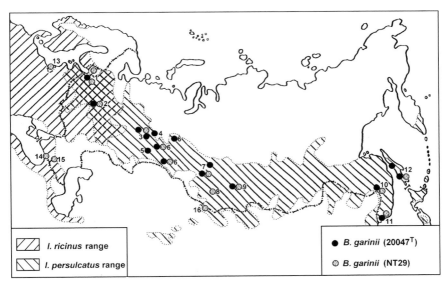

Fig. 7.1. Distribution of two *Borrelia garinii* variants in different regions of Russia (according to Postic *et al.*, 1997, with changes and supplements). (1) St Petersburg; (2) Yaroslavl; (3) Perm; (4) Sverdlovsk; (5) Kurgan; (6) Tyumen; (7) Novosibirsk; (8) Gornii Altai Republic; (9) Krasnoyarsk; (10) Khabarovsk; (11) Primor'e; (12) Sakhalin; (13) Kaliningrad; (14) Krasnodar; (15) Stavropol; (16) Kazakhstan.

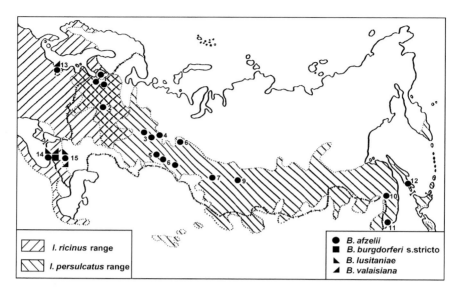

Fig. 7.2. Distribution of *Borrelia afzelii, Borrelia burgdorferi* sensu stricto, *Borrelia lusitaniae* and *Borrelia valaisiana* in Russia (according to Postic *et al.*, 1997, with changes and supplements). (1) St Petersburg; (2) Yaroslavl; (3) Perm; (4) Sverdlovsk; (5) Kurgan; (6) Tyumen; (7) Novosibirsk; (8) Gornii Altai Republic; (9) Krasnoyarsk; (10) Khabarovsk; (11) Primor'e; (12) Sakhalin; (13) Kaliningrad; (14) Krasnodar; (15) Stavropol.

extent of infection and on the epizootic and epidemiological significance of the other vector (Korenberg *et al.*, 2001a).

The tick *I. trianguliceps*, widespread in the forested zone of Eurasia from Great Britain to the southern end of the Baikal Lake in Russia, does not attack people and feeds mainly on small mammals (Korenberg and Lebedeva, 1969; Filippova, 1977). According to data on borrelial cultivation, the proportion of infected larvae and nymphs is about 10% (Gorelova *et al.*, 1996); according to the results of immunofluorescence assay (IFA) with monoclonal antibodies, the prevalence of infection is far higher (Grigoryeva and Tretyakov, 1998). This vector may be of some epizootiological significance (Gorelova *et al.*, 1996; Korenberg *et al.*, 1997), but it remains unclear whether *I. trianguliceps* can maintain borrelial circulation in the absence of *I. persulcatus* and *I. ricinus* ticks.

The tick *Ixodes pavlovskyi*, whose contemporary disjunctive range consists of the western (Altai–Sayan) and Far Eastern parts, is closely related to *I. persulcatus*, both morphologically and ecologically, and is sympatric with the latter virtually everywhere (Filippova, 1977, 1999). Borreliae were first isolated from *I. pavlovskyi* ticks collected in the southwestern Altai (Fig. 7.1, No. 16), near the Kazakh–Russian border (Gorelova *et al.*, 2001a). The role of this species in ITBB epizootiology and epidemiology requires further investigation.

Information about *B. burgdorferi* genospecies present in different species of vec-

Table 7.1. Prevalence of *Borrelia burgdorferi* sensu lato in unfed adult ixodid ticks in different regions of Russia (microscopy data).

Regions/provinces	Years	Number of ticks studied	Prevalence of *Borrelia* in ticks		Reference
			Average	Min.–Max.	
Ixodes persulcatus					
Northern and Northwest regions					
Vologda oblast	1995–1997	1868	12.0	8.4–19.8	Rybakova and Evsyukova, 1999
Leningrad oblast	1986–1992	4059	33.2	8.6–47.8	Korenberg *et al.*, 2001a
St Petersburg city	1995–1997	1559	27.0	24.2–30.7	Dubinina and Alekseev, 1999
Pskov oblast	1990	210	29.5	20.5–39.1	Korenberg *et al.*, 2001a (with supplements)
Novgorod oblast	1990–1991	484	29.8	20.0–36.5	Korenberg *et al.*, 2001a (with supplements)
Central part of European Russia					
Moscow oblast	1990–1991	174	28.7	20.0–48.0	Alekseev, 1993
Ural region					
Perm oblast	1993–2000	1324	42.7	29.5–60.8	Korenberg *et al.*, 1999 (with supplements)
Perm city	1993–1998	1311	27.2	18.4–35.6	Korenberg *et al.*, 2001b (with supplements)
Sverdlovsk oblast	1993–1995	228	37	31–42	Belyaeva, 1998
Chelyabinsk oblast	1994–1997	502	8.0	1.1–18.6	Tarasov *et al.*, 2000
Kurgan oblast	1991–1995	759	22.8	4–45	Smirnova *et al.*, 1996
West Siberia region					
Tyumen oblast	1990–1994	2046	35.6	22.6–44.6	Kolchanova, 1997
Omsk oblast	1991–1995	900	25.0	11.9–31.3	Rudakova and Matuschenko, 1996
Novosibirsk oblast	1991–1995	292	26.4	13.3–58.3	Rudakova and Matuschenko, 1996
Altai territory	1991–1995	248	21.8	17.0–46.1	Rudakova and Matuschenko, 1996

Table 7.1. Continued

Regions/provinces	Years	Number of ticks studied	Prevalence of Borrelia in ticks		Reference
			Average	Min.–Max.	
East Siberia region					
Khakassia Republic	1993–1995	3036	22.1	12.2–30.6	Naumov et al., 1994; Vasilieva and Naumov, 1996
Irkutsk oblast	1992–1995	~2500	19	12–58	Gorin et al., 1996
Irkutsk city	1994–1995	448	26.8	25.0–28.6	Apostolova et al., 1996
Far Eastern region					
Khabarovsk territory	1986–1987	288	22.5	21.2–22.9	Korenberg et al., 1989
Primorye territory	1995	99	41.4[a]	–	Sato et al., 1996
Sakhalin oblast	1995	183	25.1[a]	–	Sato et al., 1996
Ixodes ricinus					
Northwestern region					
Kaliningrad oblast	1995–1998	2470	11.1	10.0–13.6	Dubinina and Alekseev, 1999
Leningrad oblast	1986–1992	2559	16.7	5.0–21.9	Korenberg et al., 2001a
Novgorod oblast	1990–1991	38	21.0	6.3–37.5	Korenberg et al., 2001a (with supplements)
Central part of European Russia					
Kaluga oblast	1992–1998	1015	15.5	10.9–19.8	Burenkova, 2000
Moscow oblast	1991–1992	308	18.5	5.6–35.2	Alekseev, 1993
North Caucasus region					
Stavropol territory	1996–1997	207	15.9[a]	–	Korenberg et al., 1999
Krasnodar territory	2000	112	18.8[a]	–	Gorelova et al., 2001b

[a] Ticks were studied by culturing of their internal organs on BSK medium.
–, no data.

tors and reservoir hosts in different Russian regions is based on the results of identification of isolates obtained in several studies (Sato *et al.*, 1996; Korenberg *et al.*, 1997; Masuzawa *et al.*, 1997; Postic *et al.*, 1997; Saint Girons *et al.*, 1998), and on unpublished data obtained in the past few years. More than 800 isolates have been identified so far and are now kept in one of the largest collections of *B. burgdorferi* s.l. isolates in the world in the Museum of *Borrelia*, Laboratory of Vectors of Infections (Gamaleya Research Institute for Epidemiology and Microbiology, Moscow) (Gorelova, 1993).

In Russia, the spirochaetes *B. garinii* and *B. afzelii* occur virtually everywhere and jointly circulate in natural foci (Figs 7.1 and 7.2). They were isolated from three tick species (Table 7.2). The genospecies *B. garinii* is represented by two types, 20047^T and NT29. The former type occurs in all the vectors, whereas the latter has been found only in *I. persulcatus* and *I. trianguliceps*. It may be that *B. garinii* NT29 has not been found in *I. pavlovskyi* simply because only four borrelial isolates have as yet been obtained from this tick species (Gorelova *et al.*, 2001a). However, in the case of *I. ricinus*, 79 spirochaetal isolates from different Russian regions have been typed to date, and the absence of *B. garinii* NT29 among them is difficult to explain in the same way. Thus, we conclude that the distribution of *B. garinii* NT29 is associated with that of *I. persulcatus* ticks (Postic *et al.*, 1997).

The species *B. valaisiana*, *B. lusitaniae* and *B. burgdorferi* s.s. have been isolated only from *I. ricinus* ticks in the European part of Russia, with the proportion of ticks infected by *B. burgdorferi* s.s. being less than 1% (Korenberg *et al.*, 1999;

Table 7.2. Genospecies of *Borrelia burgdorferi* sensu lato isolated in Russia from vectors and reservoir hosts.

Species of vectors and reservoir hosts	*B. afzelii*	*B. garinii* Variant 20047^T	*B. garinii* Variant NT29	*B. burgdorferi* sensu stricto	*B. lusitaniae*	*B. valaisiana*
Ixodid ticks						
Ixodes persulcatus	+	+	+			
Ixodes ricinus	+	+		+	+	+
Ixodes trianguliceps	+	+	+			
Ixodes pavlovskyi		+				
Small mammals						
Clethrionomys glareolus	+	+	+			
Clethrionomys rufocanus	+	+	+			
Clethrionomys rutilus	+	+	+			
Microtus oeconomus	+	+	+			
Microtus agrestis	+					
Apodemus uralensis	+		+			
Apodemus peninsulae			+			
Sorex araneus			+			

Blank spaces indicate species not found.

Gorelova *et al.*, 2001b) although it cannot be excluded that in some localities the prevalence of this recognized human pathogen in *I. ricinus* is higher. According to Semenov *et al.* (2000), DNA of this spirochaete genospecies was detected by PCR in *I. ricinus* and *I. persulcatus* ticks from the northwest region of Russia. These findings need to be verified by isolation of the pathogen. In any case, *B. burgdorferi* s.s. is of limited significance as an aetiological agent of ITBB in Russia, and the aetiology, epidemiology and clinical manifestations of these diseases are mainly attributable to *B. garinii* and *B. afzelii* (Korenberg, 1995).

Different combinations of mixed isolates have been obtained from three tick species collected in many Russian regions (Table 7.3). For example, according to our long-term data on borrelial culturing, the proportions of mixed-infected *I. persulcatus* in the southern taiga forests of the Ural region are 1.6 ± 0.9% among adult ticks collected from the vegetation and 1.5 ± 1.1% among nymphs. Furthermore, mixed infection of *I. persulcatus* and *I. ricinus* ticks from the northwest was revealed by PCR analysis (Dubinina and Alekseev, 1999). These data support the conclusion that practically all possible combinations of mixed infections with two or more *Borrelia* spp. circulating within a certain area can occur

Table 7.3. Distribution of different *Borrelia burgdorferi* s.l. mixed isolates from vectors and reservoir hosts in Russia.

Species of vectors and reservoir hosts	Region	Variants of mixed isolates				
		B. garinii NT29 + *B. afzelii*	*B. garinii* NT29 + *B. garinii* 20047^T	*B. garinii* 20047^T + *B. afzelii*	*B. garinii* 20047^T + *B. valaisiana*	*B. garinii* NT29 + *B. garinii* 20047^T + *B. afzelii*
Ixodid ticks						
Ixodes persulcatus	Central part of European Russia			+		
	Ural region	+	+	+		+
	West Siberian region	+		+		
	Far Eastern region	+	+			+
Ixodes ricinus	North Caucasian region				+	
Ixodes trianguliceps	Ural region	+	+			
Small mammals						
Clethrionomys glareolus	Ural region	+	+	+		+
Clethrionomys rutilus	Ural region		+			
Microtus oeconomus	Ural region	+	+	+		+
Clethrionomys rufocanus and *Apodemus peninsulae*	Far Eastern region	+?	+?	+?		

+?, According to Masuzawa *et al.*, 1997. Particular variant not indicated.
Blank spaces indicate species not recorded.

in both unfed and engorged (or partially engorged) ticks at any phase of their development (Postic et al., 1997; Korenberg, 1999).

Host Preferences of Ticks

The list of vertebrates that can serve as hosts of *I. ricinus* and *I. persulcatus* comprises at least 300 species and includes virtually all terrestrial mammals occurring within the ranges of these vectors, many ground-foraging bird species and even reptiles and amphibians, which, however, have no apparent significance in this respect (Filippova, 1977; Naumov, 1985).

Larvae and nymphs parasitize more than 50 species of small mammals. The roles of these animals as hosts of larvae and nymphs are fairly similar, but insectivores are much less attractive for nymphs. Throughout the forest zone of Russia, voles of the genus *Clethrionomys* and shrews of the genus *Sorex* are of special significance as tick hosts. This universal *Clethrionomys–Sorex* type of host consists of two variants (subtypes). The first one is characteristic of European Russia (and a large part of central Europe), where most immature ticks feed on bank voles (*Clethrionomys glareolus*) and common shrews (*Sorex araneus*). The second variant is characteristic of Asian Russia, where the genus *Sorex* retains its role but the bank vole is replaced by the ruddy vole (*Clethrionomys rutilus*) and the grey large-toothed red-backed vole (*Clethrionomys rufocanus*) (Korenberg, 1979; Tupikova et al., 1980). In addition to small mammals, larger animals, such as hedgehogs, squirrels and, in Asian Russia, chipmunks, are readily attacked by larvae and nymphs. The average number of larvae and nymphs per animal can vary widely, ranging in different regions from 0.5 to 69.7 in the case of the red squirrel (*Sciurus vulgaris*) and from 0.5 to 29.3 in the case of the Siberian chipmunk (*Tamius sibiricus*) (Korenberg, 1979). Nymphs also frequently attack wild hoofed animals. Larvae of *I. ricinus* and *I. persulcatus* very rarely attack large animals in Russia.

Immature *I. persulcatus* and *I. ricinus* ticks have been found on more than 180 species of bird. However, the number of these bird species is greater in the Far East, within the optimum *I. persulcatus* range; it gradually decreases in the eastern and central regions of European Russia and increases again in the regions located farther to the west, within the *I. ricinus* optimum range (Korenberg, 1962). The only birds attacked by larvae and nymphs throughout this area are the forest species of grouse (*Tetraonidae* spp.), thrushes (*Turdus* spp.), pipits (*Anthus* spp.) and buntings (*Emberiza* spp.). It appears that, as in the case of TBE (Korenberg, 1964), the involvement of birds in the circulation of *Borrelia* spp. in most Russian regions is a relatively rare phenomenon (see below).

Hares are of special significance as hosts because ticks parasitize them at all three phases of tick development: for example, on average one mountain hare (*Lepus timidus*) can be simultaneously infested by up to 44 larvae, 24 nymphs and 85 adult *I. persulcatus* (Mishin, 1964). Adult ticks also attack virtually all larger animals, both wild and domestic, and the significance of different groups of hosts in a certain region depends on the numbers of livestock, conditions of their maintenance and pasturing, and the abundance of large wild mammals. In the southern taiga forests of the eastern East European Plain, for example, cattle are parasitized by approximately 52% of adult *I. persulcatus* ticks, hares (*L. timidus*) 34%, elks (*Alces alces*) 4%, red foxes (*Vulpes vulpes*) 2% and other species of large mammals 8% (Korenberg, 1978). In populated areas of the Far East, approximately 62% of *I. persulcatus* ticks feed on cattle, 3% on horses, 1% on dogs, 24% on red deer (*Cervus elaphus*), 3% on roe deer (*Capreolus capreolus*), 6% on hares (*L. timidus*) and only about 1% on other animal species (Savitskii, 1970). It should be noted that in contrast to *I. ricinus* ticks, which attack people as both nymphs and adults, *I. persulcatus* nymphs very rarely attack people. Larvae of both species do not as a rule attack people in Russia (Korenberg and Kovalevskii, 1981; Korenberg, 1994; Korenberg *et al.*, 2001b).

As in the case of TBE, a very important factor affecting the development of the epizootic process in natural ITBB foci is that ticks of different developmental stages, belonging to different generations, often parasitize the same host. This situation provides an especially efficient horizontal pathway of pathogen transmission, bringing together the vector generations of at least 5 successive years (Korenberg and Kovalevskii, 1994, 1999).

Host Preferences of Spirochaetes

Eleven species of small mammals from various Russian regions have been analysed for borrelial infection, but isolates were obtained and identified from only eight species (Tables 7.2 and 7.4). In all these regions animals proved to contain the same predominant *Borrelia* genospecies and their combinations as those found in ixodid ticks (Table 7.3), with *B. afzelii* and both variants of *B. garinii* having been isolated from all studied species of red-backed voles (*Clethrionomys* spp.). No preference of a certain genospecies for a particular host species was revealed, and the absence of some *Borrelia* spp. from certain species of small mammals (Table 7.4) may be due to an insufficient amount of data. In addition to the species listed in Table 7.4, we examined 13 species of small to medium-sized mammals (*Microtus arvalis*, *Mus musculus*, *Sicista betulina*, *Sorex caecutiens*, *Sorex isodon*, *Sorex minutus*, *Neomys fodiens*, *Talpa europaea*, *Erinaceus europaeus*, *Arvicola terrestris*, *Cricetus cricetus*, *L. timidus* and *Mustela nivalis*) inhabiting an active ITBB focus in the Ural region (see below), for the presence of *Borrelia* spp. in the organs. All the 164 cultures in BSKII medium (from urinary bladder, heart, spleen, kidney and, for some species, ear skin biopsies) were negative.

Our study on the presence of borreliae in birds, which is still the only work of its kind in Russia, was performed in the same natural focus in 1998. Culture medium was inoculated with 94 skin biopsies (taken from the head near the ear opening) and brain samples of 47 birds belonging to four species closely associated with the lower forest layer, namely, 34 *Turdus pilaris*, nine *Turdus musicus*, two *Emberiza citrinella* and two *Anthus trivialis*. No borreliae were found in any of these cultures, which is evidence that birds may not be involved in the circula-

Table 7.4. Distribution of different *Borrelia burgdorferi* sensu lato genospecies in small mammals in Russia.

Species of small mammals	B.garinii 20047ᵀ	B.garinii NT29	B. afzelii	B. burgdorferi sensu lato	Reference
North West region					
Clethrionomys glareolus				+	Grigoryeva and
Clethrionomys rutilus				+	Tretyakov, 1998[a]
Microtus arvalis				+	
Apodemus uralensis				+	
Sorex araneus				+	Grigoryeva, 1996[a]
Ural region					
Clethrionomys glareolus	+	+	+		Gorelova et al.,
Clethrionomys rutilus	+	+	+		1995 (with
Clethrionomys rufocanus	+	+	+		supplements);
Microtus oeconomus	+	+	+		Korenberg et al.,
Microtus agrestis			+		1997 (with
Apodemus uralensis		+	+		supplements)
Sorex araneus		+			
West Siberian region					
Clethrionomys glareolus				+	Matuschenko et
Clethrionomys rutilus				+	al., 1996
Clethrionomys rufocanus				+	
Microtus oeconomus				+	
Apodemus agrarius				+	
Micromys minutus				+	
Sorex sp.				+	
Far East region					
Clethrionomys rufocanus	+	+			Sato et al., 1996
Apodemus peninsulae	+				

[a] Borreliae were revealed by stained blood smears and imprints of kidney cuts, and also by silver-staining of histological preparations.
Blank spaces indicate species not found.

tion of *Borrelia* spp., at least in the study area, though it is also possible that some bird spirochaetes are particularly difficult to culture. In regions where birds are heavily parasitized by ticks, for example the Far East of Russia (see Miyamoto and Masozuwa, Chapter 8), some bird species appear be reservoir hosts of *Borrelia* spp.

Reservoir Hosts

More detailed data on the reservoir hosts and vectors of different *Borrelia* genospecies were obtained during long-term studies (1992–2000) on the natural ITBB foci located in the east European southern taiga forests of the Ural region.

Altogether 595 borrelial isolates were obtained and identified from seven species of small mammals and from *I. persulcatus* and *I. trianguliceps* ticks of all developmental stages.

Some researchers are of the opinion that borreliae persisting in the internal organs of small mammals are of different genospecies from those in the skin. In particular, it is maintained that *B. garinii* occurs exclusively or mainly in the internal organs, but is absent from the skin and, hence, is not transmitted to the ticks that feed on these animals. On this basis, the conclusion is drawn that small mammals do not serve as reservoir hosts of *B. garinii* and can transmit only *B. afzelii* to ticks (Humair *et al.*, 1995, 1999; Gern *et al.*, 1997; Hu *et al.*, 1997; Kurtenbach *et al.*, 1998). However, most of the isolates we obtained from ear skin biopsies of small mammals as well as from *I. persulcatus* larvae removed from these animals were identified as two variants of *B. garinii*; moreover, the incidence of these spirochaetes in the internal organs (the urinary bladder) was even slightly lower than in the skin (Table 7.5). Thus, small mammals proved to be good reservoir hosts not only for *B. afzelii*, but also for both these particular *B. garinii* genospecies variants.

The absence of any fundamental differences in the composition of *Borrelia* spp. in the internal organs and skin biopsies of small mammals makes it possible to assess the aetiological structure of natural foci by analysing all the isolates obtained (Fig. 7.3). The most abundant rodent species in the study region is the bank vole *C. glareolus*, which accounts for more than 70% of animals captured. More than 40% of *Borrelia* spp. isolates obtained from bank voles belonged to *B. garinii* NT29, approximately 25% to *B. afzelii*, less than 20% to *B. garinii* 20047^T, and the rest were mixed isolates (Fig. 7.3, histogram 2). Approximately the same ratio of *B. garinii* (both variants) to *B. afzelii* was observed among isolates obtained from the other six species of reservoir hosts (Fig. 7.3, histogram 3). Moreover, this ratio was also characteristic of partially or fully engorged immature *I. persulcatus* removed from the animals and of unfed nymphs and adult ticks collected from the vegetation. On the whole, this is evidence for the weakness or even the absence of directional selection for a certain *Borrelia* genospecies at different links of the enzootic cycle.

On the other hand, some differences in the relation of *Borrelia* spp. to their associated hosts and ticks were evident, though further study is needed to verify them. Two groups of facts are relevant in this context. Firstly, *B. garinii* 20047^T occurred more frequently in all three species of *Clethrionomys* voles, whereas none of the 35 unmixed isolates from the meadow voles of the genus *Microtus* was identified as *B. garinii* 20047^T; this variant was only found in a few mixed isolates in these voles. Meadow voles may therefore be less efficient as reservoir hosts of *B. garinii* 20047^T than the red-backed voles (*Clethrionomys* spp.). Secondly, *B. afzelii* occurred much less frequently in fully or partially engorged *I. trianguliceps* ticks removed from small mammals (Fig. 7.3, histogram 7) than in the animals proper or in immature and adult *I. persulcatus* ticks. This may be evidence that *I. trianguliceps* is less susceptible to *B. afzelii* than to *B. garinii* and, therefore, there is some selection of one of the *Borrelia* genospecies by an accessory vector.

Table 7.5. Results of identification of *Borrelia burgdorferi* sensu lato isolates, obtained in the Ural region from small mammals and *I. persulcatus* ticks collected from them (1992–2000).

Source	No. of isolates identified	*B. garinii* Total		*B. garinii* Consisting of (absolute number):			*B. afzelii*		Mixture *B. garinii* + *B. afzelii*	
		Absolute	%	Variant 20047^T	Variant NT29	Mixture 20047^T + NT29	Absolute	%	Absolute	%
Small mammals										
Ear skin biopsies	66	56	84.9	15	36	5	8	12.1	2	3.0
Urinary bladder	120	65	54.2	10	48	7	46	38.3	9	7.5
Ticks										
Larvae	40	30	75.0	16	11	3	8	20.0	2	5.0

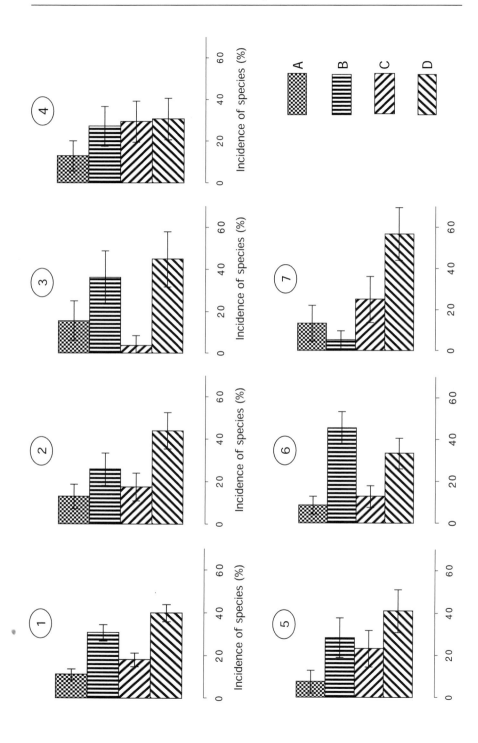

Ixodid Tick-borne Borrelioses Biotopes

As in the case of TBE (Korenberg and Kovalevskii, 1981, 1999), the range of ITBB in Russia coincides with the ranges of *I. ricinus* and *I. persulcatus* ticks (Korenberg, 1994). Within the ITBB range, large areas are free from natural foci, because of the absence of landscape-biocenotic conditions necessary for the existence of these vectors. The mosaic of ITBB biotopes is wholly determined by specific features of the biotopic preference of the main vectors, which varies in different climatic zones and landscapes because the ticks are always confined to the biotopes with hygrothermal conditions appropriate for them. The patchwork pattern of ITBB distribution is eventually formed by the effects of typical zonal, intrazonal and extrazonal conditions and their combinations.

The tick *I. ricinus* usually prefers moderately hygrophilic and mesophilic biotopes in plain and mountain broad-leaved and mixed forests. In the southern and (in places) mid-taiga subzones, it occurs in warmer areas, such as mixed coniferous and deciduous and light secondary forests, clearings, forest margins and shrubby meadows. South of the boundary of continuous zonal broad-leaved forests, this vector is confined to moister floodplain forests and shrubwoods located in depressions of the terrain (Filippova, 1977).

The tick *I. persulcatus* typically inhabits variants of taiga (boreal coniferous forests) but occurs in some other plant formations as well. In the plains its range extends over the area with at least six types of zonal vegetation: northern, mid- and southern taiga forests, mixed and broad-leaved forests, and forest-steppe. The altitudinal zonality of vegetation in the mountain regions contributes to this diversity. The list of plant formations in which this tick occurs is extensive (Korenberg and Kovalevskii, 1985).

In the plain mid-taiga forests, where humidity is considerable and heat supply (seasonal sum of daily positive temperatures) is low, *I. persulcatus* occurs only in the warmest and most well-drained areas. In the Russian northwest, for example, natural ITBB foci in such a situation are located in spruce and small-leaved forests isolated from each other by numerous lakes, bogs and wet and over-mature stands (Fig. 7.4, 1); in vast plain watershed mid-taiga forests, *I. persulcatus* occurs mainly in light forest margins and in the immediate vicinity of villages and towns (Fig. 7.4, 2); and in the mid-taiga zone of West Siberia, intrazonal conditions of large river valleys are most favourable for borrelial circulation (Fig. 7.4, 3).

Fig. 7.3. (Opposite) Incidence of different *Borrelia* genospecies and their mixtures in small mammals and ticks. The histograms (1–7) show the portion (in %) of mixed isolates (A), *Borrelia afzelii* (B), *Borrelia garinii* 20047^T (C) and *B. garinii* NT29 (D) of the number (indicated in square brackets) of isolates, obtained from: (1) total [595]; (2) *Clethrionomys glareolus* [132]; (3) small mammals of other species [58]; (4) fed and partly engorged *Ixodes persulcatus* larvae and nymphs, collected from mammals [85]; (5) unfed *I. persulcatus* nymphs, collected from vegetation [95]; (6) unfed adult *I. persulcatus* ticks, collected from vegetation [165]; (7) fed and partly engorged *Ixodes trianguliceps* ticks, collected from mammals [60].

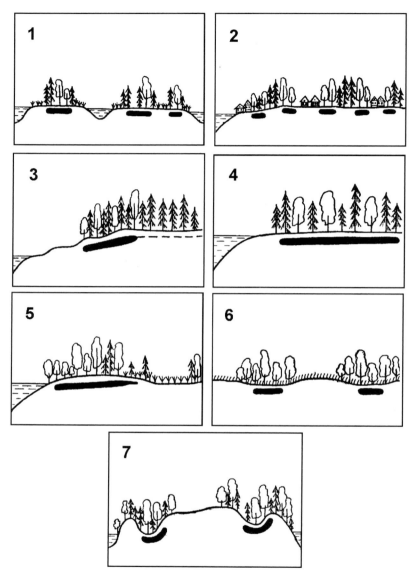

Fig. 7.4. Scheme of landscape preference of populations of *Ixodes persulcatus* ticks – plain territories (according to Korenberg and Kovalevskii, 1985). (1) North West region, middle taiga at the south of Karelia; (2) Northern region, middle taiga at the south of Komi Republic; (3) West Siberia, middle taiga at the right bank of the Ob river; (4) Volga–Vyatka region, European southern taiga forests; (5) West Siberia, southern taiga near to the Irtysh river; (6) West Siberia, northern forest–steppe; (7) Ural region, Kungur forest–steppe. Bold black lines under the landscape profiles show the places, where ticks are found.

In the southern taiga forests of the European plain, where heat supply is high, *I. persulcatus* inhabits virtually all plain watershed forest biocenoses (Fig. 7.4, 4). However, when moisture is excessive, as in many areas of the southern taiga subzone of West Siberia, this vector occurs mainly in well-drained river valleys and mixed waterless valley forests (Fig. 7.4, 5). In the forest–steppe, under conditions of some moisture deficiency but sufficient temperatures for *I. persulcatus*, the tick is only found in the central parts of birch–aspen forest islands (Fig. 7.4, 6) or on the bottoms of ravines and slopes of northern exposure (Fig. 7.4, 7).

The patterns of the altitudinal distribution of *I. persulcatus* ticks and ITBB foci vary considerably depending on such factors as the general geographical location of the mountain system, its orography and altitudinal gradient. In the mountain forests in the south of central Siberia, for example, parasitological prerequisites for the persistence of *B. burgdorferi* s.l. exist virtually everywhere, from the piedmont forest–steppes to the upper boundary of the forest belt, which is located approximately 1400 m above sea level (Fig. 7.5, 1). A similar picture is observed in the northeastern Altai mountains, but in this case the upper forest boundary is located higher and the taiga tick occurs at elevations of up to 2000 m. However, it does not occur in forest areas that, due to some orographic factors, begin at elevations above 1500 m; the same applies to woodless steppe-

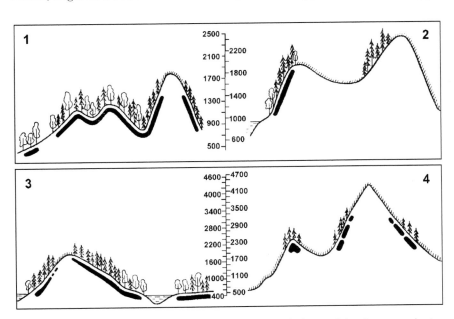

Fig. 7.5. Scheme of landscape preference of populations of *Ixodes persulcatus* ticks – mountain territories (according to Korenberg and Kovalevskii, 1985). (1) Mountains of the south of Central Siberia; (2) Central Altai; (3) mountains of the north surroundings of the Amur river (Far East); (4) Tien Shan. Bold black lines under the landscape profiles show the places where ticks are found. Vertical scale, height above sea-level.

like slopes of southern, southeastern and southwestern exposure (Fig. 7.5, 2). Near the northern boundary of the *I. persulcatus* range, in the mountain ridges located in the middle taiga subzone of Khabarovsk territory, this vector regularly occurs on the northwestern slopes at elevations of no more than 750–800 m, whereas on the opposite warmer slopes it is found up to the mountain tops at elevations of approximately 1400 m above sea level (Fig. 7.5, 3). These examples do not exhaust the entire diversity of landscape–biotic preferences of the vector that determine the distribution of ITBB biotopes over the vast and varied territory of Russia. Due to the many factors involved, essentially different conditions may form within a zone, which may be more varied than conditions between zones.

Ecological Risk Factors

In diseases with natural focality, the number of persons infected in a focus and the actual risk of infection are eventually the product of the epizootic potential of the focus and the frequency of human contact with it (Korenberg and Yurkova, 1983). In essence, the ecological factors responsible for the risk of infection are the same factors (parameters) that determine the epizootic potential of a focus. In ITBB, as in TBE (Korenberg and Kovalevskii, 1999), the principal parameter is the number of infected ticks, and this number depends on the general level of tick abundance and the prevalence of borreliae in them. The latter parameters differ considerably from one landscape subzone to another, and this is manifested especially clearly in many Russian oblasts that extend from the north to the south for a great distance. Thus, in different subzones of the Tyumen oblast in West Siberia, the average abundance of *I. persulcatus* adults and the prevalence of borreliae in them differ by factors of 9 and *c.* 2, respectively (Table 7.6). Differences between these ecological parameters in different landscape subzones of the Perm oblast in the Ural region are of similar magnitude. The frequency of visits by people to the forest is approximately the same throughout this oblast (Table 7.7), but data on the frequency of tick attachment and the parameters of

Table 7.6. Parameters of abundance of adult *I. persulcatus* ticks on vegetation and prevalence of *Borrelia burgdorferi* sensu lato in them (by dark-field microscopy of internal organs) in different landscape subzones of Tyumen oblast, situated in the West Siberian region. (After Kolchanova, 1997.)

Landscape subzones	Number of ticks per flag-hour			Prevalence of *Borrelia* spp. in ticks
	Minimum	Maximum	Average	
Middle taiga	0	24	1.6	27.5 ± 4.4
Southern taiga	0	37	2.7	32.8 ± 3.6
Subtaiga	1	57	4.6	44.6 ± 3.4
Northern forest–steppe	0	1	0.5	22.6 ± 6.2

Table 7.7. Frequency of human contacts with ticks and agents of ITBB in different landscape-geographical subdivisions of Perm oblast, situated in the Ural region. (After Alypova et al., 2002.)

Landscape subzones	% of people who visited forest	% of people who noted tick attachment	% of sera with titre 1:20 or more	Mean geometrical titre of antibodies for positive sera	Average long-term parameter of morbidity per 100,000 people
Middle taiga	90.8 ± 3.7	17.2 ± 4.8	9.2 ± 3.7	1:12.1	4.84
Southern taiga	88.1 ± 4.9	41.3 ± 2.9	27.2 ± 2.6	1:24.2	16.27
Coniferous–deciduous forests	84.0 ± 4.2	39.3 ± 5.6	22.9 ± 4.6	1:22.6	13.51
Forest–steppe	88.1 ± 3.8	38.1 ± 5.7	26.2 ± 5.2	1:19.7	8.99
Mountain taiga	88.6 ± 3.3	24.9 ± 4.5	12.7 ± 3.5	1:17.1	11.32

Table 7.8. Strength of correlations between long-term parameters of adult *Ixodes persulcatus* density, infected tick density, *Borrelia burgdorferi* sensu lato prevalence in these ticks and parameters of human ITBB morbidity (observations in the Ural region).

Year	Density of adult *I. persulcatus* (per hectare)		*Borrelia* prevalence in ticks (%)	Number of ITBB cases (per 100,000 people)
	Total	Infected ticks only		
1993	835	424	50.8	32.3
1994	663	243	36.6	16.5
1995	226	78	34.5	15.4
1996	1423	865	60.8	35.7
1997	212	67	31.6	14.6
1998	390	115	29.5	17.1
1999	1135	546	48.1	20.8
2000	463	165	35.7	11.1
Coefficient of correlation (r) with parameters of morbidity	0.79	0.85	0.90	

anti-ITBB immunity among residents indicate that people living in the subzones of southern taiga forests, coniferous/broad-leaved forests, and forest–steppe are at a higher risk of acquiring ITBB. Risk is moderate in the southern part of the mountain taiga subzone and is relatively low in the southern part of the middle taiga subzone. There is a close correlation in different landscape subzones between the proportion of people bitten by ticks and the proportion of sera with a high IFA titre of antibodies (> 1:20 and higher) (correlation coefficient $r = 0.98$) and also between long-term average parameters of ITBB morbidity ($r = 0.8$) (Alypova *et al.*, 2002).

A long-term study in the Ural region revealed that the level of ITBB morbidity and, therefore, the risk for people to acquire infection changed significantly from year to year (Table 7.8). These changes correlate well with annual fluctuations in the total density of adult ticks and the density of ticks infected by *Borrelia* spp. In contrast to the situation with TBE (Korenberg and Kovalevskii, 1999), there is also a close correlation between ITBB morbidity and *Borrelia* spp. prevalence in ticks. This may be explained by differences in the infective doses (the dose necessary for causing a clinically manifest disease) of TBE virus and borreliae and in specific features of their behaviour at the earliest stage of the infection process. Nevertheless, the density of infected ticks clearly plays the key role as the principal ecological risk factor.

Conclusion

This chapter presents an overview of the available data on the ecology of *B. burgdorferi* s.l. in Russia. The vast Russian territory, with its diverse landscape, climatic conditions and ecosystems, has been poorly studied in this respect. However, it has already become apparent that the ecology of widespread *Borrelia* genospecies has very important regional features that will be the focus of research attention in the future. In this connection, we attach special significance to studies on the specificity of the aetiological structure of natural ITBB foci in different regions, the degree of their stability and the factors determining their dynamics.

Acknowledgements

This study was supported in part by the Russian Foundation for Basic Research, grant no. 01-04-48152.

References

Alekseev, A.N. (1993) *Tick–Tick-Borne Pathogen System and its Emergent Qualities.* St Petersburg, 203 pp. (in Russian).

Alypova, I.I., Korenberg, E.I. and Vorob'eva, N.N. (2002) Risk of ixodid tick-borne borreliosis infection for people in different landscape subzones of Perm region. *Meditsinskaya Parasitologiya i Parasitarnye Bolezni* 1, 37–40 (in Russian).

Apostolova, I.Y., Borisova, T.I. and Kulikova, E.V. (1996) Infection rate of ticks with *Borrelia burgdorferi* in suburban zone of Irkutsk city. In: *Human Diseases with Natural Focality.* Omsk, pp. 190–191 (in Russian).

Assous, M., Postic, D., Paul, G., Nevot, P. and Baranton, G. (1994) Individualisation of two new genomic groups among American *Borrelia burgdorferi* sensu lato strains. *FEMS Microbiology Letters* 121, 93–98.

Baranton, G., Postic, D., Saint Girons, I., Boerlin, P., Piffaretti, J.-C., Assous, M. and Grimont, P.A.D. (1992) Delineation of *Borrelia burgdorferi* sensu stricto, *Borrelia garinii* sp. nov., and group VS461 associated with Lyme borreliosis. *International Journal of Systematic Bacteriology* 42, 378–383.

Belfaiza, J., Postic, D., Bellenger, E.,

Baranton, G. and Saint Girons, I. (1993) Genomic fingerprinting of *Borrelia burgdorferi* sensu lato by pulsed-field gel electrophoresis. *Journal of Clinical Microbiology* 31, 2873–2877.

Belyaeva, M.L. (1998) Results of studies on *Borrelia* prevalence in ticks in Sverdlovsk region. In: *Actual Problems of Infections with Natural Focality.* Izhevsk, pp. 166–167 (in Russian).

Burenkova, L.A. (2000) Seasonal changes in infection rate of ticks with agents of tick-borne borrelioses at the north of Kaluga region. *Dezinfektsionnoe Delo* 2, 14–16 (in Russian).

Burgdorfer, W., Barbour, A., Hayes, S., Peter, O. and Aeschlimann, A. (1983) Erythema chronicum migrans – a tick-borne spirochetosis. *Acta Tropica* 40, 79–83.

Dubinina, E.V. and Alekseev, A.N. (1999) The biodiversity dynamics of the pathogens being transmitted by the ticks of the genus *Ixodes*: an analysis of the long-term data. *Meditsinskaya Parasitologiya i Parasitarnye Bolezni* 2, 13–19 (in Russian).

Filippova, N.A. (1977) *Ixodid Ticks of Subfamily Ixodinae: Arachnaidae.* (Fauna of USSR. Vol. 4, iss. 4.) Nauka, Moscow, Leningrad, 396 pp. (in Russian).

Filippova, N.A. (1999) Sympatry of closely related species of *Ixodid* ticks and its possible role in parasitic systems of natural foci of transmissive diseases. *Parazitologiya* 33, 223–241 (in Russian).

Fukunaga, M., Sohnaka, M. and Yanagihara, Y. (1993) Analysis of *Borrelia* species associated with Lyme disease by rRNA gene restriction fragment length polymorphism. *Journal of General Microbiology* 139, 1141–1146.

Gern, L., Humair, P.F., Hu, C.M. and Leuba-Garcia, S. (1997) Ecology of *Borrelia burgdorferi* sensu lato in Europe. In: Suss, J. and Kahl, O. (eds) *Tick-borne Encephalitis and Lyme Borreliosis*. Pabst Science Publishers, Lengerich, pp. 271–280.

Gorelova, N.B. (1993) Museum of *Borrelia* of the Russian Center for Borrelioses. In: Korenberg, E.I. (ed.) *Problems of Tick-borne Borrelioses*. Moscow, pp. 38–44.

Gorelova, N.B., Korenberg, E.I., Kovalevskii, Y.V. and Shcherbakov, S.V. (1995) Small mammals as reservoir hosts for *Borrelia* in Russia. *Zentralblatt für Bacteriologie* 282, 315–322.

Gorelova, N.B., Korenberg, E.I., Kovalevskii, Y.V., Postic, D. and Baranton, G. (1996) Isolation of *Borrelia* from the tick *Ixodes trianguliceps* (*Ixodinae*) and the significance of this species in epizootiology of ixodid-tick-borne borrelioses. *Parazitologiya* 30, 13–18.

Gorelova, N.B., Korenberg, E.I., Filippova, N.A. and Postic, D. (2001a) First isolation of *Borrelia*, pathogenic for human, from *Ixodes pavlovskyi* Pom. ticks. *Doklady Rossiyskoyi Akademii Nauk* 378, N 4, 1–2 (in Russian).

Gorelova, N.B., Korenberg, E.I., Postic, D., Yunicheva, Y.V. and Ryabova, T.E. (2001b) First isolation of *Borrelia burgdorferi* sensu stricto in Russia. *Zhurnal Microbiologii, Epidemiologii i Immunobiologii* 3, 459–462 (in Russian).

Gorin, O.Z., Zlobin, V.I., Arbatskaya, E.V., Dzhioev, Y.P., Kozlova, I.V., Ivanova, L.V., Chernogor, L.I., Cuntsova, O.V. and Verhozina, M.V. (1996) Actual problems of epidemiology and prophylaxis of transmissible infections in Siberia. *Zhurnal Infektsionnoy Patologii* 3, 10–12 (in Russian).

Grigoryeva, L.A. (1996) Shrews as reservoir hosts for *Borrelia* in the north-west of Russia. *Parazitologiya* 30, 458–459 (in Russian).

Grigoryeva, L.A. and Tretyakov, K.A. (1998) Peculiarity of parasitic system ixodid ticks – *Borrelia* – micromammalia in the north-west of Russia. *Parazitologiya* 32, 422–430 (in Russian).

Hu, C.M., Humair, P.F., Wallich, R. and Gern, L. (1997) *Apodemus* sp. rodents, reservoir hosts for *Borrelia afzelii* in an endemic area in Switzerland. *Zentralblatt für Bacteriologie* 285, 558–564.

Humair, P.F., Peter, O., Wallich, R. and Gern, L. (1995) Strain variation of Lyme disease spirochetes isolated from *Ixodes ricinus* ticks and rodents collected in two endemic areas in Switzerland. *Journal of Medical Entomology* 32, 433–438.

Humair, P.F., Rais, O. and Gern, L. (1999) Transmission of *Borrelia afzelii* from *Apodemus* mice and *Clethrionomys* voles to *Ixodes ricinus* ticks: differential transmission pattern and overwintering maintenance. *Parasitology* 118, 33–42.

Ivanov, L.I., Liberova, R.N., Korenberg, E.I., Zdanovskaya, N.I., Vorob'eva, R.N., Volkov, V.I. and Vysochina, N.P. (1996) Natural foci of Lyme disease in Khabarovsk region. In: *International Scientific Conference 'Tick-borne Viral, Rickettsial and Bacterial Infections'*. Irkutsk, pp. 89–90.

Johnson, R.C., Schmid, G.P., Hyde, F.W., Steigerwalt, A.G., and Brenner, D.J. (1984) *Borrelia burgdorferi* sp. nov.: etiologic agent of Lyme disease. *International Journal of Systematic Bacteriology* 34, 496–497.

Kolchanova, L.P. (1997) Spontaneous *Borrelia* infection in ticks and degree of their individual infectivity in different landscape subzones of Tyumen region. *Meditsinskaya Parasitologiya i Parasitarnye Bolezni* 1, 49–50 (in Russian).

Korenberg, E.I. (1962) Role of birds in feeding of ixodid ticks in natural foci of encephalitis in forest zone. *Zoologicheskii Zhurnal* 41, 1220–1225 (in Russian).

Korenberg, E.I. (1964) Birds and the problem of natural focality of tick-borne encephalitis. *Zoologicheskii Zhurnal* 45, 245–260 (in Russian).

Korenberg, E.I. (1978) Mammals – hosts of adult taiga ticks at the east of Russian plain. In: Panteleev, P.A. (ed.) *Proceedings of the 2nd Congress of All-Union Teriological Society.* Nauka, Moscow, pp. 257–258 (in Russian).

Korenberg, E.I. (1979) Small mammals and problem of natural focus of tick-borne encephalitis. *Zoologicheskii Zhurnal* 58, 542–552 (in Russian).

Korenberg, E.I. (1993) Borrelioses. In: Pokrovsky, V. (ed.) *Manual on Epidemiology of Infectious Diseases*, Vol. 2. Meditsina, Moscow, pp. 382–392 (in Russian).

Korenberg, E.I. (1994) Comparative ecology and epidemiology of Lyme disease and tick-borne encephalitis in the former Sovet Union. *Parasitology Today* 10, 157–160.

Korenberg, E.I. (1995) Ixodid tick-borne borrelioses (ITBBs), infections of the Lyme borreliosis group, in Russia: country report. In: *Reports of WHO Workshop on Lyme Borreliosis Diagnosis and Surveillance.* Sanitati, WHO/CDC/VPH/95.141-1, Warsaw, pp. 128–136.

Korenberg, E.I. (1996) Taxonomy, phylogenetic relations and centres of origin of distinct forms of spirochetes of the genus *Borrelia* transmitted by ixodid ticks. *Uspehi Sovremennoy Biologii* 116, 389–406 (in Russian).

Korenberg, E.I. (1999) Interaction between transmissible disease agents in ixodid ticks (*Ixodidae*) with a mixed infection. *Parazitologiya* 33, 273–289 (in Russian).

Korenberg, E.I. and Kovalevskii, Y.V. (1981) *Regionalization of the Tick-borne Encephalitis Range. Advances in Science and Technology, Medical Geography*, Vol. 11. VINITI, Moscow, 148 pp.

Korenberg, E.I. and Kovalevskii, Y.V. (1985) Landscape relation. In: Filippova, N.A. (ed.) *Taiga Tick Ixodes persulcatus Schulze (Acarina, Ixodidae).* Nauka, Leningrad, pp. 193–198 (in Russian).

Korenberg, E.I. and Kovalevskii, Y.V. (1994) A model for relationships among the tick-borne encephalitis virus, its main vectors, and hosts. In: Harris, K.E. (ed.) *Advances in Disease Vector Research* 10. Springer-Verlag, New York, pp. 65–92.

Korenberg, E.I. and Kovalevskii, Y.V. (1999) Main features of tick-borne encephalitis eco-epidemiology in Russia. *Zentralblatt für Bacteriologie* 289, 525–539.

Korenberg, E.I. and Kryuchechnicov, V.N. (1996) Ixodid tick-borne borrelioses – a new group of human diseases. *Zhurnal Microbiologii, Epidemiologii i Immunobiologii* 4, 104–108 (in Russian).

Korenberg, E.I. and Lebedeva, N.N. (1969) Distribution and some general features of the ecology of *Ixodes trianguliceps* Bir. in the Soviet Union. *Folia Parasitologica* 16, 143–152.

Korenberg, E.I. and Yurkova, E.V. (1983) The problem of predicting epidemic manifestations of natural foci of human diseases. *Meditsinskaya Parasitologiya i Parasitarnye Bolezni* 3, 3–10 (in Russian).

Korenberg, E.I., Kryuchechnikov, V.N., Dekonenko, E.P., Scherbakov, S.V. and Anan'ina, Y.V. (1986) Serological verification of Lyme disease in Russia. *Zhurnal Microbiologii, Epidemiologii i Immunobiologii* 6, 111–113 (in Russian).

Korenberg, E.I., Kryuchechnikov, V.N., Kovalevskii, Y.V., Scherbakov, S.V., Kuznetsova, R.I. and Levin, M.L. (1987a) Tick *Ixodes persulcatus* Schulze – a new vector of *Borrelia burgdorferi. Doklady Academii Nauk SSSR* 297, 1268–1270 (in Russian).

Korenberg, E.I., Scherbakov, S.V. and Kryuchechnikov, V.N. (1987b) Data on distribution of Lyme disease in USSR. *Meditsinskaya Parasitologiya i Parasitarnye Bolezni* 2, 71–73 (in Russian).

Korenberg, E.I., Kovalevskii, Y.V., Kuznetsova, R.I., Fonarev, L.S., Churilova, A.A., Antykova, L.P., Kalinin, M.I., Kryuchechnikov, V.N., Mebel', B.D., Scherbakov, S.V. and Kovtunenko, S.S. (1988) Revealing and first results of studies of Lyme disease at the north-west of USSR. *Meditsinskaya Parasitologiya i Parasitarnye Bolezni* 1, 45–48 (in Russian).

Korenberg, E.I., Scherbakov, S.V., Zaharycheva, T.A., Levin, M.L., Kalinin, M.I. and Kryuchechnikov, V.N. (1989) Lyme disease in Khabarovsk region. *Meditsinskaya Parasitologiya i Parasitarnye Bolezni* 5, 74–78 (in Russian).

Korenberg, E.I., Kovalevskii,Y.V., Kryuchechnikov, V.N. and Gorelova, N.B. (1991a) Tick *Ixodes persulcatus* Schulze, 1930 as a vector of *Borrelia burgdorferi*. In: Dusbabek, F. and Bukva, V. (eds) *Modern Acarology*, Vol. 1. Academia, Prague, pp. 119–123.

Korenberg, E.I., Kuznetsova, R.I., Kovalevskii, Y.V., Vasilenko, Z.E. and Mebel', B.D. (1991b) Main features of epidemiology of Lyme disease at the north-west of USSR. *Meditsinskaya Parasitologiya i Parasitarnye Bolezni* 3, 14–17 (in Russian).

Korenberg, E.I., Gorelova, N.B., Postic, D., Kovalevskii, Y.V., Baranton, G. and Vorobyeva, N.N. (1997) Reservoir hosts and vectors of *Borrelia*, causative agents of ixodid tick-borne borrelioses in Russia. *Zhurnal Microbiologii, Epidemiologii i Immunobiologii* 6, 36–38 (in Russian).

Korenberg, E.I., Gorelova, N.B., Postic, D. and Kotti, B.K. (1999) *Borrelia* species, new for Russia, as possible causative agents of ixodid tick-borne borreliosis. *Zhurnal Microbiologii, Epidemiologii i Immunobiologii* 2, 3–5 (in Russian).

Korenberg, E.I., Kovalevskii, Y.V., Levin, M.L. and Shchyogoleva, T.V. (2001a) The prevalence of *Borrelia burgdorferi* sensu lato in *Ixodes persulcatus* and *I. ricinus* ticks in the zone of their sympatry. *Folia Parasitologica* 48, 63–68.

Korenberg, E.I., Gorban', L.J., Kovalevskii, Y.V., Frizen,V.I. and Karavanov, A.S. (2001b) Risk of human tick-borne encephalitis, borrelioses, and double infection in the Pre-Ural Region of Russia. *Emerging Infectious Diseases* (in press).

Kryuchechnikov, V.N., Korenberg, E.I., Scherbakov, S.V., Kovalevskii, Y.V. and Levin, M.L. (1988) Identification of *Borrelia*, isolated in USSR from *Ixodes persulcatus* Schulze ticks. *Zhurnal Microbiologii, Epidemiologii i Immunobiologii* 12, 41–44 (in Russian).

Kryuchechnikov, V.N., Korenberg, E.I., Sherbakov, S.V., Gorelova, N.B., Jurzicova, Z., Halouzka, Z., Hubalek, Z., Kovalevskii, Y.V., Levin M.L. and Bunikis, A.I. (1990) Identification of *Borrelia*, isolated in USSR and Czechoslovakia from *Ixodes ricinus* (L.) ticks. *Zhurnal Microbiologii, Epidemiologii i Immunobiologii* 6, 10–13 (in Russian).

Kryuchechnikov, V.N., Gorelova, N.B. and Sherbakov, S.V. (1993) Identification of *Borrelia* and results of studying isolates of Lyme disease agent from Russia and neighbouring countries. In: Korenberg, E.I. (ed.) *Problems of Tick-borne Borrelioses*. Moscow, pp. 45–55 (in Russian).

Kurtenbach, K., Peacey, M., Rijpkema, S.G.T., Hoodless, A.N., Nuttall, P.A. and Randolph, S.E. (1998) Differential transmission of the genospecies of *Borrelia burgdorferi* sensu lato by game birds and small rodents in England. *Applied and Environmental Microbiology* 64, 1169–1174.

Lisovskaya, N.D. (1958) Data on studies of chronical atrophic acrodermatitis. Candidate Sc. thesis, State Institute for Perfection of Physicians, Leningrad, USSR (in Russian).

Liveris, D., Gazumyan, A. and Schwartz, I. (1995) Molecular typing of *Borrelia burgdorferi* sensu lato by PCR– restriction fragment length polymorphism analysis. *Journal of Clinical Microbiology* 33, 589–595.

Masuzawa, T., Iwaki, A., Sato, Y., Miyamoto, K., Korenberg, E.I. and Yanagihara, Y. (1997) Genetic diversity of *Borrelia burgdorferi* sensu lato isolated in Far Eastern Russia. *Microbiology and Immunology* 41, 595–600.

Matuschenko, A.A., Venediktov, V.S., Yakimenko, V.V. and Tantsev, A.K. (1996) Role of small mammals in circulation of *Borrelia* in natural foci of tick-borne borreliosis at the south-west of Siberia. *Zhurnal Infektsionnoy Patologii* 3, 37–39 (in Russian).

Mescherskii, G.I. (1898) To the problem of atrophia cutis idiopathica progressiva acuisita. *Records of Moscow Venereologic and Dermatologic Society* 8, 135–136 (in Russian).

Mishin, A.V. (1964) Terms of existence of taiga tick in Udmurtiya ASSR. In: *Collected Works of Izhevsk Medical Institute* 20. Izhevsk, pp. 66–73 (in Russian).

Naumov, R.L. (1985) Trophic relations. In: Filippova, N.A. (ed.) *Taiga Tick Ixodes Persulcatus Schulze (Acarina, Ixodidae).* Nauka, Leningrad, pp. 277–312 (in Russian).

Naumov, R.L. (1999) Tick-borne encephalitis and Lyme disease: epizootiological parallels and monitoring. *Meditsinskaya Parasitologiya i Parasitarnye Bolezni* 2, 20–26 (in Russian).

Naumov, R.L., Gutova, V.P., Ershova, A.S. and Papel'nitskaya, N.P. (1994) Infection rate of taiga ticks with *Borrelia* in Western Sayans. *Meditsinskaya Parasitologiya i Parasitarnye Bolezni* 3, 19–20 (in Russian).

Nikol'skii, P.V. (1896) Demonstration of the case of atrophia cutis maculata. *Diary of the 6th Congress of Russian Physicians in Memory of N.I. Pirogov* 12, 89 (in Russian).

Pisemskii, N.N. (1902) A case of idiopathic acquired atrophia cutis. *Meditsinskoe obozrenie* 37, 710–713 (in Russian).

Pospelov, A.I. (1905) *Manual on Studies of Skin Diseases.* Moscow, 650 pp. (in Russian).

Pospelov, A.I. (1906) Two cases of autonomous atrophia cutis in adult humans. *Meditsinskoe obozrenie* 26, 565–574 (in Russian).

Postic, D., Assous, M.V., Grimont, P.A.D. and Baranton, G. (1994) Diversity of *Borrelia burgdorferi* sensu lato evidenced by restriction fragment length polymorphism of *rrf* (5S) –*rrl* (23S) intergenic spacer amplicons. *International Journal of Systematic Bacteriology* 44, 743–752.

Postic, D., Korenberg, E., Gorelova, N., Kovalevskii, Y.V., Bellenger, E. and Baranton, G. (1997) *Borrelia burgdorferi* sensu lato in Russia and neighbouring countries: high incidence of mixed isolates. *Research in Microbiology* 148, 691–702.

Rudakova, S.A. and Matuschenko, A.A. (1996) Results of studies of natural foci of tick-borne borreliosis at regions of the south of Western Siberia. In: *Human Diseases with Natural Focality.* Omsk, pp. 169–173 (in Russian).

Rybakova, N. and Evsyukova, N. (1999) Lyme borreliosis in Vologda province – clinical and epidemiological characteristics. In: *VIII International Conference on Lyme Borreliosis and other Emerging Tick-borne Diseases, Abstract Book.* Munich, p. 43.

Saint Girons, I., Gern, L., Gray, J.S., Guy, E.C., Korenberg, E., Nuttall, P.A., Rijpkema, S.G.T., Schönberg, A., Stanek, G. and Postic, D. (1998) Identification of *Borrelia burgdorferi* sensu lato species in Europe. *Zentralblatt für Bacteriologie* 287, 190–195.

Sato, Y., Miyamoto, K., Iwaki, A., Masuzawa, T., Yanagihara, Y., Korenberg, E., Gorelova, N.B., Volkov, V.I., Ivanov, L.I. and Liberova, R.N. (1996) Prevalence of Lyme disease spirochetes in *Ixodes persulcatus* and wild rodents in Far Eastern Russia. *Applied and Environmental Microbiology* 62, 3387–3389.

Savitskii, B.P. (1970) Types of tick-borne encephalitis foci in Khabarovsk region. *Scientific Reports of Higher School. Biological Sciences* 7, 26–30 (in Russian).

Semenov, A.V., Alekseev, A.N. and Dubinina, H.V. (2000) Electrophoretypes, genotypes and *Borrelia* agents of *Ixodes persulcatus* ticks. In: *International Symposium 'Ecological Parasitology on the Turn of Millennium'*. St Petersburg, p. 56.

Smirnova, N.N., Kilevoi, L.Y. and Kushnarev, I.G. (1996) Studies on the state of a natural focus of Lyme disease in Kurgan region. In: *Human Diseases with Natural Focality*. Omsk, pp. 177–178 (in Russian).

Tarasov, V.N., Ul'yantsenko, E.V., Kondrashova, O.N., Bobrovskaya, G.N. and Oseledchik, N.G. (2000) Prevalence of *Borrelia* in ixodid ticks in Chelyabinsk region. *Meditsinskaya Parasitologiya i Parasitarnye Bolezni* 4, 48 (in Russian).

Tupikova, N.V., Suvorova, L.G. and Korenberg, E.I. (1980) On the evaluation of the importance of different small mammals species as hosts of larvae and nymphs of *Ixodes persulcatus*. In: *Fauna and Ecology of the Rodents*, 14. Moscow University Press, Moscow, pp. 158–176 (in Russian).

Vasilieva, I.S. and Naumov, R.L. (1996) Lyme disease parasitosic system, the state of the problem. Communication 1. Pathogens and vectors. *Acarina* 4, 53–75 (in Russian).

Ecology of *Borrelia burgdorferi* sensu lato in Japan and East Asia

Kenji Miyamoto[1] and Toshiyuki Masuzawa[2]

[1]*Department of Parasitology, Asahikawa Medical College, Asahikawa, 078-8510, Japan;* [2]*Department of Microbiology, School of Pharmaceutical Sciences, University of Shizuoka, 52-1 Yada, Shizuoka 422-8526, Japan*

Introduction

The multisystemic disorder known as Lyme disease (Lyme borreliosis, LB) was first described in the USA in the 1970s (Steere *et al.*, 1977), the deer tick (*Ixodes ricinus* complex) infected with the spirochaete *Borrelia burgdorferi* sensu lato (s.l.) being the cause of this disease (Steere, 1989). *B. burgdorferi* s.l. has been classified into at least 11 genetically similar pathogenic species, including *B. burgdorferi* sensu stricto (s.s.), *Borrelia garinii* and *Borrelia afzelii* (Baranton *et al.*, 1992; Canica *et al.*, 1993). While LB is the most frequently reported arthropod-borne infection in North America and Europe (Jaenson, 1991; Centers for Disease Control, 1992; O'Connell *et al.*, 1998; Orloski *et al.*, 2000), the extent of the ecological impact of this disease in Asia has only recently been realized.

The study of LB in Japan started in 1987 after an *Ixodes persulcatus* tick bite in Nagano Prefecture (Kawabata *et al.*, 1987). This was the first recorded case of erythema migrans at the site of the tick bite along with antibodies to *B. burgdorferi*. However, cases in which there was the appearance of erythema after tick bites from forest regions in Japan, and whether these were allergic or infectious reactions, had been discussed among dermatologists before 1987.

According to Yamaguti, 15 tick species (six *Ixodes*, five *Haemaphysalis*, one *Amblyomma*, one *Rhipicephalus* and two *Argas*) have been reported from human tick bite cases across Japan (Yamaguti, 1989). Outbreaks of tick bites are observed in all seasons in the southern parts of Japan and from late spring to early autumn in northern parts. The taiga tick *I. persulcatus* has been incriminated as the potential vector of LB in the Eurasian zone including Japan. However, *Ixodes ovatus* is the most widely distributed tick in flatlands from the northernmost island of Hokkaido to the southernmost island of Kyushu. In Hokkaido, there are more areas of natural vegetation than in other areas of Japan; these areas provide the habitat for the birds and mammals that act as the reservoir for the large numbers of ixodid ticks. Recently, the population of sika deer (*Cervus nippon*) in Hokkaido has increased and has been expanding into new areas with resultant damage to agriculture (Takahashi *et al.*, 1997; Kaji *et al.*, 2000).

Numerous cases of tick bites, some accompanied by LB, occur every year, not only in the inhabitants but also in the many visitors to Hokkaido. On 1 April 1999, new legislation for infectious diseases was introduced requiring a clinical doctor to notify any LB cases to the national government.

This chapter is devoted to the natural history of the agent, *B. burgdorferi* s.l., in Japan and East Asia and to the relationships between the spirochaete, its tick vector and its host in nature.

Activity of Ticks and Outbreaks of Tick-bite Cases

The activities of host-seeking ticks have been investigated at Furano, central Hokkaido, by blanket-drag sampling over 2.4 km by two people twice a month (Fig. 8.1). Tick questing starts in mid- or late April, depending on weather conditions. Two predominant species of tick, *I. persulcatus* and *I. ovatus*, were collected from the study area and the numbers of the ticks increased as the season progressed. The maximum number of ticks was found in early June; thereafter, the number decreased towards the autumn and no ticks were collected in September. The collected ticks were almost all adults but included a small

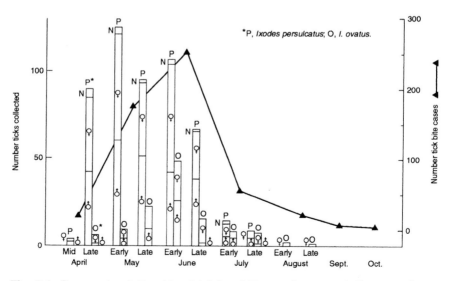

Fig. 8.1. Seasonal activity of ixodid ticks at Furano from vegetation sampling data (2.4 km) and monthly records of tick-bite cases in Hokkaido.

Table 8.1. Seasonal fluctuation of cases of tick bite in Hokkaido (1980–2000).

Month	No. of cases		Total no.
	Ixodes persulcatus	Ixodes ovatus	
March	0	0	0
April	12	6	18
May	141	33	175 (1)[a]
June	225	29	255 (1)[b]
July	71	4	75
August	17	4	21
September	4	2	6
October	1	2	3
November	0	0	0
Total	471 (85.2%)	80 (14.5%)	553 (2)

[a] One case of *Haemaphysalis flava*.
[b] One case of *Ixodes nipponensis*.

number of nymphal *I. persulcatus* between late April and early July (Miyamoto and Nakao, 1991).

All stages of *I. persulcatus* were collected from vegetation by the flag-sampling method in Niigata Prefecture including Sado Island, and peak numbers of ticks were observed between summer and early autumn (Saito, 1959). The questing activity of *I. persulcatus* in each locality is quite different. In general, the life cycle of *I. persulcatus* usually takes 2 years in Japan, but it was completed in only 4.5 months when the tick was reared in the laboratory (Saito, 1960).

Since the first tick specimen was obtained from the back of a human subject in 1980, a total of 553 ticks have been collected from cases between each April and October, from 1980 to 2000, with maximum numbers in June in Hokkaido (Table 8.1). Of these ticks, 471 (85.2%) out of 553 specimens were *I. persulcatus*, and 80 (14.5%) were *I. ovatus*, the remaining two being *Ixodes nipponensis* and *Haemaphysalis flava* (Table 8.1; unpublished data). Almost all cases were infested with female ticks, except for one male and two nymphs of *I. persulcatus*, and most were found as single specimens, although very rarely two or three occurred together.

I. persulcatus prevails on flatland from Hakodate in southern Hokkaido to Wakkanai in the northernmost part of the island, whereas on the mainland this tick species inhabits the high altitude mountainous zone (an elevation of 800–1000 m) from Aomori to Oita. There is a particularly high density of *I. persulcatus* in the mountainous areas of Nagano Prefecture (Uchikawa *et al.*, 1991).

B. burgdorferi s.l. in Japan and Relationships with Vector Species

Since the initial report of LB in Japan (Kawabata *et al.*, 1987), several *B. burgdorferi* s.l. species have been isolated from patients, wild rodents, migratory birds and five species of ixodid ticks (*I. persulcatus*, *I. ovatus*, *Ixodes tanuki*, *Ixodes turdus* and *Ixodes columnae*) (Masuzawa *et al.*, 1991a,b, 1994; Miyamoto *et al.*, 1991, 1992a,b, 1997, 2000; Nakao and Miyamoto, 1993a,b,c, 1994; Miyamoto and Nakao, 1994; Hashimoto *et al.*, 1995). Recently developments in molecular biology have provided new and powerful epidemiological tools for the accurate typing of *B. burgdorferi* s.l. Serotyping systems based on reactivity with monoclonal antibodies (mAbs), various DNA hybridization methods, as well as restriction fragment length polymorphism (RFLP) analysis using several spirochaete rRNA genes, rRNA gene intergenic spacers, the flagellin gene and genomic DNA as targets, have been successfully exploited (Baranton *et al.*, 1992; Marconi and Garon, 1992; Wilske *et al.*, 1992; Fukunaga *et al.*, 1993b; Kawabata *et al.*, 1994; Postic *et al.*, 1994; Fukunaga and Koreki, 1996). Using these powerful methods, we have begun to systematically explore the distribution of many *B. burgdorferi* s.l. isolates in the hope that such information will aid our understanding of the *Borrelia* spp. that cause LB in Japan.

Epidemiological surveys have revealed the high prevalence of LB-related *Borrelia* spp. in both *I. persulcatus* and *I. ovatus*, which are the most common tick species that infest humans in Japan (Uchikawa *et al.*, 1991; Miyamoto *et al.*, 1992a,b; Nakao *et al.*, 1992a,b). Divergent protein profiles for several Japanese *B. burgdorferi* s.l. isolates from *I. persulcatus* have been found, some of which share similar protein profile patterns with some European isolates (Masuzawa *et al.*, 1991b; Nakao *et al.*, 1992a; Fukunaga *et al.*, 1993a). RFLP patterns of the 16S rRNA gene of isolates from *I. persulcatus* can be classified into five groups (ribotypes), II–VI, whereas RFLP analysis of *B. burgdorferi* s.s., *B. garinii* and *B. afzelii* from North America and Europe reveals three ribotypes, I, II and III. While some of the Japanese isolates from *I. persulcatus* belonging to ribotypes II and III were identified as *B. garinii* and *B. afzelii*, the species identity of the other isolates generating the ribotype IV, V and VI RFLP patterns was not clear (Fukunaga *et al.*, 1993a,b; Nakao *et al.*, 1994a,b). However, because these isolates were subsequently found to have 82–85% DNA homology with *B. garinii* type strain 20047 and over 98% similarity with the *fla* gene of *B. garinii* strain Ip90, we place these isolates (ribotypes IV, V and VI), which have not been found in America and Europe, as members of *B. garinii* (Fukunaga and Koreki, 1996; Masuzawa *et al.*, 1998).

However, the picture is not yet clear-cut as genetic and immunological heterogeneity have been reported for many European *B. garinii* isolates (Wilske *et al.*, 1993; Bunikis *et al.*, 1995; Will *et al.*, 1995). At least 61 strains of *B. burgdorferi* s.l. have been classified in Japan on the basis of reactivity with 16 mAbs to outer surface protein A (OspA) (Masuzawa *et al.*, 1996b). Whereas all isolates of Japanese *B. afzelii* examined thus far can be classified into one OspA serotype, at least nine OspA serotypes have been identified as *B. garinii*. It has been reported by Wilske *et al.* (1993) that European *B. garinii* can be classified into at least five Osp serotypes. However, none of the Japanese strains identified as *B. garinii* has been found to have identical reactivity with the European *B. garinii* strains. Using a specific panel of *B. garinii* mAbs designed for this purpose, Will *et al.*

(1995) clearly established the immunological differences between the Japanese and European *B. garinii* isolates. These findings are also supported by DNA sequencing studies of the OspA genes of the spirochaetes (T. Masuzawa, unpublished data). On the other hand, all isolates identified as *B. afzelii* in Japan were homologous (and were also similar to European *B. afzelii*), showing uniformity in the molecular sizes of OspA and OspB and in mAb reactivity.

Many *B. burgdorferi* s.l. isolates worldwide have been compared, characterized and identified by analysing the RFLP patterns of the 5S–23S rRNA gene intergenic spacer, a method originally pioneered by Postic *et al.* (1994). Based on the RFLP 5S–23S spacer patterns, we have compiled data summarizing the classification of representative *B. burgdorferi* s.l. isolates from Japan and other areas in the world (Table 8.2). To our knowledge, most European *B. garinii* generated RFLP pattern B (by digestion of *Mse*I). Furthermore, *B. garinii* from Japan (Masuzawa *et al.*, 1996a), Far Eastern Russia (Masuzawa *et al.*, 1997a; Postic *et al.*, 1997), the western part of Russia (Postic *et al.*, 1997), China (Li *et al.*, 1998) and Korea (Masuzawa *et al.*, 1999) showed not only pattern B or its variation (Bv1, Bv2), but also pattern C or its variation (Cv1), which Postic *et al.* (1997) referred to as *B. garinii* variant NT29, the first isolate of which originated from *I. persulcatus* collected in the Nagano prefecture, Japan. In addition, a few isolates from China, Korea and Japan (Ishiguro *et al.*, 2000) displayed pattern R and Rv1. More importantly, isolates generating pattern C and R have not been found in any European isolates from *I. ricinus* (Postic *et al.*, 1997; T. Masuzawa *et al.*, unpublished data). It is of interest that pattern C (*B. garinii* variant NT29) was the most common *B. garinii* isolate from *I. persulcatus* in Japan as it was in most of those isolates classified in the ribotype IV group. Therefore, based on the RFLP analysis data reported here, we have tentatively classified Japanese *B. garinii* into two subtypes. These are the Eurasia-type *B. garinii* subtype (type strain 20047), which is found in Europe and Asia and generates the RFLP pattern B, and the Asia-type *B. garinii* subtype (variant strain NT29) found only in Asia, which generates pattern C by RFLP. In Japan, the Asia-type *B. garinii* is not only the most commonly found *Borrelia* spp., but also the most frequent spirochaete isolated from patients.

I. ovatus, which belongs to the subgenus *Partipalpiger* (Hoogstraal *et al.*, 1973), is another common and widely distributed ixodid tick that infests human habitats in Japan. *B. burgdorferi* s.l. isolated from *I. ovatus* shows homology in the molecular size of OspA and OspB and by reactivity to mAb H5332 specific to the *B. burgdorferi* species complex (Nakao *et al.*, 1992a; Masuzawa *et al.*, 1994). Although the classification of relapsing fever *Borrelia* spp. has been based on vector tick species, to our knowledge, the specific relationship between LB-related *Borrelia* spp. and tick species has not yet been fully characterized. Based on an extensive study comparing several *B. burgdorferi* s.l. isolates from *I. ovatus* and a variety of reservoir animal species (Masuzawa *et al.*, 1995), it appears that the isolates were homologous in protein profiles and OspC gene sequences (Fukunaga and Takahashi, 1994; Masuzawa *et al.*, 1997b). However, these isolates were different from the other LB *Borrelia* spp. such as *B. burgdorferi* s.s., *B. garinii* and *B.*

Table 8.2. *Borrelia burgdorferi* sensu lato identified by RFLP analysis of the 5S–23S intergenic rRNA gene spacer amplicons.

Species	Strain	Vector tick	Isolated in:	Amplicon size (bp)	*Dra*I RFLP pattern	*Dra*I restriction fragment sizes (bp)	*Mse*I RFLP pattern	*Mse*I restriction fragment sizes (bp)
B. burgdorferi s.s.	B31	*I. scapularis*	USA	254	A′	144, 53, 29, 28	A	108, 51, 38, 29, 28
B. garinii	20047	*I. ricinus*	Europe	253	B′	201, 52	B	108, 95, 50
	NP81	*I. persulcatus*	Japan	255	B′	203, 52	B	108, 96, 51
	HT59	*I. persulcatus*	Japan	255	NS	203, 52	B	108, 96, 51
	NT31	*I. persulcatus*	Japan	255	NS	203, 52	B	108, 96, 51
	XB190	*I. persulcatus*	Far eastern Russia	250	B′v1	201, 49	Bv1	107, 95, 48
	XB141	*I. persulcatus*	Far eastern Russia	251	B′v2	201, 50	Bv2	107, 95, 49
	NT29	*I. persulcatus*	Japan	253	C′	144, 57, 52	C	108, 57, 50, 38
	ASF	*I. persulcatus*	Japan	253	C′	144, 57, 52	C	107, 57, 51, 38
	KKU5	*I. nipponensis*	Korea	253	C′	144, 57, 52	C	107, 57, 51, 38
	XB37	*I. persulcatus*	Far eastern Russia	253	C′v1	157, 52, 44	Cv1	107, 51, 51, 44
	ChY13p	*I. persulcatus*	Northern China	237	R′	185, 52	R	105, 79, 53
	935T	*I. persulcatus*	Korea	237	R′	185, 52	Rv1	107, 79, 51
B. afzelii	VS461	*I. ricinus*	Europe	246	D′	174, 52, 20	D	108, 68, 50, 20
	KKU1	*I. nipponensis*	Korea	246	D′	174, 52, 20	D	108, 68, 50, 20
	17Y	*I. granulatus*	Korea	248	D′v1	175, 52, 21	Dv1	108, 68, 51, 21
	NT28	*I. persulcatus*	Japan	246	N′	144, 52, 30, 20	N	107, 51, 38, 30, 20
B. japonica	HO14	*I. ovatus*	Japan	236	NS	No fragment	E	108, 78, 50
B. valaisiana	VS116	*I. ricinus*	Europe	255	B″	203, 52	F	175, 50, 23, 7
B. valaisiana-related	Am501	*I. columnae*	Japan	249	NS	No fragment	Q	169, 51, 23, 6
	5MT	*I. nipponensis*	Korea	254	S′	173, 81	S	107, 58, 43, 24, 22
	KR3, KR6	*R. losea*	Taiwan	254	S′	173, 81	S	107, 58, 43, 24, 22
	CKA2b	*A. agrarius*	China	254	S′	173, 81	S	107, 58, 43, 24, 22

Species	Strain	Host	Origin					
	10MT	*I. nipponensis*	Korea	254	S'v1	173, 81	Sv1	150, 58, 24, 22
	CKA2a	*A. agrarius*	China	254	S'v1	173, 81	Sv1	150, 58, 24, 22
	TA1, TA2	*A. agrarius*	Taiwan	254	S'v1	173, 81	Sv1	150, 58, 24, 22
	CKG5	*I. granulatus*	China	254	S'v1	173, 81	Sv1	150, 58, 24, 22
	TM1	*M. formosanus*	Taiwan	254	S'v1	173, 81	Sv1	150, 58, 24, 22
	CKA3a	*A. agrarius*	China	246	S'v2	144, 102	Sv2	145, 59, 28, 14
	KR1, KR2, KR4	*R. losea*	Taiwan	246	S'v2	144, 102	Sv2	145, 59, 28, 14
	TM2	*M. formosanus*	Taiwan	246	S'v2	144, 102	Sv2	145, 59, 28, 14
	CKA1	*A. agrarius*	China	253	S'v3	144, 52, 29, 28	Sv3	145, 51, 29, 21, 7
	CKG4	*I. granulatus*	China	253	S'v3	144, 52, 29, 28	Sv3	145, 51, 29, 21, 7
B. tanukii	Hk501	*I. tanuki*	Japan	245	O'	173, 52, 20	O	174, 51, 20
B. turdi	Ya501	*I. turdus*	Japan	248	P'	144, 81, 23	P	107, 51, 38, 21, 16, 8, 7
B. andersonii	21123	*I. dentatus*	USA	266	NS	No fragment	L	120, 67, 51, 28
B. lusitaniae	PotiB1, PotiB2	*I. ricinus*	Europe	257	G'	145, 83, 29	G	108, 81, 39, 29
	PotiB3	*I. ricinus*	Europe	255	H'	158, 81, 16	H	108, 79, 52, 16
B. bissettii	DN127	*I. neotomae*	USA	257	I'	144, 53, 33, 27	I	108, 51, 38, 33, 27
	California strains	*I. neotomae*	USA	226	J'	144, 53, 29	J	108, 51, 38, 29
	25015	*I. scapularis*	USA	253	K'	173, 53, 27	K	108, 51, 34, 27, 17, 12, 4
B. sinica	CMN3	*N. confucianus*	China	235		157, 49, 29		107, 48, 38, 29, 13
	CMN1a, CMN2	*N. confucianus*	China	236		144, 50, 42		107, 49, 38, 29, 13
	CWO1	*I. ovatus*	China	236		144, 50, 42		107, 49, 38, 29, 13

Exact fragment sizes were determined from the sequences.
NS, no restriction site.

afzelii. This conclusion is based on RFLP analysis of the flagellin and 16S rRNA genes. Furthermore, DNA homology studies showed that, while the similarity of the DNA among these isolates ranged from 85 to 99%, DNA similarity with other LB-causing *Borrelia* spp. was less than 70% (Kawabata *et al.*, 1993). Thus, the unique *Borrelia* spp. that was isolated from *I. ovatus* in Japan was subsequently described in 1993 as a new species, *Borrelia japonica* (Kawabata *et al.*, 1993).

While *B. burgdorferi* s.l. spirochaetes in Japan are most commonly associated with *I. persulcatus* and *I. ovatus*, these tick species are by no means the only arthropod hosts. In a study by Nakao and Miyamoto (1993a), several rare Japanese ixodid ticks species such as *I. tunuki*, *I. turdus* and *I. columnae* were found to harbour borrelial spirochaetes. The protein profiles of isolates obtained from *I. tanuki* and *I. turdus* were different from those of known species. Of more interest perhaps is that some isolates obtained from *I. tanuki* and *I. turdus* have now been classified as new members of the *B. burgdorferi* s.l. species complex, i.e. *Borrelia tanukii* and *Borrelia turdii*, respectively (Fukunaga *et al.*, 1996a).

These conclusions were again based on RFLP analysis of 5S–23S rRNA gene intergenic spacer sequence (Masuzawa *et al.*, 1996a,d), ribotyping, PCR targeted to the 16S rRNA gene (Fukunaga *et al.*, 1996b) and DNA–DNA hybridization analysis (Masuzawa *et al.*, 1996c,d). Although strain Am501 from *I. columnae* (only one isolate thus far obtained) was also initially thought to be unique (Nakao and Miyamoto, 1993a), this strain has subsequently been classified as *Borrelia valaisiana* (Masuzawa *et al.*, 1996a, 1999; Wang *et al.*, 1997).

In summary of this section, the most common *B. burgdorferi* s.l. species found in Japan are *B. garinii* and *B. afzelii*. These two spirochaete species are most commonly isolated from *I. persulcatus* and LB patients in Japan. Moreover, Japanese *B. garinii* can be further classified into two subtypes: the Eurasian type (type strain 20047) found in Europe and Asia, and the Asian type (variant strain NT29), found in Asia. Surprisingly, *B. burgdorferi* s.s. has not yet been found in Japan. The other tick vectors, *I. ovatus*, *I. tanuki* and *I. turdus*, harbour *B. japonica*, *B. tanukii* and *B. turdii*, respectively, while the Japanese strain Am501, which has only been isolated from *I. columnae*, is now considered to be *B. valaisiana*.

Reservoir Hosts and Enzootic Transmission Cycles of *B. burgdorferi* s.l. in Japan

Eight species of mammal, two large (*C. nippon* and *Vulpes vulpes*) and six small (*Apodemus speciosus ainu*, *Apodemus argenteus*, *Clethrionomys rufocanus bedfordia*, *Clethrionomys rutilus mikado*, *Eothenmys smithii* and *Sorex unguiculatus*), and four bird species (*Turdus chrysolaus*, *Turdus pallidus*, *Emberiza spodocehala* and *Emberiza rustica*) have been shown to be infected with *B. burgdorferi* s.l. in Japan. The spirochaetes, designated as ASF strain (*B. garinii*), were first isolated in 1989 from the large Japanese field mouse, *A. speciosus ainu*, captured at Furano (Miyamoto *et al.*, 1991). In further investigations in the Furano and Nemuro areas, spirochaetes were detected from *A. speciosus ainu*, *A. argenteus*, *C. rufocanus bedfordia*, *C. rutilus mikado* in Furano, and *A. speciosus ainu* and *C. rutilus mikado* in Nemuro. Populations and infection rates were higher in *A. speciosus ainu* than in any other rodent or vole in either location. Feeding ticks were also more abundant on *A. speciosus ainu* than *A. argenteus*. In both locations immature *I. persulcatus* preferred *A. speciosus ainu* as a host to any of the other three host species in both areas. For these reasons, it has been suggested that the population of *A. speciosus ainu* constitutes the most important reservoir of *B. burgdorferi* s.l. in endemic areas of Hokkaido (Nakao and Miyamoto, 1993b). An additional survey was also carried out involving *C. rufocanus bedfordia* in Hokkaido, and *B. garinii* and *B. afzelii* were isolated from this vole (Ishiguro *et al.*, 1996). *B. garinii* was taken from *A. speciosus* in Fukushima, and both *B. garinii* and *B. afzelii* were taken from *E. smithii* in Fukui Prefecture (Ishiguro and Takada, 1996). *B. japonica* was identified from long-tailed shrews, *S. unguiculatus*, in Hokkaido (Nakao and Miyamoto, 1993c), from *A. speciosus* from sites in Aomori, Fukui and Hygo, from *A. argenteus* in Fukushima, and from *E. smithii* in Ishikawa and Fukui (Ishiguro *et al.*, 1994; Masuzawa *et al.*, 1995).

Large mammals such as red fox and sika deer are the main source of blood for female ticks. The spirochaete *B. afzelii* was isolated from the skin of a red fox that was infested with all stages of *I. persulcatus* in Sapporo, Hokkaido (Isogai *et al.*, 1994). A serosurvey of sika deer indicated a high seropositive rate (92.9%) in late October in Hokkaido (Isogai *et al.*, 1996) and *B. burgdorferi* s.l. was detected in the skin of wild sika deer from east Hokkaido by PCR (Kimura *et al.*, 1995). It is thought that horizontal transmission of LB spirochaetes could occur from infected ticks to agent-free ticks co-feeding on the skin of sika deer (Gern and Rais, 1996).

Migratory birds, especially ground-inhabiting birds, also serve as reservoir hosts, and they have been investigated at sites in Nemuro, Hokkaido and Morioka, Iwate Prefecture, in the northern part of Honshu. *B. burgdorferi* s.l. was detected in larval *I. persulcatus* feeding on black-faced bunting (*E. spodocehala*) and red-bellied thrush (*T. chrysolaus*) (Miyamoto *et al.*, 1993), and *B. garinii* was directly isolated from three skins and one liver of *T. chrysolaus* from Nemuro (Miyamoto *et al.*, 1997). Although spirochaetes have been isolated from larvae of *I. persulcatus* feeding on *E. spodocehala*, they could not be detected from 60 *E. spodocehala* from Nemuro (data not shown). Recently, larval *I. persulcatus* from rustic bunting (*E. rustica*) and *E. spodocehala* were examined at Morioka. *B. garinii* was found from larvae feeding on one *E. rustica* and on one *E. spodocehala* (Miyamoto *et al.*, 2000). Attempts to isolate the agent from ten individuals of *E. rustica* failed (data not shown). Examination of the avian reservoir was also conducted at Fukui Prefecture, in the central part of Honshu. *B. garinii* was in a larva of *I. persulcatus* infesting a pale thrush, *T. pallidus*. In rearing experiments of ticks from the same bird, 38 nymphal *H. flava* were obtained, all of which moulted to females with only one showing any indication of transstadial transmission (Ishiguro *et al.*, 2000).

These results indicate that, as in Europe (Olsen *et al.*, 1995), migratory birds may be involved as reservoirs of *B. garinii* in Japan. It is considered that these avian reservoirs serve to disperse vector ticks and *B. burgdorferi* s.l. over long distances worldwide. These birds are infested with larval and/or nymphal ticks from spring to early summer or autumn and are a source of blood for the ticks together with the spirochaetes. The immature ticks become infected with spirochaetes and transstadial transmission to the next stage occurs.

Enzootic transmission cycles between birds, rodents and *I. persulcatus* have been determined by the ribotyping system (Nakao *et al.*, 1994a). Since transovarial transmission of spirochaetes is negligible in *I. persulcatus* (Nakao and Miyamoto, 1992), enzootic transmission studies have focused on birds and rodents. *B. burgdorferi* s.l. isolated from larval *I. persulcatus* fed on migratory birds (*Emberiza* and *Turdus* spp.) and rodents (*A. speciosus* and *Clethrionomys* spp.) were identified as *B. garinii* (ribotype II) and *B. afzelii* (ribotype III), respectively. Ribotype IV, Asian-type *B. garinii*, can be found in both birds and rodents. Thus, while rodents have a potential to serve as reservoirs for *B. afzelii*, birds appear to be a specific reservoir host for Eurasian-type *B. garinii*. This latter fact of is interest when one considers the fact that the Asian-type *B. garinii* is maintained in both birds and rodents in the wild.

Pathogenicity of *Borrelia* spp. in Japan

Most of the *B. burgdorferi* s.l. isolates obtained from patients in Japan have been cultured from skin biopsy specimens. Using such specimens, *B. garinii* ribotype IV (equivalent to Asian-type *B. garinii*), a few *B. garinii* ribotype II (equivalent to Eurasian-type *B. garinii*) and *B. afzelii* have thus far been identified (Fukunaga *et al.*, 1993c). Experimental transmission studies using tick colonies maintained in the laboratory have revealed that, while *I. persulcatus* ticks can be readily infected with these Japanese patient-derived *B. burgdorferi* s.l. isolates after ingestion of the spirochaetes from experimentally infected Mongolian jirds, *I. ovatus* ticks were not infected (Nakao and Miyamoto, 1994). As mentioned earlier, *I. ovatus* is one of the most common tick species to infest human habitats in Japan. This tick also specifically transmits *B. japonica*. None the less, with the possible exception of one suspected case in 1996, clinical cases of bona fide LB due to the transmission of *B. japonica* through the bite of *I. ovatus* have not been documented in humans (Masuzawa *et al.*, 1996c). A recent infectivity and arthritis induction study by Kaneda *et al.* (1998) using *B. japonica* in severe combined immunodeficiency (scid) mice hints at this spirochaete's pathogenic potential. In this study, as expected, arthritis induction was observed 12 days after inoculation of scid mice with *B. burgdorferi* s.s. *B. japonica* was also able to induce a mild arthritis in scid mice 28 days after inoculation. However, histopathological changes were restricted to the joints. *B. japonica* also maintained its infectivity and the ability to induce arthritis, albeit with less severe consequences when compared with *B. burgdorferi* s.s. These results may help to partially explain why the incidence of LB as a result of a bite by *I. ovatus* is extremely rare in Japan. Since the other rare Japanese tick species, *I. tanuki*, *I. turdus* and *I. columnae*, seldom bite humans, the pathogenic potential of *B. tanukii*, *B. turdi* and *B. valaisiana* in Japan is still unknown.

The findings described here indicate that the principal tick vector of LB in Japan is *I. persulcatus*. Moreover, the clinical symptoms presented by patients are often limited to erythema migrans with a few people developing facial palsy, arthralgia and lymphocytoma (Nakao *et al.*, 1992b; Hashimoto *et al.*, 1995). We speculate that one of the reasons for this limited clinical spectrum is that Eurasian-type *B. garinii*, the most dominant *Borrelia* spp., may have a relatively low virulence in humans. Interestingly, although *B. afzelii*, a common pathogen in Europe and the species most often implicated in acrodermatitis chronica atrophicans (ACA), is present in Japan, no cases of ACA have been documented. The reason why most LB patients in Japan only present a mild clinical course of the disease is an important unresolved issue.

Distribution of *B. burgdorferi* s.l. in East Asian Countries

While several *B. burgdorferi* s.l. species have been fairly well characterized in Japan, the species found in other Asian countries have not been extensively studied. To clarify the current distribution of *B. burgdorferi* s.l. in Asia, we surveyed several East Asian countries.

This geographical survey was conducted from 1995 to 2001 and included Far Eastern Russia (Khabarovsk, Vladivostok and Yuzhno-Sakhalinsk), northern China (Yakeshi, Inner Mongolia), southern China (Zhejiang, Sichuan and Anhui provinces), Uighur in western China, Korea, Taiwan and Kinmen island, which is located offshore from Hisamen, Fujian province in China (Fig. 8.2). In these studies, ticks were collected by flagging vegetation in woodland areas and rodents captured with Sherman live traps were collected at the same time. RFLP analysis of the 5S–23S rRNA intergenic spacer sequence was used to determine the *Borrelia* spp. In addition, the isolates were also characterized by sequence analysis of the intergenic spacer, 16S rRNA and the flagellin gene, and hybridization analysis of genomic DNA.

B. burgdorferi s.l. isolated from *I. persulcatus* collected in the Far Eastern regions of Russia, Inner Mongolia in northeastern China, Uighur in western China and the northern part of South Korea were identified as *B. garinii* (Eurasian-type and Asian-type) and *B. afzelii* (Park *et al.*, 1993; Sato *et al.*, 1996; Masuzawa *et al.*, 1997a, 1999; Li *et al.*, 1998; Takada *et al.*, 1998). In contrast, the borrelial isolates obtained from wild rodent species such *Apodemus peninsulae* captured in Khabarovsk and Inner Mongolia, and *A. peninsulae, Apodemus agrarius* and *Crocidura lasiura* from the northern part of South Korea were identified as *B. garinii* (Asian-type) and *B. afzelii*. In these areas, *B. garinii* and *B. afzelii* appear to be naturally maintained between *I. persulcatus* and wild mammals. These observations are in good accord with the situation found in the northern part of Japan.

In the middle of south Korea (Choongju), where *I. persulcatus* is not found, *Ixodes nipponensis* is the main tick vector of these *Borrelia* spp. Although *I. nipponensis*, a member of the *I. ricinus* tick complex, is commonly found in Japan, to

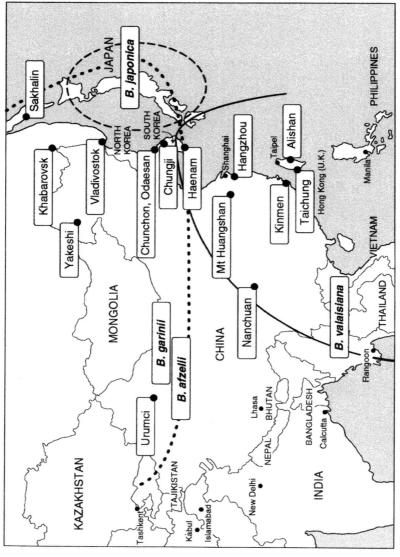

Fig. 8.2. Survey sites for vectors and reservoirs of *Borrelia burgdorferi* sensu lato in East Asian countries from 1995 to 1999 and geographical distribution of *B. burgdorferi* s.l. in East Asian countries.

our knowledge no isolation of any spirochaete from this tick species has occurred in Japan. However, the genetic differences between isolates from *I. persulcatus* and *I. nipponensis* remain to be investigated. Other Korean isolates from *I. nipponensis* collected in the southern tip of the Korean peninsula have been tentatively classified as belonging to the *B. afzelii*-related group based on 16S rRNA gene sequences (Kee *et al.*, 1994, 1996). After the initial description of *B. valaisiana* (Wang *et al.*, 1997), the DNA sequences of several *B. valaisiana* isolates have been published, so comparison between European and Asian isolates is now possible. We have characterized several strains belonging to *B. afzelii*-related genomic groups from rodents (*Rattus losea*, *Mus formosanus*) captured in Taiwan (Tachung and Kinmen Island) (Masuzawa *et al.*, 2000). *B. afzelii*-related genomic groups have also been found in *Ixodes granulatus* and wild rodents (*A. agrarius* and *Niviventer* sp.) captured in Zhejiang and Sichuan provinces located in the Yangtze river valley in China (Masuzawa *et al.*, 2001). However, on the basis of the 16S rRNA gene, the 5S–23S rRNA gene intergenic spacer and flagellin gene sequences, Am501, as isolated from the tick *I. columnae* in Japan, appears to be closely related to *B. valaisiana* isolated from *I. ricinus* in Europe (Masuzawa *et al.*, 1996a, 1999). Figure 8.3 shows the phylogenetic tree constructed on the basis of 16S rRNA gene sequences. It can be seen that Korean, Taiwanese and Chinese isolates are clustered with each other, but they are relatively divergent from *B. valaisiana* in Europe and Am501 found in Japan. Furthermore, OspA gene sequences and protein profiles of major Osps of these spirochaetes were divergent and were not identical to those of the European isolates. These types of borrelia are also present in Thailand, Nepal and the most southern island of Japan, Okinawa (Takada

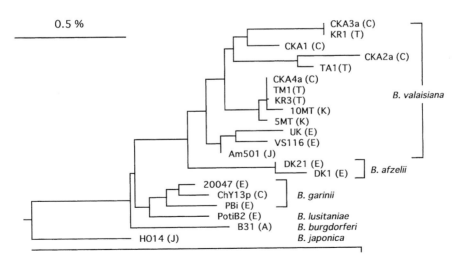

Fig. 8.3. Phylogenetic tree of *Borrelia burgdorferi* sensu lato based on 16S rRNA gene sequences. A, isolated from America; C, isolated from China; E, isolated from Europe; J, isolated from Japan; K, isolated from Korea; T, isolated from Taiwan. The bar indicates a sequence divergence of 0.5%.

et al., 2001). While it has been recently suggested that some *B. burgdorferi* s.l. isolates from *I. granulatus* ticks found in Korea formed a new genomic group (Lee *et al.*, 2000), we feel that these isolates belong to the *B. valaisiana*-related genomic groups that we have identified. None the less, the classification of *B. valaisiana*-related species found in Korea, China and Taiwan remains to be determined. On a final note, *B. valaisiana* found in various European countries is transmitted by *I. ricinus* ticks (Filipuzzi-Jenny *et al.*, 1993; Livesley *et al.*, 1995; Rijpkema *et al.*, 1995, 1996; Kirstein *et al.*, 1997). Thus, our observations further imply that *B. valaisiana* has evolved to adapt to a variety of vector ticks in that it has been found in *I. granulatus* in China, Taiwan and Korea, *I. nipponensis* in Korea and *I. columnae* in Japan. However, interesting as these findings are, the pathogenicity of *B. valaisiana* and related species has yet to be determined.

In the survey conducted in the southern part of China (Sichuan and Anhui provinces along the Yangtze river valley), a unique borrelia was isolated from *I. ovatus* and the wild rodent *Niviventer confucianus*. These isolates showed reduced similarity of the 16S rDNA sequence to the sequences of previously described *Borrelia* spp. and could be distinguished by RFLP analysis of the 5S–23S rRNA gene intergenic spacer sequences. The borrelia was morphologically similar to spirochaetes of the *B. burgdorferi* species complex, but the cells contained fewer periplasmic flagella, a maximum of four, inserted at each end. Based on these findings, the borrelia was named *Borrelia sinica*, a new member of the of LB-related *Borrelia* spp. (Masuzawa *et al.*, 2001). This species is distributed in middle China, Thailand and Nepal (unpublished results) and its pathogenicity is still unknown.

Vector competence of *Haemaphysalis* spp. was reported in China (Sheng *et al.*, 1992; Tian *et al.*, 1998). However, no isolate was obtained from *Haemaphysalis bispinosa* and *Haemaphysalis longicornis* collected in China in our survey. Furthermore, in another survey in Nepal and Thailand, cultures made from *Haemaphysalis hystricis*, *Haemaphysalis formonsensis* and *Haemaphysalis indoflava* were all negative. In contrast, a *Borrelia* sp. was isolated from a few *H. flava* collected from vegetation by flagging in the Fukui Prefecture, Japan, and was identified as *B. japonica* (Ishiguro *et al.*, 1994). One nymphal *H. flavas* fed on *T. pallidus*, transstadially transmitted *B. garinii* to the adult stage (Ishiguro *et al.*, 2000). In view of these contradictions, we suggest that *Haemaphysalis* spp. in Asia are not important as vectors for LB, but may occasionally transmit *Borrelia* spp. in nature.

In summary, the distribution of *Borrelia* spp. in East Asian countries appears to be wide and varied (Fig. 8.2). *B. garinii* (Eurasian-type and Asian-type) and *B. afzelii* transmitted by *I. persulcatus* ticks are found in Far Eastern Russia, northern and western China, Korea and Japan. On the other hand, the *B. valaisiana*-related genomic group, which has a more southern distribution in Asia, such as southern China, Taiwan, the southern tip of the Korean peninsula, Thailand and Okinawa in Japan, may be transmitted by *I. granulatus*. In addition, in Korea this spirochaete may also be transmitted by *I. nipponensis*.

The relationship between *Borrelia* spp. and vector ticks in East Asian countries is also clarified in Fig. 8.4. *I. persulcatus*, which is the vector in Russia, northern China, Uighur, the northern part of South Korea and northern Japan, carries

Eurasian-type *B. garinii*, Asian-type *B. garinii* and *B. afzelii*. However, Asian-type *B. garinii* has not been found in *I. ricinus* in Europe, and is therefore unique to Asia and *I. persulcatus*. These *Borrelia* spp. infect humans and cause LB in their respective countries. However, *I. nipponensis*, a tick found in mid-Korea (where *I. persulcatus* is not found), transmits *B. garinii* and *B. afzelii*. *B. valaisiana* is transmitted by *I. ricinus*, as are *B. burgdorferi*, *B. lusitaniae*, Eurasian-type *B. garinii* and *B. afzelii* in Europe. In southeast China, Taiwan and Korea, *B. valaisiana*-related *Borrelia* spp. are carried by *I. granulatus* while in Japan this *Borrelia* sp. is transmitted by *I. columnae*.

Conclusions

We have accumulated much data on the distribution of *B. burgdorferi* s.l. and their vectors in Japan and parts of eastern Asia. However, there remain many countries and regions in Asia that have yet to be studied. By monitoring the enzootic cycles, transmission and distribution of *Borrelia* spp. on a worldwide basis, it is hoped that the knowledge gained will lead to the design of better serodiagnosis and prophylaxis regimes. Clinical symptoms presenting in LB patients in Japan are moderate in comparison with those in Europe and the USA, and prophylaxis using vaccines against LB is not justified. At present, the raising of public awareness of this problem and establishment of early diagnosis would seem to be the most effective means of limiting the disease in Japan.

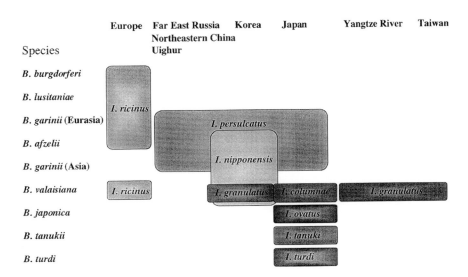

Fig. 8.4. Relationship between *Borrelia* spp. and vector tick species.

Acknowledgements

We would like to thank Drs Y. Hashimoto (Asahikawa Medical College, Japan), Y. Sato (previously at Asahikawa Medical College, Japan), F. Sato (Yamashina Institute for Ornithology, Japan), E.I. Korenberg (Gamaleya Institute, Russia), V.I. Volkov (Khabarovsk Antiplague Station, Russia), L.I. Ivanov (Khabarovsk Plague Control Station, Russia), R.N. Liberova (Khabarovsk Regional Center of Sanitary Epidemiological Administration, Russia), N. Takada and Y. Yano (Fukui Medical College, Japan), F. Ishiguro (Fukui Prefectural Institute of Public Health, Japan), H. Fujita (Oohara Institute, Japan), T. Kadosaka (Aichi Medical University, Japan), T. Itoh (Hokkaido Institute of Public Health, Japan), H. Wang and J. Wang (Chinese Medical University, China), K.H. Park (Konkuku University, Korea), M.K. Cho and W.H. Chang (Hallym University, Korea), X.-H. Ma (Zhejiang Institute of Microbiology, China), M.-J. Pan (National Taiwan University, Taiwan) and L.-K. Lin (Tunghai University, Taiwan) for conducting the survey in East Asia countries. We also thank Dr Dennis J. Grab (Johns Hopkins University, School of Medicine, USA, previously at Kurume University School of Medicine, Japan), for critical reading and editing of the final manuscript. This study was supported in part by Grants-in-Aid for International Cooperative Research (Nos 08044310, 08041181 and 10041204) and Grants-in-Aid for Scientific Research (Nos 08670312, 09670294, 11670267) from the Ministry of Education, Science and Culture of Japan.

References

Baranton, G., Postic, D., Saint Girons, I., Boerlin, P., Piffaretti, G.-C., Assous, M. and Grimont, P.A.D. (1992) Delineation of *Borrelia burgdorferi* sensu stricto, *Borrelia garinii* sp. nov., and group VS461 associated with Lyme borreliosis. *International Journal of Systematic Bacteriology* 42, 378–383.

Bunikis, J., Olsen, B., Fingerle, V., Bonnedahl, J., Wilske, B. and Bergström, S. (1995) Molecular polymorphism of the disease agent *Borrelia garinii* in Northern Europe is influenced by a novel enzootic *Borrelia* in the North Atlantic. *Journal of Clinical Microbiology* 34, 364–368.

Canica, M.M., Nato, F., du Merle, L., Mazie, J.C., Baranton, G. and Postic, D. (1993) Monoclonal antibodies for identification of *Borrelia afzelii* sp.nov. associated with late cutaneous manifestations of Lyme borreliosis. *Scandinavian Journal of Infectious Diseases* 25, 441–448.

Centers for Disease Control (1992) Notifiable diseases report. *Morbidity and Mortality Weekly Report* 40, 898–899.

Filipuzzi-Jenny, E., Blot, M., Schmid-Berger, N., Meister-Tuner, J. and Meyer, J. (1993) Genetic diversity among *Borrelia burgdorferi* isolates: more than three genospecies? *Research in Microbiology* 144, 295–304.

Fukunaga, M. and Koreki, Y. (1996) A phylogenetic analysis of *Borrelia burgdorferi* sensu lato isolates associated with Lyme disease in Japan by flagellin gene sequence determination. *International Journal of Systematic Bacteriology* 46, 416–421.

Fukunaga, M. and Takahashi, Y. (1994) Pulsed field gel electrophoresis analysis of *Borrelia burgdorferi* sensu lato isolated in Japan and taxonomic implications with Lyme disease spirochetes. *Microbiology and Immunology* 38, 747–751.

Fukunaga, M., Sohnaka, M., Takahashi, Y., Nakao, M. and Miyamoto, K. (1993a) Antigenic and genetic characterization of *Borrelia* species isolated from *Ixodes persulcatus* in Hokkaido, Japan. *Journal of Clinical Microbiology* 31, 1388–1391.

Fukunaga, M., Sohnaka, M. and Yanagihara, Y. (1993b) Analysis of *Borrelia* species associated with Lyme disease by rRNA gene restriction fragment length polymorphism. *Journal of General Microbiology* 139, 1141–1146.

Fukunaga, M., Sohnaka, M., Nakao, M. and Miyamoto, K. (1993c) Evaluation of genetic divergence of borrelial isolates from Lyme disease patients in Hokkaido, Japan, by rRNA gene probes. *Journal of Clinical Microbiology* 31, 2044–2048.

Fukunaga, M., Hamase, A., Okada, K. and Nakao, M. (1996a) *Borrelia tanukii* sp. nov. and *Borrelia turdae* sp. nov. found from ixodid ticks in Japan: rapid species identification by 16S rRNA gene-targeted PCR analysis. *Microbiology and Immunology* 40, 877–881.

Fukunaga, M., Hamase, A., Okada, K., Inoue, H., Miyamoto, K. and Nakao, M. (1996b) Characterization of spirochetes isolated from *Ixodes tanuki*, *Ixodes turdus*, and *Ixodes columnae* ticks and sequence comparison with *Borrelia burgdorferi* sensu lato strains. *Applied and Environmental Microbiology* 62, 2338–2344.

Gern, L. and Rais, O. (1996) Efficient transmission of *Borrelia burgdorferi* between cofeeding *Ixodes ricinus* (Acari: Ixodidae). *Journal of Medical Entomology* 33, 189–192.

Hashimoto, Y., Kawagishi, N., Sakai, H., Takahashi, H., Matsuo, S., Nakao, M., Miyamoto, K. and Iizuka, H. (1995) Lyme disease in Japan. Analysis of *Borrelia* species using rRNA gene restriction fragment length polymorphism. *Dermatology* 191, 193–198.

Hoogstraal, H., Clifford, M., Saito, Y. and Keirans, J.E. (1973) *Ixodes (Partipalpiger) ovatus* Neuman, subgen. Nov.: identity, hosts, ecology, and distribution (Ixodoidea:Ixodidae). *Journal of Medical Entomology* 10, 157–164.

Ishiguro, F. and Takada, N. (1996) Lyme *Borrelia* from *Ixodes persulcatus* and small rodents from northern to central parts of mainland Japan. *Medical Entomology and Zoology* 47, 183–185.

Ishiguro, F., Iida, H., Ma, X.H., Yano, Y., Fujita, H. and Takada, N. (1994) Diversity of *Borrelia* isolates found in rodent–tick relationship in Fukui Prefecture and some additional areas. *Japanese Journal of Sanitation Zoology* 45, 141–145.

Ishiguro, F., Takada, N., Nakata, K., Yano, Y., Suzuki, H., Masuzawa, T. and Yanagihara, Y. (1996) Reservoir competence of the vole, *Clethrionomys rufocanus bedfordiae*, for *Borrelia garinii* or *Borrelia afzelii*. *Microbiology and Immunology* 40, 67–69.

Ishiguro, F., Takada, N., Masuzawa, T. and Fukui, T. (2000) Prevalence of Lyme disease *Borrelia* spp. in ticks from migratory birds on the Japanese mainland. *Applied and Environmental Microbiology* 66, 982–986.

Isogai, E., Isogai, H., Kawabata, H., Masuzawa, T., Yanagihara, Y., Kimura, K., Sakai, T., Azuma, Y., Fujii, N. and Ohno, S. (1994) Lyme disease spirochetes in a wild fox (*Vulpes vulpes schrencki*) and in ticks. *Journal of Wildlife Diseases* 30, 439–444.

Isogai, E., Isogai, H., Masuzawa, T., Postic, D., Baranton, G., Kamewaka, Y., Kimura, K., Nishikawa, T., Fujii, N., Ishii, N., Ohno, S. and Yamaguti, N. (1996) *Borrelia burgdorferi* sensu lato in an endemic environment: wild sika deer (*Cervus nippon yesoensis*) with infected ticks and antibodies. *Microbiology and Immunology* 40, 13–19.

Jaenson, T.G.T. (1991) The epidemiology of Lyme borreliosis. *Parasitology Today* 7, 39–45.

Kaji, K., Miyaki, M., Saitoh, T., Ono, S. and Kaneko, M. (2000) Spatial distribution of an expanding sika deer *Cervus nippon* population on Hokkaido Island, Japan. *Wildlife Society Bulletin* 28, 699–707.

Kaneda, K., Masuzawa, T., Simon, M.M., Isogai, E., Isogai, H., Yasugami, K., Suzuki, T., Suzuki, Y. and Yanagihara, Y. (1998) Infectivity and arthritis induction of *Borrelia japonica* on SCID mice and immune competent mice: possible role of galactosylceramide binding activity on initiation of infection. *Microbiology and Immunology* 42, 171–175.

Kawabata, H., Masuzawa, T. and Yanagihara, Y. (1993) Genomic analysis of *Borrelia japonica* sp. nov. isolated from *Ixodes ovatus* in Japan. *Microbiology and Immunology* 37, 843–848.

Kawabata, H., Tashibu, H., Yamada, K., Masuzawa, T. and Yanagihara, Y. (1994) Polymerase chain reaction analysis of *Borrelia* species isolated in Japan. *Microbiology and Immunology* 38, 591–598.

Kawabata, M., Baba, S., Iguchi, K., Yamaguti, N. and Russel, H. (1987) Lyme disease in Japan and its possible incriminated tick vector, *Ixodes persulcatus*. *Journal of Infectious Diseases* 156, 854.

Kee, S., Hwang, K.J., Oh, H.B. and Park, K.S. (1994) Identification of *Borrelia burgdorferi* isolated in Korea using outer surface protein A (OspA) serotyping system. *Microbiology and Immunology* 38, 989–993.

Kee, S.H., Yoon, J.H., Oh, H.B., Park, Y.H., Kim, Y.W., Cho, M.K., Park, K.S. and Chang, W.H. (1996) Genetic analysis of *Borrelia burgdorferi* sensu lato in Korea using genomic hybridization and 16S rRNA gene sequence determination. *Microbiology and Immunology* 40, 599–605.

Kimura, K., Isogai, E., Isogai, H., Kamewaka, Y., Nishikawa, T., Ishii, N. and Fujii, N. (1995) Detection of Lyme disease spirochetes in the skin of naturally infected wild sika deer (*Cervus nippon yesoensis*). *Applied and Environmental Microbiology* 61, 1641–1642.

Kirstein, F., Rijpkema, S., Molkenboer, M. and Gray, J.S. (1997) The distribution and prevalence of *B. burgdorferi* genomospecies in *Ixodes ricinus* ticks in Ireland. *European Journal of Epidemiology* 13, 67–72.

Lee, S.H., Kim, B.J., Kim, J.H., Park, K.H., Yeo, S.J., Kim, S.J. and Kook, Y.H. (2000) Characterization of *Borrelia burgdorferi* strains isolated from Korea by 16S rDNA sequence analysis and PCR–RFLP analysis of *rrf* (5S) –*rrl* (23S) intergenic spacer amplicons. *International Journal of Systematic and Evolutionary Microbiology* 50, 857–863.

Li, M., Masuzawa, T., Takada, N., Ishiguro, F., Fujita, H., Iwaki, A., Wang, H., Wang, J., Kawabata, M. and Yanagihara, Y. (1998) Lyme disease *Borrelia* species in Northeastern China resemble those isolated from far eastern Russia and Japan. *Applied and Environmental Microbiology* 64, 2705–2709.

Livesley, M.A., Thompson, I.P., Rainey, P.B. and Nuttall, P.A. (1995) Comparison of *Borrelia* isolated from UK foci of Lyme disease. *FEMS Microbiology Letters* 130, 151–158.

Marconi, R.T. and Garon, C.F. (1992) Development of polymerase chain reaction primer sets for diagnosis of Lyme disease and for species-specific identification of Lyme disease isolates by 16S rRNA signature nucleotide analysis. *Journal of Clinical Microbiology* 30, 2830–2834.

Masuzawa, T., Okada, Y., Beppu, Y., Oku, T., Kawamori, F. and Yanagihara, Y. (1991a) Immunological properties of *Borrelia burgdorferi* isolated from the *Ixodes ovatus* in Shizuoka, Japan. *Microbiology and Immunology* 35, 913–919.

Masuzawa, T., Okada, Y., Yanagihara, Y. and Sato, N. (1991b) Antigenic properties of *Borrelia burgdorferi* isolated from *Ixodes ovatus* and *Ixodes persulcatus* in Hokkaido, Japan. *Journal of Clinical Microbiology* 29, 1568–1573.

Masuzawa, T., Kawabata, H., Beppu, Y., Miyamoto, K., Nakao, M., Sato, N., Muramatsu, K., Sato, N., Johnson, R.C. and Yanagihara, Y. (1994) Characterization of monoclonal antibodies for identification of *Borrelia japonica* isolates from *Ixodes ovatus*. *Microbiology and Immunology* 38, 393–398.

Masuzawa, T., Suzuki, H., Kawabata, H., Ishiguro, F., Takada, N., Yano, Y. and Yanagihara, Y. (1995) Identification of spirochetes isolated from wild rodents in Japan as *Borrelia japonica*. *Journal of Clinical Microbiology* 33, 1392–1394.

Masuzawa, T., Komikado, T., Iwaki, A., Suzuki, H., Kaneda, K. and Yanagihara. Y. (1996a) Characterization of *Borrelia* sp. isolated from *Ixodes tanuki*, *I. turdus*, and *I. columnae* in Japan by restriction fragment length polymorphism of rrf (5S) –rrl (23S) intergenic spacer amplicons. *FEMS Microbiology Letters* 142, 77–83.

Masuzawa, T., Wilske, B., Komikado, T., Suzuki, H., Kawabata, H., Sato, N., Muramatsu, K., Sato, N., Isogai, E., Isogai, H., Johnson, R.C. and Yanagihara, Y. (1996b) Comparison of OspA-serotypes for *Borrelia burgdorferi* sensu lato from Japan, Europe and North America. *Microbiology and Immunology* 40, 539–545.

Masuzawa, T., Yanagihara, Y. and Fujita, H. (1996c) A case of Lyme borreliosis which was suspected to be by *Borrelia japonica* infection in Shizuoka, Japan. *Journal of Japanese Association for Infectious Diseases* 70, 264–267 (in Japanese with English summary).

Masuzawa, T., Suzuki, H., Kawabata, H., Ishiguro, F., Takada, N. and Yanagihara, Y. (1996d) Characterization of *Borrelia* spp. isolated from the tick, *Ixodes tanuki*, and small rodents in Japan. *Journal of Wildlife Diseases* 32, 565–571.

Masuzawa, T., Iwaki, A., Sato, Y., Miyamoto, K., Korenberg, E.I. and Yanagihara, Y. (1997a) Genetic diversity of *Borrelia burgdorferi* sensu lato isolated in far eastern Russia. *Microbiology and Immunology* 41, 595–600.

Masuzawa, T., Komikado, T., Kaneda, K., Fukui, T., Sawaki, K. and Yanagihara, Y. (1997b) Homogeneity of *Borrelia japonica* and heterogeneity of *Borrelia afzelii* and *Borrelia tanukii* isolated in Japan, determined from OspC gene sequences. *FEMS Microbiology Letters* 153, 287–293.

Masuzawa, T., Kaneda, K., Suzuki, H., Wang, J., Yamada, K., Kawabata, H., Johnson, R.C. and Yanagihara, Y. (1998) Presence of common antigenic epitope in outer surface protein (Osp) A and OspB of Japanese isolates identified as *Borrelia garinii*. *Microbiology and Immunology* 40, 455–458.

Masuzawa, T., Fukui, T., Miyake, M., Oh, H., Cho, M., Chang, W., Imai, Y. and Yanagihara, Y. (1999) Determination of members of a *Borrelia afzelii*-related group isolated from *Ixodes nipponensis* in Korea as *Borrelia valaisiana*. *International Journal of Systematic Bacteriology* 49, 1409–1415.

Masuzawa, T., Pan, M.-J., Kadosaka, T., Kudeken, M., Takada, N., Yano, Y., Imai, Y and Yanagihara, Y. (2000) Characterization and identification of *Borrelia* isolates as *Borrelia valaisiana* in Taiwan and Kinmen Islands. *Microbiology and Immunology* 44, 1003–1009.

Masuzawa, T., Takada, N., Kudeken, M., Fukui, T., Yano, Y., Ishiguro, F., Kawamura, Y., Imai, Y. and Ezaki, T. (2001) *Borrelia sinica* sp. nov., a Lyme disease-related *Borrelia* species isolated in China. *International Journal of Systematic and Evolutionary Microbiology* 51, 1817–1824.

Miyamoto, K. and Nakao, M. (1991) Frequent occurrence of human tick bites and monthly fluctuation of ixodid ticks in Hokkaido, Japan. *Japanese Journal of Sanitary Zoology* 42, 267–269.

Miyamoto, K. and Nakao, M. (1994) A study of ixodid tick bite sites on the human body in Hokkaido. *Japanese Journal of Sanitary Zoology* 45, 79–81.

Miyamoto, K., Nakao, M., Sato, N. and Mori, M. (1991) Isolation of Lyme disease spirochetes from an ixodid tick in Hokkaido, Japan. *Acta Tropica* 49, 65–68.

Miyamoto, K., Nakao, M., Uchikawa, K. and Fujita, H. (1992a) Prevalence of Lyme borreliosis spirochetes in ixodid ticks of Japan, with special reference to a new potential vector, *Ixodes ovatus* (Acari: Ixodidae). *Journal of Medical Entomology* 29, 216–220.

Miyamoto, K., Nakao, M., Fujimoto, K., Yamaguti, N. and Hori, E. (1992b) Detection of *Borrelia burgdorferi* in ixodid ticks collected from the Chichibu mountainous region of central Honshu, Japan. *Japanese Journal of Sanitary Zoology* 43, 255–258.

Miyamoto, K., Nakao, M., Fujita, H. and Sato, F. (1993) The ixodid ticks on migratory birds in Japan and isolation of Lyme disease spirochetes from bird-feeding ticks. *Japanese Journal of Sanitary Zoology* 44, 315–326.

Miyamoto, K., Sato, Y., Okada, K., Fukunaga, M. and Sato, F. (1997) Competence of a migratory bird, red-bellied thrush (*Turdus chrysolaus*), as an avian reservoir for the Lyme disease spirochetes in Japan. *Acta Tropica* 65, 43–51.

Miyamoto, K., Masuzawa, T. and Kuddeken, M. (2000) Tick collection from wild birds and detection of Lyme disease spirochetes from a new avian reservoir in Japan. *Medical Entomology and Zoology* 51, 221–226.

Nakao, M. and Miyamoto, K. (1992) Negative finding in detection of transovarial transmission of *Borrelia burgdorferi* in Japanese ixodid ticks, *Ixodes persulcatus* and *Ixodes ovatus*. *Japanese Journal of Sanitary Zoology* 43, 343–345.

Nakao, M. and Miyamoto, K. (1993a) Isolation of spirochetes from Japanese ixodid ticks, *Ixodes tanuki*, *Ixodes turdus*, and *Ixodes columnae*. *Japanese Journal of Sanitary Zoology* 44, 49–52.

Nakao, M. and Miyamoto, K. (1993b) Reservoir competence of the wood mouse, *Apodemus speciosus ainu*, for the Lyme disease spirochetes, *Borrelia burgdorferi*, in Hokkaido, Japan. *Japanese Journal of Sanitary Zoology* 44, 69–84.

Nakao, M. and Miyamoto, K. (1993c) Long-tailed shrew, *Sorex unguiculatus*, as a potential reservoir of the spirochetes transmitted by *Ixodes ovatus* in Hokkaido, Japan. *Japanese Journal of Sanitary Zoology* 44, 237–245.

Nakao, M. and Miyamoto, K. (1994) Susceptibility of *Ixodes persulcatus* and *I. ovatus* (Acari: Ixodidae) to Lyme disease spirochetes isolated from humans in Japan. *Journal of Medical Entomology* 31, 467–473.

Nakao, M., Miyamoto, K., Uchikawa, K. and Fujita, H. (1992a) Characterization of *Borrelia burgdorferi* isolated from *Ixodes persulcatus* and *Ixodes ovatus* ticks in Japan. *American Journal of Tropical Medicine Hygiene* 47, 505–511.

Nakao, M., Miyamoto, K., Kawagishi, N., Hashimoto, Y. and Iizuka, H. (1992b) Comparison of *Borrelia burgdorferi* isolated from humans and ixodid ticks in Hokkaido, Japan. *Microbiology and Immunology* 36, 1189–1193.

Nakao, M., Miyamoto, K. and Fukunaga, M. (1994a) Lyme disease spirochetes in Japan: enzootic transmission cycles in birds, rodents, and *Ixodes persulcatus* ticks. *Journal of Infectious Diseases* 170, 878–882.

Nakao, M., Miyamoto, K., Fukunaga, M., Hashimoto, Y. and Takahashi, H. (1994b) Comparative studies on *Borrelia afzelii* isolated from a patient of Lyme disease, *Ixodes persulcatus* ticks, and *Apodemus speciosus* rodents in Japan. *Microbiology and Immunology* 38, 413–420.

O'Connell, S., Granstrom, M., Gray, J.S. and Stanek, G. (1998) Epidemiology of European Lyme borreliosis. *Zentralblatt für Bacteriologie* 287, 229–240.

Olsen, B., Jaenson, T.G.T. and Bergström, S. (1995) Prevalence of *Borrelia burgdorferi* sensu lato-infected ticks on migrating birds. *Applied and Environmental Microbiology* 61, 3082–3087.

Orloski, K.A., Hayes, E.B., Campbell, G.L. and Dennis, D.T. (2000) Surveillance for Lyme disease – United States, 1992–1998. *Morbidity and Mortality Weekly Report* 49 (SS-3), 1–12.

Park, K.H., Chang, W.H. and Schwan, T.G. (1993) Identification and characterization of Lyme disease spirochetes, *Borrelia burgdorferi* sensu lato, isolated in Korea. *Journal of Clinical Microbiology* 31, 1831–1837.

Postic, D., Assous, M.V., Grimont, P.A.D. and Baranton, G. (1994) Diversity of *Borrelia burgdorferi* sensu lato evidenced by restriction fragment length polymorphism of *rrf* (5S) –*rrl* (23S) intergenic spacer amplicons. *International Journal of Systematic Bacteriology* 44, 743–752.

Postic, D., Korenberg, E., Gorelova, N., Kovalevski, Y.V., Bellenger, E. and Baranton, G. (1997) *Borrelia burgdorferi* sensu lato in Russia and neighbouring countries: high incidence of mixed isolates. *Research in Microbiology* 148, 691–702.

Rijpkema, S.G.T., Molkenboer, M.J.C.H., Schouls, L.M., Jongejan, F. and Schellekens, J.F.P. (1995) Simultaneous detection and genotyping of three genomic groups of *Borrelia burgdorferi* sensu lato in Dutch *Ixodes ricinus* ticks by characterization of the amplified intergenic spacer region between 5S and 23S rRNA genes. *Journal of Clinical Microbiology* 33, 3091–3095.

Rijpkema, S., Golubic, D., Molkenboer, M., Verbeck-de Kruif, N. and Schellekens, J. (1996) Identification of four genomic groups of *Borrelia burgdorferi* sensu lato in *Ixodes ricinus* ticks collected in a Lyme borreliosis endemic region of north Croatia. *Experimental and Applied Acarology* 20, 23–30.

Saito, Y. (1959) Studies on ixodid ticks. Part I. On ecology, with reference to distribution and seasonal occurrence of ixodid ticks in Niigata Prefecture, Japan. *Acta Medica et Biologica* 7, 193–209.

Saito, Y. (1960) Studies on ixodid ticks. Part II. On the rearing and life history of three tick species (*Haemaphysalis flava*, *Ixodes japonica* and *Ixodes persulcatus persulcatus*) in Japan (Acarina: Ixodidae). *Acta Medica et Biologica* 7, 303–321.

Sato, Y., Miyamoto, K., Iwaki, A., Masuzawa, T., Yanagihara, Y., Korenberg, E.I., Gorelova, N.B., Ivanov, V.I. and Liberoba, R.N. (1996) Prevalence of the Lyme disease spirochete in *Ixodes persulcatus* and wild rodents in far east Russia. *Applied and Environmental Microbiology* 62, 3887–3889.

Sheng, Z., Zhang, F. and Wang, J. (1992) Lyme disease spirochetes isolated from *Haemaphysalis longicolonis* in the forest area of West-North Beijing. *Chinese Journal of Vector Biology and Control* 3 (suppl. 2), 99–100.

Steere, A.C. (1989) Lyme disease. *New England Journal of Medicine* 321, 586–596.

Steere, A.C., Malawista, S.E., Syndman, D.R., Shope, R.E., Andiman, W.A., Ross, M.R. and Steele, F.M. (1977) Lyme arthritis: an epidemic of oligoarticular arthritis in children and adults in three Connecticut communities. *Arthritis and Rheumatism* 20, 7–17.

Takada, N., Ishiguro, F., Fujita, H., Wang, H.-P. and Masuzawa, T. (1998) Lyme disease spirochetes in ticks from northeastern China. *Journal of Parasitology* 84, 499–504.

Takada, N., Fujita, H., Yano, Y., Ishiguro, F., Iwasaki, H. and Masuzawa, T. (2001) First records of tick-borne pathogens, *Borrelia*, and spotted fever group rickettsiae in Okinawajima island, Japan. *Microbiology and Immunology* 45, 163–165.

Takahashi, Y., Inukai, M., Iguchi, K., Takahashi, I. and Yamamoto, Y. (1997) A case of forest damage caused by sika deer (*Cervus nippon*) in Iwanazawa experimental plot of the Tokyo university forest in Hokkaido. *Transaction Meeting Hokkaido Branch Japan Forestry Society* 45, 84–87 (in Japanese).

Tian, W., Zhang, Z., Moldenhauer, S., Guo, Y., Yu, Q., Wang, L. and Chen, M. (1998) Detection of *Borrelia burgdorferi* from ticks (Acari) in Hebei Province, China. *Journal of Medical Entomology* 35, 295–298.

Uchikawa, K., Muramatsu, K., Miyamoto, K. and Nakao, M. (1991) An extensive prevalence of *Borrelia burgdorferi*, the etiologic agent of Lyme borreliosis, in Nagano Prefecture, Japan. *Japanese Journal of Sanitary Zoology* 42, 293–299.

Wang, G., van Dam, A.P., Le Fleche, A., Postic, D., Peter, O., Baranton, G., de Boer, R., Spajaard, L. and Dankert, J. (1997) Genetic and phylogenetic analysis of *Borrelia valaisiana* sp. nov. (*Borrelia* genomic groups VS116 and M19). *International Journal of Systematic Bacteriology* 47, 926–932.

Will, G.S., Jauris-Ileike, S., Schab, E., Busch, U., Rossler, D., Soutschek, E., Wilske, B. and Preac-Mursic, V. (1995) Sequence analysis of OspA gene shows homogeneity within *Borrelia burgdorferi* sensu stricto and *Borrelia afzelii* but reveals major subgroups within the *Borrelia garinii* species. *Medical Microbiology and Immunology* 184, 73–80.

Wilske, B., Luft, B., Schubach, W.H., Zumstein, G., Jauris, S., Preac-Mursic, V. and Kramer, M.D. (1992) Molecular analysis of the outer surface protein A (OspA) of *Borrelia burgdorferi* for conserved and variable antibody binding domains. *Medical Microbiology and Immunology* 181, 191–207.

Wilske, B., Preac-Mursic, V., Göbel, U.B., Graf, B., Jauris, S., Schwab, E. and Zumstein, G. (1993) An OspA serotyping system for *Borrelia burgdorferi* based on reactivity with monoclonal antibodies and OspA sequence analysis. *Journal of Clinical Microbiology* 31, 340–350.

Yamaguti, N. (1989) Human tick bite: variety of tick species and increase of cases. *The Saishin-Igaku* 44, 903–908 (in Japanese with English summary).

Ecology of *Borrelia burgdorferi* sensu lato in North America

9

Joseph Piesman

Division of Vector-borne Infectious Diseases, National Center for Infectious Diseases, Centers for Disease Control and Prevention, Public Health Service, US Department of Health and Human Services, PO Box 2087, Fort Collins, CO 80522, USA

Introduction

The impact of human tick-borne diseases in North America changed dramatically during the last century. Early in the 20th century, tick-borne diseases on the American continent were associated with sparsely populated remote regions of the Rocky Mountains. Rocky Mountain spotted fever, Colorado tick fever and relapsing fever were poorly known diseases associated with frontier regions (Price, 1948). The vast majority of people on the North American continent thought themselves to be free of the risk of tick-borne diseases. These perceptions were altered radically in the late 1960s, or 1969 to be precise. In 1969, Scrimenti (1970) observed an unusual case of erythema chronicum migrans in a Wisconsin resident. During the same year, another unusual case, this time of human babesiosis, was recognized in a resident of Nantucket Island, Massachusetts (Western *et al.*, 1970). Although these two reports were little noticed at the time, they were a portent of greater things to come. In the mid-1970s it became clear that several new tick-borne diseases were becoming increasingly common in highly populated areas of the northeastern USA.

In 1977, residents of Lyme, Connecticut (Fig. 9.1), and neighbouring towns were reported to be suffering from an unusual outbreak of arthritis, termed Lyme arthritis (Steere *et al.*, 1977); also in 1977, an ongoing outbreak of human babesiosis was described on Nantucket Island, Massachusetts (Ruebush *et al.*, 1977). Lyme arthritis was soon linked to erythema migrans, a condition long known in Europe, and rapidly becoming associated with the bite of the black-legged tick, *Ixodes scapularis*. By 1982, a spirochaete had been isolated from these ticks (Burgdorfer *et al.*, 1982) and the aetiologic agent of Lyme disease (Lyme borreliosis) was subsequently named *Borrelia burgdorferi* (Johnson *et al.*, 1984). This

© 2002 CAB International. *Lyme Borreliosis: Biology, Epidemiology and Control* (eds J. Gray, O. Kahl, R.S. Lane and G. Stanek)

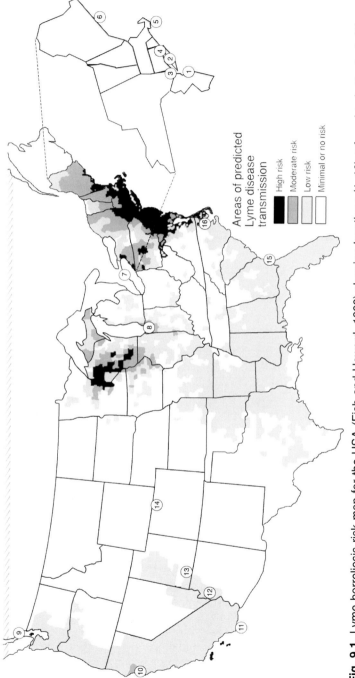

Fig. 9.1. Lyme borreliosis risk map for the USA (Fish and Howard, 1999) showing locations (1–16) referred to in the text. (1) Monmouth County, New Jersey; (2) Long Island, New York; (3) Westchester County, New York; (4) Lyme, Connecticut; (5) Nantucket Island, Massachusetts; (6) Monhegan Island, Maine; (7) Long Point, Ontario, Canada; (8) Jasper County, Indiana; (9) British Columbia, Canada; (10) Mendicino County, California; (11) Baja, Mexico; (12) Haulapai Mountain County Park, Mohave County, Arizona; (13) St George, Utah; (14) Larimer County, Colorado; (15) Sapelo Island, Georgia; (16) Camp Lejune, North Carolina.

nomenclature was modified to *B. burgdorferi* sensu stricto (s.s.) after the finding that in Europe the spirochaete occured as a complex of genospecies, known as *B. burgdorferi* sensu lato (s.l.), including *B. burgdorferi* s.s. (Wilske *et al.*, 1986). Human babesiosis was also linked to *I. scapularis* transmitting a rodent piroplasm, *Babesia microti* (Spielman, 1976). With the eye of the biomedical community closely fixed on *I. scapularis*, additional pathogens were soon found to be associated with this nefarious tick. The agent causing human granulocytic ehrlichiosis (HGE), closely related or identical to *Ehrlichia phagocytophila*, was found to be transmitted by *I. scapularis* (Pancholi *et al.*, 1995) and an arbovirus related to an agent that causes human encephalitis, Powassan virus, was also described in this same tick (Telford *et al.*, 1997). It seems prudent to review this sudden temporal and geographical clustering of human disease agents transmitted by ticks in terms of the basic ecology of Lyme borreliosis (LB) spirochaetes and their principal tick vectors in North America.

Tick ecology

Systematics

I. scapularis is a species that has long been known to science. It was described almost 200 years ago (Say, 1821). Unfortunately, Say's type specimen cannot be located and he gave as location only 'United States' (Nuttall *et al.*, 1911). In an authoritative review of the genus *Ixodes* in North America, Cooley and Kohls (1945) described the distribution of *I. scapularis* as being limited primarily to the southern USA, listing 21 records of this tick. This review does, however, cite a report by Cobb (1942) with support from a letter by the noted tick biologist C.N. Smith that placed an isolated population of *I. scapularis* in a northern location: Cape Cod, Massachusetts.

After the summary by Cooley and Kohls (1945), we essentially lost track of *I. scapularis* for approximately 30 years. Isolated scattered reports placed this tick in various locations in the southern USA from Florida (Rogers, 1953) to Virginia (Kellogg *et al.*, 1971). Northern records included Rhode Island (Hyland and Mathewson, 1961), Maryland and Massachusetts (Kellogg *et al.*, 1971), eastern Long Island, New York (Good, 1973), northern Wisconsin (Jackson and DeFoliart, 1970) and Long Point, Ontario, Canada (Watson and Anderson, 1976) (Fig. 9.1). Few reports on the distribution and biology of *I. scapularis* prior to 1976 mentioned these ticks as vectors of human or animal disease agents. One exception (Kellogg *et al.*, 1971) listed 'experimental transmission only' for bovine anaplasmosis and tularaemia for *I. scapularis* with no details or references. Then, in 1976, research attention was focused squarely on *I. scapularis* when nymphs of this tick were identified as the vector of *B. microti*, the aetiologic agent of human babesiosis on Nantucket Island, Massachusetts (Spielman, 1976). Soon thereafter, nymphal *I. scapularis* were also implicated as the vector of the yet to be described aetiologic agent of LB (Wallis *et al.*, 1978).

Controversy soon emerged when Spielman *et al.* (1979) made the bold move of describing the northern *Ixodes* spp. populations as a new species, *Ixodes dammini*, separate and distinct from the southern populations of *I. scapularis*. The grounds for this species designation were primarily based on morphology and geographical distribution. The nymphs, especially, were reported to be different morphologically; a key difference was that *I. dammini* nymphs contained large triangular posteriorly directed auriculae that arose on the margin of the ventral side of their basis capituli, whereas *I. scapularis* nymphs contained much smaller auriculae that arose somewhat mesad of the external margin. The result of this characteristic was that *I. dammini* had pointy auriculae that could be seen with a light microscope and the auriculae of *I. scapularis* were barely visible. The distribution of *I. dammini* was reported to include Massachusetts, Connecticut, New Jersey, New York, Rhode Island, Minnesota, Wisconsin and Ontario, Canada; *I. scapularis* identifications were confirmed from Florida, Georgia, Alabama, Louisiana, Texas, South Carolina, Arkansas, Illinois, Indiana and Iowa (Spielman *et al.*, 1979). An additional distinction was the host preference of immature ticks. Immature *I. dammini* were reported to feed on a variety of hosts, but to focus on rodents and *Peromyscus leucopus* (the principal reservoir of *B. microti* and *B. burgdorferi* s.s.) in particular. In contrast, *I. scapularis* immatures were reported to feed predominantly on reptiles, including lizard species that were refractory to infection with these pathogens (Spielman *et al.*, 1984).

The literature on the principal vectors of LB spirochaetes in North America generally incorporated the use of *I. dammini* through the 1980s, until Oliver *et al.* (1993b) challenged the validity of *I. dammini* as a species separate and distinct from *I. scapularis*. Their argument was based on hybridization and assortative mating experiments, morphometric analyses, isoenzymes and host-selection data. The core of their argument for conspecificity was their mating experiments. Reciprocal crosses between *I. dammini* (Massachusetts) and *I. scapularis* (Georgia) produced fertile offspring through the F_3 generation. No significant decrease in fertility or fecundity occurred in any of the *I. scapularis*–*I. dammini* crosses. In contrast, crosses between either *I. scapularis* or *I. dammini* with another *Ixodes* sp. (*Ixodes pacificus*), produced F_1 offspring, but these progeny were sterile. The assortative mating experiments suggested that no mating barriers existed between the Massachusetts (*I. dammini*) and Georgia (*I. scapularis*) colonies. Moreover, Oliver *et al.* (1993b) conducted a morphometric discriminant analysis using 65 measurable characteristics on adults and nymphs of *I. scapularis* from North Carolina and Georgia and *I. dammini* from Maryland and Massachusetts, as well as F_1 hybrids. *I. pacificus* was included as an outgroup. Conventional analysis suggested differences in the nymphs of *I. scapularis* and *I. dammini*, but these differences were considered to be size-dependent. Gross analysis of the chromosomes and isozymes also showed no significant difference between *I. dammini* and *I. scapularis*. A prior experiment (James and Oliver, 1990) was also cited to prove that both *I. dammini* and *I. scapularis* fed preferentially on mice as opposed to lizards when presented with a choice.

The debate on the species status of *I. dammini* quickly moved to include the new molecular systematic tools. The first in-depth molecular analysis of the validity of *I. dammini* as a species distinct from *I. scapularis* focused on ribosomal nucleotide sequences (rDNA), e.g. the internal transcribed spacers ITS1 and ITS2. Wesson *et al.* (1993) determined the ITS1 and 2 sequences of *I. scapularis*, *I. dammini* and *I. pacificus*. The ITS region for *I. pacificus* was quite distinct from *I. scapularis* and *I. dammini*, but the ITS regions for *I. scapularis* and *I. dammini* were similar. Phylogenetic analysis showed that the ITS sequences did not separate *I. dammini* and *I. scapularis*. The approach of using ITS sequences for phylogenetic analysis of *Ixodes* spp. ticks was subsequently challenged by Rich *et al.* (1997); these authors detected at least 4% nucleotide variation among repeated assays of individual ticks. They concluded that the ITS region should be used with great caution in phylogenetic studies of ticks.

In an exhaustive study on the population genetics of *I. scapularis* (including the putative *I. dammini*), Norris *et al.* (1996) concentrated first on the mitochondrial genes 16S and 12S. This analysis clearly distinguished *I. scapularis* from the two outgroups: *I. pacificus* and *Ixodes ricinus*. Among *I. scapularis*, two distinct clades emerged. The first clade consisted of specimens from Florida to North Carolina and was designated the southern clade. The second clade contained specimens from Mississippi to Massachusetts and was termed the American clade. Mixed populations of the southern and American clades were discovered in Oklahoma, Mississippi, Georgia and North Carolina. The authors concluded that there was 100% support for the monophyletic relationship among all haplotypes of *I. scapularis*. Interestingly, both clades were present in a single collection from a dog in North Carolina. In order to further test whether these clades were reproductively, if not geographically, isolated, Norris *et al.* (1996) went on to use random amplified polymorphic DNA PCR (RAPD–PCR) to test for reproductive isolation. Ticks from the two clades consistently grouped together in this RAPD analysis. Overall, Norris *et al.* (1996) concluded that the two *I. scapularis* clades were reproductively continuous, and that there was no reproductive isolation among *I. scapularis* populations. In contrast, a study looking solely at the 16S mitochondrial DNA of *I. scapularis* and *I. dammini* observed a similar phenomenon, the existence of two clades, but came to the opposite conclusion (Rich *et al.*, 1995), namely, that a southern clade exists that corresponded to *I. scapularis* and that a recently expanded clade including northeastern and middle-eastern regions corresponded to *I. dammini*. A study by the same group looked at the 16S sequences of nine *Ixodes* spp. from Eurasia and North America (Caporale *et al.*, 1995). They found that all the species were easily separated by > 4% divergence in their 16S nucleotide sequence analysis, except for the two clades of *I. scapularis*, where divergence was approximately 2%.

The pattern of current usage appears to have accepted the conspecificity of *I. scapularis* and *I. dammini*, with the vast majority of authors referring to the two clades simply as *I. scapularis*. An interesting argument for the continued use of the species *I. dammini* based on epidemiological grounds was presented by Telford (1998). An increased understanding of the biotic and abiotic factors that determine ecological risk of LB spirochaete transmission on a regional level is certainly needed. The species status of *I. scapularis* should, however, be determined on the basis of rigorous systematic criteria. A synthesis approach to tick systematics that involves a combination of morphological criteria and multiple genetic loci, as has been recently performed for *Rhipicephalus* and *Boophilus* spp. (Beati and Keirans, 2001), may eventually help to resolve some of the conflicts in *Ixodes* spp. systematics. For the purposes of this chapter, the conspecificity of *I. scapularis* and *I. dammini* will be accepted and these ticks will be referred to as *I. scapularis*.

Habitat

In eastern North America, *I. scapularis* is clearly associated with deciduous forest and habitat containing leaf litter. These ticks are very susceptible to desiccation (Stafford, 1994) and are therefore found in mainland forests where sufficient leaf litter provides a substrate to hold moisture and provide cover from wind, snow and other elements. In Westchester County, New York, the local distribution of *I. scapularis* was positively correlated with a greenness–wetness index calculated from geographical information system (GIS) data obtained by remote sensing (Dister *et al.*, 1997). A more down-to-earth study conducted in the same area demonstrated that all three active stages of *I. scapularis* were much more abundant in the woods than other habitats; substantial numbers of ticks were also found in the ecotonal vegetation, while fewer were recovered from ornamental vegetation and even fewer from lawns (Maupin *et al.*, 1991). Those *I. scapularis* that were found on lawns were in shaded areas that did not dry out during the day. The importance of leaf litter has been demonstrated by experiments where leaf litter was actually removed and tick populations evaluated. This backbreaking experiment showed that *I. scapularis* populations could be reduced by 72–100% by leaf litter removal (Schulze *et al.*, 1995). In summary, these ticks appear to be woodland species that are closely associated with deciduous forest and leaf litter.

The type of forested habitat where *I. scapularis* is commonly found is quite variable. The classical northeastern hardwood forest in Westchester County, New York, has been described as a mix of oak (*Quercus* spp.), sugar maple (*Acer saccharum*), grey birch (*Betula populifolia*) and tulip poplar (*Liriodendron tulipifera*) (Daniels *et al.*, 1989). Although *I. scapularis* clearly abounds in this type of habitat, it is found in many other ecosystems as well. In coastal Massachusetts, the type of vegetation where *I. scapularis* thrives is more scrub/brush-dominated than mature forest. On Nantucket Island, Massachusetts, habitats with abundant tick populations were characterized by bayberry (*Myrica pensylvanica*), wild rose (*Rosa carolina*, *Rosa virginiana*) and scrub oak (*Quercus ilicifolia*) (Piesman and Spielman, 1979). In Monmouth County, New Jersey, where mature inland forests are based on sandy soils, a mixture of hardwoods and shrubs are present in areas of high *I. scapularis* abundance: various oak and maple species, black cherry (*Prunus serotina*), tulip poplar, American beech (*Fagus grandifolia*), blackgum (*Nyssa sylvati-*

ca) and sweetgum (*Liquidambar styraciflua*) represent hardwoods, and the understorey and shrub layer are dominated by saplings of these hardwoods as well as highbush blueberry (*Vaccinium corymbosum*) and lowbush blueberry (*Vaccinium angustifolium*), as well as other bushes and greenbriar (*Smilax glauca*) (Schulze *et al.*, 1998). In New Jersey, an interesting study compared hardwood forests with pine forests containing predominantly pitch pine (*Pinus rigida*) with mixed hardwood–pine forest, and with old fields where various grasses and shrubs predominated (Schulze *et al.*, 1998). Surprisingly, *I. scapularis* was highly abundant in hardwood forest, pine forest and mixed hardwood–pine forest habitat. Old field was the only habitat where *I. scapularis* was consistently rare. In general, one may conclude that *I. scapularis* is generally associated with hardwood forests containing abundant leaf litter in the eastern USA, but these ticks are quite adaptable and can be found in other habitats such as scrub/brush and pine forests. In Europe, *I. ricinus* has been reported to thrive in certain grassland areas where sheep graze and trees are totally absent. In North America, *I. scapularis* has not been described in comparable habitats.

An interesting theory has been presented revolving around events called masting, wherein oak trees produce massive amounts of acorn crops every 2–5 years. Researchers in upstate New York conducted an experiment in which a remarkable amount of acorns were placed on small plots (Jones *et al.*, 1998). Tick, mouse and deer populations were monitored in and around these plots. Mouse populations and *I. scapularis* populations increased on these plots in subsequent years. The authors hypothesized that masting events could influence tick populations from year to year and thereby be used to predict years of high LB risk, which would occur roughly 2 years after a major masting event. Since masting events seem to occur in different areas in the northeastern forest during different years, the scale of these predictions would have to be carefully monitored and the hypothesis tested over many years. Nevertheless, it is an intriguing idea.

In the mid-western USA, *I. scapularis* is generally found in heavily wooded habitats. These areas are often surrounded by vast tracts of land cleared for agriculture in this part of the USA. In general, land dedicated to agricultural crops does not contain *I. scapularis* (Kitron and Kazmierczak, 1997). An interesting story in Indiana revealed, however, how small a patch of woodlands is needed to allow the establishment of *I. scapularis* and the natural transmission cycle of *B. burgdorferi* s.s. Despite concerted efforts to find *I. scapularis* ticks in Indiana in the 1970s and 1980s, Pinger *et al.* (1996) could not detect these ticks. A collection of three adult *I. scapularis* in 1992 and a retrospective report of an LB case in 1985 led these researchers to two sparsely populated counties in northwestern Indiana where the land was largely cleared for agriculture (Jasper and Newton Counties) (Fig. 9.1). Searches led them to a 50 ha woodlot surrounded by cropland. This woodlot had not been cleared because it sat on slightly higher ground than the surrounding flat terrain and within it they found many questing *I. scapularis*, infested and infected rodents, and *B. burgdorferi* s.s. infection in adult and nymphal ticks. From this small beginning, *I. scapularis* has now become well established in the surrounding counties of the northwest corner of Indiana.

The habitat where *I. scapularis* is found in the southeastern USA is quite diverse. It consists of a mixture of pine forest mixed with hardwoods such as oak and hickory. Dominant trees include long-leaf pine (*Pinus palustris*), loblolly pine (*Pinus taeda*), various oaks (*Quercus* spp.), hickory (*Carya* spp.) and loblolly bay (*Gordonia lasianthus*) (Apperson *et al.*, 1993). Included among the understorey are various shrubs including blueberry (*Vaccinium crassifolium*), inkberry (*Ilex glabra*), wiregrass (*Aristida stricta*) and fetterbush (*Lyonia lucida*).

The principal vector of the LB spirochaete in western North America is the western black-legged tick, *I. pacificus*. There is no systematics controversy surrounding this tick species; an examination of allozymes and mitochondrial genes demonstrated that although there was great haplotype diversity in this species' range from British Columbia down through Mexico and east to Utah (Fig. 9.1), there is no evidence of genetic isolation among these diverse populations (Kain *et al.*, 1997, 1999). The habitats where *I. pacificus* thrive are interesting. In north coastal California, the association of *I. pacificus* has been intensely studied (Li *et al.*, 2000). This area has a Mediterranean climate, with cool wet winters and warm dry summers. Numerous *I. pacificus* adults have been collected along trails in habitat characterized as Douglas fir forest, north coastal scrub, chaparral and open grasslands in one park, and redwood forest, oak forest, chaparral and open grasslands in another park. Adult *I. pacificus* were found in these habitats in areas of high brush, with uphill or downhill slope and moderate leaf litter density. Nymphs were found in areas of heavy shade in this study and can generally be found in areas of shade with leaf litter in California. Areas where nymphal *I. pacificus* were particularly abundant were characterized by the presence of trees such as black oak (*Quercus kelloggii*), Douglas fir (*Pseudotsuga menziesii*) and Pacific madrone (*Arbutus menziesii*) with an understorey of poison oak (*Toxicodendron diversiloba*) in Cloverdale, California (Clover and Lane, 1995); in Hopland, California, various oaks as well as Pacific madrone, Douglas fir, redwood (*Sequoia sempervirens*), California bay (*Umbellularia californica*) and bigleaf maple (*Acer macrophyllum*) dominated in an area of high nymphal tick abundance (Tälleklint-Eisen and Lane, 1999). Attempts to identify biotic and abiotic factors that dramatically influence the abundance of infected nymphal *I. pacificus* in Mendicino County, California, were not productive (Tälleklint-Eisen and Lane, 1999).

Although *I. pacificus* is mainly found along the Pacific coast in areas of high rainfall, it can be found in some unusual inland locations where the general environment is quite arid. A population of *I. pacificus* is extant in the Hualapai Mountain County Park, near Kingman, Mojave County, Arizona (Fig. 9.1). As the name implies, this area borders on one of the driest parts of the world, the Mojave Desert. At sufficient altitude (> 2000 m), however, sky-islands of vegetation containing oak trees and brush, such as the coffee-berry bush (*Garrya wrightii*) can be found. Snow provides ample moisture and the oak trees provide leaf litter. Numerous *I. pacificus* were collected here at an altitude of 2345 m (Olson *et al.*, 1992). Similar sky-islands are common in this region and may prove to contain isolated populations of *I. pacificus*. In another arid region, near St George, Utah, in the southwest corner of Utah, *I. pacificus* was collected in an isolated location called locally 'Oak Grove' (Kain *et al.*, 1997, 1999). This population had an extremely restricted diversity of haplotypes when tested with mitochondrial markers, indicating it may be quite isolated and derived from a small number of founder ticks. Finally, immature *I. pacificus* have been collected from lizards in extremely hot and arid locations in Washington County, in southwestern Utah (J. Piesman, unpublished data). Only the barest of leaf litter could be found along a small stream bed, but nevertheless *I. pacificus* was found to survive there.

In summary, although both *I. scapularis* and *I. pacificus* are woodland species associated with high moisture and leaf litter, these ticks can, to a limited degree, extend their distribution to relatively harsh environments.

Hosts

The one indispensable piece in the LB puzzle in North America is the presence of deer. In the eastern USA white-tailed deer, *Odocoileus virginianus*, serve as hosts for all three stages of *I. scapularis*, so much so, that the tick is generally called the 'deer tick', even though the appropriate common name is the black-legged tick. Although all three stages of *I. scapularis* feed on deer, it is the role of deer as hosts for adult *I. scapularis* that places it front and centre in the LB story. Soon after *I. scapularis* was implicated as a vector of human disease, a study was conducted in Massachusetts that sampled deer throughout the year to determine when each tick stage was present on deer. Hundreds of larval, nymphal and adult ticks were found on deer collected from Naushon and Nantucket Islands, Massachusetts (Piesman *et al.*, 1979). Close to 500 adult *I. scapularis* were removed from a single deer. Numerous publications have since detailed the close association between adult *I. scapularis* and deer. Other large animals can serve as hosts for adult *I. scapularis*, e.g. domestic pets (dogs and cats), livestock (cows, horses, sheep, goats, pigs, donkeys) and wild animals such as bears, coyotes, wolves, raccoons and opossums. But infestation levels on these other hosts do not appear to approach those frequently seen on deer. Black bears (*Ursus americanus*) may be an exception (Kazmierczak *et al.*, 1988); the density of bears, however, is quite low in eastern North America in comparison with deer. Moreover, adult *I. scapularis* rarely if ever infest smaller mammals such as rabbits, woodchucks, weasels and rodents. Adult *I. scapularis* also do not infest birds. An interesting question is whether deer are absolutely essential for the establishment and maintenance of *I. scapularis*. One approach to answering this question is to measure the abundance of *I. scapularis* in neighbouring locations and correlate the tick abundance with the density of deer. In a macrogeographical study of 13 islands in Essex County, Massachusetts, the density of deer was closely correlated with the abundance of larval *I. scapularis* on a per island basis (Wilson *et al.*, 1985). Similarly, in a study in Rhode Island, *I. scapularis* was present on two islands with deer and absent from four islands without deer (Anderson *et al.*, 1987). In a microgeographical study conducted on Long Island, New York, the frequency with which deer carrying radio-telemetry devices occupied defined quadrants was positively correlated with the number of immature *I. scapularis* on rodents the following summer (Wilson *et al.*, 1990). In natural areas and parks on Long Island, New York (Duffy *et al.*, 1994), ticks were sampled from 16 locations where deer were abundant and six locations where deer were believed to be absent (a difficult proposition to prove absolutely). The natural areas where deer were abundant had 14-fold more nymphs and 56-fold more larval *I. scapularis* compared with natural areas where deer were rare or absent. Studies on deer exclusion using fencing have also shown that deer are essential hosts (Stafford, 1993; Daniels and Fish, 1995), as have deer removal experiments (Wilson *et al.*, 1988; Deblinger *et al.*, 1993).

There is little doubt that the importance of white-tailed deer as a pivotal host for adult *I. scapularis* is due, in part, to the superabundance of these animals at the close of the 20th century. In an interesting review of the history of white-tailed deer in North America, McCabe and McCabe (1997) estimate that before European colonization (*c.* 1500) there were between 24 and 33 million white-tailed deer in what is now the USA. During the period 1850–1900, agricultural expansion and aggressive hunting practices led to the virtual demise of the species. Many counties in New England reported the shooting of the 'last deer in the county' from 1895 to 1900. In the latter half of the 20th century, a reversion of agricultural land to forest and lack of hunting pressure have returned the white-tailed deer to some of its former glory. McCabe and McCabe (1997) estimate there are between 16 and 17 million white-tailed deer in the USA at present. What is remarkable is the explosive growth of these populations over a period of 100 years. In Connecticut, for instance, there were probably only about a dozen deer left in 1896. Today there are probably > 75,000 deer in Connecticut (Gregonis, 2000) and as many as 2 million deer in New England and the adjacent mid-Atlantic states. This explosive growth in the white-tailed deer population has no doubt coincided in time with the LB epidemic in northeastern North America. The hypothesis that *I. scapularis* was once prevalent over much of the eastern deciduous forest, reduced to a few relict populations when deer were virtually eradicated around 1900 and are slowly but surely now returning to their original range as the deer population explodes is intriguing.

In western North America, *I. pacificus* is closely associated with another deer species, namely, the Columbian black-tailed deer, *Odocoileus hemionus columbianus* (Westrom *et al.*, 1985). This subspecies is among several referred to commonly as 'mule deer'. In general, mule deer are extremely abundant in areas where *I. pacificus* are abundant, and adult *I. pacificus* are found in high numbers on these animals. Although larval and nymphal *I. pacificus* can be found on deer, other hosts are considered more important for immatures (Westrom *et al.*, 1985).

Rodents also play a key role in the ecology of LB in North America, as hosts for immature *Ixodes* spp. ticks. When attention was first directed at *I. scapularis* as a vector of human pathogens, Nantucket Island, Massachusetts, was a focal point of early research. This island had a unique paucity of rodent species, with white-footed mice (*P. leucopus*), meadow voles (*M. pennsylvanicus*) and short-tailed shrews (*Blarina brevicauda*) as the only rodents present in abundance (Healy *et al.*, 1976). Furthermore, the two main ticks encountered used different hosts for their immature stages, with the American dog tick *Dermacentor variabilis* mainly associated with meadow voles and *I. scapularis* mainly associated with white-footed mice. Summarizing numerous studies conducted on Nantucket Island, Spielman *et al.* (1984) concluded that 91% of larval and 91% of nymphal *I. scapularis* fed on white-footed mice, with voles and deer feeding far fewer immature *I. scapularis*. A paradigm emerged from this body of work on Nantucket Island that featured white-tailed deer as the indispensable host for adult *I. scapularis* and white-footed mice as the indispensable host for immature *I. scapularis*.

The *P. leucopus*–*I. scapularis* association proved not to be an absolute requirement. For instance, Monhegan Island, an isolated island off the coast of Maine, was known to have deer and an abundant population of *I. scapularis* infected with *B. burgdorferi* s.s., but no white-footed mice (Smith *et al.*, 1993). It transpired that the Norway rat, *Rattus norvegicus*, had substituted efficiently as the rodent host for immature *I. scapularis*, thus demonstrating the plasticity of these ticks and the LB disease spirochaete ecological transmission cycle. Deer have now been totally removed from this island in a tick control experiment and it will be interesting to follow the survival of *I. scapularis* in these unique circumstances. In addition, as mainland locations became the sites of Lyme disease ecological studies, the predominance of *P. leucopus* as hosts for immature *I. scapularis* became less clear. In Westchester County, New York, chipmunks, shrews, skunks and raccoons have all been suggested as important alternative mammalian hosts for immature *I. scapularis* (Fish and Daniels, 1990; Daniels *et al.*, 1991). In the mid-western USA, chipmunks clearly play an important role as hosts for immature *I. scapularis*. Studies along the Rock River in Illinois revealed that the eastern chipmunk, *Tamias striatus*, had higher overall levels of infestation with *I. scapularis* than did *P. leucopus* and this was especially significant for the nymphal stage of these ticks (Mannelli *et al.*, 1993). The relative importance of these two rodents in the LB spirochaete ecological cycle should be the subject of locally based comparative studies (e.g. Slajchert *et al.*, 1997) examining tick abundance on each host, the infectivity of each host species and their relative abundance. In general, the importance of various rodents as hosts for immature *I. scapularis* and as reservoirs for *B. burgdorferi* s.s. requires intense local study.

Rodents also play a role as important hosts for immature *I. pacificus* and in an area of intensive study, the Hopland Field Station in Mendocino County, California, include the deer mouse (*Peromyscus maniculatus*), the piñon mouse (*Peromyscus truei*), the brush mouse (*Peromyscus boylii*), the California kangaroo rat (*Dipodomys californicus*), the dusky-footed woodrat (*Neotoma fuscipes*) and the western harvest mouse (*Reithrodontomys megalotis*). All of these rodents were infested with larval *I. pacificus*, but nymphal *I. pacificus* infestation on rodents was rare (Lane and Loye, 1991). The situation with the dusky-footed woodrat (*N. fuscipes*) is especially interesting. These rodents are often infested with immature *Ixodes* spp. ticks, but *I. pacificus* shares this rodent host with another species, *Ixodes spinipalpis*. For years, *I. spinipalpis* was identified as *Ixodes neotomae* until the two species were synonomized as *I. spinipalpis* (Norris *et al.*, 1997). A study conducted in southern Oregon confirmed that small numbers of immature *I. pacificus* and *I. spinipalpis* infested woodrats as well as deer mice and brush mice (Burkot *et al.*, 1999).

In general, rodents are much less significant than lizards as hosts for immature *I. pacificus*. Large numbers of both larval and nymphal *I. pacificus* regularly infest the western fence lizard, *Sceloporus occidentalis* (Lane and Loye, 1989; Tälleklint-Eisen and Eisen, 1999) and the southern alligator lizard, *Elgaria multicarinata* (Wright *et al.*, 1998). In addition, the range of *I. pacificus* and *S. occidentalis* roughly overlaps. Thus, throughout its range, the western fence lizard appears to be the primary host for immature *I. pacificus*. Clearly, they are more important hosts for immature *I. pacificus* than are rodents. As will be discussed in the section Borrelia Ecology (below), the fact that western lizards are the primary host for immature *I. pacificus* and that these reptiles are refractory to infection with *B. burgdorferi* s.s. has profound implications for the ecological cycle of LB spirochaetes in the western USA.

In the northeastern and mid-western USA, where LB is hyperendemic, lizards are generally absent. In fact, rodents are much more abundant than reptiles in the northern USA. To some degree, this helps focus the feeding of immature *I. scapularis* on rodents that are reservoir-competent for *B. burgdorferi* s.s. In the southeastern USA, however, lizards are abundant and are the predominant hosts for immature *I. scapularis*. An extensive comparison of rodents and lizards as hosts for immature *I. scapularis* was conducted on the coastal plain of North Carolina, on the Marine Corps Base in Camp Lejune (Apperson *et al.*, 1993). Four lizard species were abundantly infested with immature *I. scapularis*: the southeastern five-lined skink (*Eumeces inexpectatus*), the ground skink (*Scincella lateralis*), the broadheaded skink (*Eumeces laticeps*) and the eastern glass lizard (*Ophisaurus ventralis*). In contrast, the number of immature *I. scapularis* infesting rodents in the same location was extremely low. In fact, immature *I. scapularis* were almost 50-fold more abundant on lizards than rodents in this study. Preserved museum specimens from the southern USA were also found to have large numbers of attached *I. scapularis*; one *Eumeces* specimen had > 200 attached *I. scapularis* (Oliver *et al.*, 1993a). The fact that immature *I. scapularis* focus on rodents in the northeast and focus on lizards in the southeast has profound implications for the risk of *B. burgdorferi* s.s. transmission in these two regions (see Borrelia Ecology below).

Interest in birds as hosts for immature *I. scapularis* during the initial phases of LB research focused mainly on the potential role of migrating birds in introducing ticks to new locations (Spielman, 1988; Wilson *et al.*, 1990). Studies in Westchester County, New York, however, revealed that up to 25 resident bird species were frequently infested with immature *I. scapularis* (Battaly and Fish, 1993). House wrens (*Troglodytes aedon*), American robins (*Turdus migratorius*) and the common grackle (*Quiscalus quiscula*) had notable levels of infestation. Similar results were obtained from various passerine birds in Connecticut (Stafford *et al.*, 1995). In general, ground-inhabiting birds have higher infestation rates than birds that spend most of their time up in the canopy. The fact that immature *I. scapularis* in different locations may abundantly infest deer, rodents, medium-sized mammals, lizards or birds shows the plasticity of the host range of these ticks. Birds have also been studied as hosts for immature *I. pacificus*. Extensive studies conducted on 24 bird species captured in an oak woodland forest in Yuba County, California, at an elevation of 440–520 m revealed only five immature *I. pacificus* infesting 138 birds (Manweiler *et al.*, 1990). In contrast, a study conducted in a mixed pine–oak–madrone forest in Placer County, California, revealed that eight ground-dwelling bird species were commonly infested with *I. pacificus* (Wright *et al.*, 2000). The bird–*I. pacificus* relationship appears to vary from location to location.

Borrelia Ecology

Host complement and reservoir competence

Controversy has long surrounded the importance of various animals as reservoirs of *B. burgdorferi* s.s. in North America. For instance, the role of white-tailed deer as a reservoir serving to infect *I. scapularis* ticks that feed on them has been hotly debated (Telford *et al.*, 1988; Oliver *et al.*, 1992). Fortunately, an excellent body of work from Europe has helped to shed light on this somewhat confused aspect of LB ecology in North America. Kurtenbach *et al.* (1998) found that *B. burgdorferi* s.l. was lysed by certain host sera and that the active component was host complement. By deactivating complement in various ways, it was shown that complement acted directly via the alternative pathway to kill *B. burgdorferi* s.l., rather than through antibodies. The most borreliacidal serum was that of deer (red deer, *Cervus elaphus*), which killed *B. burgdorferi* s.s. as well as all other European genospecies of *B. burgdorferi*. Curiously, bird complement killed some European genospecies and rodent complement killed other European genospecies, but *B. burgdorferi* s.s. was relatively resistant to killing by both bird and rodent sera.

This basic observation by Kurtenbach *et al.* (1998) was quickly extended in North America. In one study, B-31, the prototype North American strain of *B. burgdorferi* s.s. (from Shelter Island, New York), was exposed to sera from white-footed mice, New Zealand white rabbits and sika deer (*Cervus nippon*). The exotic sika deer were used as surrogates for white-tailed deer in tick feeding experiments. Whereas mouse and rabbit sera demonstrated negligible borreliacidal activity, exposure to sera from sika deer resulted in almost complete killing of the spirochetes (Nelson *et al.*, 2000). This killing activity was judged to be due to the alternative complement pathway. Another study compared the killing activity of mammalian and reptilian complement. Using a California strain (CA4) of *B. burgdorferi* s.s., Kuo *et al.* (2000) demonstrated that complement from the western fence lizard and the southern alligator lizard effectively killed the spirochaetes, whereas sera from deer mice (*P. maniculatus*) and humans did not. Although definitive testing of sera from white-tailed deer, mule deer and eastern North American lizard species has not been conducted, these observations on complement are highly suggestive that these deer and lizards serve as zooprophylactic hosts, a concept first proposed by Spielman *et al.* (1985). This concept stresses the fact that such hosts remove some immature ticks from the proportion of the tick population that will acquire infection with *B. burgdorferi* s.s.; these hosts thereby serve to lower the overall risk of transmission of LB spirochaetes to people. Of course, the role of deer is a double-edged sword since they are the presumed indispensable host species for the adult stages of *I. scapularis* and *I. pacificus*. But lizards appear to be the primary force in driving down tick infection rates with *B. burgdorferi* s.s. wherever they are found, so much so that in one study the relative abundance of lizards versus mammals was considered as an inverse indicator of infection rate data for assessing risk of transmission of LB spirochaetes (Anon., 1999). If this approach turns out to be justified, it could go a long way towards explaining why risk of LB transmission is so great in the northeastern USA and extremely low in the southern USA. The only observation that contradicts this hypothesis is the finding that certain lizards found in eastern North America under isolated circumstances may actually serve as competent reservoirs for *B. burgdorferi* s.s. (Levin *et al.*, 1996).

Leaving deer and lizards aside as potential reservoir hosts for *B. burgdorferi* s.s. in North America, we are left with medium-sized mammals, rodents and birds. Although medium-sized mammals can infect *I. scapularis* with *B. burgdorferi* s.s. (Fish and Daniels, 1990), the relative scarcity of these hosts compared with rodents and birds in eastern woodlands generally rules them out as important reservoir hosts. The importance of the white-footed mouse (*P. leucopus*) as a host for immature *I. scapularis* has been firmly established. Interestingly, nymphal *I. scapularis* usually feed on these mice in late spring or early summer (May–July), whereas larvae feed on the same mice later in the year (August–September) (Piesman and Spielman, 1979; Carey *et al.*, 1980; Wilson and Spielman, 1985). This feeding pattern is thought to be essential to the efficient transfer of spirochaetes from *P. leucopus* to *I. scapularis* since juvenile mice entering the population become infected during the nymphal feeding season and are virtually all infected and maximally infectious at the time larval *I. scapularis* feed on them. Infected larval *I. scapularis* are then thought to moult to nymphs and survive the winter (Yuval and Spielman, 1990) and to transmit spirochaetes to a new crop of mice or humans the next year. This interpretation of the maintenance of *B. burgdorferi* s.s. places a heavy reliance on standard horizontal transmission of spirochaetes from nymphal ticks to rodents and back to larval ticks that then moult to nymphs and renew the cycle. It is supported by experimental observations demonstrating that laboratory-infected *P. leucopus* are extremely infectious to larval *I. scapularis* (Donahue *et al.*, 1987) and field-collected *P. leucopus* from Massachusetts were also highly infectious, infecting 46% of exposed larval *I. scapularis* in one study (Mather *et al.*, 1989b).

In Europe there has been some discussion of co-feeding as an important adjunct to the maintenance of *B. burgdorferi* s.l. in areas where sheep are the primary host for all three stages of *I. ricinus* (Ogden *et al.*, 1997). In the northeastern USA, however, the standard horizontal transmission of *B. burgdorferi* s.s.

is so efficient that adjuncts like co-feeding do not appear to be necessary to explain the commonly observed 25% infection rate in nymphal *I. scapularis* and 50% infection rate in adults (Maupin *et al.*, 1991). In the mid-western USA co-feeding may be more important because feeding of larval and nymphal *I. scapularis* tends to coincide to a greater degree (Jones and Kitron, 2000); however, the situation is not clear-cut in the mid-western USA. In a study area in Illinois, nymphal *I. scapularis* were mainly focused on chipmunks (*T. striatus*) as hosts (Mannelli *et al.*, 1993; Slajchert *et al.*, 1997), whereas larvae were mainly focused on white-footed mice (Mannelli *et al.*, 1994). Issues of horizontal versus co-feeding transmission aside, rodents are clearly the primary reservoirs of *B. burgdorferi* s.s. in eastern North America, with white-footed mice, deer mice, chipmunks, voles, short-tailed shrews and Norway rats all reported to play a role as reservoirs in specific local situations.

In California, the dusky-footed woodrat (*N. fuscipes*) is the principal reservoir of *B. burgdorferi* s.s. and the California kangaroo rat (*D. californicus*) is also frequently infected (Brown and Lane, 1992, 1994). *Peromyscus* spp., however, seem to be less frequent reservoirs of *B. burgdorferi* s.s. in this region. The genetic diversity of *Borrelia* spp. found in these rodents makes it difficult to determine the primary reservoir of those spirochaetes causing LB in the western USA (see section on Diversity of *Borrelia* genospecies in this chapter).

Birds clearly have an auxiliary role to play as reservoirs of *B. burgdorferi* s.s. in North America. Several reports suggest that larval *I. scapularis* feeding on a variety of naturally infected birds acquired infection with *B. burgdorferi* s.s. (Stafford *et al.*, 1995); but at least the grey catbird (*Dumatella carolinensis*) appeared to be incompetent as a reservoir of *B. burgdorferi* s.s. (Mather *et al.*, 1989a). In contrast, American robins (*T. migratorius*) infected almost all the *I. scapularis* experimentally fed on them, although this infectivity waned after 2 months (Richter *et al.*, 2000). A primary importance of birds may prove to be that, when specific control measures are directed towards targeted killing of ticks feeding on reservoir-competent rodents, birds may prove to be a troubling alternative source of spirochaetes to the tick population.

Diversity of *Borrelia* genospecies

B. burgdorferi s.l. has been split into numerous genospecies in Europe and Asia based on minor differences in the 5S–23S and 16S–23S intergenic spacer regions in the ribosomal genes. Spirochaetes in North America were for a long time thought to be more uniform. In fact, *B. burgdorferi* s.s. was the sole genospecies discussed in the North American literature until Marconi *et al.* (1995) described isolates from cottontail rabbits and a tick species closely associated with these rabbits (*Ixodes dentatus*) as a unique genospecies, *Borrelia andersonii*. It is unknown whether *B. andersonii* can commonly infect ticks that feed on people, e.g. *I. scapularis*, and whether human infection with *B. andersonii* ever occurs. Subsequent to the description of *B. andersonii*, a diverse collection of spirochaetes isolated in California were characterized as an additional new species, *Borrelia bissettii* (Postic *et al.*, 1998). It is now clear that *B. bissettii* can be found in many regions spanning the USA, including California, Colorado, Illinois and Connecticut. In Larimer County, Colorado, *B. bissettii* is most commonly found in woodrats (*Neotoma mexicana*) and the associated tick *I. spinipalpis* (Schneider *et al.*, 2000). In California, *B. bissettii* is also closely associated with woodrats and *I. spinipalpis*, although it occasionally infects the human-biting *I. pacificus*. Moreover, several strains of *Borrelia* spp. isolated from *Ixodes* spp. ticks and rodents in California are clearly separate and distinct from both *B. burgdorferi* s.s. and *B. bissettii*. These spirochaetes may be derived from a separate unique cycle involving the rare tick *Ixodes jellisoni* and its interesting host, the California kangaroo rat (*D. californicus*) (Lane *et al.*, 1999). Curiously, diversity of *B. burgdorferi* s.l. found in *I. pacificus* increases as one heads from north to south, even though the prevalence of infection decreases (T. Schwan, personal communication). Although a large number of isolates from *I. scapularis* are similarly available along a north–south cline along the Atlantic coast, these isolates are just beginning to be subject to molecular characterization (Lin *et al.*, 2001). Although it is interesting to study and discuss the great diversity of *B. burgdorferi* s.l. in North America, it is important to remember that the vast majority of isolates from *I. scapularis* and *I. pacificus* are indeed *B. burgdorferi* s.s. Moreover, to my knowledge all well-characterized autochthonous human isolates of *B. burgdorferi* s.l. from North American patients have proved to be *B. burgdorferi* s.s. (Mathiesen *et al.*, 1997). Current evidence strongly suggests that interrupting the transmission of *B. burgdorferi* s.s. is where public health effort is needed.

Other *Borrelia* species in ixodid ticks

Argasid ticks in North America are infected with a variety of *Borrelia* spp., some of which cause relapsing fever. Among the hard ticks, *Ixodes* spp. are infected with a variety of genospecies of *B. burgdorferi* s.l. For many years a variety of observers have reported seeing live spirochaetes in an ixodid tick that is extremely common in North America, namely the Lone Star tick, *Amblyomma americanum*. All three stages of *A. americanum* aggressively bite people in the southern USA, and an erythema-like rash is commonly associated with the bite of this tick (Campbell *et al.*, 1995). Approximately 1–2% of *A. americanum* were infected with live spirochaetes across the range of this tick, but these spirochaetes could not be cultured in BSK medium. Finally, a PCR-based molecular study, sequencing the flagellin gene and 16S gene of spirochaetes from an *A. americanum* from Texas and one from New Jersey, revealed that these ticks were infected with a spirochaete that was intermediate between *B. burgdorferi* s.l. spirochaetes and relapsing fever spirochaetes (Barbour *et al.*, 1996). The tentative name of *Borrelia lonestari* was proposed for the *A. americanum* spirochaete. *B. lonestari* is most closely related to *Borrelia miyamotoi*, a spirochaete isolated from *Ixodes* spp. ticks in Japan, and *Borrelia theileri*, a spirochaete transmitted by *Rhipicephalus* and *Boophilus* genera in Africa and the Americas. Rich *et al.* (2001) have made the interesting suggestion that *B. lonestari* evolved from *B. theileri*, which was introduced into the Americas along with *Boophilus* spp. ticks. Presumably, these spirochaetes then made the jump from *Boophilus* spp. to *Amblyomma* spp. when these ticks fed on the same animals, such as deer. It is difficult to test this hypothesis since the spirochaetes involved have not been cultured, the reservoir hosts are unknown and the pathogenicity of the spirochaetes is undetermined. Much work needs to be done in this field.

Another closely related spirochaete has been described recently in North America. This spirochaete apparently infects *I. scapularis*, can be passed from tick to mouse to tick, and is efficiently passed transovarially from female *I. scapularis* to larval offspring. Although the spirochaete has not been cultured, it has been characterized based on sequence analysis and is most closely related to *B. miyamotoi* (Scoles *et al.*, 2001). One of the more important consequences of this discovery is the apparent need to re-examine all reports of transovarial transmission of *B. burgdorferi* s.s. in larval *I. scapularis*. Several observers have reported collecting questing larvae infected with spirochaetes and have suggested that these spirochaetes were *B. burgdorferi* s.s. Transovarial transmission of *B. burgdorferi* s.s. in the laboratory, however, is difficult to reproduce. Molecular characterization of spirochaetes derived from questing larval *I. scapularis* may soon clarify this situation.

Human Risk

Geographical distribution

The area of LB transmission risk in North America encompasses areas where *I. scapularis* and *I. pacificus* are extant. In the USA, county-by-county surveys established a rough 'roadmap' of where these ticks had been found by the early 1990s (Dennis *et al.*, 1998). There is little doubt that *I. scapularis* is expanding its geographical range. Studies in New York State have shown that these ticks are moving northward up the Hudson River valley. Whereas the tick was only commonly found within New York State on Long Island before the 1980s, it has now established itself virtually throughout New York State. In the mid-western USA, dense populations of *I. scapularis* were mainly restricted to an area along the border of Minnesota and Wisconsin. Infected ticks are now found in a slowly expanding radius from this central focus. The suggestion that *I. scapularis* was once found throughout eastern deciduous forests in North America and is now regaining its former range with the re-establishment of deer populations will require close monitoring.

Although this review has focused on the ecology of LB in the USA, there are some areas of transmission in Canada. In eastern Canada isolated adult *I. scapularis* can be found in several places along the Atlantic coast and as far inland as Saskatchewan (Lindsay, R. *et al.*, 1999; Artsob *et al.*, 2000). Self-sustaining dense populations of *I. scapularis* infected with *B. burgdorferi* s.s. appear to be limited, however, to one location along the shore of Lake Erie. This location is Long Point, Ontario, and the *I. scapularis* population has long been established here (Watson and Anderson, 1976). In a series of excellent ecological studies, the presence of *B. burgdorferi*-infected ticks in various habitats at Long Point has been documented (Barker *et al.*, 1992; Lindsay, L.R. *et al.*, 1999) and the risk of transmission of *B. burgdorferi* s.s. to people visiting parkland on Long Point is real. The potential for the spread of *I. scapularis* to neighbouring locations in Ontario is being closely monitored. Along the Pacific Coast, *I. pacificus* extends from British Columbia down to Baja, Mexico (Kain *et al.*, 1997, 1999). The relative transmission risk of *B. burgdorferi* s.s. to people from the northern edge of the distribution of *I. pacificus* in Canada to the southern edge in Mexico requires further study. In addition, *Ixodes uriae* infected with *Borrelia garinii* have been found in association with seabirds in Alaska, but the public health significance of this observation is unknown (Olsen *et al.*, 1995).

Importance of nymphs

There seems little doubt that the nymphal stage of *I. scapularis* is the stage that transmits *B. burgdorferi* s.s. to the vast majority of LB victims in North America. In the USA, LB is clearly a summer disease with the vast majority of cases reported in July and August (Dennis, 1995). The feeding of nymphal *I. scapularis* generally starts in May and continues through early July. If one assumes an incubation period for erythema migrans of up to 3 weeks, the time of transmission of *B. burgdorferi* s.s. in the northeastern and mid-western USA clearly coincides with the activity period of the nymphal tick (Falco *et al.*, 1999). In contrast, adult *I. scapularis* feed primarily in the colder months, November–April, when few people acquire LB. This may be due in part to reduced outdoor activity during the winter months, but collections of ticks from tick-bite victims in Westchester County, New York, document that people are frequently bitten by both nymphal and adult *I. scapularis* in regions hyperendemic for LB (Falco *et al.*, 1996). An important reason why nymphs are such good vectors of *B. burgdorferi* s.s. to tick-bite victims but adult ticks are not lies in the duration of attachment required for successful spirochaete transmission. *B. burgdorferi* s.s. must first multiply in the midgut of an infected tick and then migrate through the haemolymph to the salivary glands before a sufficient dose of infectious spirochaetes can be transmitted (Ohnishi *et al.*, 2001). This process generally takes about 2 days in both nymphal and adult *I. scapularis* (Piesman *et al.*, 1991; des Vignes *et al.*, 2001). Since nymphal *I. scapularis* are quite small, < 2 mm, whereas female *I. scapularis* are five- to tenfold larger, it is thought that nymphal ticks often escape detection and feed for a sufficient duration of attachment to transmit *B. burgdorferi* s.s., whereas female ticks are detected and removed quite early in the feeding process. In Westchester County, New York, approximately three-quarters of adult ticks removed from tick-bite victims were removed in the first day of feeding, whereas only half of nymphs were detected during the first day (Falco et al., 1996). This difference may be even more pronounced in ticks that are not detected and complete feeding successfully.

Infection in nymphs

Infection with *B. burgdorferi* s.s. among nymphal *I. scapularis* in the northeastern USA generally varies from 15 to 30% and an estimate of 25% infected is fairly constant when large samples are taken (Piesman et al., 1986; Maupin et al., 1991). Although this spirochaete population shows some genetic variability (Qiu et al., 1997), they are virtually all *B. burgdorferi* s.s., the genospecies known to be capable of causing human infection in North America. Infection rates in nymphal *I. scapularis* in the mid-western USA tend to be slightly lower, generally around 15% (Walker et al., 1994). It is clear, however, that in endemic regions in the northeastern and mid-western USA, nymphal *I. scapularis* are commonly infected with *B. burgdorferi* s.s., frequently bite people and are extremely competent vectors of spirochaetes to residents and visitors to these areas. In the southern USA the situation is less clear. A large-scale effort to collect ticks removed from people in Georgia and South Carolina turned up approximately 1000 adult ticks, but only a single *I. scapularis* nymph (Felz et al., 1996). The infection rate of southern populations of nymphal *I. scapularis* is still unknown because questing nymphal *I. scapularis* are so hard to collect that sufficient samples for testing are not available. The author spent 1 week dragging for ticks on two islands (Sapelo and St Catherines) off the coast of Georgia where adult *I. scapularis* are known to be common, at the time when nymphal *I. scapularis* infesting lizards were at their peak. He managed to collect a dozen questing nymphs during this week and return ten alive to the laboratory; none of them was infected with *B. burgdorferi* s.l. The author knows of no larger sample of questing nymphal *I. scapularis* tested from the southern USA.

The majority of testing of *I. pacificus* for *B. burgdorferi* s.l. has utilized adults. Generally, the infection rate is quite low, 1–2% positive. Surprisingly, when large-scale testing of nymphal *I. pacificus* was conducted in a location in Mendocino County, 14% of nymphal *I. pacificus* proved to be infected with *B. burgdorferi* s.l. (Clover and Lane, 1995). Clearly, the majority of LB cases along the Pacific coast are acquired during the spring and summer, when nymphal *I. pacificus* are active (March–September). But human cases of LB in California do occur throughout the year and some of these may be associated with adult ticks (Lane and Lavoie, 1988; Clover and Lane, 1995; Monsen et al., 1999).

A recent study in Mendocino County demonstrated that infection rates in nymphal *I. pacificus* from two nearby locations varied from 2 to 12% (Tälleklint-Eisen and Lane, 1999). Basic ecological studies are needed at a local and regional level to determine the factors controlling rates of infection with *B. burgdorferi* s.l. in nymphal tick populations. Such studies are essential to predict risk of LB and to develop a plan to prevent it.

Summary

Over the last 25 years our knowledge of the biology and ecology of the ticks that transmit LB spirochaetes in North America has grown tremendously. We have a much better concept of the habitats in which these ticks thrive, the hosts they feed on, the manner in which they acquire and transmit spirochaetes, and the diversity of *Borrelia* spp. in North America. However, there remains much to be learnt. We do not know the limits of the distribution of vector ticks, whether the removal of essential hosts will eradicate tick populations, and whether *B. burgdorferi* s.s. is the only *Borrelia* sp. causing human LB in North America. As we seek the answers to these questions, we must keep in mind that the ultimate goal in accumulating knowledge on the biology and ecology of the ticks that transmit LB spirochaetes is to devise prevention strategies that will stop the transmission of the spirochaetes to people and thus aid the public's health.

References

Anderson, J.F., Johnson, R.C., Magnarelli, L.A., Hyde, F.W. and Myers, J.E. (1987) Prevalence of *Borrelia burgdorferi* and *Babesia microti* in mice on islands inhabited by white-tailed deer. *Applied and Environmental Microbiology* 53, 892–894.

Anon. (1999) Recommendations for the use of Lyme disease vaccine. Recommendations of the advisory committee on immunization practice (ACIP). *Morbidity and Mortality Weekly Report* 48 (RR-7), 1–25.

Apperson, C.S., Levine, J.F., Evans, T.L., Braswell, A. and Heller, J. (1993) Relative utilization of reptiles and rodents as hosts by immature *Ixodes scapularis* (Acari: Ixodidae) in the coastal plain of North Carolina, USA. *Experimental and Applied Acarology* 17, 719–731.

Artsob, H., Maloney, R., Conboy, G. and Horney, B. (2000) Identification of *Ixodes scapularis* in Newfoundland, Canada. *Canadian Communicable Disease Report* 26, 133–134.

Barbour, A.G., Maupin, G.O., Teltow, G.J., Carter, C.J. and Piesman, J. (1996) Identification of an uncultivable *Borrelia* species in the hard tick *Amblyomma americanum*: possible agent of a Lyme disease-like illness. *Journal of Infectious Diseases* 173, 403–409.

Barker, I.K., Surgeoner, G.A., Artsob, H., McEwen, S.A., Elliott, L.A., Campbell, G.D. and Robinson, J.T. (1992) Distribution of the Lyme disease vector, *Ixodes dammini* (Acari: Ixodidae) and isolation of *Borrelia burgdorferi* in Ontario, Canada. *Journal of Medical Entomology* 29, 1011–1022.

Battaly, G.R. and Fish, D. (1993) Relative importance of bird species as hosts for immature *Ixodes dammini* (Acari: Ixodidae) in a suburban residential landscape of southern New York State. *Journal of Medical Entomology* 30, 740–747.

Beati, L. and Keirans, J.E. (2001) Analysis of the systematic relationships among ticks of the genera *Rhipicephalus* and *Boophilus* (Acari: Ixodidae) based on mitochondrial 12S ribosomal DNA gene sequences and morphological characters. *Journal of Parasitology* 87, 32–48.

Brown, R.N. and Lane, R.S. (1992) Lyme disease in California: a novel enzootic transmission cycle of *Borrelia burgdorferi*. *Science* 256, 1439–1442.

Brown, R.N. and Lane, R.S. (1994) Natural and experimental *Borrelia burgdorferi* infections in woodrats and deer mice from California. *Journal of Wildlife Diseases* 30, 389–398.

Burgdorfer, W., Barbour, A.G., Hayes, S.F., Benach, J.L., Grunwaldt, E. and Davis, J.P. (1982) Lyme disease – a tickborne spirochetosis? *Science* 216, 1317–1319.

Burkot, T.R., Clover, J.R., Happ, C.M., DeBess, E. and Maupin, G.O. (1999) Isolation of *Borrelia burgdorferi* from *Neotoma fuscipes*, *Peromyscus maniculatus*, *Peromyscus boylii*, and *Ixodes pacificus* in Oregon. *American Journal of Tropical Medicine and Hygiene* 60, 453–457.

Campbell, G.L., Paul, W.S., Schriefer, M.E., Craven, R.B., Robbins, K.E. and Dennis, D.T. (1995) Epidemiologic and diagnostic studies of patients with suspected early Lyme disease, Missouri, 1990–1993. *Journal of Infectious Diseases* 172, 470–480.

Caporale, D.A., Rich, S.M., Spielman, A., Telford, S.R. and Kocher, T.D. (1995) Discriminating between *Ixodes* ticks by means of mitochondrial DNA sequences. *Molecular Phylogenetics and Evolution* 4, 361–365.

Carey, A.B., Krinsky, W.L. and Main, A.J. (1980) *Ixodes dammini* (Acari: Ixodidae) and associated ixodid ticks in south-central Connecticut, USA. *Journal of Medical Entomology* 17, 89–99.

Clover, J.R. and Lane, R.S. (1995) Evidence implicating nymphal *Ixodes pacificus* (Acari: Ixodidae) in the epidemiology of Lyme disease in California. *American Journal of Tropical Medicine and Hygiene* 53, 237–240.

Cobb, S. (1942) Tick parasites on Cape Cod. *Science* 95, 503.

Cooley, R.A. and Kohls, G.M. (1945) The genus *Ixodes* in North America. *National Institute of Health Bulletin* 184, 1–246.

Daniels, T.J. and Fish, D. (1995) Effect of deer exclusion on the abundance of immature *Ixodes scapularis* (Acari: Ixodidae) parasitizing small and medium-sized mammals. *Journal of Medical Entomology* 32, 5–11.

Daniels, T.J., Fish, D. and Falco, R.C. (1989) Seasonal activity and survival of adult *Ixodes dammini* (Acari: Ixodidae) in southern New York State. *Journal of Medical Entomology* 26, 610–614.

Daniels, T.J., Fish, D. and Falco, R.C. (1991) Evaluation of host-targeted acaricide for reducing risk of Lyme disease in southern New York State. *Journal of Medical Entomology* 28, 537–543.

Deblinger, R.D., Wilson, M.L., Rimmer, D.W. and Spielman, A. (1993) Reduced abundance of immature *Ixodes dammini* (Acari: Ixodidae) following incremental removal of deer. *Journal of Medical Entomology* 30, 144–150.

Dennis, D.T. (1995) Lyme disease. *Dermatology Clinics* 13, 537–551.

Dennis, D.T., Nekomoto, T.S., Victor, J.C., Paul, W.S. and Piesman, J. (1998) Reported distribution of *Ixodes scapularis* and *Ixodes pacificus* (Acari: Ixodidae) in the United States. *Journal of Medical Entomology* 35, 629–638.

Dister, S.W., Fish, D., Bros, S.M., Frank, D.H. and Wood, B.L. (1997) Landscape characterization of peridomestic risk for Lyme disease using satellite imagery. *American Journal of Tropical Medicine and Hygiene* 57, 687–692.

Donahue, J.G., Piesman, J. and Spielman, A. (1987) Reservoir competence of white-footed mice for Lyme disease spirochetes. *American Journal of Tropical Medicine and Hygiene* 36, 92–96.

Duffy, D.C., Campbell, S.R., Clark, D., DiMotta, C. and Gurney, S. (1994) *Ixodes scapularis* (Acari: Ixodidae) deer tick mesoscale populations in natural areas: effects of deer, area, and location. *Journal of Medical Entomology* 31, 152–158.

Falco, R.C., Fish, D. and Piesman, J. (1996) Duration of tick bites in a Lyme disease-endemic area. *American Journal of Epidemiology* 143, 187–192.

Falco, R.C., McKenna, D.F., Daniels, T.J., Nadelman, R.B., Nowakowski, J., Fish, D. and Wormser, G.P. (1999) Temporal relation between *Ixodes scapularis* abundance and risk of Lyme disease associated with erythema migrans. *American Journal of Epidemiology* 149, 771–776.

Felz, M., Durden, L.A. and Oliver, J.H. Jr (1996) Ticks parasitizing humans in Georgia and South Carolina. *Journal of Parasitology* 82, 505–508.

Fish, D. and Daniels, T.J. (1990) The role of medium-sized mammals as reservoirs of *Borrelia burgdorferi* in southern New York. *Journal of Wildlife Diseases* 26, 339–445.

Fish, D. and Howard, C. (1999) Methods used for creating a national Lyme disease risk map. *Morbidity and Mortality Weekly Report* 48 (RR07), 21–24.

Good, N.E. (1973) Ticks of eastern Long island: notes on host relations and seasonal distribution. *Annals of the Entomological Society of America* 66, 240–243.

Gregonis, M. (2000) Aerial deer survey results indicate population is increasing. *Connecticut Wildlife* 20, 12–13.

Healy, G.R., Spielman, A. and Gleason, N. (1976) Human babesiosis: reservoir of infection on Nantucket Island. *Science* 192, 479–480.

Hyland, K.E. and Mathewson, J.A. (1961) The ectoparasites of Rhode Island mammals I. The tick fauna. *Journal of Wildlife Diseases* 11, 1–14.

Jackson, J.O. and DeFoliart, G.R. (1970) *Ixodes scapularis* Say in northern Wisconsin. *Journal of Medical Entomology* 7, 124–125.

James, A.M. and Oliver, J.H. Jr (1990) Feeding and host preference of immature *Ixodes dammini*, *I. scapularis*, and *I. pacificus* (Acari: Ixodidae). *Journal of Medical Entomology* 27, 324–330.

Johnson, R.C., Schmid, G.P., Hyde, F.W., Steigerwalt, A.G. and Brenner, D.J. (1984) *Borrelia burgdorferi* sp. nov.: etiologic agent of Lyme disease. *International Journal of Systematic Bacteriology* 34, 496–497.

Jones, C.G., Ostfeld, R.S., Richard, M.P., Schauber, E.M. and Wolff, J.O. (1998) Chain reactions linking acorns to gypsy moth outbreaks and Lyme disease risk. *Science* 279, 1023–1026.

Jones, C.J. and Kitron, U. (2000) Populations of *Ixodes scapularis* (Acari: Ixodidae) are modulated by drought at a Lyme disease focus in Illinois. *Journal of Medical Entomology* 37, 408–415.

Kain, D.E., Sperling, F.A.H. and Lane, R.S. (1997) Population genetic structure of *Ixodes pacificus* (Acari: Ixodidae) using allozymes. *Journal of Medical Entomology* 34, 441–450.

Kain, D.E., Sperling, F.A., Daly, H.V. and Lane R.S. (1999) Mitochondrial DNA sequence variation in *Ixodes pacificus* (Acari: Ixodidae). *Heredity* 83, 378–386.

Kazmierczak, J.J., Amundson, T.E. and Burgess, E.C. (1988) Borreliosis in free-ranging black bears from Wisconsin. *Journal of Wildlife Diseases* 244, 366–368.

Kellogg, F.E., Kistner, T.P., Strickland, R.K. and Gerrish, R.B. (1971) Arthropod parasites collected from white-tailed deer. *Journal of Medical Entomology* 8, 495–498.

Kitron, U. and Kazmierczak, J.J. (1997) Spatial analysis of the distribution of Lyme disease in Wisconsin. *American Journal of Epidemiology* 145, 558–566.

Kuo, M.M., Lane, R.S. and Gicias, P.C. (2000) A comparative study of mammalian and reptilian alternative pathway of complement-mediated killing of the Lyme disease spirochete (*Borrelia burgdorferi*). *Journal of Parasitology* 86, 1223–1228.

Kurtenbach, K., Sewell, H.S., Ogden, N.H., Randolph, S.E. and Nuttall, P.A. (1998) Serum complement sensitivity as a key factor in Lyme disease ecology. *Infection and Immunity* 66, 1248–1251.

Lane, R.S. and Lavoie, P.E. (1988) Lyme borreliosis in California: acarological, clinical, and epidemiological studies.

Annals of the New York Academy of Sciences 539, 192–203.

Lane, R.S. and Loye, J.E. (1989) Lyme disease in California: interrelationship of *Ixodes pacificus* (Acari: Ixodidae), the western fence lizard (*Sceloporus occidentalis*), and *Borrelia burgdorferi*. *Journal of Medical Entomology* 26, 272–278.

Lane, R.S. and Loye, J.E. (1991) Lyme disease in California: interrelationship of ixodid ticks (Acari), rodents, and *Borrelia burgdorferi*. *Journal of Medical Entomology* 28, 719–725.

Lane, R.S., Peavey, C.A., Padgett, K.A. and Hendson, M. (1999) Life history of *Ixodes* (*Ixodes*) *jellisoni* (Acari: Ixodidae) and its vector competence for *Borrelia burgdorferi* sensu lato. *Journal of Medical Entomology* 36, 329–340.

Levin, M., Levine, J.F., Yang, S., Howard, P. and Apperson, C.S. (1996) Reservoir competence of the southeastern five-line skink (*Eumeces inexpectatus*) and the green anole (*Anolis carolinensis*) for *Borrelia burgdorferi*. *American Journal of Tropical Medicine and Hygiene* 54, 92–97.

Li, X.L., Peavey, C.A. and Lane, R.S. (2000) Density and spatial distribution of *Ixodes pacificus* (Acari: Ixodidae) in two recreational areas in north coastal California. *American Journal of Tropical Medicine and Hygiene* 62, 415–422.

Lin, T., Oliver, J.H. Jr, Gao, L., Kollars, T.M. Jr and Clark, K.L. (2001) Genetic heterogeneity of *Borrelia burgdorferi* sensu lato in the southern United States based on restriction fragment length polymorphism and sequence analysis. *Journal of Clinical Microbiology* 39, 2500–2507.

Lindsay, L.R., Mathison, S.W., Barker, I.K., McEwen, S.A. and Surgeoner, G.A. (1999) Abundance of *Ixodes scapularis* (Acari: Ixodidae) larvae and nymphs in relation to host density and habitat on Long Point, Ontario. *Journal of Medical Entomology* 36, 243–254.

Lindsay, R., Artsob, H., Galloway, T. and Horsman, G. (1999) Vector of Lyme borreliosis, *Ixodes scapularis*, identified in Saskatchewan. *Canadian Communicable Disease Report* 25, 81–83.

Mannelli, A., Kitron, U., Jones, C.J. and Slajchert, T.L. (1993) Role of the eastern chipmunk as a host for immature *Ixodes dammini* (Acari: Ixodidae) in northwestern Illinois. *Journal of Medical Entomology* 30, 87–93.

Mannelli, A., Kitron, U., Jones, C.J. and Slajchert, T.L. (1994) Influence of season and habitat on *Ixodes scapularis* infestation on white-footed mice in northwestern Illinois. *Journal of Parasitology* 80, 1038–1042.

Manweiler, S.A., Lane, R.S., Block, W.M. and Morrison, M.L. (1990) Survey of birds and lizards for ixodid ticks (Acari) and spirochetal infection in northern California. *Journal of Medical Entomology* 27, 1011–1015.

Marconi, R.T., Liveris, D. and Schwartz, I. (1995) Identification of novel insertion elements, restriction fragment length polymorphism patterns, and discontinuous 23S rRNA in Lyme disease spirochetes: phylogenetic analyses of rRNA genes and their intergenic spacers in *Borrelia japonica* sp. nov. and genomic group 21038 (*Borrelia andersonii* sp. nov.). *Journal of Clinical Microbiology* 33, 2427–2434.

Mather, T.N., Telford, S.R. III, MacLachlan, A.B. and Spielman, A. (1989a) Incompetence of catbirds as reservoirs for the Lyme disease spirochetes (*Borrelia burgdorferi*). *Journal of Parasitology* 75, 66–69.

Mather, T.N., Wilson, M.L., Moore, S.I., Ribeiro, J.M.C. and Spielman, A. (1989b) Comparing the relative potential of rodents as reservoirs of the Lyme disease spirochete (*Borrelia burgdorferi*). *American Journal of Epidemiology* 130, 143–150.

Mathiesen, D.A., Oliver, J.H. Jr, Kolbert, C.P., Tullson, E.D., Johnson, B.J., Campbell, G.L., Mitchell, P.D., Reed, K.D., Telford, S.R. III, Anderson, J.F., Lane, R.S. and Persing, D.H. (1997)

Genetic heterogeneity of *Borrelia burgdorferi* in the United States. *Journal of Infectious Diseases* 175, 98–107.

Maupin, G.O., Fish, D., Zultowsky, J., Campos, E.G. and Piesman, J. (1991) Landscape ecology of Lyme disease in a residential area of Westchester County, New York. *American Journal of Epidemiology* 133, 1105–1113.

McCabe, T.R. and McCabe, R.E. (1997) Recounting whitetails past. In: McShea, W.J., Underwood, H.B. and Rappole, J.H. (eds) *The Science of Overabundance: Deer Ecology and Population Management*. Smithsonian Institution Press, Washington, DC, pp. 11–26.

Monsen, S.E., Bronson, L.R., Tucker, J.R. and Smith, C.R. (1999) Experimental and field evaluations of two acaricides for control of *I. pacificus* (Acari: Ixodidae) in northern California. *Journal of Medical Entomology* 36, 660–665.

Nelson, D.R., Rooney, S., Miller, N.J. and Mather, T.R. (2000) Complement-mediated killing of *Borrelia burgdorferi* by nonimmune sera from sika deer. *Journal of Parasitology* 86, 1232–1238.

Norris, D.E., Klompen, J.S.H., Keirans, J.E. and Black, W.C. IV (1996) Population genetics of *Ixodes scapularis* (Acari: Ixodidae) based on mitochondrial 16S and 12S genes. *Journal of Medical Entomology* 33, 78–89.

Norris, D.E., Klompen, J.S.H., Keirans, J.E., Lane, R.S., Piesman, J. and Black, W.C. IV (1997) Taxonomic status of *Ixodes neotomae* and *I. spinipalpis* (Acari: Ixodidae) based on mitochondrial DNA evidence. *Journal of Medical Entomology* 34, 696–703.

Nuttall, G.H.F., Warburton, C., Cooper, W.F. and Robinson, L.E. (1911) *Ticks. A Monograph of the Ixodoidea*, Part II. Cambridge University Press, London, pp. 133–348.

Ogden, N.H., Nuttall, P.A. and Randolph, S.E. (1997) Natural Lyme disease cycles maintained via sheep by co-feeding ticks. *Parasitology* 115, 591–599.

Ohnishi, J., Piesman, J. and de Silva, A.M. (2001) Antigenic and genetic heterogeneity of *Borrelia burgdorferi* populations transmitted by ticks. *Proceedings of the National Academy of Sciences, USA* 98, 670–675.

Oliver, J.H. Jr, Stallknecht, D., Chandler, F.W., James, A.M., McGuire, B.S. and Howerth, E. (1992) Detection of *Borrelia burgdorferi* in laboratory-reared *Ixodes dammini* (Acari: Ixodidae) fed on experimentally inoculated white-tailed deer. *Journal of Medical Entomology* 29, 980–984.

Oliver, J.H. Jr, Cummins, G.A. and Joiner, M.S. (1993a) Immature *Ixodes scapularis* (Acari: Ixodidae) parasitizing lizards from the southeastern U.S.A. *Journal of Parasitology* 79, 684–689.

Oliver, J.H. Jr, Owsley, M.R., Hutcheson, H.J., James, A.M., Chen, C., Irby, W.S., Dotson, E.M. and McLain, D.K. (1993b) Conspecificity of the ticks *Ixodes scapularis* and *I. dammini* (Acari: Ixodidae). *Journal of Medical Entomology* 30, 54–63.

Olsen, B., Duffy, D.C., Jaenson, T.G., Gylfe, A., Bonnedahl, J. and Bergstrom, S. (1995) Transhemispheric exchange of Lyme disease spirochetes by seabirds. *Journal of Clinical Microbiology* 33, 3270–3274.

Olson, C.A., Cupp, E.W., Luckhart, S., Ribeiro, J.M.C. and Levy, C. (1992) Occurrence of *Ixodes pacificus* (Parasitoformes: Ixodidae) in Arizona. *Journal of Medical Entomology* 29, 1060–1062.

Pancholi, P., Kolbert, C.P., Mitchell, P.D., Reed, K.D., Dumler, J.S., Bakken, J.S., Telford, S.R. and Persing, D.H. (1995) *Ixodes dammini* as a potential vector of human granulocytic ehrlichiosis. *Journal of Infectious Diseases* 172, 1007–1012.

Piesman, J. and Spielman, A. (1979) Host-associations and seasonal abundance of immature *Ixodes dammini* in southeastern Massachusetts. *Annals of the Entomological Society of America* 72, 829–832.

Piesman, J., Spielman, A., Etkind, P.,

Ruebush, T.K. II and Juranek, D.D. (1979) Role of deer in the epizootiology of *Babesia microti* in Massachusetts, USA. *Journal of Medical Entomology* 15, 537–540.

Piesman, J., Mather, T.N., Donahue, J.G., Levine, J., Campbell, J.D., Karakashian, S.J. and Spielman, A. (1986) Comparative prevalence of *Babesia microti* and *Borrelia burgdorferi* in four populations of *Ixodes dammini* in eastern Massachusetts. *Acta Tropica* 43, 263–270.

Piesman, J., Maupin, G.O., Campos, E.G. and Happ, C.M. (1991) Duration of adult female *Ixodes dammini* attachment and transmission of *Borrelia burgdorferi*, with description of a needle aspiration isolation method. *Journal of Infectious Diseases* 163, 895–897.

Pinger, R.R., Timmons, L. and Karris, K. (1996) Spread of *Ixodes scapularis* (Acari: Ixodidae) in Indiana: collections of adults in 1991–1994 and description of a *Borrelia burgdorferi*-infected population. *Journal of Medical Entomology* 33, 852–855.

Postic, D., Ras, N.M., Lane, R.S., Hendson, M. and Baranton, G. (1998) Expanded diversity among Californian borrelia isolates and description of *Borrelia bissettii* sp. nov. (formerly *Borrelia* group DN127). *Journal of Clinical Microbiology* 36, 3497–3504.

Price, E.G. (1948) *Fighting Spotted Fever in the Rockies*. Naegele Printing Company, Helena, Montana, 269 pp.

Qiu, W.G., Bosler, E.M., Campbell, J.R., Ugine, G.D., Wang, I.N., Luft, B.J. and Dykhuizen, D.E. (1997) A population genetic study of *Borrelia burgdorferi* sensu stricto from eastern Long Island, New York, suggested frequency-dependent selection, gene flow and host adaptation. *Hereditas* 127, 203–216.

Rich, S.M., Caporale, D.A., Telford, S.R., Kocher, T.D., Hartl, D.L. and Spielman, A. (1995) Distribution of the *Ixodes ricinus*-like ticks of eastern North America. *Proceedings of the National Academy of Sciences, USA* 92, 6284–6288.

Rich, S.M., Rosenthal, B.M., Telford, S.R., Spielman, A., Hartl, D.A. and Ayala, F.J. (1997) Heterogeneity of the internal transcribed spacer (ITS-2) region within individual deer ticks. *Insect Molecular Biology* 6, 123–129.

Rich, S.M., Armstrong, P.M., Smith, R.D. and Telford, S.R. III (2001) Lone Star tick-infecting borreliae are most closely related to the agent of bovine borreliosis. *Journal of Clinical Microbiology* 39, 494–497.

Richter, D., Spielman, A., Komar, N. and Matuschka, F.R. (2000) Competence of American Robins as reservoir hosts for Lyme disease spirochetes. *Emerging Infectious Diseases* 6, 133–138.

Rogers, A.J. (1953) A study of the ixodid ticks of northern Florida, including the biology and life history of *Ixodes scapularis* Say (Ixodidae: Acarina). PhD thesis, University of Maryland, USA.

Ruebush, T.K., Juranek, D.D., Chisholm, E.S., Snow, P.C., Healy, G.R. and Sulzer, A.J. (1977) Human babesiosis on Nantucket Island. Evidence for self-limited and subclinical infections. *New England Journal of Medicine* 297, 825–827.

Say, T. (1821) An account of the Arachnides of the United States. *Journal of the Academy of Natural Science, Philadelphia* 2, 59–83.

Schneider, B.S., Zeidner, N.S., Burkot, T.R., Maupin, G.O. and Piesman, J. (2000) *Borrelia* isolates in northern Colorado identified as *Borrelia bissettii*. *Journal of Clinical Microbiology* 38, 3103–3105.

Schulze, T.L., Jordan, R.A. and Hung, R.W. (1995) Suppresion of subadult *Ixodes scapularis* (Acari: Ixodidae) following removal of leaf litter. *Journal of Medical Entomology* 32, 730–733.

Schulze, T.L., Jordan, R.A. and Hung, R.W. (1998) Comparison of *Ixodes scapularis* (Acari: Ixodidae) populations and their habitats in established and emerging Lyme disease areas in New Jersey. *Journal of Medical Entomology* 35, 64–70.

Scoles, G.A., Papero, M., Beati, L. and Fish, D. (2001) A relapsing fever group spirochete transmitted by *Ixodes scapularis* ticks. *Vector Borne and Zoonotic Diseases* 1, 19–31.

Scrimenti, R.J. (1970) Erythema chronicum migrans. *Archives of Dermatology* 102, 104–105.

Slajchert, T., Kitron, U.D., Jones, C.J. and Mannelli, A. (1997) Role of the eastern chipmunk (*Tamias striatus*) in the epizootiology of Lyme borreliosis in northwestern Illinois, USA. *Journal of Wildlife Diseases* 33, 40–46.

Smith, R.P. Jr, Rand, P.W., Lacombe, E.H., Telford, S.R. III, Rich, S.M., Piesman, J. and Spielman, A. (1993) Norway rats as reservoir hosts for Lyme disease spirochetes on Monhegan Island, Maine. *Journal of Infectious Diseases* 168, 687–691.

Spielman, A. (1976) Human babesiosis on Nantucket Island: transmission by nymphal *Ixodes* ticks. *American Journal of Tropical Medicine and Hygiene* 25, 784–787.

Spielman, A. (1988) Lyme disease and human babesiosis: evidence incriminating vector and reservoir hosts. In: Englund, P.T. and Sher, A. (eds) *Biology of Parasitism*. A.R. Liss, New York, pp. 147–165.

Spielman, A., Clifford, C.M., Piesman, J. and Corwin, M.D. (1979) Human babesiosis on Nantucket Island: description of the vector, *Ixodes* (*Ixodes*) *dammini*, n. sp. (Acarina: Ixodidae). *Journal of Medical Entomology* 15, 218–234.

Spielman, A., Levine, J.F. and Wilson, M.L. (1984) Vectorial capacity of North American *Ixodes* ticks. *Yale Journal of Biology and Medicine* 57, 507–513.

Spielman, A., Wilson, M.L., Levine, J.F. and Piesman, J. (1985) Ecology of *Ixodes dammini*-borne human babesiosis and Lyme disease. *Annual Review of Entomology* 30, 439–460.

Stafford, K.C. III (1993) Reduced abundance of *Ixodes scapularis* (Acari: Ixodidae) with exclusion of deer by electric fencing. *Journal of Medical Entomology* 30, 986–996.

Stafford, K.C. III (1994) Survival of immature *Ixodes scapularis* (Acari: Ixodidae) at different relative humidities. *Journal of Medical Entomology* 31, 310–314.

Stafford, K.C. III, Bladen, V.C. and Magnarelli, L.A. (1995) Ticks (Acari: Ixodidae) infesting wild birds (Aves) and white-footed mice in Lyme, CT. *Journal of Medical Entomology* 32, 453–466.

Steere, A.C., Malawista, S.E., Snydman, D.R., Shope, R.E., Andiman, W.A., Ross, M.R. and Steele, F.M. (1977) Lyme arthritis: an epidemic of oligoarticular arthritis in children and adults in three Connecticut communities. *Arthritis and Rheumatology* 20, 7–17.

Tälleklint-Eisen, L. and Eisen, R.J. (1999) Abundance of ticks (Acari: Ixodidae) infesting the western fence lizard, *Sceloporus occidentalis*, in relation to environmental factors. *Environmental and Applied Acarology* 23, 731–740.

Tälleklint-Eisen, L. and Lane, R.S. (1999) Variation in the density of questing *Ixodes pacificus* (Acari: Ixodidae) nymphs infected with *Borrelia burgdorferi* at different spatial scales in California. *Journal of Parasitology* 85, 824–831.

Telford, S.R. III (1998) The name *Ixodes dammini* epidemiologically justified. *Emerging Infectious Diseases* 4, 132–134.

Telford, S.R. III, Mather, T.N., Moore, S.I., Wilson, M.L. and Spielman, A. (1988) Incompetence of deer as reservoirs of the Lyme disease spirochete. *American Journal of Tropical Medicine and Hygiene* 39, 105–109.

Telford, S.R. III, Armstrong, P.M., Katavolos, P., Foppa, I., Garcia, A.S.O., Wilson, M.L. and Spielman, A. (1997) A new tick-borne encephalitis-like virus infecting New England deer ticks, *Ixodes dammini*. *Emerging Infectious Diseases* 3, 165–170.

des Vignes, F., Piesman, J., Heffernan, R., Schulze, T.L., Stafford, K.C. III and Fish, D. (2001) Effect of tick removal on transmission of *Borrelia burgdorferi* and *Ehrlichia phagocytophila* by *Ixodes scapularis* nymphs. *Journal of Infectious Diseases* 183, 773–778.

Walker, E.D., Smith, T.W., DeWitt, J., Braudo, D.C. and McLean, R.G. (1994) Prevalence of *Borrelia burgdorferi* in host-seeking ticks (Acari: Ixodidae) from a Lyme disease endemic area in Northern Michigan. *Journal of Medical Entomology* 31, 524–528.

Wallis, R.C., Brown, S.E., Kloter, K.O. and Main, A.J. (1978) Erythema chronicum migrans and Lyme arthritis: field study of ticks. *American Journal of Epidemiology* 108, 322–327.

Watson, T.G. and Anderson, R.C. (1976) *Ixodes scapularis* Say on white-tailed deer (*Odocoileus virginianus*) from Long Point, Ontario. *Journal of Wildlife Diseases* 12, 66–71.

Wesson, D.M., McLain, D.K., Oliver, J.H., Piesman, J. and Collins, F.C. (1993). Investigation of the validity of species status of *Ixodes dammini* (Acari: Ixodidae) using rDNA. *Proceedings of the National Academy of Sciences, USA* 90, 10221–10225.

Western, K.A., Benson, G.D., Gleason, N.N., Healy, G.R. and Schultz, M.G. (1970) Babesiosis in a Nantucket resident. *New England Journal of Medicine* 283, 854–856.

Westrom, D.R., Lane, R.S. and Anderson, J.R. (1985) *Ixodes pacificus* (Acari: Ixodidae): population dynamics and distribution on Columbian black-tailed deer (*Odocoileus hemionus columbianus*). *Journal of Medical Entomology* 22, 507–511.

Wilske, B., Preac-Mursic, V., Schierz, G. and von Busch, K. (1986) Immunochemical and immunological analysis of European *Borrelia burgdorferi* strains. *Zentralblatt für Bakteriologie und Hygiene A* 263, 92–102.

Wilson, M.L. and Spielman, A. (1985) Seasonal activity of immature *Ixodes dammini* (Acari: Ixodidae). *Journal of Medical Entomology* 22, 408–414.

Wilson, M.L., Adler, G.H. and Spielman, A. (1985) Correlation between deer abundance and that of the deer tick *Ixodes dammini* (Acari: Ixodidae). *Annals of the Entomological Society of America* 78, 172–176.

Wilson, M.L., Telford, S.R. III, Piesman, J. and Spielman, J. (1988) Reduced abundance of immature *Ixodes dammini* (Acari: Ixodidae) following removal of deer. *Journal of Medical Entomology* 25, 224–228.

Wilson, M.L., Ducey, A.M., Litwin, T.S., Gavin, T.A. and Spielman, A. (1990) Microgeographic distribution of immature *Ixodes dammini* ticks correlated with that of deer. *Medical and Veterinary Entomology* 4, 151–159.

Wright, S.A., Lane, R.S. and Clover, R.S. (1998) Infestation of the southern alligator lizard (Squamata: Anguidae) by *Ixodes pacificus* (Acari: Ixodidae) and its susceptibility to *Borrelia burgdorferi*. *Journal of Medical Entomology* 35, 1044–1049.

Wright, S.A., Thompson, M.A., Miller, M.J., Knerl, K.M., Elms, S.L., Karpowicz, J.C., Young, J.F. and Kramer, V.L. (2000) Ecology of *Borrelia burgdorferi* in ticks (Acari: Ixodidae), rodents, and birds in the Sierra Nevada foothills, Placer County, California. *Journal of Medical Entomology* 37, 909–918.

Yuval, B. and Spielman, A. (1990) Duration and regulation of the development cycle of *Ixodes dammini* (Acari: Ixodidae). *Journal of Medical Entomology* 27, 196–201.

10 Epidemiology of Lyme Borreliosis

David T. Dennis and Edward B. Hayes

Division of Vector-borne Infectious Diseases, National Center for Infectious Diseases, Centers for Disease Control and Prevention, Public Health Service, US Department of Health and Human Services, PO Box 2087, Fort Collins, CO 80522, USA

Introduction

Lyme borreliosis (LB) is the most commonly reported arthropod-borne disease in North America and Europe, accounting for tens of thousands of new cases yearly in both regions (O'Connell *et al.*, 1998; Orloski *et al.*, 2000). The disease has long been endemic in Europe, where its principal clinical syndromes were described more than 50 years ago, and where early indirect evidence suggested a tick-transmitted infection responsive to antibiotics (Weber, 2001). In the USA, cases of erythema migrans (EM) were initially recognized in the 1960s and 1970s, and epidemiological evidence suggests that LB became endemic at several foci at that time (Steere, 1989; Spielman, 1994). Molecular examinations of archived materials show that natural cycles of *Borrelia burgdorferi* sensu lato (s.l.), the agent of LB, were present in northeastern coastal areas in the late 19th century, and they have probably been present in North America for millennia (Persing *et al.*, 1990; Marshall *et al.*, 1994).

The seminal investigations of an outbreak of arthritis in coastal Connecticut in the mid-1970s quickly established the principal epidemiological and ecological features of LB: a multi-system inflammatory disease affecting children and adults of both sexes; clustering of cases within townships and even by road and household; onsets of illness mostly occurring in summer months; association of cases with residence in wooded areas where tick-infested mice and deer were abundant; and linkage of illness to antecedent tick bites (Steere *et al.*, 1977, 1978).

In the USA, the annual number of reported cases of LB has increased more than 30-fold since *B. burgdorferi* was first identified in 1982 (Centers for Disease Control and Prevention, 2001). The increase has been associated with an increasing density and distribution of vector ticks (Godsey *et al.*, 1987; White *et al.*, 1991;

Stafford *et al.*, 1998) and changing demographics, which place more persons at risk of infective tick bites (Bowen *et al.*, 1984). The consequence has been a rise in case rates in established endemic foci (Petersen *et al.*, 1989; White *et al.*, 1991) and an expanding endemic range that now includes parts or all of more than 15 states (Dennis *et al.*, 1998; Orloski *et al.*, 2000). The early rapid rise in cases and the geographical expansion have justified the designation of LB as one of the most rapidly emerging infections in the USA (Institute of Medicine, 1992). More recently, spread of the disease has been characterized by a slower but steady increase of incidence among counties on the inland border of LB-endemic areas (Centers for Disease Control and Prevention, unpublished). In Eurasia, disease incidence and geographical range have been relatively stable, and increases in numbers of reported cases are thought to be due mainly to a greater awareness and enhanced surveillance (O'Connell *et al.*, 1998; Strle, 1999).

An understanding of the epidemiology of LB requires a thorough knowledge of the complex natural history of *B. burgdorferi* s.l., its tick vectors and vertebrate hosts, as well as those human activities and behaviours that place persons at risk of infection. LB is, epidemiologically, firstly a disease of place, and maps of the distribution of cases clearly identify endemic areas and their relationship to the distribution of known vector ticks and other biological variables, such as availability of suitable vertebrate reservoirs of infection and maintenance hosts for ticks (Spielman *et al.*, 1985; Fish, 1995; Wilson, 1998; Centers for Disease Control and Prevention, 1999). Descriptive studies have yielded consistent demographic profiles of cases and trends in disease distribution (Petersen *et al.*, 1989; White *et al.*, 1991; Alpert *et al.*, 1992; Berglund *et al.*, 1995; de Mik *et al.*, 1997; Huppertz *et al.*, 1999; Strle, 1999; Orloski *et al.*, 2000). Analytical epidemiological studies show associations between incident disease and various environmental and behavioural factors that result in increased risk of exposure to infective ticks (Ley *et al.*, 1995; de Mik *et al.*, 1997; Cromley *et al.*, 1998; Orloski *et al.*, 1998; Belongia *et al.*, 1999). Epidemiological studies have, however, poorly characterized various factors that place some persons with endemic exposures at greater risk than others, and have not clearly shown beneficial effects of various measures of personal protection.

This chapter provides a synopsis of biological factors relevant to the epidemiology of LB (for details, see Chapters 6, 7, 8 and 9); it outlines the known distribution and determinants of the disease, using information provided by surveillance data, surveys and epidemiological studies, and it briefly addresses some cost- and prevention-effectiveness evaluations.

The Agent and its Vector Associations

LB is caused by infection with several spirochaete genospecies of the grouping *B. burgdorferi* s.l. – a diversity that has important epidemiological and ecological implications (Gern and Humair, 1998; Wang *et al.*, 1999). Transmission occurs as a result of bites by infective hard ticks of the *Ixodes ricinus* species complex. In

North America, *B. burgdorferi* sensu stricto (s.s.) is the only known cause of human LB and, as far as is known, it is transmitted to humans solely by *Ixodes scapularis* (the black-legged tick) in eastern regions and *Ixodes pacificus* (the western black-legged tick) in the far western region (Spielman *et al.*, 1985; Lane *et al.*, 1991). These ticks have a discrete, non-overlapping distribution (Dennis *et al.*, 1998). In Europe, there is a wider diversity of human-infecting *B. burgdorferi* s.l. The several genospecies known to cause disease are, in order of decreasing disease-causing frequency, *Borrelia afzelii, Borrelia garinii* and *B. burgdorferi* s.s. (Saint Girons *et al.*, 1998). Eurasian strains are transmitted by *I. ricinus* (the European sheep tick) and *Ixodes persulcatus* (the taiga tick), which have different but overlapping ranges. In eastern Russia and in Asia, only *B. afzelii* and *B. garinii* are recognized human pathogens, transmitted in these regions by *I. persulcatus* ticks (Korenberg, 1994; Postic *et al.*, 1997; Li *et al.*, 1998).

Other related *Borrelia* spp. are less easily classified. In the USA, *Borrelia lonestari* sp. nov., a hitherto uncultivable borrelia, was identified by DNA analysis in the midgut contents of *Amblyomma americanum* (lone star) ticks (Barbour *et al.*, 1996), and it has been shown that bites by these ticks are associated in some instances with an illness characterized by mild constitutional symptoms and a rash similar to EM (Campbell *et al.*, 1995; Barbour *et al.*, 1996; Kirkland *et al.*, 1997; Felz *et al.*, 1999; James *et al.*, 2001). An unnamed agent newly identified in *I. scapularis* ticks in the eastern USA (Scoles *et al.*, 2001) is, like *B. lonestari*, more closely related to *Borrelia theileri* and the relapsing fever borreliae than to *B. burgdorferi* s.l. The pathogenic status of *Borrelia bissettii*, an agent associated with *Ixodes spinipalpis* ticks and woodrats in the western USA (Schneider *et al.*, 2000), is ambiguous. Although genetically close to *B. burgdorferi* s.s., *B. bissettii* has not been related epidemiologically or clinically with human infection in the USA, but has been found in *I. ricinus* ticks in Europe and linked to human LB in Slovenia (Picken *et al.*, 1996). *Borrelia andersoni*, a spirochaete associated with *Ixodes dentatus* ticks and rabbits in the southern and eastern USA, is not known to cause human borreliosis. In Eurasia, there are several *Borrelia* spp. vectored by *Ixodes* spp. ticks that are closely related to *B. burgdorferi* s.s. but have separate natural histories and are not known to infect humans; examples include *Borrelia valaisiana* and *Borrelia lusitaniae* in Europe, and *Borrelia japonica* and *Borrelia miyamotoi* in Japan (see Gern and Humair, Chapter 6 and Miyamoto and Masuzawa, Chapter 8).

Reservoirs of Infection (see Kurtenbach *et al.*, Chapter 5)

In North America, various rodents serve as the principal reservoirs of infection of *B. burgdorferi* s.s, predominantly *Peromyscus leucopus*, the white-footed mouse, in the eastern LB-endemic regions, and wood rats (*Neotoma* spp.) and kangaroo rats in the far-western endemic regions (Spielman *et al.*, 1985; Brown and Lane, 1996). Several bird species are competent reservoirs in eastern regions, such as robins (*Turdus migratorius*) (Richter *et al.*, 2000). In Europe, a recent review identified nine small mammals (including several mice, the bank vole and shrews), seven medium-sized mammals (especially squirrels) and a number of birds as competent hosts of human pathogenic strains (Gorelova *et al.*, 1995; Gern *et al.*, 1998). The Norway rat, *Rattus norvegicus*, has been shown to be a competent reservoir host in both North America and Europe (Smith *et al.*, 1993; Matuschka *et al.*, 1997). Various feral rodents are known reservoirs of infection in eastern regions of the former USSR and in China, and rodents and birds have been shown to be important reservoir hosts in Japan (Nakao *et al.*, 1996).

Transmission of Infection to Humans

Mode of transmission

Vector ticks transmit infection by cutaneous inoculation of spirochaete-infected saliva. Experimental and epidemiological studies indicate that this occurs only after the tick begins taking a blood meal, which for *I. scapularis* and *I. pacificus* usually begins after 36–48 h of tick attachment (Sood *et al.*, 1997; des Vignes *et al.*, 2001). There is evidence that transmission may sometimes occur earlier after *I. ricinus* or *I. persulcatus* attachment (Kahl *et al.*, 1998). In both North America and Europe, the nymphal stage of vector ticks is thought to be responsible for the majority of LB cases (Falco *et al.*, 1999; Robertson *et al.*, 2000a). Nymphal ticks often have high infection rates (15–40% in highly endemic areas); they quest in the spring and summer months when people are most likely to engage in outdoor activities and they are so small that they are difficult to detect on clothing or body surfaces. Questing larvae, on the other hand, are rarely infected, principally seek small vertebrate hosts in leaf litter and are unimportant in transmitting infection to humans. Although adult ticks may have a high prevalence of infection, in the USA they usually quest in the period of late autumn to early spring, when people are less apt to be exposed to tick habitat and more likely to wear protective clothing. Furthermore, adult ticks are large enough to be often noticed and removed before feeding. However, in the former USSR and Asia, where the main vector is *I. persulcatus*, most cases are probably caused by adults because nymphs of this species rarely bite people (Korenberg, 1994).

Although *B. burgdorferi* s.l. has been isolated from the mouthparts of various biting insects (Magnarelli and Anderson, 1988), there is no epidemiological evidence that arthropods other than *Ixodes* spp. ticks are important in transmission of infection to humans.

LB is not contagious; there is no good evidence that infection spreads between humans by sexual contact, through breast milk or through blood transfusions. Transplacental infection in humans occurs readily in other spirochaetal diseases, and has been suggested for LB. Rare silver-stained spirochaetal structures have been described in fetal and neonatal tissues in association with adverse outcomes of pregnancy in residents of LB-endemic areas (Schlesinger *et al.*, 1985); however, histological materials stained with silver are notoriously difficult to interpret, and transplacental transmission of *B. burgdorferi* s.l. in humans has not been supported by cultural isolation, by immunohistochemical staining or by DNA characterization of spirochaetes obtained from fetal tissues. In the one report of cultural recovery of *B. burgdorferi* s.s. from fetal tissue, the culture was contaminated by a *Bacillus* sp. and the identification of *B. burgdorferi* s.l. was made indirectly by immunofluorescence staining (MacDonald *et al.*, 1987). If maternal infections do result in transplacental infection and fetal or neonatal death, these occurrences must be rare.

Dynamics of infection in humans and classification of disease

LB of humans is a multi-stage, multi-system, inflammatory disease, and patients can usually be categorized as having early localized, early disseminated or late persistent infection (Steere, 1989) (see Stanek *et al.*, Chapter 1).

The incubation period of LB is days to weeks from infective exposure to earliest manifestations, such as EM rash, lymphadenitis and mild constitutional symptoms. More than 80% of patients develop a recognized EM rash that appears 3–30 days after a tick bite (median 7–10 days) (Nadelman and Wormser, 1995). Not all patients, however, recognize EM or other manifestations of early acute infection. In a large prospective study of the efficacy of vaccine against LB, persons receiving placebo experienced approximately one incident of asymptomatic infection for every four definite cases (Steere *et al.*, 1998). In patients who do not recognize early disease, a latency period of weeks or months may elapse before the diagnosis is triggered by the appearance of later-stage manifestations such as arthritis, various neurological syndromes or acrodermatitis chronica atrophicans (ACA).

The period from exposure to the development of laboratory evidence of infection may be several days or weeks, depending on the test employed. *B. burgdorferi* s.l. can be isolated in culture from early EM lesions (up to 80% success from skin biopsies in untreated patients), or from the plasma of untreated EM patients (up to 50% success) (Wormser *et al.*, 2000). The usual duration of infection in untreated patients is unknown; occasional isolations have been made of *B. burgdorferi* s.l. from skin, cerebrospinal fluid, myocardium and synovial fluid, months and even years after onset of symptoms (Snydman *et al.*, 1986; Stanek *et al.*, 1990; Pfister *et al.*, 1991; Strle *et al.*, 1995). IgM antibodies directed against *B. burgdorferi* s.l. antigens usually appear within 3 weeks of infective exposure, and peak between 4 and 6 weeks post-exposure, when IgG antibodies are also likely to be present. Although antibodies tend to wane, they may remain detectable for months or years, with or without treatment, and serology has no role in measuring the response to treatment (Nadelman and Wormser, 1998).

Considerable controversy exists over whether chronic active infection can persist in the absence of detectable antibodies directed against *B. burgdorferi* s.l. Seronegative, disseminated infection has been reported, in which T lymphocyte reactivity was the only immunological indicator of infection (Dattwyler *et al.*, 1988); this phenomenon is, however, probably quite rare. Similarly controversial is the question of persistent infection following one or more appropriate courses of antimicrobial drugs. In the USA, reports describe patients with objective evidence of chronic neuroborreliosis who require more than one course of treatment to bring about a satisfactory resolution of illness, but these patients appear to be the exception rather than the rule (Logigian *et al.*, 1990). Persons with chronic and refractory arthritic manifestations of LB are more likely than controls to have HLA-DRB1*0401 or related alleles, and increased outer surface protein A (OspA) reactivity (Kalish *et al.*, 1993; Steere *et al.*, 1994; Steere, 2001).

A recent controlled trial of antimicrobial treatment of patients with persistent symptoms following well-documented, previously treated LB found no difference in outcome between patients who received antibiotics and those who received a placebo (Klempner *et al.*, 2001). The study evaluated the efficacy of treatment in both seropositive and seronegative patients. Patients received either intravenous ceftriaxone at 2 g daily for 30 days, followed by oral doxycycline at 200 mg daily for 60 days, or matching intravenous and oral placebos. The cause of the patients' symptoms was unknown; however, the evidence suggested that it was not persistent infection, and argued against the validity of a condition described as 'chronic (refractory) Lyme disease'. The occurrence of continuing symptoms following appropriate treatment for LB has several possible explanations other than persistent infection, including initial misdiagnosis, permanent tissue damage not responsive to antibiotics, sterile inflammation in response to poorly degraded antigens, autoimmunity and triggering of conditions such as chronic fatigue syndrome or fibromyalgia (Sigal, 1994).

Although it is possible that persons can acquire a degree of natural immunity against *B. burgdorferi* s.l. infection, cases of reinfection have been well documented (Stobierski *et al.*, 1994), and repeated infection is probably common in highly endemic areas (Lastavica *et al.*, 1989). It is not known whether acquired or inherited factors can lead to an increased susceptibility to infection. Studies of immunocompromised patients with early localized infection indicated that they were more likely than immunocompetent persons to develop early disseminated infection before treatment and to require more than one course of antibiotics, although a favourable outcome was noted after 1 year in both groups (Marasapin *et al.*, 1999).

The complications of untreated LB can be severe and disabling, but reports of LB as an immediate cause of death are rare and mostly unsubstantiated; in one fatal case associated with culture-confirmed Lyme carditis, the patient had concomitant babesiosis (Marcus *et al.*, 1985).

Distribution of Lyme Borreliosis

Global patterns

The distribution of LB closely matches the worldwide distribution of ticks of the *I. ricinus* complex (see Fig. 4.1, Eisen and Lane, Chapter 4) (Schmid, 1985;

Korenberg *et al.*, 1993; Dennis *et al.*, 1998; Li *et al.*, 1998). LB is endemic in certain temperate areas of the USA and Canada; it is not known to occur in Mexico. In Europe, the disease occurs in localized areas of the British Isles and throughout continental Europe, with relatively high frequency in southern Scandinavia, The Netherlands, parts of Germany and in eastern European states, such as Austria and Slovenia. Endemic levels are relatively low in France, Switzerland, Italy, Spain and Portugal, and the Balkan states. The rates of reported cases decrease from south to north in Scandinavia, and from north to south in Italy (O'Connell *et al.*, 1998). Among states of the former USSR, it occurs from Baltic countries eastward through northern forest areas of Russia to the Pacific Coast. In China, the disease is most highly endemic in forested northeastern regions, and occurs only sporadically in widely scattered foci elsewhere (Wanchun *et al.*, 1998). In Japan, the disease is rare, and most cases arise from exposures in limited subalpine forest areas of Nagano Prefecture of central Japan and on Hokkaido Island in northern Japan (Yanagihara and Masuzawa, 1997). The risk to humans from enzootic cycles of infection in Korea and Taiwan is unknown.

Although cases of suspected LB have been reported from sub-Saharan Africa, Latin America and Australia, these are unsubstantiated: epidemiological patterns of these cases have not been defined, transmission cycles have not been described and individual cases have not been confirmed by bacterial isolation or molecular characterization of the agent. Established populations of *I. ricinus* do occur in some mountainous areas of North Africa and the Near East, but neither human cases nor established enzootic cycles of *B. burgdorferi* s.l. strains known to infect humans have been confirmed there.

Focal patterns and ecological correlates in North America

The distribution of LB is focal within endemic regions and geopolitical areas such as provinces, states, counties, districts, townships and even within neighbourhoods. In the USA, cases occur mostly among persons living in coastal northeastern and mid-Atlantic regions, from Maine to Maryland, and the north-central states of Wisconsin and Minnesota (Orloski *et al.*, 2000; Centers for Disease Control and Prevention, 2001). High-risk areas are characterized by a combination of forest and forest-edge habitats that support white-tailed deer (*Odocoileus virginianus*), humid, temperate microclimatic ground-level conditions that favour *Ixodes* spp. ticks in all stages of their development (conditions most often provided by shade and leaf litter), and habitats that are suitable for rodent reservoirs such as *P. leucopus* and *Tamius* (chipmunk) spp. (Spielman *et al.*, 1985; Fish, 1995; Kitron and Kazmierczak, 1997; Wilson, 1998). Much smaller numbers of LB cases occur at Pacific coastal sites where the human-biting tick *I. pacificus* is a bridge vector from a natural cycle of *B. burgdorferi* s.s. involving woodrats, mice and *I. spinipalis* ticks, and where lizards serve as preferred hosts of subadult ticks (Lane *et al.*, 1991). The favoured habitat in California is open oak and bay tree woodland with dense underlying leaf litter. The black-tailed deer (*Odocoileus hemionus columbianus*) is an important maintenance host for adult *I. pacificus* ticks. In eastern Canada, LB is a known enzoosis only in southeastern Ontario Province where there are established populations of *I. scapularis* ticks, although these ticks have been collected in scattered foci elsewhere in eastern maritime Canada (Canada Health and Welfare, 1998). In western Canada, scarce *I. pacificus* infected with *B. burgdorferi* s.s. have been reported from British Columbia, as have rare human cases (Canada Health and Welfare, 1993).

Focal patterns and ecological correlates in Europe

The distribution of LB in Europe and Asia is also discrete. In Europe, risk is associated with residence in rural areas, but not those characterized by intensive agriculture, with recreational and leisure activities in forested natural areas set aside for public use, and with occupational activities in forested areas (O'Connell *et al.*, 1998; Gray, 1999). Urban parks may also pose a significant risk (Guy and Farquhar, 1991; Matuschka *et al.*, 1997; Junttila *et al.*, 1999). The tick vectors of LB in Eurasia parasitize a great number and diversity of vertebrate hosts, and the various genospecies of *B. burgdorferi* s.l. are associated with particular groups of competent reservoir hosts, such as *B. garinii* and *B. valaisiana* with birds, *B. afzelii* with small mammals (such as various mice, voles, rats and shrews), and *B. burgdorferi* s.s. and *B. afzelii* with squirrels (Gern and Humair, 1998). *Ixodes uriae*, a tick that feeds on seabirds, maintains transmission cycles of *B. garinii*, and infected *I. uriae* ticks have been reported in both northern and southern hemispheres, suggesting global avian dispersal of infection (Olsen *et al.*, 1995). Passerine birds have been suggested as sources of dispersal of infection in both North America and in Europe (Anderson *et al.*, 1986). The usual maintenance hosts for *I. ricinus* and *I. persulcatus* ticks are ungulates, such as sheep and deer, especially the roe deer, red deer and sika deer, and various other large and medium-sized vertebrates (Gern *et al.*, 1998). In some restricted urban parks, however, robust enzootic cycles are maintained in the absence of these animals (Juntilla *et al.*, 1999).

Descriptive Population Data on Lyme Borreliosis in North America

Emergence of Lyme borreliosis in the USA

In the mid-1970s, Steere and co-workers described an outbreak of arthritis and EM in the town of Lyme and adjacent Connecticut communities (Steere *et al.*, 1977). They named the disease after the town of its discovery, determined that the disease first appeared 3–4 years earlier, and that its period prevalence was 4.3 per 1000 persons in the study communities. Other studies identified similar small communities with rapidly emerging LB (Hanrahan *et al.*, 1984; Steere *et al.*, 1986; Lastavica *et al.*, 1989; Alpert *et al.*, 1992). A review of reported cases in Connecticut for the period 1977–1985 suggested a threefold to eightfold increase in incidence in the communities where LB was first described, and an extension of disease from one county to all counties in Connecticut (Petersen *et al.*, 1989). These patterns of emergence continue unabated in this state (Connecticut Department of Public Health, 2001). Rapid emergence has also been well documented in New York State. Prior to 1979, vector ticks were known only from eastern Long Island, but by 1989 they were recognized in 22 New York counties, and enzootic cycles of *B. burgdorferi* s.s. were described in eight adjacent counties in southern New York (White *et al.*, 1991). Since then, spread has continued in New York mostly as a contiguous extension of established foci (Glavanakov *et al.*, 2001). In the mid-western USA, extension of LB follows riparian fringe habitat (Kitron and Kazmierczak, 1997; Guerra *et al.*, 2001).

LB in the far-western USA is considered a relatively low risk and appears to be stable. The highest reported incidence occurs in several coastal counties of northern California. An epidemiological investigation of 230 residents of a semi-rural subdivision in Sonoma County, considered to one of the most endemic areas in California, found that only 1.4% of persons had serological evidence of infection with *B. burgdorferi* s.s., although 12.6% of residents gave a history of illness thought by them to be LB (Fritz *et al.*, 1997). Active surveillance for LB in several of the most highly endemic counties in California, using physician-diagnosed EM as the case criterion, yielded an annual incidence of only 5.5 per 100,000 population, a case rate substantially lower than in the northeastern USA. The authors concluded that the disease was being overdiagnosed in the area (Ley *et al.*, 1994).

National surveillance for Lyme borreliosis in the USA

In the USA, an informal system of LB reporting was established in the early 1980s, and was standardized after a uniform case definition was adopted for national surveillance, beginning in 1991 (Centers for Disease Control and Prevention, 1998). This definition consists of clinical criteria for early localized borreliosis (EM) and for disseminated borreliosis, and employs serological testing to substantiate cases presenting with signs other than EM (Table 10.1). State and local health authorities employ various procedures for passive and active surveillance in their jurisdictions, and surveillance, although mandatory in all states, is carried out with varying intensity from one place to the next. Increasingly, states have relied on laboratory-based surveillance to identify suspected cases, which then undergo various validation processes before being reported as confirmed cases. In Canada, a modified version of the USA clinical case definition was established for surveillance purposes in 1991. Clinical case reporting is mandatory in Ontario and British Columbia Provinces, the only Canadian provinces with known established enzootic cycles of *B. burgdorferi* s.s., and public health laboratory-based reporting occurs throughout Canada.

Table 10.1. Case definition of Lyme borreliosis for national surveillance in the USA. (Adapted from recommendations made by the Centers for Disease Control and Prevention (CDC, 1997), and modified by Steere (2001).)

Skin	Erythema migrans observed by a physician. This skin lesion expands slowly over a period of days or weeks to form a large, round lesion, often with central clearing. To be counted for surveillance purposes, a solitary lesion must reach a size of at least 5 cm.
Nervous system	Lymphocytic meningitis, cranial neuritis, radiculoneuropathy or, rarely, encephalomyelitis, alone or in combination. For encephalomyelitis to be counted for surveillance purposes, there must be evidence in cerebrospinal fluid of the intrathecal production of antibodies against *Borrelia burgdorferi*.
Cardiovascular system	Acute-onset, high-grade (second or third degree) atrioventricular conduction defects that resolve in days or weeks and are sometimes associated with myocarditis.
Musculoskeletal system	Recurrent, brief attacks (lasting weeks to months) of objectively confirmed joint swelling in one or a few joints, sometimes followed by chronic arthritis in one or a few joints.
Laboratory evidence	Isolation of *B. burgdorferi* from tissue or body fluid or detection of diagnostic levels of antibody against the spirochete by the two-test approach of ELISA and Western blotting, interpreted according to the criteria of the Centers for Disease Control and Prevention and the Association of State and Territorial Public Health Laboratory Directors.[a]

[a] In a person with acute disease of less than 1 month's duration, IgM and IgG antibody responses should be measured in serum samples obtained during the acute and convalescent phases. A Western blot for IgM antibodies is considered positive if at least two of the following three bands are present: 23, 39 and 41 kDa. A blot IgG antibodies is considered positive if at least five of the following ten bands are present: 18, 23, 28, 30, 39, 41, 45, 58, 66 and 93 kDa. Only the IgG response should be used to support the diagnosis after the first month of infection; after that time, an IgM response alone is likely to represent a false positive result.

As is the case with other infectious diseases, LB reporting in the USA is unenforced, and poorly and irregularly complied with. In 1991–1992, in Connecticut, more than 80% of all cases of LB registered with the state were reported by physicians from one of four primary care specialities, even though only 7% of physicians in these specialities had reported cases (Centers for Disease Control and Prevention, 1993). Follow-up interviews with a sample of these physicians suggested that at best only 16% of all cases diagnosed and treated as incident cases of LB had been reported (Meek *et al.*, 1996). A study in Maryland suggested that the disease was under-reported by 10- to 12-fold in that state, and that many more patients were seen and treated for presumptive LB and for tick-bite alone than patients meeting the case criteria for reporting (Coyle *et al.*, 1996).

Overdiagnosis and misclassification of cases are also important surveillance problems, especially in areas of low or questionable endemicity (Steere *et al.*, 1993; Seltzer and Shapiro, 1996). Too much reliance is placed on serological tests for making the diagnosis of LB, and testing is often applied inappropriately, especially in persons with suspected EM, and in persons with vague symptoms. ELISA tests are, however, generally highly sensitive in early disseminated and late LB, and they can be helpful in these circumstances in ruling out the diagnosis. The two-tiered approach of an ELISA followed by immunoblotting of ELISA-positive sera is highly specific, and can be usefully applied to 'rule in' infection for individual case evaluations and for seroepidemiological studies (Fritz *et al.*, 1997; Trevejo *et al.*, 2001).

In the USA, LB accounts for more than 90% of all reports of vector-borne infectious disease. In 1999, 16,273 cases were reported by 45 states (overall incidence of 6.0 per 100,000 population) (Centers for Disease Control and Prevention, 2001). This compares with 497 cases by 11 states in 1982, and continues a secular trend of rising case numbers (Fig. 10.1). The map of the distribution of cases by county clearly shows geographical concentrations in the northeastern, mid-Atlantic and north-central regions, and in northern

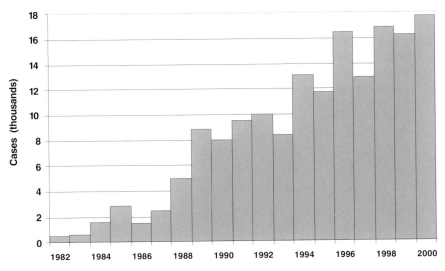

Fig. 10.1. Reported cases of Lyme borreliosis in the USA, 1982–2000.

California (Fig. 10.2). Nine states reporting incidences higher than the national rate (i.e. Connecticut, 98.0; Rhode Island, 55.1; New York, 24.2; Pennsylvania, 23.2; Delaware, 22.2; New Jersey, 21.1; Maryland, 17.4; Massachusetts, 12.7; Wisconsin, 9.3 per 100,000 population) accounted for 92.0% of the nationally reported cases. Although cases are regularly reported by about 45 states, reporting is by residence of case, not by place of exposure. Epidemiological data suggest that only about 15 states are endemic for LB, with sporadic cases occurring in some bordering states, and evidence is lacking for transmission of infection in about 25 other states. Cases reported from the latter states are likely to have had exposures elsewhere or to represent case misclassifications.

In 1999, county of residence was available for 16,214 (99.6%) nationally reported cases. Among the 3143 USA counties, 713 (22.7%) had at least one case reported during 1999; 90% of these cases were from just 109 (15.3%) reporting counties in ten states. The highest county-specific incidence (950.7 per 100,000) occurred in Nantucket County, which comprises a small population residing on the offshore Massachusetts island of Nantucket. Incidence exceeded 100 cases per 100,000 persons in 24 counties. Several counties in northeastern states routinely report annual incidence rates of 400 per 100,000 population or greater.

LB affects persons in all age groups, but the rate of disease incidence is bimodal with the highest rates found in children less than 15 years of age and in adults 35 years of age and older. Consistently, over time and from one geographical area to another, surveillance statistics show a pronounced downturn in the case rates in the age group 15–35 years, with least rates in the third decade of life (Fig. 10.3). In the USA from 1992–1998, the median age of reported cases was 30 years (Centers for Disease Control and Prevention, 2001). Overall, slightly

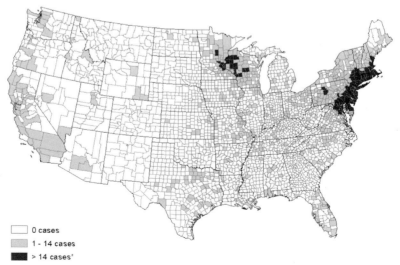

Fig. 10.2. Reported cases of Lyme borreliosis in the USA by county, in 2000.

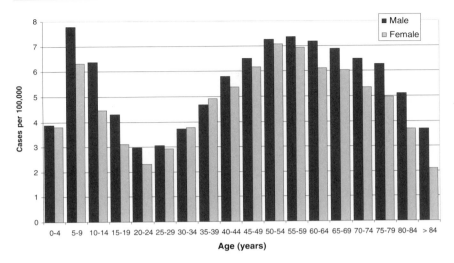

Fig. 10.3. Mean annual incidence of reported cases of Lyme borreliosis by age group and sex in the USA, 1992–2000.

more males than females were reported as cases (51.9%); the difference was found to be greatest in children aged 5–14 years, and in adults greater than 60 years of age.

The majority (about 70%) of cases experience disease onset in June, July or August, although cases occur in all months of the year (Fig. 10.4). Statistics on clinical presentations of reported cases are not available nationally, but are recorded by some states. Connecticut, the state with the highest reported case

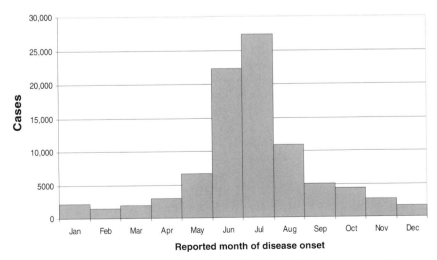

Fig. 10.4. Month of Lyme borreliosis onset for reported cases in the USA, 1992–2000.

rates and the most complete surveillance analyses, recorded nearly 4000 cases in 2000; 65% were reports of cases with EM alone, 7% had EM and a systemic manifestation of LB, and 28% had one or more systemic manifestations and a positive serological test for antibodies to *B. burgdorferi* s.s. Of the more than 1000 disseminated LB cases, arthritis was reported in 69%, neurological manifestations in 36% and cardiac complications in only 2% of cases (Connecticut Department of Public Health, 2001). Published clinical case series, which represent a different patient population from nationally reported cases, describe higher proportions of patients with EM rashes (Nadelman and Wormser, 1995).

Descriptive Population Data on Lyme Borreliosis in Europe

Surveillance statistics in Europe are based on varying case criteria and on disparate systems of data collection. LB is nationally reportable only in Scotland, Slovenia and Denmark (neuroborreliosis); however, a number of jurisdictions below the state level have developed surveillance systems that allow estimations of incidence and its trends. Nearly all countries have made efforts to determine the national scope and magnitude of the problem, including the total number of new cases annually, and the mapping of the cases by place of residence or report.

A report of a 1995 World Health Organization workshop on LB diagnosis and surveillance provided estimates of the occurrence of LB in several European countries. Estimates of annual incidence ranged from 0.3 and 0.6 per 100,000 in the UK and Ireland, respectively, 16–39 in several central European countries, 55 in Bulgaria, and 120 and 130 per 100,000 in Slovenia and Austria, respectively. Annual numbers of cases ranged from 30 in Ireland to 20,000 in Germany, with a total estimate of 60,000 for the ten reporting countries (O'Connell *et al.*, 1998).

The existence of several pathogenic *B. burgdorferi* genospecies complicates surveillance. These organisms are not evenly distributed geographically and they demonstrate differing tick and vertebrate host preferences; serological testing needs to consider different antigenic expressions of the various strains, especially the variability in expression of OspC, and clinical manifestations differ somewhat depending on which genospecies is responsible for infection (van Dam *et al.*, 1993; Wang *et al.*, 1999). In general, *B. afzelii* infection is associated more with cutaneous disease (especially ACA), *B. garinii* more with neurological manifestations and *B. burgdorferi* s.s. more with arthritis (Wang *et al.*, 1999).

The case definition for LB developed by the European Union Concerted Action on Risk Assessment in Lyme Borreliosis (EUCALB) is more clinically detailed than that used in the USA, and incorporates criteria for lymphocytoma and ACA (Stanek *et al.*, 1996), which are very rare manifestations of infection in North America. To support the clinical diagnosis, an attempt has been made by EUCALB to develop standardized criteria for the interpretation of immunoblots, but internationally useful criteria have been hard to achieve (Robertson et al., 2000b).

Investigators in Sweden conducted a prospective population-based survey of cases in seven southern counties in a 1 year period, 1992–1993 (Berglund et al., 1995). The study area was considered the most highly endemic region in Sweden and comprised 24% (2.13 million) of the population. Primary care physicians were asked to report all cases of illness with diagnosed EM or one or more objective clinical manifestations of disseminated disease plus serological confirmation. The study identified 1471 cases giving an overall annual incidence of 69 per 100,000 population (range 26–160). EM was seen in 77% of cases, followed by neuroborreliosis (16%) and arthritis (7%). Carditis was extremely rare (only two cases). The incidence varied according to area, and was highest in the eastern coastal area. Cases occurred in all months of the year, but tick bites were most frequent in July, and onsets of EM peaked in August. As in the USA, the incidence by age group was bimodal, with highest rates in children in the age group 5–9 years and in adults aged 60–75 years. Nearly 80% of patients remembered an antecedent tick bite; bites on the head and neck region were more common in children than adults, and were associated with an increased risk of neuroborreliosis. The epidemiological picture in southern Sweden was very similar to that seen in endemic areas of North America.

In a retrospective study in The Netherlands in 1995 (de Mik et al., 1997), all general practitioners in the country were asked to provide the number of patients in their practice who had been seen in the previous year for tick bite or for EM rash illness. The response rate was 80% and using extrapolations it was estimated that the practitioners saw approximately 33,000 patients with tick bites and 6500 patients with EM (approximately 43 per 100,000 population). Practices were geocoded by local government area ($n = 633$) and linked to geographical information system data on several major ecological variables. Ecological factors positively associated with both tick bites and EM were the proportion of the area covered by woods, sandy soil, dry uncultivated land, the number of tourist-nights per inhabitant and sheep population density. Despite the limitations of a retrospective, non-validated survey, this innovative study provided a quick estimate of approximate national rates of incident disease and some ecological correlates of risk important for developing educative messages.

Slovenia health authorities have required reporting of LB since 1988. A review of the epidemiological situation there noted that the annual incidence of reported cases has, since 1994, regularly exceeded 100 per 100,000 population overall, and is more than 200 per 100,000 in some parts of the country (Strle, 1999). Under-reporting is an acknowledged problem.

Since 1994, the northeastern Brandenburg region of Germany, which surrounds Berlin, has had an established reporting system using uniform case definitions for early localized and disseminated LB. The incidence of reported cases has increased over time, but it is unclear how much of this increase may be due to reporting artefact such as increased recognition and compliance. The annual incidence in 1997 was as high as 48 per 100,000 in some districts (Talaska, 1998).

To assess the incidence of LB in central Europe, a 12 month (1996–1997) population-based surveillance study was conducted in a population of 279,000 persons in the Wurzburg region of Germany (Huppertz et al., 1999). All physicians were requested to report cases meeting a case definition developed for the study. This definition required the presence of an EM rash of 5 cm in diameter or greater, lymphocytoma, or an objective manifestation of later disseminated infection with serological confirmation. A total of 313 cases (111 per 100,000) were reported. The highest rates occurred in children and elderly adults living in wooded as opposed to agricultural or urban areas. EM was the only manifestation in 89% of patients.

In a prospective serological survey in southeastern Bavaria, Germany, nearly 5000 adults having no history of LB were enrolled (Reimer et al., 1999). Blood samples obtained at times 0, 1 year and 2 years were tested using IgG ELISA and substantiated by Western immunoblotting. In the follow-up period, which included more than 55,000 person-observation months, 33 incident cases of LB were diagnosed, giving an incidence of 7.1 per 1000 person-observation years; additionally, the authors reported an asymptomatic seroconversion of 7.6 per 1000 person-observation years. On this basis, it was concluded that the incidence of infection in this poorly defined sample was 1.5% per year, and the incidence of diagnosed LB was 0.7% per year.

Several seroprevalence studies for LB in selected European population groups considered to be at high risk have been reported, with rates varying from 5 to 20%, and exceeding by twofold or more the rates of reported LB in these participants (O'Connell et al., 1998). The largest number of studies have examined populations thought to be at high risk because of recreational or occupational exposures to infective ticks, including orienteers, foresters, military personnel, farmers and other classes of outdoor workers (O'Connell et al., 1998). Although these studies reported elevated levels of seropositivity in the at-risk groups, and increasing levels with a longer duration of exposure, the pretest probabilities of LB in the study and comparison populations were generally poorly defined, the serodiagnostic methods used were unstandardized and the confidence intervals around the prevalence estimates were not given.

In a sero-epidemiological study of 950 Swiss orienteers, diagnostic levels of IgG ELISA antibodies against B31 strain whole-cell sonicated antigens were detected in 26% of bloods taken in the spring of the year from persons in the study group, and in 3.9% and 6.0% of two comparison groups (Fahrer et al., 1991). Using questionnaire responses, only 1.9% and 3.1% of the study group had a past history compatible with confirmed or probable LB. Seroconversions occurred in 45 (8.1%) of 755 orienteers between the spring bleed and a second bleed 6 months later; however, questionnaire data indicated that only one (2.2%) of those who seroconverted developed clinically recognizable LB.

A study in Austria showed that the seasonal peak of EM cases occurred in July, with the peak of neuroborreliosis (NB) cases occurring approximately 1 month later (Fig. 10.5). This is very similar to the seasonal pattern of disease onset in the USA (Fig. 10.4), despite differences in the life cycles of *I. ricinus* in

Europe and *I. scapularis* in the USA (Gray, 1998), though in both regions the nymphal stage is thought to be responsible for most cases (Falco *et al.*, 1999; Robertson *et al.*, 2000a).

Determinants of Infection

Risk factors for Lyme borreliosis

In the USA, the principal risk factor for LB is permanent or seasonal residence in an area with high infestation by infected ticks (Spielman, 1994; Fish, 1995; Cromley *et al.*, 1998). Exposures to infected ticks are most likely to occur in peri-residential areas that are relatively undisturbed, i.e. wooded tracts, and tall grass and brushy scrub at edges of gardens (yards), and are considerably less likely to occur on lawns and other maintained areas (Maupin *et al.*, 1991).

A case–control study was conducted in Hunterdon County, New Jersey, to determine risk factors for LB in reported cases. Comparisons of age-matched cases and controls showed that rural residence, clearing of peri-residential brush during spring and summer months, and the presence of stone walls, woods, deer or a bird feeder on residential property were each associated with incident disease (Orloski *et al.*, 1998).

A study of *I. scapularis*-borne illness (LB and ehrlichiosis) in Wisconsin compared exposure factors in cases and their matched controls (Belongia *et al.*, 1999).

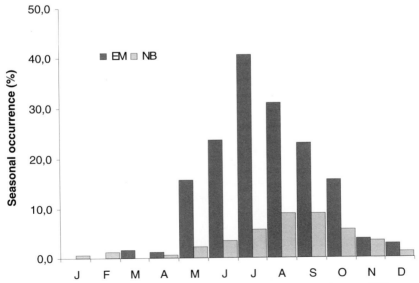

Fig. 10.5. Comparison of seasonal occurrence of erythema migrans (EM, *n* = 566) and neuroborreliosis (NB, *n* = 148) in northeastern Austria, 1994–1995 (G. Stanek *et al.*, unpublished results).

Results of multivariate logistic regression modelling identified residence in a rural neighbourhood and a history of camping as risk factors for illness. Using conditional logistic regression, recognized tick bite in the month before symptom onset, jogging, clearing brush off property, camping and residential property size > 2 acres were significantly associated with tick-borne illness. The study was unable to document the effectiveness of insect repellent use or other personal protective measures.

In California, a case–control study of persons with physician-diagnosed EM and matched controls found that the observation of deer and lizards around the home, a recognized exposure to ticks, and outdoor leisure activities were risk factors there for LB (Ley et al., 1995).

In the original epidemiological studies of Lyme arthritis in coastal Connecticut, USA, ownership of cats was found to be weakly associated with an increased risk of acquiring LB (Steere et al., 1978), and domestic cats have been observed to bring unattached nymphal ticks into homes (Curran and Fish, 1989). The association of cat ownership and LB in their owners has not been substantiated in other studies. A study of dogs and persons living in the same households in two highly endemic areas in Massachusetts showed that dogs were more likely to have serological evidence of *B. burgdorferi* s.s. infection than their human co-residents (Lindenmayer et al., 1991), and that dog ownership was not associated with an increased risk to their owners (Eng et al., 1988). Canine serosurveys have been useful in identifying geographical areas of risk (Daniels et al., 1993; Guerra et al., 2001).

A number of studies have shown increased risk among certain occupational groups. A study of employees in New Jersey showed that outdoor workers were considerably more likely to have had LB than indoor workers (Bowen et al., 1984; Schwartz and Goldstein, 1990). In New York, workers in endemic counties with a history of outdoor employment were twice as likely to be seropositive as persons without such a history. Although the difference was not statistically significant, the seroprevalence rate of outdoor employees was 5.9 times higher than a comparison group of anonymous blood donors. The most important risk in New York workers was a history of spending 30 or more hours of leisure a week outdoors (Smith et al., 1988).

Studies in Europe and Asia have shown that foresters are at high risk of infection: serological evidence of *B. burgdorferi* s.l. exposure in forest workers has been positively associated with workers' age, history of tick bite and history of EM (Kuiper et al., 1991; Ai et al., 1994; Nakama et al., 1994; Rath et al., 1996).

Risk for adverse outcomes of pregnancy

Several epidemiological studies have examined whether *B. burgdorferi* s.l. infection during pregnancy results in adverse outcomes to the fetus, including stillbirth, abortion and various congenital defects. A prospective study of 2000 prenatal patients in a highly endemic area of New York State concluded that maternal

LB before conception and during pregnancy was not associated with fetal death, birth weight, length of gestation at delivery or congenital malformations (Strobino et al., 1993). A retrospective case–control study in the same population determined that women who had been bitten by a tick or treated for LB during or before pregnancy were not at an increased risk for giving birth to a child with a congenital heart defect (Strobino et al., 1999). Cardiac defects were selected as the outcome measure because they comprise a large proportion of congenital defects in general, and because previous observational studies had suggested a possible association with maternal LB. A large population-based study using an ecological design showed no positive associations between adverse outcomes of pregnancy and endemic levels of LB in the county of residence of the mothers (Centers for Disease Control and Prevention, unpublished data).

Cost-effectiveness Analyses

Decision trees have been used to evaluate the cost per case averted (cost-effectiveness) of vaccinating persons against LB. One study, assuming a 0.80 probability of diagnosing and treating early LB, a 0.005 probability of contracting LB and a vaccination cost of $50 per year, yielded a cost to society of $4466 per case averted. Increasing the probability of contracting LB to 0.03 and the cost of vaccination to $100 per year resulted in a mean net societal savings per case averted of $3377 (Meltzer et al., 1999). Of the variables examined, the incidence of LB in the population at risk had the greatest impact on the cost-effectiveness of vaccination. The likelihood of early diagnosis and treatment also had a substantial impact on the model because of the potential for reduced incidence of complications of infection. In a second study, a decision-analytic model was developed that measured cost-effectiveness in terms of quality-adjusted life years (QALY) (Shadick et al., 2001). Vaccinating 10,000 residents in endemic areas with an expected annual incidence of 0.01 averted 202 cases of LB during a 10 year period. The additional cost per QALY gained compared with no vaccination was $62,300. If a yearly booster shot is required for persisting efficacy, the marginal cost-effectiveness ratio increased to $72,700 per QALY. The authors of both studies concluded that vaccination was economically attractive only when the annual incidence of infection was greater than 0.01. These cost-effectiveness estimates were important in the development of national recommendations for vaccine use based on expected risk of exposure to infective ticks, advising that vaccination be considered only for persons who reside, work or take recreation in areas of high or moderate risk and who also engage in activities that result in exposure to tick-infested habitats.

The cost-effectiveness for treating tick bite with 2 weeks of oral doxycyline was examined by Magid et al. (1992). The authors concluded that empirical treatment is indicated when the probability of *B. burgdorferi* s.l. infection after a bite is 0.036 or higher, and that this treatment may be preferred when the probability ranges between 0.01 and 0.035, since at these levels the value of preventing major complications of illness outweigh minor complications of treatment; however, at probabilities less than 0.01, the model predicted that the costs of the excess minor complications incurred would exceed the costs of major disease complications averted.

A study of the costs of serological diagnostic testing for LB in Maryland, an endemic state in the mid-Atlantic region of the USA having a population of about 5 million persons, showed that from 1992 to 1995 approximately 30,000 tests were performed annually at a cost to Maryland residents of about $2 million in direct medical costs per year (Strickland *et al.*, 1997). The percentage of positive ELISA and Western immunoblot tests did not exceed 7.9% for any one year, emphasizing the misuse of tests for screening and other inappropriate purposes and the risk of misleading test results, especially of ELISA results. In a separate study, it was determined that more patients were seen and treated for tick bites and for presumptive LB than for substantiated cases (Coyle *et al.*, 1996).

A diagnosis of LB is best made by evaluating the patient's clinical findings, combined with an epidemiological assessment of exposure and supported by serological testing (Tugwell *et al.*, 1997). A two-tier serodiagnostic approach, using the ELISA as a first test, followed by immunoblotting of sera found to be positive or equivocal, is recommended by authorities both in the USA (Centers for Disease Control and Prevention, 1995) and in Europe (Robertson *et al.*, 2000b) as the strategy providing the greatest accuracy and reliability. Referring to this algorithm, the American College of Physicians concluded that testing for LB was in general not useful if the pretest probabilities were less than 0.20 or greater than 0.80 (American College of Physicians, 1997; Tugwell *et al.*, 1997). A cost-effectiveness analysis of differing test-treatment strategies concluded that the most economically attractive strategy for persons with myalgia alone was no testing and no treatment; empirical treatment was most attractive for patients with EM, and two-tier testing was most attractive for persons with oligoarticular arthritis with one or no other features suggestive of LB (Nichol *et al.*, 1998). These findings complement earlier cost-effectiveness studies by Lightfoot *et al.* (1993), who determined that empirical treatment with intravenous antibiotics of seropositive patients who have only non-specific symptoms incurs costs and risks that far exceed any benefits of such treatment. A follow-up cost-effectiveness study by Eckman *et al.* (1997) used a simulation model to analyse decision trees and perform sensitivity analyses in a consideration of oral compared with intravenous antibiotic therapy of two groups of patients: those with early LB and those with Lyme arthritis who had no clinical evidence of neurological involvement. These analyses showed that, for both groups of patients, treatment with oral doxycyline was as effective as intravenous ceftriaxone and was less costly.

Prevention of Lyme Borreliosis

Infection with *B. burgdorferi* s.l. can be prevented by reducing exposure to infective ticks or by reducing the likelihood that ticks will transmit infection when exposure is unavoidable. Strategies to reduce tick densities in the environment and to reduce infection rates in ticks are discussed by Stafford and Kiken in Chapter 12. The use of repellents, personally applied acaricides and clothing barriers to tick attachment, as well as the prompt removal of ticks once they are attached, should theoretically reduce the risk of acquiring LB, but studies to evaluate the efficacy of such methods have not provided conclusive evidence of their effectiveness (Smith *et al.*, 1988; Ley *et al.*, 1995; Belongia *et al.*, 1999; Orloski *et al.*, 2000). A vaccine recently developed for use in the USA was nearly 80% efficacious in preventing LB (Steere *et al.*, 1998) and recommendations for its use were published (Centers for Disease Control and Prevention, 1999). The vaccine and its use are discussed in detail by Hayes and Schriefer in Chapter 11. Using a computer simulation model and basic epidemiological concepts, Hayes *et al.* (1999) estimated the effectiveness of several interventions for preventing LB in a hypothetical endemic community where the disease is transmitted by *I. scapularis* ticks. The model examined such interventions as the spreading of granular acaricides, topical application of acaricides to deer, reduction in deer density, deer exclusion, reduction of tick habitat through landscape modification, education to increase the use of personal protective measures and administration of vaccine. The effectiveness of these interventions was measured by numbers of LB cases prevented. The model highlighted the concept that the effectiveness of any intervention is determined by its theoretical efficacy and by the level of engagement of the target population.

Populations living in LB-endemic areas may be exposed to large numbers of bites by vector ticks (Campbell *et al.*, 1998). Several early studies, including a meta-analysis, did not show that treatment after *I. scapularis* tick bites is efficacious in preventing LB (Warshafsky *et al.*, 1996). Recently, Nadelman *et al.* (2001) conducted a large, randomized, double-blind, placebo-controlled trial in an area of New York where LB is endemic, and reported that a single 200 mg dose of doxycyline given within 72 h after an *I. scapularis* tick bite prevented the development of the disease. The ease of administration, safety and low cost of this regimen may justify its use as an adjunct to other prevention practices in selected high-risk circumstances.

The development and implementation of effective prevention measures requires an in-depth understanding of LB epidemiology, which is in turn dependent on knowledge of the complex biology of the pathogens, *B. burgdorferi* s.l., and of their interface with humans. The increasingly evident genetic diversity of members of this species complex, and their ecological adaptations and changing human–parasite interactions, has implications for pathogenicity, laboratory diagnosis, exposure risks, and for the development of effective vaccines and other methods of prevention and control. This great breadth of diversity presents a formidable public health challenge in both Old and New World circumstances.

References

Ai, C., Zhang, W. and Zhao, J. (1994) Seroepidemiology of Lyme disease in an endemic area in China. *Microbiology and Immunology* 38, 505–509.

Alpert, B., Esin, J., Sivak, S.L. and Wormser, G.P. (1992) Incidence and prevalence of Lyme disease in a suburban Westchester County community. *New York State Journal of Medicine* 92, 5–8.

American College of Physicians (1997) Clinical guideline, Part 1. Guidelines for laboratory evaluation in the diagnosis of Lyme disease. *Annals of Internal Medicine* 127, 1106–1108.

Anderson, J.F., Johnson, R.C., Magnarelli, L.A. and Hyde, F.W. (1986) Involvement of birds in the epidemiology of the Lyme disease agent *Borrelia burgdorferi*. *Infection and Immunity* 51, 394–396.

Barbour, A.G. (1996) Does Lyme disease occur in the south? A survey of emerging tick-borne infections in the region. *American Journal of Medical Science* 311, 34–40.

Barbour, A.G., Maupin, G.O., Teltow, G.J., Carter, C.J. and Piesman, J. (1996) Identification of an uncultivable *Borrelia* species in the hard tick *Ambloymma americanum*: possible agent of a Lyme disease-like illness. *Journal of Infectious Diseases* 173, 403–409.

Belongia, E.A., Reed, K.D., Mitchell, P.D., Chyou, P.H., Mueller-Rizner, N., Finkel, M.F. and Schriefer, M.E. (1999) Clinical and epidemiological features of early Lyme disease and human granulocytic ehrlichiosis in Wisconsin. *Clinical Infectious Diseases* 29, 1472–1477.

Berglund, J., Eitrem, R., Ornstein, K., Lindberg, A., Ringner, A., Elmrud, H., Carlsson, M., Runehagen, A., Svanborg, C. and Norrby, R. (1995) An epidemiologic study of Lyme disease in southern Sweden. *New England Journal of Medicine* 333, 1319–1324.

Bowen, G.S., Schulze, T.L., Hayne, C. and Parkin, W.E. (1984) A focus of Lyme disease in Monmouth County, New Jersey. *American Journal of Epidemiology* 120, 387–394.

Brown, R.N. and Lane, R.S. (1996) Reservoir competence of four chaparral-dwelling rodents for *Borrelia burdorferi* in California. *American Journal of Tropical Medicine and Hygiene* 54, 84–91.

Campbell, G.L., Paul, W.S., Schriefer, M.E., Craven, R.B., Robbins, K.E. and Dennis, D.T. (1995) Epidemiologic and diagnostic studies of patients with suspected early Lyme disease, Missouri, 1990–1993. *Journal of Infectious Diseases* 172, 470–480.

Campbell, G.L., Fritz, C.L., Fish, D., Nowakowski, J., Nadelman, R.B. and Wormser, G.P. (1998) Estimation of the incidence of Lyme disease. *American Journal of Epidemiology* 148, 1018–1026.

Canada Health and Welfare (1993) Isolation of *Borrelia burgdorferi* in British Columbia. In: *Canada Communicable Disease Report*, pp. 19–24, 204–205.

Canada Health and Welfare (1998) Distribution of *Ixodes pacificus* and *Ixodes scapularis* re concurrent babesiosis and Lyme disease. In: *Canada Communicable Disease Report*, pp. 24–15.

Centers for Disease Control and Prevention (1993) Physician reporting of Lyme disease – Connecticut, 1991–1992. *Morbidity and Mortality Weekly Report* 42, 348–350.

Centers for Disease Control and Prevention (1995) Recommendations for test performance and interpretation from the Second National Conference on Serologic Diagnosis of Lyme Disease. *Mortality and Morbidity Weekly Report* 44, 590–591.

Centers for Disease Control and Prevention (1997) Case definitions for infectious conditions under public health surveillance. *Morbidity and Mortality Weekly Report* 46, 20–21.

Centers for Disease Control and Prevention (1998) Preventing emerging infectious diseases: a strategy for the 21st century. *Morbidity and Mortality Weekly Report* 47, 1–14.

Centers for Disease Control and Prevention (1999) Recommendations for the use of Lyme disease vaccine: recommendations of the Advisory Committee on Immunization Practices (ACIP). *Morbidity and Mortality Weekly Report* 48, 1–25.

Centers for Disease Control and Prevention (2001) Lyme disease – United States, 1999. *Morbidity and Mortality Weekly Report* 50, 181–185.

Connecticut Department of Public Health (2001) Lyme disease – Connecticut, 2000. *Connecticut Epidemiology* 21, 9–12.

Coyle, B.S., Strickland, G.T., Liang, Y.L., Pena, C., McCarter, R. and Israel, E. (1996) The public health impact of Lyme disease in Maryland. *Journal of Infectious Diseases* 173, 1260–1262.

Cromley, E.K., Cartter, M.L., Mrozinski, R.D. and Ertel, S.H. (1998) Residential setting as a risk factor for Lyme disease in a hyperendemic region. *American Journal of Epidemiology* 147, 472–477.

Curran, K.L. and Fish, D. (1989) Increased risk of Lyme disease for cat owners. *New England Journal of Medicine* 320, 183.

van Dam, A.P., Kuiper, H., Vos, K., Widjojokusumo, A., de Jongh, B.M., Spanjaard, L., Ramselaar, A.C.P., Kramer, M.D. and Dankert, J. (1993) Different genospecies of *Borrelia burgdorferi* are associated with distinct clinical manifestations of Lyme borreliosis. *Clinical Infectious Diseases* 17, 708–717.

Daniels, T.J., Fish, D., Levine, J.F., Greco, M.A., Eaton, A.T., Padgett, P.J. and LaPointe, D.A. (1993) Canine exposure to *Borrelia burgdorferi* and prevalence of *Ixodes dammini* (Acari: Ixodidae) on deer as a measure of Lyme disease risk in the northeastern United States. *Journal of Medical Entomology* 30, 171–178.

Dattwyler, R.J., Volkman, D.J., Luft, B.J., Halperin, J.J., Thomas, J. and Golightly, M.G. (1988) Seronegative Lyme disease: dissociation of specific T- and B-lymphocyte to *Borrelia burgdorferi*. *New England Journal of Medicine* 319, 1441–1446.

Dennis, D.T., Nekomoto, T.S., Victor, J.C., Paul, W.S. and Piesman, J. (1998) Reported distribution of *Ixodes scapularis* and *Ixodes pacificus* (Acari: Ixodidae) in the United States. *Journal of Medical Entomology* 35, 629–638.

Eckman, M.H., Steere, A.C., Kalish, R.A. and Parker, S.G. (1997) Cost-effectiveness of oral as compared with intravenous antibiotic therapy for patients with early Lyme disease or Lyme arthritis. *New England Journal of Medicine* 337, 357–363.

Eng, T.R., Wilson, M.I., Spielman, A. and Lastavica, C.C. (1988) Greater risk of *Borrelia burgdorferi* infection in dogs than in people. *Journal of Infectious Diseases* 158, 1410–1411.

Fahrer, H., van der Linden, S.M., Sauvain, M.J., Gern, L., Zhioua, E. and Aeschlimann, A. (1991) The prevalence and incidence of clinical and asymptomatic Lyme borreliosis in a population at risk. *Journal of Infectious Diseases* 163, 305–310.

Falco, R.C., McKenna, D.F., Daniels, T.J., Nadelman, R.B., Nowakowski, J., Fish, D. and Wormser, G.P. (1999) Temporal relation between *Ixodes scapularis* abundance and risk for Lyme disease associated with erythema migrans. *American Journal of Epidemiology* 149, 771–776.

Felz, M.W., Chandler, F.W., Oliver, J.H., Rahn, D.W. and Schriefer, M.E. (1999) Solitary erythema migrans in Georgia and South Carolina. *Archives of Dermatology* 135, 1317–1326.

Fish, D. (1995) Environmental risk and prevention of Lyme disease. *American Journal of Medicine* 98, 4A-2S-9S.

Fritz, C.L., Kjemtrup, A.M., Conrad, P.A., Flores, G.R., Campbell, G., Schriefer, M.E., Gallo, D. and Vugia, D.J. (1997)

Seroepidemiology of emerging tick-borne infectious diseases in a northern California community. *Journal of Infectious Diseases* 175, 1432–1439.

Gern, L. and Humair, P.-F. (1998) Natural history of *Borrelia burgdorferi* sensu lato. *Weiner Klinische Wochenschrift* 110, 856–858.

Gern, L., Estrada-Pena, A., Frandsen, F., Gray, J.S., Jaenson, T.G.T., Jongejan, F., Kahl, O., Korenberg, E., Mehl, R. and Nuttall, P.A. (1998) European reservoir hosts of *Borrelia burgdorferi* sensu lato. *Zentralblatt für Bakteriologie* 287, 196–204.

Glavanakov, S., White, D.J., Caraco, T., Lapenis A., Robinson, G.R., Szymanski, B.K. and Maniatty, W.A. (2001) Lyme disease in New York State: spatial pattern at a regional scale. *American Journal of Tropical Medicine and Hygiene* 65, 538–545.

Godsey, M.S., Amundson, T.E., Burgess, E.C., Schell, W., Davis, J.P., Kaslow, R. and Edelman, R. (1987) Lyme disease ecology in Wisconsin: distribution and host preferences of *Ixodes dammini*, and prevalence of antibody to *Borrelia burgdorferi* in small mammals. *American Journal of Tropical Medicine and Hygiene* 37, 180–187.

Gorelova, N.B., Korenberg, E.I., Kovalevskii, Y.V. and Shcherbakov, S.V. (1995) Small mammals as reservoir hosts for *Borrelia* in Russia. *Zentrablatt für Bakteriologie* 282, 315–322.

Gray, J.S. (1998) The ecology of ticks transmitting Lyme borreliosis. *Experimental and Applied Acarology* 22, 249–258.

Gray, J. (1999) Risk assessment in Lyme borreliosis. *Weiner Klinische Wochenschrift* 111, 990–993.

Guerra, M.A., Walker, E.D. and Kitron, U. (2001) Canine surveillance for Lyme borreliosis in Wisconsin and northern Illinois: geographic distribution and risk factor analysis. *American Journal of Tropical Medicine and Hygiene* 65, 546–552.

Guy, E.C. and Farquhar, R.G. (1991) *Borrelia burgdorferi* in urban parks. *Lancet* 338, 253.

Hanrahan, J.P., Benach, J.L., Coleman, J.L., Bosler, E.M., Morse, D.L., Cameron, D.J., Edelman, R. and Kaslow, R.A. (1984) Incidence and cumulative frequency of endemic Lyme disease in a community. *Journal of Infectious Diseases* 150, 489–496.

Hayes, E.B., Maupin, G.O., Mount, G.A. and Piesman, J. (1999) Assessing the prevention effectiveness of local Lyme disease control. *Journal of Public Health Management Practice* 5, 84–92.

Huppertz, H.I., Bohme, M., Standaert, S.M., Karch, H. and Plotkin, S.A. (1999) Incidence of Lyme borreliosis in the Wurzburg region of Germany. *European Journal of Clinical Microbiology and Infectious Diseases* 18, 697–703.

Institute of Medicine (1992) *Infections: Microbial Threats of Health in the United States*. National Academy Press, Washington, DC.

James, A.M., Liveris, D., Wormser, G.P., Schwartz, I., Montecalvo, M.A. and Johnson, B.J.B. (2001) *Borrelia lonstari* infection after a bite by an *Amblyomma americanum* tick. *Journal of Infectious Diseases* 183, 1810–1814.

Junttila, J., Peltomaa, M., Soini, H., Marjamaki, M. and Viljanen, M.K. (1999) Prevalence of *Borrelia burgdorferi* in *Ixodes ricinus* ticks in urban recreational areas of Helsinki. *Journal of Clinical Microbiology* 37, 1361–1365.

Kahl, O., Janetzki-Mittmann, C., Gray, J.S., Jonas, R., Stein, J. and de Boer, R. (1998) Risk of infection with *Borrelia burgdorferi* sensu lato for a host in relation to the duration of nymphal *Ixodes ricinus* feeding and the method of tick removal. *Zentralblatt für Bakteriologie* 287, 41–52.

Kalish, R.A., Leong, J.M. and Steere, A.C. (1993) Association of treatment-resistant chronic Lyme arthritis with HLA-DR4 and antibody reactivity to OspA and

OspB of *Borrelia burgdorferi*. *Infection and Immunity* 61, 2774–2779.

Kirkland, K.B., Klimko, T.B., Meriweather, R.A., Schriefer, M., Levin, M., Levine, J., MacKenzie, W.R. and Dennis, D.T. (1997) Erythema migrans-like rash illness at a camp in North Carolina. *Archives of Internal Medicine* 157, 2635–2641.

Kitron, U. and Kazmierczak, J.J. (1997) Spatial analysis of the distribution of Lyme disease in Wisconsin. *American Journal of Epidemiology* 145, 558–566.

Klempner, M.S., Hu, L.T., Evans, J., Schmid, C.H., Johnson, G.M., Trevino, R.P., Norton, D., Levy, L., Wall, D., McCall, J., Kosinski, M. and Weinstein, A. (2001) Two controlled trials of antibiotic treatment in patients with persistent symptoms and a history of Lyme disease. *New England Journal of Medicine* 345, 85–91.

Korenberg, E.I. (1994) Comparative ecology and epidemiology of Lyme disease and tick-borne encephalitis in the former Soviet Union. *Parasitology Today* 10, 157–160.

Korenberg, E.I., Kryuchechnikov, V.N. and Kovalevsky, Y.V. (1993) Advances in investigations of Lyme borreliosis in the territory of the former USSR. *European Journal of Epidemiology* 9, 86–91.

Kuiper, H., de Jong, B.M., Nauta, A.P., Houweling, H., Wiessing, L.G., Moll van Charante, A.W. and Spanjaard, L. (1991) Lyme borreliosis in Dutch forestry workers. *Journal of Infection* 23, 279–286.

Lane, R.S., Piesman, J. and Burgdorfer, W. (1991) Lyme borreliosis: relation of its causative agent to its vectors and host in North America and Europe. *Annual Review of Entomology* 36, 587–609.

Lastavica, C.C., Wilson, M.L., Berardi, V.P., Spielman, A. and Deblinger, R.D. (1989) Rapid emergence of a focal epidemic of Lyme disease in coastal Massachusetts. *New England Journal of Medicine* 320, 133–137.

Ley, C., Le, C., Olshen, E.M. and Reingold, A.L. (1994) The use of serologic tests for Lyme disease in a prepaid health plan in California. *Journal of the American Medical Association* 271, 460–463.

Ley, C., Olshen, E.M. and Reingold, A.L. (1995) Case–control study of risk factors for incident Lyme disease in California. *American Journal of Epidemiology* 142, S39–S47.

Li, M., Masuzawa, T., Takada, N., Ishiguro, F., Fujita, H., Iwaki, A., Wang, H., Wang, J., Kawabata, M. and Yanagihara, Y. (1998) Lyme disease *Borrelia* species in northeastern China resemble those isolated from far eastern Russia and Japan. *Applied and Environmental Microbiology* 64, 2705–2709.

Lightfoot, R.W. Jr, Luft, B.J., Rahn, D.W., Steere, A.C., Sigal, L.H., Zoschke, D.C., Gardner, P., Britton, M.C. and Kaufman, R.L. (1993) Empiric parenteral antibiotic treatment of patients with fibromyalgia and fatigue and a positive result for Lyme disease: a cost-effectiveness analysis. *Annals of Internal Medicine* 119, 503–509.

Lindenmayer, J.M., Marshall, D. and Onderdonk, A.B. (1991) Dogs as sentinels for Lyme disease in Massachusetts. *American Journal of Public Health* 81, 1448–1454.

Logigian, E.L., Kaplan, R.F. and Steere, A.C. (1990) Chronic neurologic manifestations of Lyme disease. *New England Journal of Medicine* 323, 1438–1444.

MacDonald, A.B., Benach, J.L. and Burgdorfer, W. (1987) Stillbirth following maternal Lyme disease. *New York State Journal of Medicine* 87, 615–616.

Magid, D., Schwartz, B., Craft, J. and Schwartz, J.S. (1992) Prevention of Lyme disease after tick bites: a cost-effectiveness analysis. *New England Journal of Medicine* 324, 534–541.

Magnarelli, L.A. and Anderson, J.F. (1988) Ticks and biting insects infected with the etiologic agent of Lyme disease,

Borrelia burgdorferi. *Journal of Clinical Microbiology* 26, 1482–1486.

Maraspin, V., Lotric-Furlan, S., Cimperman, J., Ruzic-Sabljic, E. and Strle, F. (1999) Erythema migrans in the immunocompromised host. *Wiener Klinische Wochenschrift* 111, 923–932.

Marcus, L.C., Steere, A.C., Duray, P.H., Anderson, A.E. and Mahoney, E.B. (1985) Fatal pancarditis in a patient with coexistent Lyme disease and babesiosis: demonstration of spirochetes in the myocardium. *Annals of Internal Medicine* 103, 374–376.

Marshall, W.F. III, Telford, S.R. III, Rys, P.N., Rutledge, B.J., Mathiesen, D., Malawista, S.E., Spielman, A. and Persing, D.H. (1994) Detection of *Borrelia burgdorferi* DNA in museum specimens of *Peromyscus leucopus*. *Journal of Infectious Diseases* 170, 1027–1032.

Matuschka, F.R., Endepols, S., Richter, D. and Spielman, A. (1997) Competence of urban rats as reservoir hosts for Lyme disease spirochetes. *Journal of Medical Entomology* 34, 489–493.

Maupin, G.O., Fish, D., Zultowsky, J., Campos, E.G. and Piesman, J. (1991) Landscape ecology of Lyme disease in a residential area of Westchester county, New York. *American Journal of Epidemiology* 133, 1105–1113.

Meek, J.I., Roberts, C.L., Smith, E.V. Jr and Cartter, M.L. (1996) Underreporting of Lyme disease by Connecticut physicians, 1992. *Journal of Public Health Management Practice* 2, 61–65.

Meltzer, M.J., Dennis, D.T. and Orloski, K.A. (1999) The cost effectiveness of vaccinating against Lyme disease. *Emerging Infectious Diseases* 5, 321–328.

de Mik, E.L., van Pelt, W., Docters-van Leeuwen, B., van der Veen, A., Schellekens, J.F.P. and Borgdorff, M.W. (1997) The geographical distribution of tick bites and erythema migrans in general practice in the Netherlands. *International Journal of Epidemiology* 26, 451–457.

Nadelman, R.B. and Wormser, G.P. (1995) Erythema migrans and early Lyme disease. *American Journal of Medicine* 98 (suppl. 4A), 15S–24S.

Nadelman, R.B. and Wormser, G.P. (1998) Lyme borreliosis. *Lancet* 352, 557–565.

Nadelman, R.B., Nowakowski, J., Fish, D., Falco, R.C., Freeman, K., McKenna, D., Welch, P., Marcus, R., Aguero-Rosenfeld, M.E., Dennis, D.T. and Wormser, G.P. (2001) Prophylaxis with single-dose doxycycline for the prevention of Lyme disease after an *Ixodes scapularis* tick bite. *New England Journal of Medicine* 345, 79–84.

Nakama, H., Muramatsu, K., Uchikawa, K. and Yamagishi, T. (1994) Possibility of Lyme disease as an occupational disease – seroepidemiological study of regional residents and forestry workers. *Asia-Pacific Journal of Public Health* 7, 214–217.

Nakao, M., Uchikawa, K. and Dewa, H. (1996) Distribution of *Borrelia* species associated with Lyme disease in the subalpine forests of Nagano prefecture, Japan. *Microbiology and Immunology* 40, 307–311.

Nichol, G., Dennis, D.T., Steere, A.C., Lightfoot, R., Wells, G., Shea, B. and Tugwell, P. (1998) Test-treatment strategies for patients suspected of having Lyme disease: a cost-effectiveness analysis. *Annals of Internal Medicine* 128, 37–48.

O'Connell, S., Granstrom, M., Gray, J.S. and Stanek, G. (1998) Epidemiology of European Lyme borreliosis. *Zentralblatt für Bakteriologie* 287, 229–240.

Olsen, B., Duffy, D.C., Jaenson, T.G.T., Gylfe, A., Bonnedahl, J. and Bergstrom, S. (1995) Transhemispheric exchange of Lyme disease spirochetes by seabirds. *Journal of Clinical Microbiology* 33, 3270–3274.

Orloski, K.A., Campbell, G.L., Genese, C.A., Beckley, J.W., Schriefer, M.E., Spitalny, K.C. and Dennis, D.T. (1998) Emergence of Lyme disease in

Hunterdon County, New Jersey, 1993: a case-control study of risk factors and evaluation of reporting patterns. *American Journal of Epidemiology* 147, 391–397.

Orloski, K.A., Hayes, E.B., Campbell, G.L. and Dennis, D.T. (2000) Surveillance for Lyme disease – United States, 1992–1998. *Morbidity and Mortality Weekly Report* 49, 1–11.

Persing, D.H., Telford, S.R. III, Rys, P.N., Dodge, D.E., White, T.J., Malawista, S.E. and Spielman, A. (1990) Detection of *Borrelia burgdorferi* DNA in museum specimens of *Ixodes dammini* ticks. *Science* 249, 1420–1423.

Petersen, L.R., Sweeney, A.H., Checko, P.J., Magnarelli, L.A., Mshar, P.A., Gunn, R.A. and Hadler, J.L. (1989) Epidemiological and clinical features of 1,149 persons with Lyme disease identified by laboratory-based surveillance in Connecticut. *Yale Journal of Biology and Medicine* 62, 253–262.

Pfister, H.-W., Preac-Mursic, V., Wilske, B., Schielke, E., Sorgel, F. and Einhaupl, K.M. (1991) Randomized comparison of ceftriaxone and cefotaxime in Lyme neuroborreliosis. *Journal of Infectious Diseases* 163, 311–318.

Picken, R.N., Cheng, Y., Strle, F. and Picken, M.M. (1996) Patient isolates of *Borrelia burgdorferi* sensu lato with genotypic and phenotypic similarities to strain 25015. *Journal of Infectious Diseases* 174, 1112–1115.

Postic, D., Korenberg, E., Gorelova, N., Kovalevski, Y.V., Bellenger, E. and Baranton, G. (1997) *Borrelia burgdorferi* sensu lato in Russia and neighbouring countries: high incidence of mixed isolates. *Research in Microbiology* 148, 691–702.

Rath, P.M., Ibershoff, B., Mohnhaupt, A., Albig, J., Eljaschewitsch, B., Jurgens, D., Horbach, I., and Fehrenbach, F.J. (1996) Seroprevalence of Lyme borreliosis in forestry workers from Brandenburg, Germany. *European Journal of Clinical Microbiology and Infectious Diseases* 15, 372–377.

Reimer, B., Marschang, A., Fingerle, V., Wilske, B., van Sonnenburg, F. and van Hoecke, C. (1999) Epidemiology of Lyme borreliosis in south-eastern Bavaria (Germany). *Zentralblatt für Bakterologie* 289, 653–654.

Richter, D., Spielman, A., Komar, N., and Matuschka, F.R. (2000) Competence of American robins as reservoir hosts for Lyme disease spirochetes. *Emerging Infectious Disease* 6, 133–138.

Robertson, J.N., Gray, J.S. and Stewart, P. (2000a) Tick bite and Lyme borreliosis risk at a recreational site in England. *European Journal of Epidemiology* 16, 647–652.

Robertson, J., Guy, E., Andrews, N., Wilske, B., Anda, P., Granstrom, M., Hauser, U., Moosmann, Y., Sambri, V., Schellekens, J., Stanek, G. and Gray, J. (2000b) A European multicenter study of immunoblotting in serodiagnosis of Lyme borreliosis. *Journal of Clinical Microbiology* 38, 2097–2102.

Saint Girons, I., Gerns, L., Gray, J.S., Guy, E.C., Korenberg, E., Nuttall, P.A., Rijpkema, S.G.T., Schonberg, A., Stanek, G. and Postic, D. (1998) Identification of *Borrelia burgdorferi* sensu lato species in Europe. *Zentralblatt für Bakteriologie* 287, 190–195.

Schlesinger, P.A., Duray, P.H., Burke, P.A., Steere, A.C. and Stillman, T. (1985) Maternal–fetal transmission of the Lyme disease spirochete, *Borrelia burgdorferi*. *Annals of Internal Medicine* 103, 67–68.

Schmid, G.P. (1985) The global distribution of Lyme disease. *Review of Infectious Diseases* 7, 41–50.

Schneider, B.S., Zeidner, N.S., Burkot, T.R., Maupin, G.O. and Piesman, J. (2000) *Borrelia* isolates in northern Colorado identified as *Borrelia bissetti*. *Journal of Clinical Microbiology* 38, 3103–3105.

Schwartz, B.S. and Goldstein, M.D. (1990)

Lyme disease in outdoor workers: risk factors, preventive measures, and tick removal methods. *American Journal of Epidemiology* 131, 877–884.

Scoles, G.A., Papero, M., Beati, L. and Fish, D. (2001) A relapsing fever group spirochete transmitted by *Ixodes scapularis* ticks. *Vector Borne and Zoonotic Diseases* 1, 19–32.

Seltzer, E.G. and Shapiro, E.D. (1996) Misdiagnosis of Lyme disease: when not to order serologic tests. *Pediatric Infectious Diseases Journal* 15, 762–763.

Shadick, N.A., Liang, M.H., Phillips, C.B., Fossel, K. and Kuntz, K.M. (2001) The cost-effectiveness of vaccination against Lyme disease. *Archives of Internal Medicine* 161, 554–561.

Sigal, L.H. (1994) Persisting complaints attributed to chronic Lyme disease: possible mechanisms and implications for management. *American Journal of Medicine* 96, 365–374.

Smith, P.F., Benach, J.L., White, D.J., Stroup, D.F. and Morse, D.L. (1988) Occupational risk of Lyme disease in endemic areas of New York State. *Annals of the New York Academy of Sciences* 539, 289–301.

Smith, R.P., Rand, P.W., Lacombe, E.H., Telford, S.R., Rich, S.M., Piesman, J. and Spielman, A. (1993) Norway rats as reservoir hosts for Lyme disease spirochetes on Monhegan island, Maine. *Journal of Infectious Diseases* 168, 687–691.

Snydman, D.R., Schenkein, D.P., Berardi, V.P., Lastavica, C.C. and Pariser, K.M. (1986) *Borrelia burgdorferi* in joint fluid in chronic Lyme arthritis. *Annals of Internal Medicine* 104, 798–800.

Sood, S.K., Salzman, M.B., Johnson, B.J.B., Happ, C.M., Fieg, K., Carmody, L., Rubin, L.G., Hilton, E. and Piesman, J. (1997) Duration of tick attachment as a predictor of the risk of Lyme disease in an area in which Lyme disease is endemic. *Journal Infectious Diseases* 175, 996–999.

Spielman, A. (1994) The emergence of Lyme disease and human babesiosis in a changing environment. *Annals of the New York Academy of Sciences* 740, 146–156.

Spielman, A., Wilson, M.L., Levine, J.F. and Piesman, J. (1985) Ecology of *Ixodes dammini*-borne human babesiosis and Lyme disease. *Annual Review of Entomology* 30, 439–460.

Stafford, K.C., Cartter, M.L., Magnarelli, L.A., Ertel, S.H. and Mshar, P.A. (1998) Temporal correlations between tick abundance and prevalence of ticks infected with *Borrelia burgdorferi* and increasing incidence of Lyme disease. *Journal of Clinical Microbiology* 36, 1240–1244.

Stanek, G., Klein, J., Bittner, R. and Glogar, D. (1990) Isolation of *Borrelia burgdorferi* from the myocardium of a patient with longstanding cardiomyopathy. *New England Journal of Medicine* 322, 249–252.

Stanek, G., O'Connell, S., Cimmino, M., Aberer, E., Kristoferitsch, W., Granstrom, M., Guy, E. and Gray, J. (1996) European Union Concerted Action on risk assessment in Lyme borreliosis: clinical case definitions for Lyme borreliosis. *Weiner Klinische Wochenschrift* 108, 741–747.

Steere, A.C. (1989) Medical progress: Lyme disease. *New England Journal of Medicine* 321, 586–596.

Steere, A.C. (2001) Lyme disease. *New England Journal of Medicine* 345, 115–125.

Steere, A.C., Malawista, S.E., Snydman, D.R., Shope, R.E., Andiman, W.A., Ross, M.R. and Steele, F.M. (1977) Lyme arthritis: an epidemic of oligoarticular arthritis in children and adults in three Connecticut communities. *Arthritis and Rheumatism* 20, 7–17.

Steere, A.C., Broderick, T.F. and Malawista, S.E. (1978) Erythema chronicum migrans and Lyme arthritis: epidemiological evidence for a tick vector. *American Journal Epidemiology* 108, 685–695.

Steere, A.C., Taylor, E., Wilson, M.L., Levine, J.F. and Spielman, A. (1986) Longitudinal assessment of the clinical and epidemiological features of Lyme disease in a defined population. *Journal of Infectious Diseases* 154, 295–300.

Steere, A.C., Taylor, E., McHugh, G.L. and Logigian, E.L. (1993) The overdiagnosis of Lyme disease. *Journal of the American Medical Association* 269, 1812–1816.

Steere, A.C., Levin, R.E., Molloy, P.J., Kalish, R.A., Abraham, J.H., Liu, N.Y. and Schmid, C.H. (1994) Treatment of Lyme arthritis. *Arthritis and Rheumatism* 37, 878–888.

Steere, A.C., Sikand, V.K., Meurice, F., Parenti, D.L., Fikrig, E., Schoen, R.T., Nowakowski, J., Schmid, C.H., Laukamp, S., Buscarino, C., Krause, D.S. and the Lyme Disease Vaccine Study Group (1998) Vaccination against Lyme disease with recombinant *Borrelia burgdorferi* outer-surface lipoprotein A with adjuvant. *New England Journal of Medicine* 339, 209–215.

Stobierski, M.G., Hall, W.N., Robinson-Dunn, B., Stiefel, H., Shiflet, S. and Carlson, V. (1994) Isolation of *Borrelia burgdorferi* from two patients in Michigan. *Clinical Infectious Diseases* 19, 944–946.

Strickland, G.T., Karp, A.C., Mathews, A. and Pena, C.A. (1997) Utilization and cost of serological tests for Lyme disease in Maryland. *Journal of Infectious Diseases* 176, 819–821.

Strle, F. (1999) Lyme borreliosis in Slovenia. *Zentralblatt für Bakteriologie* 289, 643–652.

Strle, F., Cheng, Y., Cimperman, J., Maraspin, V., Lotric-Furlan, S., Nelson, J.A., Picken, M.M., Ruzic-Sabljic, E. and Picken, R.N. (1995) Persistence of *Borrelia burgdorferi* sensu lato in resolved erythema migrans lesions. *Clinical Infectious Diseases* 21, 380–389.

Strobino, B.A., Williams, C.L., Abid, S., Chalson, R. and Spierling, P. (1993) Lyme disease and pregnancy outcome: a prospective study of two thousand prenatal patients. *American Journal of Obstetrics and Gynecology* 169, 367–374.

Strobino, B.A., Abid, S. and Gewitz, M. (1999) Maternal Lyme disease and congenital heart disease: a case–control study in an endemic area. *American Journal of Obstetrics and Gynecology* 180, 711–716.

Talaska, T. (1998) Borreliose-Epidemiologie am Beispiel des Bundeslandes Brandenburg. *Für Die Praxis: Lyme-Borreliose* 40–41.

Trevejo, R.T., Krause, P.J., Schriefer, M.E. and Dennis, D.T. (2001) Evaluation of a two-test serodiagnostic method for community assessment of Lyme disease in an endemic area. *American Journal of Tropical Medicine and Hygiene* 65, 563–566.

Tugwell, P., Dennis, D.T., Weinstein, A., Wells, G., Shea, B., Nichol, G., Hayward, R., Lightfoot, R., Baker, P. and Steere, A.C. (1997) Laboratory evaluation in the diagnosis of Lyme disease. *Annals of Internal Medicine* 127, 1109–1123.

des Vignes, F., Piesman, J., Heffernan, R., Schulze, T.L., Stafford, K.C. and Fish, D. (2001) Effect of tick removal on transmission of *Borrelia burgdorferi* and *Ehrlichia phagocytophila* by *Ixodes scapularis* nymphs. *Journal of Infectious Diseases* 183, 773–778.

Wanchun, T., Zhikun, Z., Moldenhauer, S., Yixiu, G., Qiuli, G., Qiuli, Y., Lina, W. and Mei, C. (1998) Detection of *Borrelia burgdorferi* from ticks (Acari) in Hebei Province, China. *Journal of Medical Entomology* 35, 95–98.

Wang, G., van Dam, A.P., Schwartz, I. and Dankert, J. (1999) Molecular typing of *Borrelia burgdorferi* sensu lato: taxonomic, epidemiological, and clinical implications. *Clinical Microbiology Reviews* 12, 633–653.

Warshafsky, S., Nowakowski, J., Nadelman, R.B., Kamer, R.S., Peterson, S.J. and

Wormser, G.P. (1996) Efficacy of antibiotic prophylaxis for prevention of Lyme disease. *Journal of General Internal Medicine* 11, 329–333.

Weber, K. (2001) Aspects of Lyme borreliosis in Europe. *European Journal of Clinical Microbiology and Infectious Diseases* 20, 6–13.

White, D.J., Chang, H., Benach, J.L., Bosler, E.M., Meldrum, S.C., Means, R.G., Debbie, J.G., Birkhead, G.S. and Morse, D.L. (1991) The geographic spread and temporal increase of the Lyme disease epidemic. *Journal of the American Medical Association* 266, 1230–1236.

Wilson, M.L. (1998) Distribution and abundance of *Ixodes scapularis* (Acari: Ixodidae) in North America: ecological processes and spatial analysis. *Journal of Medical Entomology* 35, 446–457.

Wormser, G.P., Bittker, S., Cooper, D., Nowakowski, J., Nadelman, R.B. and Pavia, C. (2000) Comparison of the yields of blood cultures using serum or plasma from patients with early Lyme disease. *Journal of Clinical Microbiology* 38, 1648–1650.

Yanagihara, Y. and Masuzawa, T. (1997) Lyme disease (Lyme borreliosis). *FEMS Immunology and Medical Microbiology* 18, 249–261.

Vaccination Against Lyme Borreliosis

11

Edward B. Hayes and Martin E. Schriefer

Division of Vector-borne Infectious Diseases, Centers for Disease Control and Prevention, Public Health Service, US Department of Health and Human Services, PO Box 2087, Fort Collins, CO 80522, USA

Introduction

Within 20 years of the identification of the tick vectors of Lyme borreliosis (LB) and isolation of the North American causative agent, *Borrelia burgdorferi* sensu stricto (s.s.), an efficacious vaccine to prevent LB in humans was commercially available. Vaccine development began with the demonstration that both passive and active immunization of hamsters protected these animals against intraperitoneal challenge with *B. burgdorferi* s.s. By 1990, a chemically inactivated whole-cell vaccine was being administered to dogs, although the efficacy of the vaccine had not been convincingly demonstrated. Whole-cell vaccines were not developed for humans because of concerns that such preparations might induce deleterious autoimmune reactions. Development of vaccines for humans focused on immunization with outer surface protein A (OspA) following the discovery that this protein was relatively homogenous among *B. burgdorferi* s.s. strains isolated from North America, and that immunization with OspA was highly protective against *B. burgdorferi* s.s. infection in mice. Two pharmaceutical companies developed recombinant OspA vaccines in the early 1990s and conducted independent phase III trials with over 10,000 participants in each trial. One of these vaccines, LYMErix™, is currently licensed for use and until recently was commercially available in the USA*. The efficacy of the vaccine in protecting against LB in Europe or Asia is unknown.

*Note: The manufacturer of LYMErix™, the only commercially available vaccine against Lyme disease, discontinued sales of the vaccine in February 2002 because of low commercial demand for the vaccine. The authors are not aware of any current plans to resume manufacture or distribution of the vaccine. The recommendations for use of the vaccine at the time of discontinuation of sales were unchanged from those described in this chapter.

© 2002 CAB *International. Lyme Borreliosis: Biology, Epidemiology and Control* (eds J. Gray, O. Kahl, R.S. Lane and G. Stanek)

The impetus for development of a vaccine against LB for humans arose from the increasing numbers of LB cases reported to health authorities in the USA, and the increased recognition of LB as an important vector-borne disease in Europe and temperate Asia. In addition, although the disease is rarely if ever fatal and responds well to treatment if diagnosed in its early stages, failure to receive early treatment can sometimes result in disabling sequelae, which may be refractory to antibiotic therapy. Efforts to prevent LB by educating people to wear protective clothing and repellents, and to check for and remove attached ticks, while theoretically efficacious, have not been effective in reducing the incidence of disease in endemic areas. Proposals to reduce vector tick populations in endemic areas by applying acaricides to tick habitats have generated concern about potentially toxic effects of the acaricides on the environment. The elimination of deer can reduce tick densities in isolated endemic foci, but this intervention has not been adopted on a wide scale. Thus, in the early 1990s, efforts were focused on developing an effective vaccine to prevent LB.

Epidemiology of Lyme Borreliosis in Relation to Vaccination

Reliable estimates of LB case numbers are difficult to obtain but approximately 17,000 new cases are reported annually in the USA and it has been estimated that in ten European countries nearly 60,000 cases may occur each year (see Dennis and Hayes, Chapter 10). In the USA, a geographical analysis of LB risk based on reported cases of LB, distribution of vector ticks and impact of host distribution on the prevalence of infection in ticks identified 115 (4%) counties in the USA as being high risk for LB, 146 (5%) as moderate risk, 1143 (36%) as low risk and 1736 (55%) as minimal or no risk (Centers for Disease Control and Prevention, 1999). The high-risk counties are all within the northeast, mid-Atlantic and north-central regions of the country. This focal distribution of LB risk within the USA provided the basis for public health recommendations for vaccine use that target the vaccine to people who are likely to be exposed to tick habitat in high or moderate risk areas (Centers for Disease Control and Prevention, 1999). LB risk in Europe appears to increase from west to east, paralleling the prevalence of infected ticks, with the highest reported incidence in Austria and Slovenia, but foci of high risk have also been described in western Europe (O'Connell *et al.*, 1998).

The incidence of LB in the USA is highest in children aged 5–9 years and adults aged 45–54 years (Orloski *et al.*, 2000). A study in Sweden found the highest incidence in children aged 5–9 years and adults aged 60–74 years (Berglund *et al.*, 1995). The relatively high risk of LB in children, as well as the potential for children to fail to notice or report to their parents an erythema migrans (EM) rash, have prompted efforts to develop a vaccine that will be safe and efficacious in children as well as adults. Males and females appear to be at approximately equal risk in the USA, but in Europe both equal risk by gender and higher risk in males have been reported (Berglund *et al.*, 1995; O'Connell *et al.*, 1998). LB risk may be higher than average in some occupational and recreational groups, which have more frequent contact with infected ticks (Smith *et al.*, 1988; O'Connell *et al.*, 1998).

Another spirochaete, *B. lonestari*, has been associated with the occurrence of a rash, strongly resembling the characteristic EM of LB, in humans in southern and central areas of the USA that are not endemic for LB (see Dennis and Hayes, Chapter 10). An *ospA* homologue has not been identified in *B. lonestari*, despite attempts to amplify *ospA* from ticks (*Amblyomma americanum*) and from EM biopsies (Felz *et al.*, 1999; James *et al.*, 2001), which suggests that vaccination with recombinant OspA (rOspA) is unlikely to provide protection against this spirochaete.

Development of Recombinant OspA Vaccines

Despite the early development of whole-cell vaccines for dogs, there was concern that whole cell vaccines in humans would be highly reactogenic and might elicit immune responses that would cross-react with human antigens and provoke inflammatory tissue damage (Wormser, 1995). Research towards developing vaccines in humans therefore focused on subunit vaccines. Several proteins expressed by *B. burgdorferi* s.s. were considered as candidates for vaccine development, primarily OspA, OspB and OspC. These outer surface proteins were shown to be protective in animal models (Probert and LeFebvre, 1994). There was some indication that both OspA and OspB conferred protection against infection and overt disease (Fikrig *et al.*, 1990; Telford and Fikrig, 1995). This was seen as a desirable property for a candidate vaccine, because of concerns that subclinical infection might result in late-stage sequelae such as those seen in untreated naturally acquired LB. Because of the relative homogeneity of OspA in *B. burgdorferi* s.s. isolates (Fikrig *et al.*, 1990), efforts to develop a human subunit vaccine focused on this protein.

Despite the finding that immunization with OspA could provide protection against infection in mice, protective immunity was achieved only against homologous *B. burgdorferi* s.s. strains. Two important protective and non-overlapping B cell epitopes defined by the monoclonal antibodies LA-2 and 336 are located at the C-terminal end of OspA (Kalish *et al.*, 1995; Wormser, 1995; Ding *et al.*, 2000). Protective immunity is more highly correlated with levels of LA-2-specific antibody than with total levels of OspA antibody. LA-2 recognizes three exposed loops on the surface of *B. burgdorferi* s.s. OspA, but does not bind to this region of *Borrelia garinii* or *Borrelia afzelii* OspA. Amino acid variation within these loops may account for the failure of *B. burgdorferi* s.s. OspA to elicit protection against heterologous *Borrelia* spp. and strains. None the less, variation within this looped epitope of *B. garinii* and *B. afzelii* strains is limited, suggesting that a multivalent OspA vaccine that provides broader protection against all three borrelial genospecies may be feasible (Telford and Fikrig, 1995; Ding *et al.*, 2000). An N-terminal lipid moiety enhances immunogenicity of OspA (Erdile *et al.*, 1993; van Hoecke *et al.*, 1996).

While in the vector tick midgut, *B. burgdorferi* sensu lato (s.l.) expresses OspA, but after migration from the tick's gut to the salivary glands and on entering the mammalian host, the expression of OspA is suppressed and most spirochaetes preferentially express OspC (de Silva *et al.*, 1996). Vaccination with OspA elicits antibody that destroys or inactivates the spirochaetes while they are still inside the tick's gut. This protective mechanism is unique among human vaccines. Presumably, any spirochaetes that enter host tissue while still expressing OspA can be neutralized through a more traditional antibody–antigen interaction within the vaccinated host's tissues (Telford and Fikrig, 1995). The actual immune mechanism by which antibody to OspA provides protection is not known, but may involve enhanced phagocytosis through Fc receptor binding (Montgomery *et al.*, 1994), immobilizing, killing or inhibiting the growth of spirochaetes (Pavia *et al.*, 1991; Keller *et al.*, 1994; Wormser, 1995), or blocking of spirochaete attachment to cell surfaces (Wormser, 1995).

In the early 1990s, Kalish *et al.* (1993) determined that individuals with treatment-resistant Lyme arthritis following naturally acquired infection were more likely to have detectable antibody to OspA and a higher prevalence of certain HLA-DR4 haplotypes than those patients whose Lyme arthritis responded to conventional antibiotic therapy. An autoimmune mechanism was proposed as the cause for the persistent joint inflammation and these findings generated concerns that OspA vaccines could induce autoimmune reactions. Further research indicated that certain epitopes of OspA were structurally similar to epitopes on human leukocyte function-associated antigen-1 and that this similarity might provide the basis for an autoimmune-mediated release of inflammatory cytokines in joint tissue (Gross *et al.*, 1998). Whether vaccination with OspA could provoke or exacerbate such a reaction remains unknown, but the results of several clinical trials in humans, described below, suggest that immunization alone does not cause arthritis.

Clinical Trials in Humans

The immunogenicity and safety of the vaccine in humans was first evaluated in 24 healthy volunteers, with no known exposure to LB, aged 18–65 in the non-endemic state of New Mexico (Keller *et al.*, 1994). Twelve participants received 10 μg recombinant OspA without adjuvant and 12 received 10 μg rOspA with 0.5 mg aluminium adjuvant. An additional 12 participants received a placebo. The most common adverse event was tenderness or pain at the injection site, which occurred in 22 (46%) persons after the first dose of rOspA, in 20 (42%) persons after the second dose and in eight (67%) of 12 persons who were given a third dose of vaccine without adjuvant. Local reactions occurred more frequently in vaccinated persons than among placebo recipients, but the study lacked sufficient power to detect a significant difference. No serious adverse events were reported. Systemic reactions such as headache within 72 h, fatigue within 6 h and fever within 24 h were reported after each injection by fewer than 20%

of vaccine and placebo recipients. Seven vaccine recipients and one placebo recipient reported joint pain. In all cases the joint pain resolved without treatment and the investigators believed it to be causally unrelated to vaccination. Joint pain was more frequently reported by vaccine and placebo recipients who were HLA-DR4-positive, but this difference was not statistically significant. None of the participants developed a clinical picture consistent with Lyme arthritis. Both vaccine formulations produced high titres of antibodies that inhibited *B. burgdorferi* s.s. growth *in vitro*. The geometric mean titres of antibody elicited by the two formulations were similar and were significantly higher than in the group given placebo.

The safety and immunogenicity of three different dosage strengths of rOspA vaccine with adjuvant was evaluated in an uncontrolled clinical trial in adults with previous LB (Schoen *et al.*, 1995). Five participants received 5 µg of the vaccine, five received 10 µg, and 20 received 30 µg. All but two of the participants received three doses of vaccine at 0, 1 and 2 months. No serious adverse events and no sustained systemic adverse effects were reported. Local reactions at the injection site were reported frequently, but participants rated 96% of these reactions as mild. The most common reaction was soreness, reported by 40–85% of vaccine recipients after any given dose. Systemic reactions after injection included fever and rash (each reported by no more than 20% of participants after any given dose), headache and fatigue (each reported by up to 40% of participants after any given dose) and arthralgia (reported by no more than 33% of participants after any given dose). Eighty-one per cent of these reactions were classified as mild. All adverse events reported in the study were of short duration. All dosage strengths of the vaccine resulted in increased geometric mean titres of OspA antibody, but those receiving the 30 µg dose developed higher titres than those receiving the lower dose formulations. The occurrence of adverse reactions was not related to antibody titre.

The safety and immunogenicity of three different formulations of rOspA vaccine was assessed in 300 seronegative volunteers in a randomized trial without a placebo group (van Hoecke *et al.*, 1996). The three vaccine formulations were non-structural protein (NS1)–OspA, NS1–OspA with monophosphoryl lipid A (MPL) and lipoprotein OspA, all with 10 µg of rOsp A adsorbed to aluminium hydroxide. Vaccines were administered on days 0, 28 and 56. In the first component of the study, each formulation was given to 20 different participants. At least one symptom, mostly mild local reactions, was reported by 35–47% of vaccine recipients in each group. All symptoms resolved within 4 days. Fever, headache and malaise were reported by fewer than 12% of participants and none of these general symptoms was felt to be related to the vaccine. One person developed pneumonia that was unrelated to the vaccine. No other severe adverse events were reported. In the second component of the study, each formulation was administered to 80 different participants. Pain at the injection site was the most frequent local reaction, reported after 24–49% of injections. General symptoms such as headache, fever or malaise were reported by 10% or fewer participants after any given dose, and were reported as either mild or moderate. Two participants reported mild arthralgia that resolved within 3 days of receiving their first vaccine dose. No severe adverse events were reported. The lipoprotein formulation produced a higher and more persistent antibody response than the other two formulations.

Following the generally favourable results of these phase I and phase II trials, two vaccine manufacturers sponsored large phase III trials of OspA vaccine to assess safety, immunogenicity and efficacy in preventing LB. Pasteur Merieux Connaught sponsored a trial of rOspA lipoprotein vaccine without adjuvant (Sigal *et al.*, 1998). The lipoprotein was derived from a B31 strain of *B. burgdorferi* s.s. that was expressed by *Escherichia coli*. The vaccine was given to 5156 people 18 years of age or older and an additional 5149 participants received placebo. Sigal *et al.* (1998) reported a vaccine efficacy of 68% (95% CI 36%–85%) after two doses of vaccine in the first year. However, the criteria for confirming the diagnosis of LB in study participants were unclear and only a subset of suspected cases were reviewed in depth to confirm diagnosis. Data on efficacy after a third dose were reported but administration of the third dose was not designed into the original protocol and was only given to a subset of participants selected through unspecified criteria. There were no serious adverse events associated with the vaccine in this trial. Vaccine recipients reported pain and inflammation at the injection site significantly more frequently than did placebo recipients.

SmithKline Beecham sponsored a randomized clinical trial to assess the safety, immunogenicity and efficacy of rOspA lipoprotein vaccine with aluminium hydroxide adjuvant (Steere *et al.*, 1998). The *ospA* gene of *B. burgdorferi* s.s. strain ZS7 was expressed in transformed *E. coli*. The vaccine was administered to 5469 study participants aged 15–70 years, and 5467 participants of the same age range received a placebo consisting of adjuvant in buffered saline. Doses of vaccine and placebo were administered at 0, 1 and 12 months and timed to provided high antibody levels just before the late spring increase in LB transmission. The following criteria were used to diagnose LB in both study groups: the presence of investigator-observed EM or objective neurologic, musculoskeletal or cardiovascular manifestations of LB, with laboratory confirmation of infection by culture, PCR, or seroconversion documented by Western blot. The estimated vaccine efficacy in reducing laboratory-confirmed LB was 49% (95% CI 15%–69%, NNT 256, where NNT = the number needed to treat, or in this case, the number of persons that would have to be vaccinated to prevent one case of LB) after two doses in the first year and 76% (95% CI 58%–86%, NNT 109) after a third dose administered 1 year later. The efficacy in preventing asymptomatic infection was 83% (95% CI 32%–97%, NNT 500) after two doses, and 100% (95% CI 26%–100%, NNT 370) after three doses. The development of OspA antibody was evaluated in a subset of participants from a single study site (SmithKline Beecham Pharmaceuticals, 1998, LYMErix™ product label). Of approximately 460 participants who received vaccine at the site, serologic results were reported for 267 participants. The criteria for selection of these participants were not specified. The geometric mean titre (GMT) was 1227 ELISA units ml^{-1} at month 2 ($n = 264$, 1 month after the second dose), 116 ELISA units ml^{-1} at month 12 before the third dose ($n = 241$), 6006 ELISA units ml^{-1} at month 13 ($n = 267$, 1 month after the third dose) and 1991 ELISA units ml^{-1} at month 20 ($n = 267$). ELISA units ml^{-1} were defined based on an evaluation of pooled sera from people who had been vaccinated with OspA (van Hoecke et al., 1996). A subsequent analysis by the manufacturer concluded that a titre greater than 1400 ELISA units ml^{-1} was associated with protection from disease (van Hoecke et al., 1999).

Data on adverse events in this study were collected in two separate ways. At one site 938 participants were asked to complete diary cards recording adverse events for the first 3 days after each dose. The remaining 9998 participants at other study sites were simply asked to report any adverse events to the investigators. Among the 9998, the most frequently described adverse event that was determined to be related or possibly related to the vaccine was soreness at the injection site, which was reported by 24.1% of vaccine recipients compared with 7.6% of placebo recipients ($P < 0.001$, NNH 6, where NNH = number needed to harm, or in this case, the number of persons that would be vaccinated for every single occurrence of the specified adverse event). Systemic events were reported by 19.4% of vaccine recipients compared with 15.1% of placebo recipients ($P < 0.001$, NNH 23), but no single systemic event was reported by more than 3.9% of vaccine recipients. The individual systemic events that were significantly associated with the vaccine were myalgias (3.2% of vaccine recipients versus 1.8% of placebo recipients, NNH 71), influenza-like illness (2.0% versus 1.1%, NNH 111), fever (2.0% versus 0.8%, NNH 83) and chills (1.8% versus 0.5%, NNH 77). Among the 938 participants from whom adverse event data were solicited by diary card, systemic events were reported by 63% of vaccine recipients, compared with 53% of placebo recipients ($P = 0.004$, NNH 10). For all 10, 936 participants, there was no significant difference between vaccine and placebo recipients in the proportions reporting adverse events 30 days or more after receiving a dose at the $P = 0.05$ significance level. The incidence of arthritis was not significantly different between vaccine and placebo recipients. No hypersensitivity reactions were noted in vaccine recipients.

A recombinant bacille Calmette-Guérin vaccine expressing OspA lipoprotein was evaluated in a dose-escalation study in 24 adults (Edelman et al., 1999). The vaccine elicited no detectable antibody to OspA in any of the volunteers and induced skin ulcers at the higher dose strength.

Clinical trials in children

The safety and immunogenicity of rOspA lipoprotein vaccine were assessed in a clinical trial that randomly assigned 250 children aged 5–15 years in the Czech Republic to receive either 15 μg or 30 μg of the vaccine (Feder et al., 1999). There was no placebo group. Each study group received three doses of vaccine on a 0, 1 and 2 month schedule. None of the 250 children had detectable antibody to OspA at the time they were enrolled in the study. One month after the second injection, the GMT of the group that received 15 μg was 2346 ELISA units ml^{-1}, and the GMT of the group that received 30 μg was 4259 ELISA units ml^{-1}. At month 3, after the third injection, the GMTs were 5967 ELISA units ml^{-1} in the 15 μg group and 10267 ELISA units ml^{-1} in the 30 μg group. Comparing these results with those of the phase III trial in adults (Steere et al., 1998) suggests that immunization with 30 μg of OspA tends to elicit higher OspA titres in children than in adults.

Participants in this children's trial were asked to record adverse events on diary cards for 3 days after each vaccination, and cards were returned for all children in the study after all doses. Soreness at the injection site was the most frequently reported adverse event, reported by 72% of those receiving their first 30 μg dose and 69% of those receiving their first 15 μg dose. Soreness was less frequently reported in both groups after subsequent doses. Redness at the injection site was reported by, at most, 43% of participants after any given dose. Headache and malaise were the most frequently reported systemic symptoms, with the highest proportion reported by those who received the first 15 μg dose (18% reported headache and 20% reported malaise). All systemic symptoms resolved within 6 days of vaccination and no serious adverse events were reported.

The safety and immunogenicity of OspA vaccine were evaluated in 91 children aged 2–5 years in the Czech Republic, who were randomly assigned to receive two doses, 1 month apart, of either 15 μg or 30 μg of lipoprotein OspA with adjuvant (Beran et al., 2000). Pain at the injection site was the most frequently reported symptom, reported by 64% of children who received the 15 μg formulation and 69% of those receiving 30 μg. There was no indication that the 30 μg dose was substantially more reactogenic than the 15 μg dose, but the GMTs of antibody to OspA elicited by the 30 μg dose were significantly higher (9877 at 1 month after two 30 μg doses compared with 4366 after two 15 μg doses).

In a randomized placebo-controlled trial to evaluate the safety and immunogenicity of OspA vaccine in children 4 years of age and older in the USA, 3063 children were vaccinated with 30 μg of rOspA vaccine and 1024 were given placebo (Sikand et al., 2001). Doses of vaccine and placebo were administered on a 0, 1 and 12 month schedule. Vaccine recipients reported a higher incidence of local reactions (78% with vaccine versus 55% with placebo; NNH 4) and general symptoms (30% with vaccine versus 22% with placebo; NNH 13) such as fatigue, fever, headache and arthralgia. The mean duration of these adverse events was 2–3 days. Study participants were asked about any rheumatological adverse events, and reports of arthritis were not any higher among vaccine recipients than among placebo recipients. The GMT of antibody to OspA 1 month after the third dose was 27,485 ELISA units ml^{-1} compared with a GMT of 8216 ELISA units ml^{-1} found in retested sera from adults in the adult Phase III efficacy trial. After three doses of vaccine, all of the children in the pediatric trial had an antibody level greater than or equal to 1400 ELISA units ml^{-1} compared with 90% of vaccine recipients from the adult trial.

Evaluation of alternative dosing schedules and booster doses

A randomized trial of alternative dosing schedules involving 800 healthy adult volunteers found that the GMT of OspA antibody 1 month after a 0, 1 and 6 month dosing schedule was 7205 ELISA units ml^{-1}, compared with 10,659 ELISA units ml 1 month after the 0, 1, 12-month dosing schedule (van Hoecke et al., 1999). There were no serious adverse events thought to be related to vaccination in either group. The frequency of local reactions and systemic symptoms after vaccination was similar between the two groups. Headache and malaise were the most frequently reported systemic symptoms, each reported by less than 12% of people in each group. Ninety-two per cent of participants who received the vaccine on a 0, 1 and 6 month dose schedule had titres greater than or equal to 1400 ELISA units ml^{-1} after the third dose, compared with 93% of those on the 0, 1 and 12 month schedule. Despite the substantial differences in the GMTs, the authors concluded that the two vaccination schedules were equivalent, because of the similarity in proportions with titres greater than or equal to 1400 ELISA units ml^{-1} and because the relative difference in GMTs was less than 100%.

Another trial randomly assigned 956 volunteers aged 17–72 years to receive 30 μg of rOspA vaccine on either a 0, 1 and 12 month schedule or a 0, 1, 2 and 12 month schedule (Schoen et al., 2000). Ninety per cent of volunteers who were vaccinated by the 0, 1, 2 and 12 month schedule had antibody levels greater than or equal to 1400 ELISA units ml^{-1} after the third dose at 2 months. This group had higher GMTs after the 12 month dose than participants who were vaccinated by the 0, 1 and 12 month schedule. The adverse event profile of the two schedules after the 12 month dose was similar, with only one volunteer in each group experiencing a serious adverse event possibly related to vaccination (shaking chills in one person and syncope with full recovery in another).

Studies of the immunogenicity of OspA vaccines have indicated that antibody levels to OspA decline with increasing time following vaccination. The trial by Schoen et al. (2000) indicated that a booster dose after initial immunization on a 0, 1 and 2 month schedule was needed for protection in the second year, and that the booster dose was well tolerated and provided over 99% of recipients with presumably protective antibody levels. An unpublished evaluation of the need for booster doses after a 0, 1 and 12 month initial immunization schedule suggested that booster doses were needed 1 year after the third dose (D. Parenti et al., unpublished data). The optimal schedule for booster doses in persons vaccinated on accelerated (0, 1 and 2 month, or 0, 1 and 6 month) schedules is uncertain, but it seems likely that such individuals would require a booster dose just before the tick transmission season that follows the year of their initial vaccination. Subsequent booster doses may need to be given annually or perhaps every other year, but data to answer this question and to evaluate the safety of such repeated boosters have not been published.

Post-licensure safety and effectiveness evaluations

In the USA, the Centers for Disease Control and Prevention (CDC) and the Food and Drug Administration (FDA) jointly monitor reports of adverse events following vaccinations through the Vaccine Adverse Events Reporting System (VAERS). This is a passive surveillance system that provides for collection of adverse events following vaccinations reported by patients, healthcare providers and vaccine manufacturers. The data in VAERS do not directly allow for determining causal relationships between reported adverse events and the vaccines administered, but periodic review of the data can show unexpected patterns of adverse events associated with a given vaccine and may identify rare adverse reactions to vaccines that were not seen in the limited scope of clinical trials. The reports to VAERS associated with the commercially licensed LB vaccine, LYMErix™, from the date of its licensure in December 1998 through to July 2000, were reviewed by the CDC and FDA (Lathrop *et al.*, 2002). During this time over 1,400,000 doses were distributed and 905 adverse events were reported. The review detected no unexpected pattern of adverse events compared with the clinical trials except that rare hypersensitivity reactions including urticaria and occasional cases of dyspnoea were reported. Such reactions had not been reported during use of the vaccine in the phase III clinical trial. The numbers of reports of arthritis were well below what would have been expected based on rates of arthritis reported in the phase III trial.

A phase IV study to evaluate safety and effectiveness of the vaccine among members of managed care organizations in the northeastern USA is being conducted by the manufacturer. At least two case–control studies to evaluate post-licensure effectiveness of the vaccine in endemic states of the USA are in progress.

Utility and effectiveness of the vaccine

The clinical trials of OspA vaccine have demonstrated its efficacy in preventing LB and indicated that it has an acceptable safety profile. For any given individual, the question of whether or not to be vaccinated depends on the individual's risk of acquiring LB, the likelihood of obtaining appropriate treatment if LB is contracted, the cost and relative convenience of vaccination, and the individual's level of concern about the consequences of acquiring LB compared with concern about possible adverse events from the vaccine. The public health benefit of vaccination to prevent LB is somewhat limited compared with many other vaccine-preventable diseases. Because LB is not contagious and humans are dead-end hosts of *B. burgdorferi* s.s., vaccination of some humans does not decrease the risk of LB for others. Furthermore, vaccination does not protect against other diseases, such as human granulocytic ehrlichiosis and babesiosis, which are transmitted by the tick vectors of *B. burgdorferi* s.l. If vaccination were to induce complacency towards exposure to ticks, then vaccinated persons could potentially be at increased risk of other tick-borne diseases. The efficacy of the vaccine in protecting against LB outside North America is unknown, but would probably be lower because of the heterogeneity of the *Borrelia* spp. that cause LB in Europe and Asia. Despite these limitations, vaccination is currently one of the most effective measures available to prevent LB in the USA.

Optimal protection against LB requires three doses of the vaccine, which, under current recommendations, are given over the course of a year, as they were administered in the phase III clinical trial (Centers for Disease Control and Prevention, 1999). The vaccine is administered as a 0.5 ml dose of 30 µg rOspA given by injection into the deltoid muscle. A second dose is administered 1 month after the first, and a third dose 1 year after the first dose. The administration of the vaccine should be timed such that the second dose in year 1 and the third dose in year 2 are given several weeks before the onset of the *B. burgdorferi* s.s. transmission season in endemic areas, which usually begins in April. This will ensure that the vaccine recipient's OspA antibody levels will be highest during the months when the risk of acquiring LB is highest. The currently recommended schedule is somewhat awkward since it requires three doses of vaccine given in 2 separate years to achieve optimal protection. The safety and immunogenicity trial of the vaccine in Czech children indicated that high antibody levels could be achieved after three doses of vaccine given over 2 months (a 0, 1 and 2 month schedule). In adults, the 0, 1 and 2 month and 0, 1 and 6 month schedules appear to provide higher levels of antibody in the first vaccination year than the 0, 1 and 12 month schedule (van Hoecke *et al.*, 1999; Schoen *et al.*, 2000). Implementation of these accelerated schedules would enhance the clinical and public health utility of the vaccine since most vaccine recipients would not be exposed to a transmission season with suboptimal levels of protective antibody. However, the safety and protective efficacy of such accelerated schedules have not been fully evaluated by the FDA.

The FDA has licensed LYMErix™ for use in individuals aged 15–70 years. The Advisory Committee on Immunization Practices (ACIP) of the CDC has published the following recommendations for the use of LB vaccine (CDC, 1999):

1. Vaccination should be considered for persons aged 15–70 years who engage in activities that result in frequent or prolonged exposure to tick-infested habitats in areas of high or moderate risk of acquiring LB.
2. Vaccination may be considered for persons aged 15–70 years who are exposed to tick-infested habitats in areas of high or moderate LB risk, but whose exposure is neither frequent nor prolonged.
3. Vaccination is not recommended for persons who have minimal or no exposure to tick-infested habitats, for persons in areas of low or no risk of LB, or for pregnant women. In addition, the ACIP states that the vaccine should not be given to persons with treatment-resistant Lyme arthritis because of the association of this condition with immune reactivity to OspA.

The cumulative incidence of LB among placebo recipients during the second year of the phase III trial of LYMErix™ was 1.2%. Among a population with comparable incidence, and assuming a vaccine efficacy after three doses of 76%, about 110 people would need to be vaccinated to prevent one case of LB.

The effectiveness of vaccination has been compared with other strategies for preventing LB through a model involving a hypothetical community (Hayes *et al.*, 1999). Under the baseline assumptions of the model (which assumed a vaccine efficacy of 68% after the first dose and 92% after the third dose), vaccination was among the three most effective interventions in preventing LB. An evaluation of the cost-effectiveness of vaccination from a societal perspective indicated that at an assumed cost of vaccination of $100 per person per year, an effectiveness of 85%, a probability of 0.85 of correctly identifying and treating early LB, and an assumed cumulative incidence of LB of 1% per year, the net cost of vaccination was $5692 per case of LB prevented (Meltzer *et al.*, 1999). Under the same assumptions the cost of vaccination exceeded the cost of not vaccinating unless the cumulative incidence of LB was above 1.97% per year. Variations in LB incidence and in the likelihood of early diagnosis and treatment each had a substantial impact on the relative cost-effectiveness of vaccination. In a separate analysis that assumed a vaccination cost of $150, vaccine efficacy of 63% after two doses and 87% after three doses, and an LB incidence of 1% per year, vaccination of 10,000 individuals prevented 202 cases over a 10 year period, and cost $62,300 per quality-adjusted life year (QALY) gained compared with no vaccination (Shadick *et al.*, 2001). An accelerated 2 month dosing schedule reduced the cost of vaccination to $53,700 per QALY gained.

Development of Alternative Vaccines

As noted in the above discussion, the recombinant OspA vaccine appears to be efficacious and have an acceptable safety profile but for several reasons it is not an ideal vaccine to prevent LB. First, the published efficacy is less than 80%, leaving 20% of vaccinated individuals susceptible to disease. Efficacy against infection with *B. garinii* or *B. afzelii* would probably be even lower because of the heterogeneity of outer surface proteins. Secondly, the theoretical concerns about autoimmune reactions that could be induced by the vaccine have limited the acceptance of the vaccine by both healthcare providers and people at risk of the disease. A substantial amount of publicity in the lay press has been generated by public concern about the safety of the vaccine despite the reassuring scientific evidence. Thirdly, the currently recommended 0, 1 and 12 month dosing schedule and apparent need for continued boosters, limit the usefulness and effectiveness of the vaccine. Over 100 distinct *B. burgdorferi* s.l. proteins have been identified and, beyond OspA, a handful of these, including OspC, OspF, decorin-binding protein (Dbp) A, BBK32 and BBK50, have been studied as potential vaccine candidates.

OspC vaccines

OspC is a major outer surface lipoprotein and immunodominant antigen. This protein is upregulated in the feeding tick and expressed early in infection. Vaccination of outbred mice with recombinant OspC provided protection against infection when the mice were challenged with *B. burgdorferi* s.s. by tick bite (Gilmore *et al.*, 1996). In addition, passive transfer of OspC antibody resulted in a dose-dependent resolution of fully established infection, arthritis and carditis (Zhong *et al.*, 1999). Unlike OspA antibody, the protective action of OspC antibody appears to take place in the immunized host (Gilmore *et al.*, 1996), which should benefit from an anamnestic response during each exposure. However, significant OspC heterogeneity exists among *B. burgdorferi* spp. and strains at both antigenic and genetic levels. Over 30 distinct OspC types have been described by restriction fragment length polymorphisms and sequencing (Livey *et al.*, 1995). The N-terminal and C-terminal regions of the mature OspC molecule form relatively conserved domains. A highly variable domain is located in the central portion of the molecule. Comparison of the conserved regions of *ospC* genes among different isolates indicates that this gene and 16S RNA sequences have evolved at similar rates. However, comparison of the variable region of *ospC* suggests that *ospC* diversity has evolved at a greater rate than diversity in other *B. burgdorferi* s.l. genes (Livey *et al.*, 1995). It has been suggested that immune pressure exerted by an active antibody response to early expression of *ospC* might play a role in its diversity. However, studies in OspC-hyperimmunized mice do not support this hypothesis; mice infected with clonal *B. burgdorferi* s.s. or *B. afzelii* and then immunized with homologous recombinant OspC failed to elicit any sequence changes in *ospC* over a 6 month period of immune pressure (Hodzic *et al.*, 2000). These findings support the alternative hypothesis that lateral transfer of *ospC* genes between organisms is the major mechanism for generating diversity of OspC. Mixed infections within tick vectors and mammalian hosts are well documented but proof of a lateral transfer hypothesis awaits tracking of multiple defining markers within such infections.

In an attempt to account for antigenic variation within OspC, Baxter developed a pentavalent recombinant OspC vaccine (Livey *et al.*, 1999). In one unpublished, non-randomized trial, 40 seronegative human volunteers were given three doses of either 10, 25 or 50 µg of this vaccine. No serious adverse events were reported. Both reactogenicity and antibody response were more prominent at higher vaccine doses (Livey *et al.*, 1999).

Anti-adhesin vaccines

Bacterial adhesion to and colonization of host tissues is a critical step in the establishment of infection (Patti *et al.*, 1994). *B. burgdorferi* s.l. probably possesses several mechanisms of host and tissue specific adherence. The fibronectin-binding protein, BBK32, and the decorin-binding proteins A and B (DbpA and DbpB) are several proteins that mediate spirochaete attachment to the extracellular matrix (Guo *et al.*, 1995; Probert and Johnson, 1998) and have been investigated as vaccine candidates and therapeutic agents (Hanson *et al.*, 1998; Fikrig *et al.*, 2000).

Expression of *bbk32* in the tick is detectable during engorgement and it appears to be upregulated after the spirochaete enters the mammalian host at the cutaneous site of inoculation, and in the heart and spleen. BBK32 immunization of mice, or passive transfer of antiserum, provides partial protection from tick-vectored infection but does not clear established infections. Interestingly, acquisition of infection by ticks and transstadial transmission of spirochaetes in ticks are both reduced in the presence of antibody to BBK32 (Fikrig *et al.*, 2000). Together, these findings suggest that fibronectin-binding protein is important for bacterial adhesion in both the arthropod and mammalian host.

Genes for the *Borrelia* decorin-binding proteins, DbpA and DbpB, have been partially characterized. Immunization of mice with DbpA, as well as passive transfer of serum antibody to DbpA, protects mice from heterologous *B. burgdorferi* s.s. challenge via needle inoculation. Immunization with DbpB is less protective than immunization with DbpA. Passive transfer of DpbA antibody is therapeutic in mice up to 4 days after infection. Combined immunization with OspA and DbpA has produced protection against 100-fold higher challenges than injection with either protein alone (Hanson *et al.*, 2000). Recent studies have also shown that decorin-deficient mice are more resistant to disease and suggest that this adhesin molecule is important in the spirochaetal colonization of joint tissue (Brown *et al.*, 2001).

Other protein vaccines

OspE and OspF often elicit an antibody response during *B. burgdorferi* s.l. infection of mammals. However, immunization with OspF provides only partial protection against tick-borne or low-dose needle challenge, and immunization with OspE is ineffective (Nguyen *et al.*, 1994). Other antigens that have been tested as vaccine candidates but fail to elicit a protective immune response include the 41 kDa subunit of flagellin, P30, P39, P55 and P83 (Fikrig *et al.*, 1992; Das *et al.*, 1996; Feng *et al.*, 1996; Gilmore *et al.*, 1996).

DNA vaccines

Several studies in the early 1990s demonstrated that injection of DNA from pathogenic organisms into experimental animals resulted in humoral and cellular immune responses to antigens encoded by the DNA. DNA vaccines have been shown to be protective against specific viral, bacterial and parasitic infections in experimental animals (Wolff *et al.*, 1990, 1991; Vogel and Sarver, 1995). When mice were immunized with plasmid DNA encoding OspA, they produced

OspA-specific antibodies, which inhibited growth of homologous *B. burgdorferi* s.s. strains *in vitro* and protected the mice from needle challenge with two heterologous strains (Luke *et al.*, 1997).

In contrast to conventional protein antigen immunization, DNA vaccines may be engineered to direct antigen processing and presentation, depending on whether or not specific signal sequences are incorporated into the vaccine (Weiss *et al.*, 1999). The addition of a human tissue–plasminogen–activator signal sequence to DNA encoding OspC enhanced both the humoral and cellular immune responses in vaccinated mice (Weiss *et al.*, 1999). These results suggest that a signal that directs antigen presentation through the endoplasmic reticulum may optimize both the antibody and cellular immune responses to OspC-based plasmid vaccines (Weiss *et al.*, 1999).

Other DNA vaccines have been engineered to produce antibody against multiple spirochaetal antigens. In one study a DNA vaccine expressing a fusion product of OspA and OspC was shown to elicit antibody responses to both antigens. However, only OspA antibodies were protective against needle challenge infection and neither antibody response was capable of clearing an established infection (Wallich *et al.*, 2001). No DNA vaccine has yet been shown to protect against tick-transmitted infection of *B. burgdorferi* s.l.

Anti-tick vaccines

Ixodes spp. complex ticks are common vectors of LB, tick-borne encephalitis, ehrlichiosis and babesiosis, and mixed infections can occur within a single tick. Vaccination against any one disease may theoretically lead to a false perception among vaccinees of protection against a greater range of tick-borne diseases and thus minimize the use of other risk-reducing practices or result in delayed diagnosis and treatment.

Reduction in vector-tick populations has been shown to have significant impact on LB incidence. However, current control efforts are limited largely to acaricide use and habitat modification. Vaccination against ticks is a relatively new alternative approach. Two targets of this approach are salivary gland antigens that are secreted into the host by the feeding tick (exposed antigens) and others that remain within the tick (concealed antigens) (Willadsen and Jongejan, 1999). The latter do not naturally prime the host's immune system but are potential targets of antibody ingested by the tick while feeding. Antibody elicited against a recombinant midgut antigen of *Boophilus microplus*, a tick vector of babesiosis in cattle, has been shown to be effective in controlling infestation by this and other *Boophilus* spp. as well as providing partial protection against other tick genera (Willadsen and Jongejan, 1999). Other concealed antigen targets might include tick hormones that regulate moulting or reproduction.

Repeated feeding by ticks induces a number of immune responses in the host that may lead to inflammation at the attachment site or cause the tick to detach. Identification of exposed antigens from the tick mouth-parts or salivary gland secretions that elicit a protective rejection of the tick may result in a vaccine that would be boosted by natural exposure.

Summary

The development of an efficacious vaccine against LB followed shortly after the identification and characterization of the *Borrelia* spp. that cause the disease. Immunization with OspA was demonstrated to be protective against infection through the unusual mechanism of eliciting antibodies that presumably destroy or inactivate the spirochaete inside the attached vector tick. In clinical trials a recombinant OspA vaccine with adjuvant has been shown to be nearly 80% efficacious and has an acceptable safety profile. Long-term concerns about the vaccine's safety centre around evidence that OspA may be a contributing component to possible autoimmune reactions after naturally acquired infection, but there is no evidence that the vaccine directly causes arthritis or other serious adverse events in humans. Rare hypersensitivity reactions to the vaccine have been reported. Despite its limitations, the OspA vaccine is a rare example of a successful transmission-blocking vaccine. A better OspA vaccine might be produced by incorporating antibody epitopes of *B. garinii* and *B. afzelii* and removing T cell epitopes that are possibly involved in autoimmune responses. Alternative vaccines could incorporate a combination of different protein antigens, pursue a DNA-directed immune response, or target the vector tick itself in order to provide safe and more complete protection against the variety of *Borrelia* genospecies that cause LB in North America, Europe and Asia.

References

Beran, J., de Clercq, N., Dieussaert, I. and van Hoecke, C. (2000) Reactogenicity and immunogenicity of a Lyme disease vaccine in children 2–5 years old. *Clinical Infectious Diseases* 31, 1504–1507.

Berglund, J., Eitrem, R., Ornstein, K., Lindberg, A., Ringer, A., Elmrud, H., Carlsson, M., Runehagen, A., Svanborg, C. and Norrby, R. (1995) An epidemiologic study of Lyme disease in southern Sweden. *New England Journal of Medicine* 333, 1319–1327.

Brown, E.L., Wooten, R.M., Johnson, B.J., Iozzo, R.V., Smith, A., Dolan, M.C., Guo, B.P., Weis, J.J. and Hook, M. (2001) Resistance to Lyme disease in decorin-deficient mice. *Journal of Clinical Investigation* 107, 845–852.

Centers for Disease Control and Prevention (1999) Recommendations for the use of Lyme disease vaccine: recommendations of the Advisory Committee on Immunization Practices (ACIP). *Morbidity and Mortality Weekly Report* 48, 1–24.

Das, S., Shraga, D., Gannon, C., Lam, T.T., Feng, S., Brunet, L.R., Telford, S.R., Barthold, S.W., Flavell, R.A. and Fikrig, E. (1996) Characterization of a 30-kDa *Borrelia burgdorferi* substrate-binding protein homologue. *Research in Microbiology* 147, 739–751.

Ding, W., Huang, X., Yang, X., Dunn, J.J.,

Luft, B.J., Koide, S. and Lawson, C.L. (2000) Structural identification of a key protective B-cell epitope in Lyme disease antigen OspA. *Journal of Molecular Biology* 302, 1153–1164.

Edelman, R., Palmer, K., Russ, K.G., Secrest, H.P., Becker, J.A., Bodison, S.A., Perry, J.G., Sills, A.R., Barbour, A.G., Luke, C.J., Hanson, M.S., Stover, C.K., Burlein, J.E., Bansal, G.P., Connor, E.M. and Koenig, S. (1999) Safety and immunogenicity of recombinant Bacille Calmette–Guérin (rBCG) expressing *Borrelia burgdorferi* outer surface protein A (OspA) lipoprotein in adult volunteers: a candidate Lyme disease vaccine. *Vaccine* 17, 904–914.

Erdile, L.F., Brandt, M.A., Warakomski, D.J., Westrack, G.J., Sadziene, A., Barbour, A.G. and Mays, J.P. (1993) Role of attached lipid in immunogenicity of *Borrelia burgdorferi* OspA. *Infection and Immunity* 61, 81–90.

Feder, H.M. Jr, Beran, J., van Hoecke, C., Abraham, B., de Clercq, N., Buscarino, C. and Parenti, D.L. (1999) Immunogenicity of a recombinant *Borrelia burgdorferi* outer surface protein A vaccine against Lyme disease in children. *Journal of Pediatrics* 135, 575–579.

Felz, M.W., Chandler, F.W. Jr, Oliver, J.H. Jr, Rahn, D.W. and Schriefer, M.E. (1999) Solitary erythema migrans in Georgia and South Carolina. *Archives of Dermatology* 135, 1317–1326.

Feng, S., Barthold, S.W., Telford, S.R. III and Fikrig, E. (1996) P55, an immunogenic but nonprotective 55-kilodalton *Borrelia burgdorferi* protein in murine Lyme disease. *Infection and Immunity* 64, 363–365.

Fikrig, E., Barthold, S.W., Kantor, F.S. and Flavell, R.A. (1990) Protection of mice against the Lyme disease agent by immunizing with recombinant OspA. *Science* 250, 553–556.

Fikrig, E., Barthold, S.W., Marcantonio, N., Deponte, K., Kantor, F.S. and Flavell, R.A. (1992) Roles of OspA, OspB, and flagellin in protective immunity to Lyme borreliosis in laboratory mice. *Infection and Immunity* 60, 657–661.

Fikrig, E., Feng, W., Barthold, S.W., Telford, S.R. III and Flavell, R.A. (2000) Arthropod- and host-specific *Borrelia burgdorferi* bbk32 expression and the inhibition of spirochete transmission. *Journal of Immunology* 164, 5344–5351.

Gilmore, R.D. Jr, Kappel, K.J., Dolan, M.C., Burkot, T.R. and Johnson, BJ. (1996) Outer surface protein C (OspC), but not P39, is a protective immunogen against a tick-transmitted *Borrelia burgdorferi* challenge: evidence for a conformational protective epitope in OspC. *Infection and Immunity* 64, 2234–2239.

Gross, D.M., Forsthuber, T., Tary-Lehmann, M., Etling, C., Ito, K., Nagy, Z.A., Field, J.A., Steere, A.C. and Huber, B.T. (1998) Identification of LFA-1 as a candidate autoantigen in treatment-resistant Lyme arthritis. *Science* 281, 703–706.

Guo, B.P., Norris, S.J., Rosenberg, L.C. and Hook, M. (1995) Adherence of *Borrelia burgdorferi* to the proteoglycan decorin. *Infection and Immunity* 63, 3467–3472.

Hanson, M.S., Cassatt, D.R., Guo, B.P., Patel, N.K., McCarthy, M.P., Dorward D.W. and Hook, M. (1998) Active and passive immunity against *Borrelia burgdorferi* decorin binding protein A (DbpA) protects against infection. *Infection and Immunity* 66, 2143–2153.

Hanson, M.S., Patel, N.K., Cassatt, D.R. and Ulbrandt, N.D. (2000) Evidence for vaccine synergy between *Borrelia burgdorferi* decorin binding protein A and outer surface protein A in the mouse model of Lyme borreliosis. *Infection and Immunity* 68, 6457–6460.

Hayes, E.B., Maupin, G.O., Mount, G.A. and Piesman, J. (1999) Assessing the prevention effectiveness of local Lyme disease control. *Journal of Public Health Management and Practice* 5, 84–92.

Hodzic, E., Feng, S. and Barthold, S.W. (2000) Stability of *Borrelia burgdorferi* outer surface protein C under immune selection pressure. *Journal of Infectious Diseases* 181, 750–753.

van Hoecke, C., Comberbach, M., de Grave, D., Desmons, P., Fu, D., Hauser, P., Lebacq, E., Lobet, Y. and Voet, P. (1996) Evaluation of the safety, reactogenicity and immunogenicity of three recombinant outer surface protein (OspA) Lyme vaccines in healthy adults. *Vaccine* 14, 1620–1626.

van Hoecke, C., Lebacq, E., Beran, J. and Parenti, D. (1999) Alternative vaccination schedules (0, 1, and 6 months versus 0, 1, and 12 months) for a recombinant OspA Lyme disease vaccine. *Clinical Infectious Diseases* 28, 1260–1264.

James, A.M., Liveris, D., Wormser, G.P., Schwartz, I., Montecalvo, M.A. and Johnson, B.J. (2001) *Borrelia lonestari* infection after a bite by an *Amblyomma americanum* tick. *Journal of Infectious Diseases* 183, 1810–1814.

Kalish, R.A., Leong, J.M. and Steere, A.C. (1993) Association of treatment-resistant chronic Lyme arthritis with HLA-DR4 and antibody reactivity to OspA and OspB of *Borrelia burgdorferi*. *Infection and Immunity* 61, 2774–2779.

Kalish, R.A., Leong, J.M. and Steere, A.C. (1995) Early and late antibody responses to full-length and truncated constructs of outer surface protein A of *Borrelia burgdorferi* in Lyme disease. *Infection and Immunity* 63, 2228–2235.

Keller, D., Koster, F.T., Marks, D.H., Hosbach, P., Erdile, L.F. and Mays, J.P. (1994) Safety and immunogenicity of a recombinant outer surface protein A Lyme vaccine. *Journal of the American Medical Association* 271, 1764–1768.

Lathrop, S.L., Ball, R., Haber, P., Mootrey, G.T., Braun, M.M., Shadomy, S.V., Chen, R.T. and Hayes, E.B. (2002) Adverse event reports following vaccination for Lyme disease: December 1998–July 2000. *Vaccine* 20, 1603–1608.

Livey, I., Gibbs, C.P., Schuster, R. and Dorner, F. (1995) Evidence for lateral transfer and recombination in OspC variation in Lyme disease *Borrelia*. *Molecular Microbiology* 18, 257–269.

Livey, I., Crowe, B.A., Eder, G., Barrett, N., Gerencer, M., Low-Baselli, A. and Dorner, F. (1999) Clinical studies of a multivalent OspC Lyme disease vaccine. In: *VII International Conference on Lyme Borreliosis and other Emerging Tickborne Diseases*. Munich, Germany, Abstract S02.

Luke, C.J., Carner, K., Liang, X. and Barbour, A.G. (1997) An OspA-based DNA vaccine protects mice against infection with *Borrelia burgdorferi*. *Journal of Infectious Diseases* 175, 91–97.

Meltzer, M.I., Dennis, D.T. and Orloski, K.A. (1999) The cost effectiveness of vaccinating against Lyme disease. *Emerging Infectious Diseases* 5, 321–328.

Montgomery, R.R., Nathanson, M.H. and Malawista, S.E. (1994) Fc- and non-Fc-mediated phagocytosis of *Borrelia burgdorferi* by macrophages. *Journal of Infectious Diseases* 170, 890–893.

Nguyen, T.P., Lam, T.T., Barthold, S.W., Telford, S.R. III, Flavell, R.A. and Fikrig, E. (1994) Partial destruction of *Borrelia burgdorferi* within ticks that engorged on OspE- or OspF-immunized mice. *Infection and Immunity* 62, 2079–2084.

O'Connell, S., Granstrom, M., Gray, J.S. and Stanek, G. (1998) Epidemiology of European Lyme borreliosis. *Zentralblatt für Bacteriologie* 287, 229–240.

Orloski, K.A., Hayes, E.B., Campbell, G.L. and Dennis, D.T. (2000) Surveillance for Lyme disease – United States, 1992–1998. *Morbidity and Mortality Weekly Report* 49, 1–12.

Patti, J.M., Allen, B.L., McGavin, M.J. and Hook, M. (1994) MSCRAMM-mediated adherence of microorganisms to host tissues. *Annual Review of Microbiology* 48, 585–617.

Pavia, C.J., Kissel, V., Bittker, S., Cabello, F. and Levine, S. (1991) Antiborrelial activity of serum from rats injected with the Lyme disease spirochete. *Journal of Infectious Diseases* 163, 656–659.

Probert, W.S. and Johnson, B.J. (1998) Identification of a 47 kDa fibronectin-binding protein expressed by *Borrelia burgdorferi* isolate B31. *Molecular Microbiology* 30, 1003–1015.

Probert, W.S. and LeFebvre, R.B. (1994) Protection of C3H/HeN mice from challenge with *Borrelia burgdorferi* through active immunization with OspA, OspB, or OspC, but not with OspD or the 83-kilodalton antigen. *Infection and Immunity* 62, 1920–1926.

Schoen, R.T., Meurice, F., Brunet, C.M., Cretella, S., Krause, D.S., Craft, J.E. and Fikrig, E. (1995) Safety and immunogenicity of an outer surface protein A vaccine in participants with previous Lyme disease. *Journal of Infectious Diseases* 172, 1324–1329.

Schoen, R.T., Sikand, V.K., Caldwell, M.C., van Hoecke, C., Gillet, M., Buscarino, C. and Parenti, D.L. (2000) Safety and immunogenicity profile of a recombinant outer-surface protein A Lyme disease vaccine: clinical trial of a 3-dose schedule at 0, 1, and 2 months. *Clinical Therapeutics* 22, 315–325.

Shadick, M.A., Liang, M.H., Phillips, C.B., Fossel, K. and Kuntz, K.M. (2001) The cost-effectiveness of vaccination against Lyme disease. *Archives of International Medicine* 161, 554–561.

Sigal, L.H., Zahradnik, J.M., Lavin, P., Patella, S.J., Bryant, G., Haselby, R., Hilton, E., Kunkel, M., Adler-Klein, D., Doherty, T., Evans, J., Molloy, P.J., Seidner, A.L., Sabetta, J.R., Simon, H.J., Klempner, M.S., Mays, J., Marks, D. and Malawista, S.E. (1998) A vaccine consisting of recombinant *Borrelia burgdorferi* outer-surface protein A to prevent Lyme disease. *New England Journal of Medicine* 339, 216–222.

Sikand, V.K., Halsey, N., Krause, P.J., Sood, S.K., Geller, R., van Hoecke, C., Buscarino, C. and Parenti, D. (2001) Safety and immunogenicity of a recombinant *Borrelia burgdorferi* outer surface protein A vaccine against Lyme disease in healthy children and adolescents: a randomized controlled trial. *Pediatrics* 108, 123–128.

de Silva, A.M., Telford, S.R. III, Brunet, L.R., Barthold, S.W. and Fikrig, E. (1996) *Borrelia burgdorferi* OspA is an arthropod-specific transmission-blocking Lyme disease vaccine. *Journal of Experimental Medicine* 183, 271–275.

Smith, P.F., Benach, J.L., White, D.J., Stroup, D.F. and Morse, D.L. (1988) Occupational risk of Lyme disease in endemic areas of New York state. *Annals of the New York Academy of Sciences* 539, 289–301.

Steere, A.C., Sikand, V.K., Meurice, F., Parenti, D.L., Fikrig, E., Schoen, R.T., Nowakowski, J., Schmid, C.H., Laukamp, S., Buscarino, C., Krause, D.S. and the Lyme Disease Vaccine Study Group (1998) Vaccination against Lyme disease with recombinant *Borrelia burgdorferi* outer surface lipoprotein A with adjuvant. *New England Journal of Medicine* 339, 209–215.

Telford, S.R. and Fikrig, E. (1995) Progress towards a vaccine for Lyme disease. *Clinical Immunotherapy* 4, 49–60.

Vogel, F.R. and Sarver, N. (1995) Nucleic acid vaccines. *Clinical Microbiology Reviews* 8, 406–410.

Wallich, R., Siebers, A., Jahraus, O., Brenner, C., Stehle, T. and Simon, M.M. (2001) DNA vaccines expressing a fusion product of outer surface proteins A and C from *Borrelia burgdorferi* induce protective antibodies suitable for prophylaxis but not for resolution of Lyme disease. *Infection and Immunity* 69, 2130–2136.

Weiss, R., Durnberger, J., Mostbock, S.,

Scheiblhofer, S., Hartl, A., Breitenbach, M., Strasser, P., Dorner, F., Livey, I., Crowe, B. and Thalhamer, J. (1999) Improvement of the immune response against plasmid DNA encoding OspC of *Borrelia* by an ER-targeting leader sequence. *Vaccine* 18, 815–824.

Willadsen, P. and Jongejan, F. (1999) Immunology of the tick–host interaction and the control of ticks and tick-borne diseases. *Parasitology Today* 15, 258–262.

Wolff, J.A., Malone, R.W., Williams, P., Chong, W., Acsadi, G., Jani, A. and Felgner, P.L. (1990) Direct gene transfer into mouse muscle *in vivo*. *Science* 247, 1465–1468.

Wolff, J.A., Williams, P., Acsadi, G., Jiao, S., Jani, A. and Chong, W. (1991) Conditions affecting direct gene transfer into rodent muscle *in vivo*. *Biotechniques* 11, 474–485.

Wormser, G.P. (1995) Prospects for a vaccine to prevent Lyme disease in humans. *Clinical Infectious Diseases* 21, 1267–1274.

Zhong, W., Gern, L., Stehle, T., Museteanu, C., Kramer, M., Wallich, R. and Simon, M.M. (1999) Resolution of experimental and tick-borne *Borrelia burgdorferi* infection in mice by passive, but not active immunization using recombinant OspC. *European Journal of Immunology* 29, 946–957.

Environmental Management for Lyme Borreliosis Control

Kirby C. Stafford[1] and Uriel Kitron[2]

[1]*Connecticut Agricultural Experiment Station, PO Box 1106, New Haven, CT 06540, USA;* [2]*Department of Veterinary Pathobiology, College of Veterinary Medicine, University of Illinois, 2001 S. Lincoln, Urbana, IL 61802, USA*

Introduction

Ticks are obligate haematophagous ectoparasites of humans, livestock and wildlife, associated with many zoonotic disease pathogens. Tick management has historically been more of a concern in the veterinary realm, where control options such as rotational grazing, dipping and quarantine are also available. Control of human tick-borne diseases may target the vector or the pathogen (Jaenson *et al.*, 1991; Tatchell, 1992; Wilson and Deblinger, 1993; Schmidtmann, 1994; Sonenshine, 1994) with the purpose of reducing tick–human contact or reducing the prevalence of infection in the tick vector and/or the vertebrate hosts. Strategies against the vector include personal protection from tick bite and approaches that reduce tick introduction into tick-free areas, tick abundance or tick reproduction. The pathogen is targeted through vaccination or, indirectly, through methods to reduce the prevalence of infection in the vector or reservoir host. Methods that have been explored to control ticks include biological control, area chemical applications, acaricidal treatment of small and large mammalian hosts, habitat modification, reduction in host availability and integrated pest management (IPM) strategies. Other approaches, such as an anti-tick vaccine, remain experimental or speculative in terms of human disease. The commercial availability of a recombinant vaccine against *Boophilus microplus* is an important new step in the control of ticks on livestock (Willadsen *et al.*, 1995).

Four tick species of the genus *Ixodes* that readily feed on humans, *Ixodes scapularis*, *Ixodes pacificus*, *Ixodes ricinus* and *Ixodes persulcatus*, are competent vectors of the *Borrelia* spp. that cause Lyme borreliosis (LB). They are distributed over broad areas of North America, Europe and Asia, placing many people at risk of not only LB, but also other tick-borne infections carried by these ticks. Control of LB has relied heavily on personal protection behaviours, antibiotic and supportive therapy after infection, and to a lesser extent the application of acaricides to tick habitats. The environmental management of tick populations and reduction in risk of the diseases they transmit requires knowledge of tick ecology and the unique habitats of each tick species, the transmission dynamics of LB (which may vary by *Borrelia* genospecies) and a proper assessment of risk. The wide geographical distribution, abundance and host range of these ticks, and the diversity of known reservoirs for *Borrelia burgdorferi* sensu lato (s.l.) and other genospecies complicate the management or control of tick-borne disease.

Tick control over large areas was the leading strategy for the prevention of tick-borne encephalitis (TBE) in Russia from about 1950 through to the 1970s (Uspensky, 1996, 1999). The use of DDT in a large-scale campaign successfully suppressed *I. persulcatus* in high-risk areas of taiga forest resulting in substantial reductions in human contacts with ticks and decreased incidence of TBE. With the gradual reduction and cessation of tick control programmes in Russia, TBE and the more recently recognized LB has increased (Korenberg, 1994). Although acaricides are still an important element of tick management, environmental and toxicological safety concerns have restricted the availability and wide-scale use of these chemicals. Nevertheless, acaricides are still commonly used for short-term control of *I. scapularis* around private residences and other localized high-risk areas (Stafford, 1997). While a variety of effective acaricides, mainly pyrethroids, are still available, there is a need for newer, less toxic alternatives and the integration of these materials with other methodologies. Targeting the acaricides to tick hosts is one approach. As the prevalence of LB continues to increase, environmentally acceptable approaches are needed to manage tick populations and reduce the risk of acquiring disease. Human tick-borne illnesses require an ecologically based integrated management approach based on an assessment of risk. The ecology of the various *Ixodes* spp., which feed on a diverse range of wildlife and are broadly distributed geographically, is complex. No single methodology will address the cycle of tick vector, vertebrate reservoirs, spirochaetes and exposure of the susceptible human population, and focused efforts to reduce the incidence of LB are few. In this chapter, we address the assessment of risk for LB, current strategies in managing tick populations, approaches that are experimental, and future prospects and strategies for tick control and the management of human disease.

Habitat Assessment and Risk

While individual ticks operate in the microenvironment and encounter only one host at each stage, risk assessment and control strategies to reduce the probability of tick–human encounters and the transmission of tick-borne diseases need to be considered over a range of spatial scales (Cortinas *et al.*, 2002). Thus, potential individual human hosts apply measures of personal protection and targeted acaricides are directed towards ticks infesting individual reservoir hosts. In contrast, landscape and host management and area application of acaricides operate on a wider geographical range and affect a large number of hosts and vectors.

Tick habitat can be viewed on the micro-, meso- and macroscales ranging from the ground level microenvironment to the country- or continent-wide epidemiological picture (Kitron, 1998, 2000; Cortinas et al., 2002). Satellite imagery and geographical information systems (GIS) have been used to map tick and LB risk distribution (Daniel and Kolar, 1990; Nicholson and Mather, 1996; Kitron and Kazmierczak, 1997; Daniel et al., 1998), and can be used as surveillance tools to highlight risk areas and periods and to determine recommendations for application of control measures.

Risk maps for transmission of LB (and other diseases transmitted by *Ixodes* spp. ticks) have been developed primarily based on habitat suitability for tick distribution in the USA (Kitron et al., 1992; Glass et al., 1994, 1995; Kitron and Kazmierczak, 1997), although the zooprophylactic effect of lizards and the effect of deer exclusion have also been considered (Fish and Howard, 1999). In Europe, where the variability in *B. burgdorferi* s.l. genotypes is well documented, tick distribution has also been associated with environmental features (Daniel and Kolar, 1990; Merler et al., 1996; Gray et al., 1998), but the role of alternative hosts in transmission of the various *B. burgdorferi* subspecies has also been considered and contrasted with the requirements for transmission of tick-borne encephalitis (Randolph, 2000; Randolph and Rogers, 2000; Randolph et al., 2000).

On a very fine scale, Dister et al. (1997) used Landsat Thematic Mapper (TM) remotely sensed data to categorize residential properties in Westchester County, New York, as no, low or high risk based on vegetation structure, moisture and abundance. Thus, residential mapping was used to determine habitat suitability for ticks and risk of transmission of LB within communities. Such an analysis can be used to derive landscape maps that can be used to target intervention techniques (e.g. deer exclusion, area insecticide application to specific properties). TM data were also used by Daniel et al. (1998) to detect micro-foci in Bohemia, Czech Republic.

In Maryland and Rhode Island, USA, risks at county and state level were assessed using satellite data and GIS (Glass et al., 1995; Nicholson and Mather, 1996). In Wisconsin, the distribution of human LB cases on the county level was associated with forest cover (Kitron and Kazmierczak, 1997). A risk map for Wisconsin and Illinois was developed based on canine serology and tick distribution associated with a range of environmental features (Cortinas et al., 2002; Guerra et al., 2001a,b) and can be used to target surveillance efforts and control priorities. On a continental scale, Fish and Howard (1999) provided a coarse continental map for LB risk in the USA that can be used to determine whether vaccination should be recommended or even considered. In Europe, a collaborative effort provided descriptions of LB habitats from 16 countries and these were used to associate high-risk situations with heterogeneous deciduous forest, recreational activities and diverse fauna (Gray et al., 1998).

In a recent review, Randolph (2000) provided a comprehensive summary and analysis of different approaches to identify tick habitat, and highlighted the biological basis underlying critical satellite signals. Randolph *et al.* (2000) used coarse resolution NOAA-AVHRR (National Oceanic and Atmospheric Administration Advanced Very High Resolution Radiometer) satellite imagery and considered both NDVI (Normalized Difference Vegetation Index) and LST (Land Surface Temperature). They processed monthly images using temporal Fourier analysis, and contrasted the risk of LB transmission with the much more restricted geographical risk of TBE transmission. Much of the biological knowledge necessary to successfully apply satellite data to understand and produce risk maps was based on field and laboratory studies that determined the necessary condition for transmission of the various tick-borne pathogens.

In contrast, efforts to produce continental risk maps, based on meta-analysis of published data (Estrada-Peña, 1998, 2001) and on cokriging of simple habitat features (e.g. NDVI), can result in maps that are discordant with known distribution of ticks (e.g. habitat suitability versus known distribution in Wisconsin, USA, and in Britain and Ireland). In Europe, in particular, the biological diversity of LB transmission systems (due to genetic diversity of the spirochaete and variations in reservoir competence) renders predictive risk maps less precise.

The ability to incorporate large amounts of environmental data (some of them through remote sensing) with biological information allows for effective targeting of control measures on a range of spatial scales. Accurate maps of tick distribution and disease risk are based on good understanding of tick biology and pathogen transmission, as are effective control strategies. In particular, spatial tools for habitat assessment can play an important role in environmental management and integration of control measures where consideration of habitat heterogeneity on multiple spatial scales is necessary for successful implementation.

Measures for Control of *Ixodes* spp. Ticks and Lyme Borreliosis

Control measures can be divided into measures applied on an individual basis (personal protection and vaccination) and measures that target the tick population (on a range of spatial and temporal scales) (Table 12.1). In this chapter, we emphasize the latter measures, which are applied in the environment. After a brief review of measures for prevention of tick bites, treatment and vaccination (including host immunity and anti-tick vaccines), we concentrate on management of landscape and hosts and area application of acaricides. We also discuss the actual and potential use of biological control agents and pheromones, whose application is part of an environmentally based strategy, even though they may target individual ticks. We conclude by discussing the use of various strategies in an IPM programme and the future of tick and LB control.

Personal protection and behaviours to prevent or reduce tick bite are the first

Table 12.1. Existing and potential management strategies for the control of Lyme borreliosis. These approaches may reduce exposure to the tick vector, target the tick vector or target the pathogen.

Personal protection	Tick-bite prevention to minimize exposure: protective clothing, use of repellents.
	Tick removal and tick-bite prophylaxis: full examination of the body and prompt removal of ticks, prophylactic treatment with antimicrobials.
	Vaccination against *Borrelia burgdorferi*.
	Early diagnosis and treatment of infection.
Immunologically based approaches	Research into immunological interactions of the tick–host–pathogen interface.
	Development of anti-tick vaccines.
Landscape management	Vegetative modifications (mowing, clearing leaves, xeric substrates) and burning to render the environment less suitable for tick survival and for tick hosts.
	Use of deer-resistant plants to reduce host visitation.
Management host abundance and availability	Exclusion of hosts from areas by fencing.
	Host reduction by management, elimination or alteration of the environment.
Area application of acaricides	Treatment of white-tailed deer with an acaricide through passive topical application devices.
Host-targeted acaricides	Treatment of white-footed mice with permethrin-treated cotton or fipronil-based bait boxes.
Biological control	Parasitoids: potential of encrytid wasp *Ixodiphagus hookeri* for tick management.
	Pathogens: potential of fungal, nematode and other pathogens for tick management.
	Predators: role of predators in reducing tick abundance.
Pheromones	Use of pheromone-based lures and decoys with acaricides for tick control, either on or off host.
Integrated pest management (IPM)	Concepts of IPM related to ticks and vector-borne disease and application of IPM strategies.
	Use of computer models to simulate various strategies for the prevention of Lyme borreliosis.

line of defence in the prevention of LB. These include avoidance and reduction of time spent in tick-infested habitats, using protective clothing and tick repellents, checking the entire body for ticks and promptly removing attached ticks before transmission of borrelial spirochaetes can occur. However, the increasing incidence of disease and surveys in the northeastern USA (Cartter et al., 1989; Brown et al., 1992; Shadick et al., 1997; Phillips et al., 2001) suggest that few people practise these measures consistently enough for them to be effective. Despite the efficacy of tick repellents, particularly with N,N-diethyl-3-methylbenzamide (DEET) applied to skin and permethrin applied to clothing, they are grossly underused. Preventive measures are often considered inconvenient; those who believe their chance of acquiring LB is high are more likely to take precautions and frequent detection of ticks appears to be a major incentive for engaging in prevention behaviours.

When the tick feeds, spirochaetes residing in the gut of an infected tick multiply; some migrate through the gut into the haemolymph, disseminate to the salivary glands and are introduced into the host during the later stages of feeding (Ribeiro et al., 1987; Piesman, 1995; Gern et al., 1996). The risk of infection is low during the first 24–48 h of attachment by *I. scapularis* (des Vignes et al., 2001; Ohnishi et al., 2001) and *I. ricinus* (Kahl et al., 1998) and increases after attachment for 48 h or more. Consequently timely tick removal and tick-bite prevention are effective control measures. However, the prophylactic use of antibiotics following a tick bite has not generally been recommended in the USA (Warshafsky et al., 1996), as the chance of infection from a known tick bite with *I. scapularis* appears low (< 5%, about equal to the chance of a reaction to the antibiotic) and the majority of patients diagnosed with LB do not detect the tick (Costello et al., 1989; Shapiro et al., 1992). Nevertheless, a large proportion (55%) of patients with tick bite apparently receive antibiotic therapy (Fix et al., 1998). Risk is higher with engorged ticks and, although tick testing is not a reliable predictor of infection, prophylactic treatment may be justified in such cases (Sood et al., 1997). Transmission rates with *I. ricinus*, and particularly with adult *I. persulcatus*, may be higher and treatment may reduce the risk of LB (Alekseev et al., 1996; Maiwald et al., 1998). Although not strictly a prevention strategy, early diagnosis and treatment are important in the management of LB. Early treatment has an excellent prognosis and can prevent the development of later stages of disease (Wormser et al., 2000).

The use of recombinant outer surface protein A (OspA) vaccines was found to prevent human LB in large-scale safety and efficacy trials (Sigal et al., 1998; Steere et al., 1998), and one vaccine (LYMErix™, GlaxoSmithKline), consisting of lipidated OspA with an adjuvant, was approved in the USA in December 1998 (Centers for Disease Control and Prevention, 1999a). European isolates of borrelia show great variation in their outer surface proteins and may express OspA poorly or not at all. Vaccines that utilize a complex of other candidate antigens, such as OspC or a cocktail of various OspAs, may be needed for an effective European borreliosis vaccine (Gern et al., 1997).

Recommendations for use of the vaccine are based on the risk of exposure to infected ticks (Centers for Disease Control and Prevention, 1999b). The economic benefit of vaccinating against LB is dependent mainly on the probability of contracting the disease, followed by the cost of treating cases and the probability of early detection and treatment of LB (Meltzer *et al.*, 1999). In modelling the effectiveness of various prevention measures, Hayes *et al.* (1999) found that use of the vaccine was among the more effective options in preventing cases in the simulated scenarios. For example, vaccination would prevent 148 cases of LB over a 4 year period in a hypothetical population of 10,000 individuals, assuming 65% of the population were targeted, 74% of those were actually vaccinated and vaccine efficacy was 60% with two doses and 90% for the first year after three doses. The future role that an LB vaccine could play, as a component of an integrated programme for the prevention of LB, is unclear as the first LB vaccine (LYMErix^{TM}, GlaxoSmithKline) was withdrawn from the market by the manufacturer in February 2002 due to a lack of sales. Future improvements should focus on developing second-generation vaccines, possibly through antigens expressed in the human host. Vaccination is discussed in depth by Hayes and Schriefer in Chapter 11.

Another promising approach in the control of ticks is immunologically based, focusing on the development of anti-tick vaccines (Wikel *et al.*, 1996; Willadsen and Jongejan, 1999). Ixodid ticks attach and feed on their hosts for many days, introducing through the saliva a variety of pharmacologically active compounds that aid the feeding process, improve engorgement success, suppress innate and acquired immunity of the host and affect pathogen transmission (reviewed by Bowman *et al.*, 1997). These elements also function as antigens that stimulate the host immune response. There is a balance between immune-mediated resistance by the host and strategies employed by the tick to circumvent these mechanisms. Vaccination with salivary gland-derived or other antigens can induce resistance similar to naturally acquired resistance. An anti-tick feeding vaccine could inhibit tick-feeding success and potentially reduce transmission of multiple pathogens from that vector.

Vaccination with a commercial recombinant vaccine against ticks has proved effective as a component in the control of cattle ticks and of bovine babesiosis (Willadsen *et al.*, 1995) and offers potential for reducing transmission of tick-borne infections in reservoir hosts or the prevention of disease in humans. Research currently focuses on a better understanding of the biological basis by which ticks modulate host immune responses and finding tick-derived proteins that could be used for an anti-tick vaccine.

A similar approach, vaccination or treatment of wild animal reservoirs with antimicrobials, may also have potential to reduce the transmission of *B. burgdorferi* s.l. and subsequently the risk of human infection. The vaccination of *B. burgdorferi*-infected *Peromyscus leucopus* (white-footed mouse) with recombinant OspA reduced transmission of spirochaetes to ticks fed upon vaccinated animals (Taso *et al.*, 2001).

Landscape management

Modification of the tick microenvironment to reduce relative humidity and soil moisture has long been considered effective in reducing the abundance of ticks. Free-living ixodid ticks spend the majority of their life off the host in the cover vegetation, leaf litter, duff or other accumulated vegetative debris containing a suitable microclimate for survival and search for hosts. Survival is a function of maintaining a balance between energy use and body water homoeostasis, which is affected by permeability of the cuticle, water mass, sensitivity to desiccation and ability to absorb atmospheric water vapour (Needham and Teel, 1986; Kahl and Knülle, 1988). Therefore, humidity is crucial to the survival of many ticks, including *I. ricinus* and *I. scapularis*: with their small water mass and relatively permeable cuticle, they cannot survive as long as some other mesic tick species (Lees, 1946; Yoder and Spielman, 1992; Stafford, 1994).

Modifications to the landscape and alterations to the vegetation to produce an unfavourable microclimate for tick oviposition, egg hatch, moulting and host seeking, or for vertebrate hosts may be particularly effective against the hydrophilic subadult stages of *Ixodes* spp. The influence of habitat structure on populations of ticks and tick hosts determines in turn the risk of human exposure to LB. Adler *et al.* (1992), for example, found that larval burdens on white-footed mice were positively associated with woody vegetation density and negatively with grassy and herbaceous vegetation. Nymphal burdens were also negatively correlated with herbaceous vegetation, but less so than the larvae as nymphs are transported and distributed (as engorged larvae) by birds and mammals. Mowed lawns with relatively little shade have fewer blacklegged tick nymphs despite high tick densities in adjacent ornamental ground cover and wooded areas, and a barrier of xeric landscaping materials at the lawn–ecotone interface could reduce residential exposure (Maupin *et al.*, 1991; Stafford and Magnarelli, 1993).

A number of studies have examined the impact of vegetation management on ticks. Mechanical clearing of the vegetation, clearing with herbicides, removal of leaf litter and controlled burns have been used to modify the vegetative environment. The mechanical or chemical clearing of vegetation has been shown in numerous studies to suppress populations of the lone star tick, *Amblyomma americanum* (e.g. Mount, 1981; Barnard, 1986). Vegetation management has also been shown to affect *I. scapularis*. Mowing and burning produced a dramatic, but temporary decline in adult tick abundance (Wilson, 1986), while removal of leaf litter in wooded residential plots or at lawn perimeters reduced the abundance of *I. scapularis* nymphs significantly (Schulze *et al.*, 1995; K. Stafford, unpublished data). Since untreated landscape stones and pinebark woodchips were found to deter the movement of *I. scapularis* nymphs in the laboratory (Patrican and Allan, 1995b), a field study on residential lawns found that the use of a woodchip border (Fig. 12.1) reduced the number of *I. scapularis* nymphs on the lawn perimeter by approximately 50% on average (K. Stafford, unpublished data).

Fire has been used extensively as a tool for forest and range management. It

Fig. 12.1. Landscape management using a woodchip barrier at the woodland–residential lawn interface to reduce movement of ticks into the lawn; a perimeter application of an acaricide can increase the efficacy of the barrier treatment. (Photograph by K. Stafford.)

was also one of the earliest tools advocated to control ticks, and grass burns were used in the attempt to control the Rocky Mountain wood tick and spotted fever (Fricks, 1915). Controlled burns were reported to reduce the abundance of lone star ticks (Hoch *et al.*, 1972). Significantly lower numbers of *I. scapularis* were found in pine palmetto flatwoods that had been burned 1–2 years prior to sampling than an area last burned 14 years previously (Rogers, 1955). Stafford (1998) found that a single, light to moderate controlled spring burn in a northeastern deciduous forest reduced nymphal ticks by 74%. A moderate to severe burn resulted in a reduction of nymphs by 97% in comparison with an adjacent unburned woodland. However, the effect appeared temporary judging by the comparable number of adult *I. scapularis* recovered in the burned and unburned tracts the following autumn. Similar results were obtained in a state park in northern Illinois (Fig. 12.2; C. Jones and U. Kitron, unpublished data). Vegetation burns on Great Island, Massachusetts, reduced adult *I. scapularis* by 70 and 80% following the burns (Wilson, 1986). Again, the effect was temporary and no effect on adult tick abundance could be detected 1.5 years later. In a single burn on Shelter Island, New York, 49% fewer nymphs were observed compared with an unburned tract, but the fire did not penetrate the humus layer (Mather *et al.*, 1993). The impact on larval and adult ticks later in the season was not reported. The potential use of the destruction of vegetation by controlled burns is limited as the main areas for risk of LB in the USA are in relatively densely populated residential areas and burning woodlands can improve deer browsing and the density of *Peromyscus* spp., which would increase host densities.

Fig. 12.2. Controlled burning in Castle Rock State Park in northern Illinois, USA. Although intended for vegetation control and landscape management, the effects of the burn on tick and small mammal densities were monitored as part of a CDC-sponsored tick control study. (Photograph by C. Jones and U. Kitron.)

Host management, reduction and exclusion

Members of the *I. ricinus* species complex associated with LB are three-host ticks with a broad host range, attacking many avian and mammalian hosts (see Chapters 6–9 on the ecology of *B. burgdorferi* s.l.). Many of these animals are reservoirs for *B. burgdorferi*, and white-footed mice, *P. leucopus*, in particular, are important hosts for the spirochaete where *I. scapularis* is found. White-tailed deer (*Odocoileus virginianus*) are the preferred host for the adult stage of *I. scapularis* and are crucial for the reproduction of this tick, but not for the transmission of *B. burgdorferi*. These vertebrate hosts have been closely linked to the abundance and distribution of *I. scapularis* and indirectly to the rising incidence of LB.

Management of tick hosts has been an important part of controlling cattle ticks (*Boophilus* spp.) by denying a primary host to free-living larvae of these one-host ticks through pasture spelling, sometimes conducted with rotational grazing (Wilkinson, 1957). Pasture spelling has also been effective against *I. ricinus* by denying this tick access to sheep (Randolph and Steele, 1985). The presence of *B. burgdorferi* and *Babesia microti*, an agent of human babesiosis, and their vector-ticks was examined on two islands inhabited by deer and mice and four islands without deer in Rhode Island (Anderson *et al.*, 1987). Both pathogens were isolated from tick-infested white-footed mice and meadow voles (*Microtus pennsylvanicus*) on the two deer-inhabited islands. Neither of the pathogens nor the blacklegged tick was detected in mice on the four islands that had no deer. This suggested that, in the absence of deer, other mammalian hosts could not sustain the tick in sufficient numbers to transmit the LB or babesiosis agents. However, a population of the bird-feeding *Ixodes uriae* and *Borrelia garinii* were maintained on a Baltic island and on the Faeroe Islands where marine birds were the only vertebrate host, demonstrating that deer and/or other mammals are not a prerequisite for completion of the life cycle and transmission of *B. burgdorferi* s.l. (Olsen *et al.*, 1993; Gylfe *et al.*, 1999).

The exclusion or reduction of white-tailed deer could potentially reduce vector abundance and affect the dynamics of disease transmission. When engorged female ticks drop off deer, they seek protected microclimates under the leaf litter or grass to lay their eggs. Hatching larvae actively disperse only a few metres from the egg mass (Lees and Milne, 1951; Stafford, 1992a), relying upon a passing host for survival and dispersal. Therefore, the exclusion of deer, with the corresponding reduction in egg-laying female ticks, would be expected to have the greatest effect on the number of larvae. Indeed, the exclusion of deer by electric fencing from two relatively small areas ($\approx$ 3.5 and 7.4 ha) in Connecticut was shown to reduce the abundance of *I. scapularis*, particularly larvae, within the fenced areas (Stafford, 1993). A seven-wire, slanted, high-tensile electric deer fence resulted in reductions in larvae of 88.5–97.8% and nymphs of 47.4–55.8% within the fence compared with outside the deer exclosures. Daniels *et al.* (1993) reported 83% fewer nymphs and 90% fewer larvae inside five exclosures with traditional fencing. These reductions were similar to those obtained for *A. americanum* larvae (98%) and nymphs (53%) in a 2.4 ha deer exclosure at the Land Between the Lakes recreational area in Tennessee (Bloemer *et al.*, 1986, 1990). Similarly, the exclusion of fallow deer, *Dama dama*, by a permanent fence resulted in significantly lower densities of *I. ricinus* nymphs compared with an adjacent forested area inhabited by deer in Ireland (Gray *et al.*, 1992). The size of the exclusion area is a primary determinant of the impact of exclusion on the tick population. Small and medium-sized mammals and birds are important to the passive dispersal of immature *I. scapularis* and these animals may reintroduce ticks within a deer exclosure, particularly near the edges (Daniels and Fish, 1995). Therefore, deer fencing alone can significantly reduce, but not eliminate, nymphal and adult tick populations and the risk of LB. There may also be local restrictions on the use, size and type of fencing permitted. Electric fences require regular maintenance, suitable terrain and proper grounding, and may be inappropriate in areas with a potential for frequent human contact (Ellingwood and Caturano, 1988). Another strategy for individual residences is the use of repellents and browse-resistant plantings to discourage deer. However, deer repellents may be ignored and less palatable vegetation browsed when alternative food is scarce or deer densities are high.

The incremental removal of deer has been shown to reduce tick abundance (Deblinger *et al.*, 1993; K. Stafford, unpublished data). A gradual reduction in deer densities over a 7 year period at a 567 ha site in Ipswich, Massachusetts, reduced the abundance of *I. scapularis* larvae and nymphs on white-footed mice by roughly one-half, although the density of female ticks on deer increased four- to sixfold during the 7 year reduction of deer at Crane's Beach (Deblinger et al., 1993). Similarly, the reduction of deer from 97.3 km^{-2} to 13.1 km^{-2} over a 7 year period on a 176 ha forested, fenced property in Connecticut, resulted in an almost fivefold reduction in nymphal tick abundance (K. Stafford, unpublished data). The virtual elimination of deer from Great Island, a 240 ha coastal peninsula near Cape Cod, Massachusetts, produced a dramatic reduction in the subadult population of this tick (Wilson et al., 1988a,b), although adult tick abundance initially seemed to increase after deer became scarce. This is due to the 2 year life cycle of the tick and the fact that a reduction in deer density (or exclusion), in the short term, may increase apparent host-seeking activity of adult ticks, as fewer deer will remove fewer adult ticks from the environment (Ginsberg and Zhioua, 1999). Adult tick populations eventually decline with the removal of deer. While deer removal has not eliminated the tick or interrupted the enzootic cycle on Great Island (adult ticks still occur on dogs, raccoons and foxes), the elimination of deer has apparently reduced or prevented human cases of LB on Great Island and observations suggest a deer density of 3 km^{-2} will reduce tick densities sufficiently to interrupt tick–human transmission of LB (Telford, 1993).

The computer simulation model LYMESIM was developed to assess the population dynamics of *I. scapularis* (Mount et al., 1997a). The model was used to examine the relationship between the density of white-footed mice and white-tailed deer in coastal Connecticut and the density and *B. burgdorferi* infection rates in the tick. Tick density thresholds for the maintenance of *B. burgdorferi* were also examined. There was a linear relationship between the density of free-living nymphal *I. scapularis* and mice, with tick density and infection rate influenced by the density of alternative hosts. In contrast, there was a curvilinear relationship between deer and tick density, with ticks disappearing when deer density was reduced to zero. The infection rate in nymphs increased with moderate decreases in deer density. In the short term, the reduction in numbers of a major host species may result in increased feeding on alternative hosts, minimizing the impact on the tick population and the prevalence of infection in the tick population. Actual reduction (or exclusion) of deer has resulted in a notable increase in the prevalence of *B. burgdorferi* in host-seeking nymphs (Gray et al., 1992; K. Stafford, unpublished data). In the simulation, however, infection rates dropped with less than 0.75 deer km^{-2} and went to zero at 0.3 deer km^{-2}. There was a reduced density of ticks with each decrease in deer density. In the model, a density of *I. scapularis* below a threshold of 87 nymphs ha^{-1} does not maintain *B. burgdorferi* in the ecosystem. Mount et al. (1997b) extended the model to simulations of several management strategies on tick density, including the impact of reduced deer density on the abundance of *I. scapularis* infected with *B. burgdorferi*. In a 10 year simulation, an initial deer density of 25 animals km^{-2} was reduced to 7.5, 2.5 and 0.25 deer km^{-2} (70, 90 and 99%, respectively). With only 0.25 deer km^{-2}, infected nymphs declined by 74% by year 3 and 98% by year 10.

The removal of most of the deer or complete elimination of the animal presents some severe logistical and political obstacles. Deer management has traditionally relied on regulated seasonal or controlled hunting and implementation can be controversial, particularly on public lands (Kilpatrick *et al.*, 1997). With the exception of some islands, it would be difficult to reduce deer sufficiently in most environments to affect tick populations or the dynamics of disease transmission. Conversely, unmanaged deer populations can result in superabundant tick populations as evident on several islands and other isolated tracts along the New England coast. Within a community, the reduction or elimination of deer requires the approval of the majority of community residents. The successful implementation of a deer reduction programme will require clear communication to the public and press of the biological basis and objectives for any wildlife reduction (i.e. protection of biodiversity, reduction in vehicle collisions or tick-associated disease), the support of stakeholders (local residents, politicians and the public) and good estimates of the size of the deer population (DeNicola *et al.*, 2000). However, the need for such a drastic reduction in deer population to achieve satisfactory control levels is likely to render this strategy unrealistic or unacceptable in many, if not most, areas.

Area application of acaricides

Although the large-scale application of acaricides against ticks is precluded by current environmental and safety concerns, these chemicals have been shown to provide effective short-term suppression of *I. scapularis* in applications targeting primarily the nymphal stage associated with most cases of LB. In early trials against *I. scapularis*, autumn application of carbaryl or diazinon was shown to reduce adult *I. scapularis* populations in an oak-dominated forest for 1 year and a single application of carbaryl to lawns and adjacent grassy and woodland areas of residential properties provided 90.8% control of *I. scapularis* nymphs for almost 2 months (Schulze *et al.*, 1987; Stafford, 1991). Curran *et al.* (1993) reported that various concentrations of liquid and granular formulations of carbaryl and chlorpyrifos, and a single spray application of cyfluthrin provided 63.4–85.7%, 84.9–95.6% and 93.1% reductions, respectively, in the density of nymphal *I. scapularis* for as long as 41 days after application. Granular formulations of diazinon, chlorpyrifos and carbaryl also provide effective control of *I. scapularis* (Schulze *et al.*, 1991, 1994, 2000). Similar results with chlorpyrifos and carbaryl were reported for adult *I. pacificus* in northern California foothills (Monsen *et al.*, 1999). The effectiveness of materials is dependent on contact with the litter layer, depth of the leaf litter and sufficient pressure to penetrate vegetation and the leaf litter (Schulze and Harrison, 1995).

Chemical control as a strategy to reduce tick abundance has principally been applied in residential settings in the USA. In a survey of licensed commercial pesticide applicators in Connecticut, treatment of residential properties comprised a larger proportion (90.1%) of tick control services than commercial properties (9.9%) (Stafford, 1997). The principal acaricides used for the control of *I. scapularis* were cyfluthrin, chlorpyrifos and carbaryl. A more recent survey in the heavily populated southeastern portion of the state found that the pyrethroid cyfluthrin was used by over half of the commercial applicators (K. Stafford, unpublished data). It is unknown how many homeowners apply acaricides to control ticks, but only 22.5% of 402 respondents in an LB survey of two Connecticut towns indicated that they had used acaricides and of those 16.9% indicated someone in their household had applied the chemical, while 75.6% used a commercial business to perform the acaricide application (Connecticut Department of Public Health, unpublished data). With the phase-out of residential use of the organophosphates chlorpyrifos and diazinon by the US Environment Protection Agency, fewer products will be available to individual homeowners for the control of ticks. With the effectiveness of synthetic pyrethroids, these chemicals will increasingly become the acaricide of choice in area tick control.

The synthetic pyrethroid cyfluthrin is particularly effective at a rate of application (0.1 kg a.i. ha^{-1}) 6–45 times less than organophosphate and carbamate materials (Curran *et al.*, 1993; K. Stafford, unpublished data). Solberg *et al.* (1992) also found spring applications of granular and liquid cyfluthrin (87% and 100% control, respectively, at 2 months) effective against *I. scapularis*. A single barrier application of granular deltamethrin at woodland edges effectively reduces *I. scapularis* nymphal abundance in residential settings (Schulze *et al.*, 2001; K. Stafford, unpublished data). However, the application of synthetic chemicals is unacceptable to a substantial proportion of the public in some areas. Non-residual, less toxic chemical control products such as desiccants and insecticidal soaps have been found to significantly reduce nymphal and adult *I. scapularis* in laboratory and small field trials when containing pyrethrin (Allan and Patrican, 1995; Patrican and Allan, 1995a,b). Nymphs crawling on to landscape stones treated with pyrethrin-based desiccants and insecticidal soaps suffer high (> 88%) mortality (Patrican and Allan, 1995b). While a residential application of natural pyrethrin did not provide effective control of *I. scapularis* nymphs, K. Stafford (unpublished data) found that a mixture of pyrethrin, the synergist piperonyl butoxide and insecticidal soap was effective in suppressing these ticks. However, more frequent applications of pyrethrin-based products may be required because of a lack of residual activity. Properly timed and targeted applications (i.e. single application as a barrier treatment or application to xeric landscaping substrates in maintained beds and borders near wooded areas) can minimize human exposure, reduce non-target and other potential environmental effects, and yet provide significant levels of tick control.

Host-targeted acaricides

The application of acaricides to small mammalian hosts has also been examined for the control of ticks. The relatively small amount of active chemical needed for host-targeted approaches to control LB is an attractive alternative to traditional chemical control because of decreased area-wide pesticide applications and decreased risk of exposure to non-target species. Originally used for the control of plague vectors (Kartman, 1958), Sonenshine and Haines (1985) found that an acaricide bait-box treatment station for the control of immature American dog ticks, *Dermacentor variabilis*, reduced infestations of rodent populations by > 90%. Frequently, a single species or small complex of species serve as principal hosts for the immature stages or as reservoirs for the bacterial agents of LB. The commercial development of host-targeted permethrin-treated cotton was the first approach for a vector of LB, *I. scapularis*.

Permethrin-treated cottonballs are aimed at immature stages of *I. scapularis* on white-footed mice, which collect the cotton as nesting material from cardboard tubes distributed approximately every 9–10 m throughout the mouse habitat. Several studies documented significant reductions in tick infestations on white-footed mice in treated sites (Mather *et al.*, 1987b; Deblinger and Rimmer, 1991; Stafford, 1992b). Mixed results on mice were obtained in studies in New York (Daniels *et al.*, 1991; Ginsberg, 1992). Nymphal reductions on both mice and in host-seeking ticks were also reported in a Massachusetts study with the treatment of one 7.3 ha site (Deblinger and Rimmer, 1991). However, studies in Connecticut and New York State failed to show any reduction in the number of infected, host-seeking *I. scapularis* nymphs when this product was used for a 3 year period in woodland and residential areas of about 1.6 ha or less (Daniels *et al.*, 1991; Stafford, 1992b). The failure of the tubes of permethrin-treated cotton to significantly affect the tick population in some areas may be due to the presence of alternative nesting materials, alternative tick hosts and/or temporal limitations on the effectiveness of the permethrin. Immigration of untreated tick-infested hosts may also reduce the impact of the permethrin-treated cotton in small, residential applications of the tubes. Mejlon *et al.* (1995) noted that the application of permethrin-treated cotton reduced numbers of larval *I. ricinus* on bank voles in Sweden and that significantly fewer nymphs were subsequently collected in the treated areas. However, no impact on larval infestations on the yellow-necked field mouse, *Apodemus flavicollis*, was observed and the authors noted that, with the diverse *I. ricinus* host species capable of transmitting *B. burgdorferi* s.l., the usefulness of rodent-targeted acaricides on nesting material would be limited.

Another recent approach, using bait boxes for the topical treatment of rodents with fipronil, is being evaluated for the control of *I. scapularis* on wild white-footed mice (patents pending, Aventis Environmental Science and Centers for Disease Control and Prevention). Discovered in 1987, fipronil is the first of a new class of insecticides, the phenyl pyrazoles, labelled for a broad range of insect pests on more than 50 crops and various non-crop uses (Aventis, 2000). Fipronil is a reversible inhibitor of the gamma-aminobutyric acid (GABA)-regulated chloride channel that displays selectively tighter binding in insects than in mammals. Percutaneous passage of the compound is low, with the compound concentrating in the epidermis, hair follicle and sebaceous glands (Cochet *et al.*, 1997). Fipronil is commercially available for the control of fleas and ticks on dogs and cats, with tick control lasting at least a month with a single application. A single topical application of fipronil to a mouse can kill ticks on the animal for up to 42 days (Maupin, 1999), reducing the need for frequent reapplication to the host. Theoretically, a single treatment at a bait box could render a white-footed mouse virtually tick-free and continue killing any ticks contacting the mouse for over a month. At the time of this writing, trials with a fipronil bait box have been expanded in Connecticut and the northeastern USA and a commercial version of the bait box may be available within a year or two.

The acaricide treatment of large wild animal hosts to kill feeding female ticks and thereby control free-living ticks is a relatively recent approach (Pound *et al.*, 1994; Sonenshine *et al.*, 1996). Ivermectin, an avermectin derivative from the soil bacterium *Streptomyces avermitilis*, is a broad-spectrum systemic parasiticide effective against nematodes, insects, mites and ticks, including *A. americanum*, *B. microplus* and *I. scapularis*. The potential of acaricide treatment of white-tailed deer was demonstrated by the control of *A. americanum* with ivermectin (Miller *et al.*, 1989; Pound *et al.*, 1996). Although an attempt to control *I. scapularis* by providing ivermectin-treated maize to free-ranging white-tailed deer on an isolated island did not succeed in reducing free-living tick abundance, > 90% control was obtained on animals with a sufficient serum concentration of ivermectin (Rand *et al.*, 2000). Difficulties in estimating the size of the island's deer population resulted in insufficient treatment of the host population. The use of a systemic agent in the USA and Europe is unlikely because of the withdrawal period required for clearance of a drug in food animals prior to human consumption. Adult *I. scapularis* activity and required timing of deer treatments corresponds with the autumn deer-hunting season in the USA.

A pesticide self-application device for the control of ticks on white-tailed deer has been developed and patented by scientists at the US Department of Agriculture Livestock Insects Research Laboratory in Kerrville, Texas (US patent no. 5,367,983). The topical pesticide self-application device, termed the 4-poster, consists of a central bin that contains bait to attract deer with one trough and two vertical pesticide impregnated rollers on each end (Pound *et al.*, 1994). Deer contact one of the rollers as they feed, treating the head, neck and ears with an acaricide (Fig. 12.3). Ninety-two and 94% control of nymphal and larval lone star ticks on deer in a large (22.3 ha) deer-fenced pasture in comparison with an untreated pasture was obtained with 2% amitraz (an acaricide with no withdrawal period) applied to deer during the tick season over 3 consecutive years (Pound *et al.*, 2000). The efficacy of the 4-poster and amitraz in controlling *I. scapularis* feeding on deer and free-living populations of the tick is being evaluated in restricted and unrestricted forested, suburban settings in several states in the northeastern USA. Although adjustments had to be made to ensure sufficient delivery of product to northern deer with their heavy hair coats and larger appetites, and availability of acorns in the autumn at some of the sites affected usage of the device during varying years of the study, generally over 90% of deer in the targeted areas visited the 4-poster (K. Stafford, unpublished data). In Maryland, preliminary analysis has shown that nymphal black-legged

Fig. 12.3. (a) The '4-poster' passive topical treatment device for controlling ticks on white-tailed deer with a central bait bin and two feeding and acaricide application stations on either side of the bin. Three of the four vertical applicator rollers are visible. (b) Close-up view of a feeding station showing an adjustable overhanging plate to force deer to rub the sides of the head and neck against the rollers. Only a small amount of bait maize exits from the bait port and deer must place their heads in the required predetermined position to feed (Pound *et al.*, 2000).

ticks were down by 68–71% after the third year of treatment (McGraw and McBride, 2001). Similar results have been observed with lone star ticks in New Jersey (McGraw and McBride, 2001) and with *I. scapularis* in Connecticut (K. Stafford, unpublished data). The application of acaricides to white-tailed deer is the only approach at this time that can conceivably reduce ticks and have an impact on LB on a community-wide scale, with minimal participation by most residents and with minimal environmental concerns. Whether this approach would be applicable to common European roe deer, *Capreolus caprelolus*, is unknown and implementation may be more difficult in Europe. The *I. ricinus*–roe deer system in Europe is less well understood and roe deer are a smaller species, which are broadly dispersed and therefore probably harder to target with a host-treatment system.

Biological control

Ticks have few natural enemies and a number of studies have explored the occurrence and potential use of predators, parasites and pathogens for the control of ticks. There have been very few attempts to control ticks with natural enemies (reviewed by Mwangi *et al.*, 1991). Hu *et al.* (1998), Samish and Rehacek (1999) and Samish and Alekseev (2001) have reviewed the role of parasitoids, pathogens and predators and arthropod predators on ticks, respectively.

Arthropod predators are generally non-specific feeders. Most reports consist of sporadic observations and many observations are based on laboratory observations within a container. Ants, beetles and spiders appear to be the primary arthropods preying on ticks and engorged ticks appear to be a more common target than unfed stages. Samish and Alekseev (2001) reported that 33% of the publications on arthropod predators of ticks were on *Ixodes* spp., mainly on work from the former USSR, and in contrast with some other genera, the genus *Ixodes* was preyed upon mainly by beetles rather than ants. Fire ants (*Solenopsis geminata*) feed on engorged ticks and may have an impact on populations of *Boophilus* (Barre *et al.*, 1991), and the red imported fire ant, *Solenopsis invicta*, has severely affected populations of the lone star tick (Harris and Burns, 1972; Burns and Melancon, 1977; Fleetwood *et al.*, 1984). Rodents, lizards, birds and chickens have all been reported to prey on ticks. Milne (1950) noted that predation of preovipositional female *I. ricinus* was high, with shrews and, to a lesser extent, birds the major predators. The best-known bird predation is probably by the oxpeckers in Africa (Moreau, 1933). Poultry readily eat *I. ricinus* dropped in sheep pens and chickens will consume attached ticks from cattle and unfed ticks from the vegetation (Milne, 1950; Hassan *et al.*, 1991). Anecdotal reports suggest that chickens and guineafowl may have an impact on local abundance of *I. scapularis*, but it remains unclear what effect these predators have on the tick population.

Ticks are parasitized by seven known species of minute (2 mm) chalcid wasps in the family Encyrtidae, which have been recovered from many species of ticks worldwide (Davis, 1986; Hu *et al.*, 1998). Two species of these insects were first recognized and described in the USA, *Ixodiphagus hookeri* (Howard) and *Ixodiphagus texanus* Howard. The first species was mass-released in the late 1920s and 1930s on the Elizabeth Islands and Martha's Vineyard, Massachusetts, to control the American dog tick, *D. variabilis*, and in Montana and other western states to control the Rocky Mountain wood tick, *Dermacentor andersoni*, a vector of Rocky Mountain spotted fever (Larrouse *et al.*, 1928; Cooley and Kohls, 1934; Smith and Cole, 1943). No reduction in tick abundance due to the parasitoids was observed and few wasps were recovered following these releases, although the wasp apparently became established in *I. scapularis* on Naushon Island (Larrouse *et al.*, 1928; Cobb, 1942; Mather *et al.*, 1987a). The failures of the previous releases decades ago may be attributed to some ecological constraint on the parasitoid, timing of the release, climatic incompatibility, or limitation on parasitoid reproduction in the targeted tick species.

The report by Mather *et al.* (1987a) that the prevalence of *B. burgdorferi* or

B. microti, the cause of human babesiosis, was nearly 40% lower in host-seeking nymphs of *I. scapularis* parasitized by *I. hookeri* renewed interest in these parasitoids in vector control, and more recently a theoretical case has been made for the control of *I. scapularis* by parasitoid augmentation, with a call for a major research initiative (Knipling and Steelman, 2000). Mwangi *et al.* (1997) reported that nymphal *Amblyomma variegatum* declined by 95% following the monthly release over a year of 150,000 *I. hookeri* on ten head of cattle in a 4 ha pasture. A parasitism rate of 51% in the nymphs was obtained with the high parasitoid release rate, but few (9%) nymphs were parasitized during the post-release period of the study. These wasps are established at several insular locations in New England and New York where deer and tick populations are extremely high (Hu *et al.*, 1993; Stafford *et al.*, 1996), suggesting that the parasitoid is strongly dependent on densities of ticks or tick hosts. The reduction of deer at two insular populations of the black-legged tick in Connecticut has resulted in dramatic declines in the presence of *I. hookeri* (K. Stafford, unpublished data), suggesting that the usefulness of this parasitoid to control *I. scapularis* may be limited in any integrated control programme.

The application of entomopathogenic fungi to tick habitat or vertebrate hosts to control ticks is another environmentally acceptable approach, and fungi are one of the more promising biological control agents for ticks. A number of fungi have been isolated from ticks, primarily engorged females from deer, although some fungal species have been found on larvae, nymphs and unfed females. These include *Paecilomyces farinosus*, *Verticillium lecanii*, *Beauveria bassiana* and some other species from these three genera in the *Hyphomycetes* family. Recovery rates are low, however, generally less than 2–5%. *V. lecanii* appears to be a common fungal pathogen of *I. scapularis* and *I. ricinus* in nature (Kalsbeek *et al.*, 1995; Zhioua *et al.*, 1999). Entomopathogenic fungi such as *B. bassiana* and *Metarhizium anisopliae* have been reported to cause mortality in several tick species such as *Rhipicephalus appendiculatus*, *A. variegatum* (Kaaya *et al.*, 1996), *I. ricinus* (Samsinakova *et al.*, 1974) and *I. scapularis* (Zhioua *et al.*, 1997). The role of these fungi in the control of natural populations is unknown, but appears low. Even though local strains isolated from ticks may be more pathogenic, existing commercial formulations of entomopathogenic fungi might be used to control ticks. Two products, BotaniGard (Mycotech, Butte, Montana) and Naturalis T & O (Troy Biosciences, Phoenix, Arizona), are labelled in the USA for ornamental and turf applications. Perimeter treatments in trials at residential sites with these two strains of *B. bassiana* reduced nymphal *I. scapularis* by 59–89% in one trial and 80–90% of a sample of nymphs developed mucoses (K. Stafford and S. Allan, unpublished data).

Although ticks do not support a full nematode life cycle, nematode parasites of insects have been proposed as possible biological control agents for ticks (Samish and Glazer, 1991; Samish and Rehacek, 1999). The engorged females of several tick species have been found to be susceptible to infection with entomopathogenic steinernematid and heterorhabiditid nematodes (Samish *et al.*, 1999, 2000). Juvenile nematodes penetrate the host by the genital pore in female ticks, and release symbiotic bacteria, which kill the host. Susceptibility to nematodes varies with the species of tick, developmental stage of the tick and nematode strain. Zhioua *et al.* (1995) and Hill (1998) demonstrated that engorged *I. scapularis* females, but not unfed females, engorged nymphs or engorged larvae, are susceptible to steinernematid nematodes. The application of nematodes to engorged female ticks will kill these ticks prior to laying eggs and could potentially reduce tick reproduction. However, sensitivity to autumn temperatures and lack of susceptibility of unfed female ticks limit the potential application of nematodes for the vectors of LB at present. Microsporidian parasites (*Nosema* spp.) have been recovered from *I. ricinus* and, although microsporidia have potential as biocontrol agents for a number of pests, their role in ticks remains unknown (Weiser and Rehacek, 1975).

Pheromones

Tick pheromones and related chemicals are potential tools that could be applied to tick management by disrupting tick behaviour and reproduction. Pheromones are messenger chemicals used by animals to modify behaviour in members of the same species. A messenger chemical that benefits only the recipient, such as a host odour attractive to ticks or to parasitoids, is referred to as a kairomone. Pheromones used by various tick species to control and modulate tick movement and mating behaviour are broadly classified as assembly, aggregation-attachment or sex-attractant pheromones. For a more detailed review of pheromone types and methods for the identification of tick pheromones, readers are referred to Sonenshine (1991).

Assembly pheromones are present in many, although not all, tick species and induce clustering of free-living ticks in their environment. Aggregation-attachment pheromones are produced by males of several *Amblyomma* spp. and strongly attract unfed adult, sometimes nymphal, ticks. They induce attachment in areas where male ticks are feeding, and permit identification of infested hosts by questing ticks, clustering of ticks on the host and facilitation of mating (Norval *et al.*, 1989). The mating process in ticks is governed by several sex pheromones in a discrete series of phases. The hierarchy of the mating sequence in ticks other than the genus *Ixodes* is relatively consistent as egg or sperm development proceeds only following attachment and commencement of feeding (for a full review of tick reproduction, see Sonenshine, 1994). In *Ixodes* spp., mating may occur off the host in the vegetation or on the host, because male gamete development is not dependent on tick feeding. Little is known about pheromonal regulation of behaviour in *Ixodes* spp. ticks. The chemical 2,6-dichlorophenol is the sex attractant secreted by at least 14 species of ticks (Hamilton, 1992), but is not the sex attractant for *Ixodes* spp. ticks, which still need to be characterized. The existence of assembly and sex pheromones has been reported for *I. ricinus* (Graf, 1975; Hajkova and Leahy, 1982). The presence of assembly pheromones, which appear to be purines or purine-like substances, attractive to nymphal and adult

I. scapularis, has recently been documented using bioassays of cast skins and exudates from moulted ticks (Allan, 1999; Allan and Sonenshine, 2002). Carroll (2001; Carroll *et al.*, 1996) has reported that adult *I. scapularis* exhibit an arrestant response to host-derived chemicals (e.g. tarsal and interdigital gland) secreted by white-tailed deer in the laboratory and the field. While the distance at which ticks can detect and respond to these substances is unknown, kairomones could serve to concentrate ticks in areas with increased likelihood of finding a host.

Research has focused on use of pheromones as a lure to attract ticks to a toxicant, to disrupt mating behaviour, or in a baited decoy system to attract and kill male ticks (Sonenshine, 1994). A combination of a pheromone and an acaricide could kill ticks more effectively and selectively because ticks are attracted by the pheromone to the acaricide. The combination of 2,6-dichlorophenol with an acaricide kills mate-seeking male *D. variabilis* and disrupts the mating process (Sonenshine and Haines, 1985; Sonenshine *et al.*, 1992). An attractant decoy tag containing an acaricide–pheromone mixture on collars and/or the tails of cattle has been shown to provide excellent control of bont ticks (*Amblyomma hebraeum* and *A. variegatum*) for many weeks (Norval *et al.*, 1996; Allan *et al.*, 1998). A controlled-release formulation containing an assembly pheromone or a kairomone and an acaricide could potentially be used against free-living ticks and improve area control of LB vectors with better efficacy and less toxicity than current applications.

Integrated tick management

There are many definitions of IPM, but it basically involves the selection, integration and implementation of several pest control actions based on predicted ecological, economic and sociological consequences (Axtell, 1981) with the goal of reduction, rather than elimination, of the pest population. Two of the basic concepts of IPM are the economic injury level, the pest population density at which economic losses exceed the cost of control, and the economic threshold, the pest density at which control measures should be applied to prevent the pest population from reaching the injury level. From a public health perspective, this implies an acceptable density of vector ticks and risk of LB before management practices are implemented. In the case of LB, eradication of the tick vector is not feasible and acceptable thresholds would be different for residential areas, recreational areas and less visited parts of woodland parks, and tied to the perception of risk as well as the actual density of ticks required for transmission to humans. In terms of human nuisance, for example, an economic threshold of 0.65 ticks h^{-1} in a carbon dioxide sample was proposed for the lone star tick based on an attack rate of less than one tick per human visitor per day (Mount and Dunn, 1983). It is difficult to put a value on a human disease like LB and, consequently to define an acceptable 'economic' level for managing the disease. Nevertheless, individuals make this decision when they elect to implement or obtain tick control services, alter their landscape or seek vaccination. As noted earlier, the use of personal protection measures is influenced by both the perception of risk and the degree of inconvenience involved in the protective activity.

The basic aim of any LB prevention programme is to reduce the number of human cases within a community or larger geographical region. On a practical basis, reducing the risk of tick bite, and consequently disease, has been limited to personal protection measures and the residential use of acaricides. The use of many technologies (e.g. host immunity, anti-tick vaccines, pheromones, parasites and pathogens) is being researched or remains speculative and others (e.g. acaricide on mouse nesting material) have had mixed or negative results (Daniels *et al.*, 1991; Ginsberg, 1992; Stafford, 1992b). The combination of methods offers the potential for the greatest level of tick management. For example, results of a project to manage lone star tick populations in a recreation area demonstrated that combinations of three tick control methods – acaricides, vegetation management and exclusion of white-tailed deer by a fence – provided greater tick reductions than each method alone (Bloemer *et al.*, 1990). All three methods together produced the greatest reduction, with 92–99% control of lone star larvae, nymphs and adults. Computer models of the life cycle of a number of tick species (Haile and Mount, 1987; Mount *et al.*, 1997a) have been used to study various combinations of tick management strategies, which could assist in the development of tick management programmes (Haile *et al.*, 1990; Mount *et al.*, 1997b).

The LYMESIM model has been extended to the simulation of several management strategies on *I. scapularis* densities (Mount *et al.*, 1997a,b). Area-wide acaricide, vegetation reduction or a combination of the two was the most effective in small recreational or residential sites. Simulated 50 and 100% reductions in vegetation were highly effective in reducing infected nymphal abundance (89 and > 99% by year 3, respectively). Simulations of area-wide acaricides applied annually to each stage reduced larvae, nymphs and adults by 82, 91 and 93%, respectively. The combination of acaricides and vegetation reduction in the LYMESIM simulation produced a 96% reduction in the seasonal abundance of infected nymphal *I. scapularis*, comparable with that observed with *A. amercainum* nymphs in experiments in Kentucky, Tennessee and Oklahoma (Mount and Whitney, 1984; Bloemer *et al.*, 1990). The acaricidal treatment of deer was the most cost-effective, long-term strategy for larger areas, but the model also suggested deer density reductions to be an important element. The simulations suggested that it might take 4–10 years to reduce the population of *I. scapularis* below the threshold necessary to maintain spirochaetal transmission (Mount *et al.*, 1997b).

LYMESIM was also used to assess the effectiveness of eight interventions in a hypothetical community of 10,000 people over 4 years with an annual incidence rate of 0.01 cases per person per year, i.e. 1000/100,000 population (Hayes *et al.*, 1999), a high rate that has been noted in some Connecticut communities (Stafford *et al.*, 1998; Connecticut Department of Public Health, 2001). The interventions simulated were area-wide acaricide, acaricide to nesting mice, acaricide to white-tailed deer, removal of white-tailed deer, fencing out deer, reduction of habitat vegetation, repellent and tick checks and the Lyme vaccine. Various estimates of intervention use were based, in part, on published surveys on LB-related behaviours of people and experiment efficacy of the intervention. The number of LB cases prevented was determined under varying assumptions for a baseline, worst-case and best-case scenario for each intervention. The impact of the interventions was dependent upon the level of engagement. For example, if only 50% of the population is vaccinated, 74 cases of LB would be prevented, but with 95% vaccinated, 361 cases would be prevented. Reduction in tick habitat vegetation by 40% of residents with 90% habitat reduction on lawns, 80% habitat reduction in ecotone and 10% reduction in forested areas resulted in the prevention of only 94 cases. Acaricidal treatment of nesting mice prevented the fewest cases, while the acaricidal treatment of deer prevented the most cases (113–306) of LB under all scenarios except the best use of the Lyme vaccine. Use of an LB vaccine only protects vaccinated individuals and is dependent on a high level of engagement, while some other interventions may protect neighbours who do not engage in these interventions. Removal or exclusion of deer also decreases disease incidence, although risk increased for the first 2 years, reflecting an increase in the proportion of infected nymphs.

The Future for Management of Lyme Borreliosis

An integrated approach to the control of LB is at an early stage of development, with the objective of lowering tick densities and reducing the proportion of ticks infected with *B. burgdorferi* s.l. The choice of intervention strategies used in an integrated programme within a disease-endemic community will depend on the level of engagement by that community. Not all methods may be applicable or acceptable to a specific target population. Some interventions only require the participation of individual households while others will require the cooperation of the entire community. Interventions such as the vaccine or personal protection measures will only reduce risk for the individual practising such behaviours. The application of acaricides to the environment or tick hosts may benefit neighbouring properties and visitors to the treated properties. Current technologies, properly applied, clearly could reduce the incidence of LB. Nevertheless, utilization of existing technology, particularly acaricides, is low and novel strategies for the control of ticks and the pathogens they transmit are needed. Existing technologies need to be more intensively applied in areas endemic for LB.

Geographical areas of high risk for LB can be identified by human case reports, records of tick bite and mapping of natural foci for the pathogen–vector–host cycle based on a sound understanding of the epidemiology and ecology of the disease and known human activity patterns. Remote sensing and GIS technologies can be used to identify and map the risk of disease and target intervention efforts. Satellite imagery, GIS and spatial statistics are tools that can contribute more to the design and implementation of LB management programmes. For example, a detailed geo-referenced database can be used to determine which control strategies will be applied, and to what extent, in different parts of a control area.

As long as deer and other hosts remain abundant, tick vectors increase in abundance or spread geographically and more people either live by or visit tick-infested areas, the incidence of LB and other tick-associated diseases will increase. LB is, or could be, a largely preventable disease. Several experimental community or population-based prevention projects have recently been initiated in the northeastern USA, but LB prevention studies or tick control programmes in Europe or Russia are few or non-existent. In the absence of an effective LB programme, wider acceptance of the current LB vaccine (in the USA), or development of more effective broadly applicable vaccines or better tick management techniques, LB will continue to be an emerging disease problem. The application of tick-management programmes for the control of LB will ultimately be a policy choice based on the risk or perceived risk of LB and the cost and effectiveness of an intervention programme. The environmental management of LB will require government-based programmes supported by public funds, similar to mosquito and mosquito-borne disease programmes that have existed in the USA for decades, and will integrate education, vaccine and various strategies for tick control. Public health initiatives to reduce LB in selected, targeted communities in the USA are an important step in that direction. At the current time, the reduction of LB and other tick-associated diseases will require a multifaceted approach utilizing a variety of measures for tick management and disease prevention by both the individual and the community.

References

Adler, G.H., Telford, S.R. III, Wilson, M.L. and Spielman, A. (1992) Vegetation structure influences the burden of immature *Ixodes dammini* on its main host, *Peromyscus leucopus*. *Parasitology* 105, 105–110.

Alekseev, A.N., Burenkova, L.A., Vasilieva, I.S., Dubinina, H.V. and Chunikhin, S.P. (1996) Preliminary studies on virus and spirochete accumulation in the cement plug of ixodid ticks. *Experimental and Applied Acarology* 20, 713–723.

Allan, S.A. (1999) Pheromonal attraction in ticks. In: *Proceedings of the National Conference on the Prevention and Control of Tick-borne Disease*, 7–9 March, 1999, New York. American Lyme Disease Foundation, p. 18.

Allan, S.A. and Patrican, L.A. (1995) Reduction of immature *Ixodes scapularis* (Acari: Ixodidae) in woods by application of desiccant and insecticidal soap formulations. *Journal of Medical Entomology* 32, 16–20.

Allan, S.A. and Sonenshine, D.E. (2002) Evidence of an assembly pheromone in the black-legged deer tick, *Ixodes scapularis*. *Journal of Chemical Ecology* 28, 15–27.

Allan, S.A., Barre, N., Sonenshine, D.E. and Burridge, M.J. (1998) Efficacy of tags impregnated with pheromone and acaricide for the control of *Amblyomma variegatum*. *Medical Veterinary Entomology* 12, 141–150.

Anderson, J.F., Johnson, R.C., Magnarelli,

L.A., Hyde, F.W. and Myers, J.E. (1987) Prevalence of *Borrelia burgdorferi* and *Babesia microti* in mice on islands inhabited by white-tailed deer. *Applied and Environmental Microbiology* 53, 892–894.

Aventis (2000) *Fipronil Worldwide Technical Bulletin*. Aventis Crop Science, Lyon, France, 27 pp.

Axtell, R.C. (1981) Livestock Integrated Pest Management (IPM): principals and prospects. In: Knapp, F.W. (ed.) *Systems Approach to Animal Health and Production: a Symposium*. University of Kentucky, pp. 31–40.

Barnard, D.R. (1986) Density perturbation in populations of *Amblyomma americanum* (Acari: Ixodidae) in beef cattle forage areas in response to two regimens of vegetation management. *Journal of Economic Entomology* 79, 122–127.

Barre, N., Maulson, H., Harris, G.I. and Kermarrec, A. (1991) Predators of the tick *Amblyomma variegatum* (Acari: Ixodidae) in Guadeloupe, French West Indies. *Experimental and Applied Acarology* 12, 163–170.

Bloemer, S.R., Snoddy, E.L., Cooney, J.C. and Fairbanks, K. (1986) Influence of deer exclusion on populations on lone star ticks and American dog ticks (Acari: Ixodidae). *Journal of Economic Entomology* 79, 679–683.

Bloemer, S.R., Mount, G.A., Morris, T.A., Zimmerman, R.H., Barnard, D.R. and Snoddy, E.L. (1990) Management of lone star ticks (Acari: Ixodidae) in recreational areas with acaricide applications, vegetative management, and exclusion of white-tailed deer. *Journal of Medical Entomology* 27, 543–550.

Bowman, A.S., Coons, L.B., Needham, G.R. and Sauer, J.R. (1997) Tick saliva: recent advances and implications for vector competence. *Medical and Veterinary Entomology* 11, 277–285.

Brown, S.W., Cartter, M.L., Hadler, J.L. and Hooper, P.F. (1992) Lyme disease knowledge, attitudes, and behaviors – Connecticut, 1992. *Morbidity and Mortality Weekly Report* 41, 505–507.

Burns, E.C. and Melancon, D.G. (1977) Effect of imported fire ant (Hymenoptera: Formicidae) invasion on lone star tick (Acari: Ixodidae) populations. *Journal of Medical Entomology* 14, 247–249.

Carroll, J.F. (2001) Interdigital gland substances of white-tailed deer and the response of host-seeking ticks (Acari: Ixodidae). *Journal of Medical Entomology* 38, 114–117.

Carroll, J.F., Mills, G.D. Jr. and Schmidtmann, E.T. (1996) Field and laboratory responses of adult *Ixodes scapularis* (Acari: Ixodidae) to kairomones produced by white-tailed deer. *Journal of Medical Entomology* 33, 640–644.

Cartter, M.L., Farley, T.A., Ardito, H.A. and Hadler, J.L. (1989) Lyme disease prevention – knowledge, beliefs, and behaviors among high school students in an endemic area. *Connecticut Medicine* 53, 354–356.

Centers for Disease Control and Prevention (1999a) Notice to readers: availability of Lyme disease vaccine. *Morbidity and Mortality Weekly Report* 48, 35–36, 43.

Centers for Disease Control and Prevention (1999b) Recommendations for the use of Lyme disease vaccine: recommendations of the Advisory Committee on Immunization Practices (ACIP). *Morbidity and Mortality Weekly Report* 48 (Suppl. RR-7), 1–26.

Cobb, S. (1942) Tick parasites on Cape Cod. *Science* 95, 503.

Cochet, P., Birckel, P., Bromet-Petit, M., Bromet, N. and Weil, A. (1997) Skin distribution of fipronil by microautoradiography following topical administration to the beagle dog. *European Journal of Drug Metabolism and Pharmacokinetics* 22, 211–216.

Connecticut Department of Public Health (2001) *Lyme Disease Statistics, 1996–2000*. www.state.ct.us/dph, 4.

Cooley, R.A. and Kohls, G.M. (1934) A summary of tick parasites. *5th Pacific Science Congress Proceedings* 5, 3375–3381.

Cortinas, M.R., Guerra, M.A., Jones, C.J. and Kitron, U. (2002) Characterization and prediction of tick-borne disease foci. *International Journal of Medical Microbiology* 291(33) (Suppl.).

Costello, C.M., Steere, A.C., Pinkerton, R.E. and Feder, H.M. (1989) A prospective study of tick bites in an endemic area for Lyme disease. *Journal of Infectious Diseases* 159, 136–139.

Curran, K.L., Fish, D. and Piesman, J. (1993) Reduction of nymphal *Ixodes dammini* (Acari: Ixodidae) in a residential suburban landscape by area application of insecticides. *Journal of Medical Entomology* 30, 107–113.

Daniel, M. and Kolar, J. (1990) Using satellite data to forecast the occurrence of the common tick *Ixodes ricinus* (L.). *Journal of Hygiene, Epidemiology, Microbiology and Immunology* 34, 243–252.

Daniel, M., Kolar, J., Zeman, P., Pavelka, K. and Sadlo, J. (1998) Predictive map of *Ixodes ricinus* high-incidence habitats and a tick-borne encephalitis risk assessment using satellite data. *Experimental and Applied Acarology* 22, 417–433.

Daniels, T.J. and Fish, D. (1995) Effect of deer exclusion on the abundance of immmature *Ixodes scapularis* (Acari: Ixodidae) parasitizing small and medium-sized mammals. *Journal of Medical Entomology* 32, 5–11.

Daniels, T.J., Fish, D. and Falco, R.C. (1991) Evaluation of host-targeted acaricide for reducing risk of Lyme disease in southern New York state. *Journal of Medical Entomology* 28, 537–543.

Daniels, T.J., Fish, D. and Schwartz, I. (1993) Reduced abundance of *Ixodes scapularis* (Acari: Ixodidae) and Lyme disease risk by deer exclusion. *Journal of Medical Entomology* 30, 1043–1049.

Davis, A.J. (1986) Bibilography of the Ixodiphagini (Hymenoptera, Chalcidoidea, Encyrtidae), parasites of ticks (Acari, Ixodidae), with notes on their biology. *Tijdschrift voor Entomologie* 129, 181–190.

Deblinger, R.D. and Rimmer, D.W. (1991) Efficacy of permethrin-based acaricide to reduce the abundance of *Ixodes dammini* (Acari: Ixodidae). *Journal of Medical Entomology* 28, 708–711.

Deblinger, R.D., Wilson, M.L., Rimmer, D.W. and Spielman, A. (1993) Reduced abundance of immature *Ixodes dammini* (Acari: Ixodidae) following incremental removal of deer. *Journal of Medical Entomology* 30, 144–150.

DeNicola, A.J., VerCauteren, K.C., Curtis, P.D. and Hygnstrom, S.E. (2000) *Managing White-tailed Deer in Suburban Environments*. Cornell Cooperative Extension, Ithaca, New York, 52 pp.

Dieter, S.W., Fish, D., Bros, S.M., Frank, D.H. and Wood, B.L. (1997) Landscape characterization of peridomestic risk for Lyme disease using satellite imagery. *American Journal of Tropical Medicine and Hygiene* 57, 687–692.

Ellingwood, M.R. and Caturano, S.L. (1988) *An Evaluation of Deer Management Options*. Publication no. DR-11, Connecticut Department of Environmental Protection, Wildlife Division, Hartford, Connecticut, 12 pp.

Estrada-Peña, A. (1998) Geostatistics and remote sensing as predictive tools of tick distribution: a cokriging system to estimate *Ixodes scapularis* (Acari: Ixodidae) habitat suitability in the United States and Canada from advanced very high resolution radiometer satellite imagery. *Journal of Medical Entomology* 35, 989–995.

Estrada-Peña, A. (2001) Forecasting habitat suitability for ticks and prevention of tick-borne diseases. *Veterinary Parasitology* 98, 111–132.

Fish, D. and Howard, C.A. (1999) Methods used for creating a national Lyme disease risk map. *Morbidity and Mortality Weekly Report* 48, 21–24.

Fix, A.D., Strickland, G.T. and Grant, J.

(1998) Tick bites and Lyme disease in an endemic setting: problematic use of serologic testing and prophylactic antibiotic therapy. *Journal of the American Medical Association* 279, 206–210.

Fleetwood, S.C., Teel, P.D. and Thompson, G. (1984) Impact of imported fire ant on lone star tick mortality in open and canopied pastured habitats of east central Texas. *Southwest Entomologist* 9, 158–162.

Fricks, L.D. (1915) Rocky Mountain spotted fever. *Public Health Reports* 30, 148–165.

Gern, L., Lebet, N. and Moret, J. (1996) Dynamics of *Borrelia burgdorferi* infection in nymphal *Ixodes ricinus* ticks during feeding. *Experimental and Applied Acarology* 20, 649–658.

Gern, L., Hu, C.M., Voet, P., Hauser, P. and Lobet, Y. (1997) Immunization with a polyvalent OspA vaccine protects mice against *Ixodes ricinus* tick bites infected by *Borrelia burgdorferi* ss, *Borrelia garinii*, and *Borrelia afzelii*. *Vaccine* 15, 1551–1557.

Ginsberg, H.S. (1992) *Ecology and Management of Ticks and Lyme Disease at Fire Island National Seashore and Selected National Parks*. Science Monograph NPS/NRSUNJ/NRSM-92/20, US Department of the Interior, National Park Service, 77 pp.

Ginsberg, H.S. and Zhioua, E. (1999) Influence of deer abundance on the abundance of questing adult *Ixodes scapularis* (Acari: Ixodidae). *Journal of Medical Entomology* 36, 379–381.

Glass, G.E., Amerasinghe, F.P., Morgan, J.M. III and Scott, T.W. (1994) Predicting *Ixodes scapularis* abundance on white-tailed deer using geographic information systems. *American Journal of Tropical Medicine and Hygiene* 51, 538–544.

Glass, G.E., Schwartz, B.S., Morgan, J.M. III, Johnson, D.T., Noy, P.M. and Israel, E. (1995) Environmental risk factors for Lyme disease identified with geographic information systems. *American Journal of Public Health* 85, 944–948.

Graf, J.F. (1975) Ecologie and ethologie d'*Ixodes ricinus* L. en Suisse (Ixodoidea: Ixodidae). Cinquième note: mise en évidence d'une phéromone sexuelle chez *Ixodes ricinus*. *Acarologia* 17, 436–441.

Gray, J.S., Kahl, O., Janetzki, C. and Stein, J. (1992) Studies on the ecology of Lyme disease in a deer forest in County Galway, Ireland. *Journal of Medical Entomology* 29, 915–920.

Gray, J.S., Kahl, O., Robertson, J.N., Daniel, M., Estrada-Peña, A., Gettinby, G., Jaenson, T.G.T., Jensen, P., Jongejan, F., Korenburg, E., Kurtenbach, K. and Zeman, P. (1998) Lyme borreliosis habitat assessment. *Zentralblatt für Bakteriologie* 287, 211–228.

Guerra, M., Walker, E., Jones, C., Paskewitz, S., Cortinas, M., Stancil, A., Beck, L., Bobo, M. and Kitron, U. (2001a) Predicting the risk of Lyme disease: habitat suitability for *Ixodes scapularis* in the north-central U.S. *Emerging Infectious Diseases* 8, 289–297.

Guerra, M.A., Walker, E.D. and Kitron, U. (2001b) Canine surveillance system for Lyme borreliosis in Wisconsin and northern Illinois: geographic distribution and risk factor analysis. *American Journal of Tropical Medicine and Hygiene* 65, 546–592.

Gylfe, A., Olsen, B., Strasevicius, D., Ras, N.M., Weihe, P., Noppa, L., Ostberg, Y., Baranton, G. and Bergstrom, S.J. (1999) Isolation of Lyme disease *Borrelia* from puffins (*Fratercula arctica*) and seabird ticks (*Ixodes uriae*) on the Faeroe Islands. *Journal of Clinical Microbiology* 37, 890–896.

Haile, D.G. and Mount, G.A. (1987) Computer simulation of population dynamics of the lone star tick, *Amblyomma americanum* (Acari: Ixodidae). *Journal of Medical Entomology* 24, 356–369.

Haile, D.G., Mount, G.A. and Cooksey,

L.M. (1990) Computer simulation of management strategies for American dog ticks (Acari: Ixodidae) and Rocky Mountain spotted fever. *Journal of Medical Entomology* 27, 686–696.

Hajkova, Z. and Leahy, M.G. (1982) Pheromone-regulated aggregation in larvae, nymphs, and adults of *Ixodes ricinus* (L.) (Acarina: Ixodidae). *Folia Parasitologica* 29, 61–67.

Hamilton, J.G.C. (1992) The role of pheromones in tick biology. *Parasitology Today* 8, 130–133.

Harris, W.G. and Burns, E.C. (1972) Predation on the lone star tick by the imported fire ant. *Environmental Entomology* 1, 362–365.

Hassan, S.M., Dipeolu, O.O., Amoo, A.O. and Odhiambo, T.R. (1991) Predation of livestock ticks by chickens. *Veterinary Parasitology* 38, 199–204.

Hayes, E.B., Maupin, G.O., Mount, G.A. and Piesman, J. (1999) Assessing the prevention effectiveness of local Lyme disease control. *Journal of Public Health Management and Practice* 5, 84–92.

Hill, D.E. (1998) Entomopathogenic nematodes as control agents of developmental stages of the black-legged tick, *Ixodes scapularis*. *Journal of Parasitology* 84, 1124–1127.

Hoch, A.L., Semtner, P.J., Baker, R.W. and Hair, J.A. (1972) Preliminary observation on controlled burning for lone star tick (Acarina: Ixodidae) control in woodlots. *Journal of Medical Entomology* 9, 446–451.

Hu, R., Hyland, K.E. and Mather, T.N. (1993) Occurence and distribution in Rhode Island of *Hunterellus hookeri* (Hymenoptera: Encyrtidae), a wasp parasitoid of *Ixodes dammini*. *Journal of Medical Entomology* 30, 277–280.

Hu, R., Hyland, K.E. and Oliver, J.H. (1998) A review on the use of *Ixodiphagus* wasps (Hymenoptera: Encyrtidae) as natural enemies for the control of ticks (Acari: Ixodidae). *Systematic and Applied Acarology* 3, 19–28.

Jaenson, T.G.T., Fish, D., Ginsberg, H.S., Gray, J.S., Mather, T.N. and Piesman, J. (1991) Workshop summary: methods for control of tick vectors of Lyme borreliosis. *Scandinavian Journal of Infectious Disease* Suppl. 77, 151–157.

Kaaya, G.P., Mwangi, E.N. and Ouna, E.A. (1996) Prospects for biological control of livestock ticks, *Rhipicephalus appendiculatus* and *Amblyomma variegatum*, using the entomogenous fungi *Beauveria bassiana* and *Metarhizium anisopliae*. *Journal of Invertebrate Pathology* 67, 15–20.

Kahl, O. and Knülle, W. (1988) Water vapour uptake from subsaturated atmospheres by engorged immature ixodid ticks. *Experimental and Applied Acarology* 4, 73–83.

Kahl, O., Janetzki-Mittmann, C., Gray, J.S., Jonas, R., Stein, J. and de Boer, R. (1998) Risk of infection with *Borrelia burgdorferi* sensu lato for a host in relation to the duration of nymphal *Ixodes ricinus* feeding and the method of tick removal. *Zentralblatt für Bakteriologie* 287, 41–52.

Kalsbeek, V., Frandsen, F. and Steenberg, T. (1995) Entomopathogenic fungi associated with *Ixodes ricinus* ticks. *Experimental and Applied Acarology* 19, 45–51.

Kartman, L. (1958) An insecticide bait-box method for the control of sylvatic plague vectors. *Journal of Hygiene* 56, 455–465.

Kilpatrick, H.J., Spohr, S.M. and Chasko, G.G. (1997) A controlled deer hunt on a state-owned coastal reserve in Connecticut: controversies, strategies, and results. *Wildlife Society Bulletin* 25, 451–456.

Kitron, U. (1998) Landscape ecology and epidemiology of vector-borne diseases: tools for spatial analysis. *Journal of Medical Entomology* 35, 435–445.

Kitron, U. (2000) Risk maps: transmission and burden of vector-borne diseases. *Parasitology Today* 16, 324–325.

Kitron, U. and Kazmierczak, J. (1997) Spatial analysis of the distribution of

Lyme disease in Wisconsin. *American Journal of Epidemiology* 145, 558–566.

Kitron, U., Jones, C.J., Bouseman, J.K., Nelson, J.A. and Baumgartner, D.L. (1992) Spatial analysis of the distribution of *Ixodes dammini* (Acari: Ixodidae) on white-tailed deer in Ogle County, Illinois. *Journal of Medical Entomology* 29, 259–266.

Knipling, E.F. and Steelman, C.D. (2000) Feasibility of controlling *Ixodes scapularis* ticks (Acari: Ixodidae), the vector of Lyme disease, by parasitoid augmentation. *Journal of Medical Entomology* 37, 645–652.

Korenberg, E.I. (1994) Comparative biology and epidemiology of Lyme disease and tick-borne encephalitis in the former Soviet Union. *Parasitology Today* 10, 157–160.

Larrouse, F., King, A.G. and Wolbach, S.B. (1928) The overwintering in Massachusetts of *Ixodiphagus caucurteri*. *Science* 67, 351–353.

Lees, A.D. (1946) The water balance in *Ixodes ricinus* L. and certain other species of ticks. *Parasitology* 37, 1–20.

Lees, A.D. and Milne, A. (1951) The seasonal and diurnal activities of individual sheep ticks (*Ixodes ricinus* L.). *Parasitology* 41, 189–208.

Maiwald, M., Oehme, R., March, O., Petney, T.N., Kimmig, P., Naser, K., Zappe, H.A., Hassler, D. and Doeberitz, M.V.K. (1998) Transmission risk of *Borrelia burgdorferi* sensu lato from *Ixodes ricinus* ticks to humans in southwest Germany. *Epidemiological Infections* 121, 103–108.

Mather, T.N., Piesman, J. and Spielman, A. (1987a) Absence of spirochetes (*Borrelia burgdorferi*) and piroplasms (*Babesia microti*) in deer ticks (*Ixodes dammini*) parasitized by chalcid wasps (*Hunterellus hookeri*). *Medical and Veterinary Entomology* 1, 3–8.

Mather, T.N., Ribeiro, J.M.C. and Spielman, A. (1987b) Lyme disease and babesiosis: acaricide focused on potentially infected ticks. *American Journal of Tropical Medicine and Hygiene* 36, 609–614.

Mather, T.N., Duffy, D.C. and Campbell, S.R. (1993) An unexpected result from burning vegetation to reduce Lyme disease transmission risks. *Journal of Medical Entomology* 30, 642–645.

Maupin, G.O. (1999) Innovations in acaricides and delivery methods against ticks. In: *Proceedings of the National Conference on the Prevention and Control of Tick-borne Disease*, 7–9 March, 1999, New York. American Lyme Disease Foundation, p. 16.

Maupin, G.O., Fish, D., Zultowsky, J., Campos, E.G. and Piesman, J. (1991) Landscape ecology of Lyme disease in a residential area of Westchester County, New York. *American Journal of Epidemiology* 133, 1105–1113.

McGraw, L. and McBride, J. (2001) Out of the Lyme-light. *Agricultural Research* 49, 4–7.

Mejlon, H.A., Jaenson, T.G.T. and Mather, T.N. (1995) Evaluation of host-targeted applications of permethrin for control of *Borrelia*-infected *Ixodes ricinus* (Acari: Ixodidae). *Medical Veterinary Entomology* 9, 207–210.

Meltzer, M.I., Dennis, D.T. and Orloski, K.A. (1999) The cost effectiveness of vaccinating against Lyme disease. *Emerging Infectious Diseases* 5, 321–328.

Merler, S., Furlanello, C., Chemini, C. and Nicolini, G. (1996) Classification tree methods for analysis of mesoscale distribution of *Ixodes ricinus* (Acari: Ixodidae) in Trentino, Italian alps. *Journal of Medical Entomology* 33, 888–893.

Miller, J.A., Garris, G.I., George, J.E. and Oehler, D.D. (1989) Control of lone star ticks (Acari: Ixodidae) on Spanish goats and white-tailed deer with orally administered ivermectin. *Journal of Economic Entomology* 82, 1650–1656.

Milne, A. (1950) The ecology of the sheep tick, *Ixodes ricinus* L.: microhabitat economy of the adult tick. *Parasitology* 40, 14–34.

Monsen, S.E., Bronson, L.R., Tucker, J.R. and Smith, C.R. (1999) Experimental and field evaluations of two acaracides for control of *I. pacificus* (Acari: Ixodidae) in northern California. *Journal of Medical Entomology* 36, 660–665.

Moreau, R.E. (1933) The food of the redbilled oxpecker, *Buphagus erythrorhynchus* (Stanley). *Bulletin of Entomological Research* 24, 325–335.

Mount, G.A. (1981) Control of the lone star tick in Oklahoma parks through vegetative management. *Journal of Economic Entomology* 74, 173–175.

Mount, G.A. and Dunn, J.E. (1983) Economic thresholds for lone star ticks (Acari: Ixodidae) in recreational areas based on a relationship between CO_2 human subject sampling. *Journal of Economic Entomology* 76, 327–329.

Mount, G.A. and Whitney, R.W. (1984) Ultralow volume mists of chlorpyrifos from a tractor-mounted blower for area control of the lone star tick (Acari: Ixodidae). *Journal of Economic Entomology* 77, 1219–1223.

Mount, G.A., Haile, D.G. and Daniels, E. (1997a) Simulation of blacklegged tick (Acari: Ixodidae) population dynamics and transmission of *Borrelia burgdorferi*. *Journal of Medical Entomology* 34, 461–484.

Mount, G.A., Haile, D.G. and Daniels, E. (1997b) Simulation of management strategies for the blacklegged tick (Acari: Ixodidae) and the Lyme disease spirochete, *Borrelia burgdorferi*. *Journal of Medical Entomology* 34, 672–683.

Mwangi, E.N., Dipeolu, O.O., Newson, R.M., Kaaya, G.P. and Hassan, S.M. (1991) Predators, parasitoids and pathogens of ticks: a review. *Biocontrol Science and Technology* 1, 147–156.

Mwangi, E.N., Hassan, S.M., Kaaya, G.P. and Essuman, S. (1997) The impact of *Ixodiphagus hookeri*, a tick parasitoid, on *Amblyomma variegatum* (Acari: Ixodidae) in a field trial in Kenya. *Experimental and Applied Acarology* 21, 117–126.

Needham, G.R. and Teel, P.D. (1986) Water balance by ticks between bloodmeals. In: Sauer, J.R. and Hair, J.A. (eds) *Morphology, Physiology, and Behavioral Ecology of Ticks*. Ellis Horwoord, Chichester, pp. 100–164.

Nicholson, M.C. and Mather, T.N. (1996) Methods for evaluating Lyme disease risks using geographic information systems and geospatial analysis. *Journal of Medical Entomology* 33, 711–720.

Norval, R.A.I., Andrew, J.H.R. and Yunker, C.E. (1989) Pheromone-mediation of host selection in bont ticks (*Amblyomma hebraeum*). *Science* 243, 364–365.

Norval, R.A.I., Sonsenshine, D.E., Allan, S.A. and Burridge, M.J. (1996) Efficacy of pheromone–acaricide-impregnated tail-tag decoys for controlling the bont tick, *Amblyomma hebraeum* (Acari: Ixodidae), on cattle in Zimbabwe. *Experimental and Applied Acarology* 20, 31–46.

Ohnishi, J., Piesman, J. and de Silva, A.M. (2001) Antigenic and genetic heterogeneity of *Borrelia burgdorferi* populations transmitted by ticks. *Proceedings of the National Academy of Sciences, USA* 98, 670–675.

Olsen, B., Jaenson, T.G., Noppa, L., Bunikis, J. and Bergstrom, S. (1993) A Lyme borreliosis cycle in seabirds and *Ixodes uriae* ticks. *Nature* 362, 340–342.

Patrican, L.A. and Allan, S.A. (1995a) Application of desiccant and insecticidal soap treatments to control *Ixodes scapularis* (Acari: Ixodidae) nymphs and adults in a hyperendemic woodland site. *Journal of Medical Entomology* 32, 859–863.

Patrican, L.A. and Allan, S.A. (1995b) Laboratory evaluation of desiccants and insecticidal soap applied to various substrates to control the deer tick *Ixodes scapularis*. *Medical and Veterinary Entomology* 9, 293–299.

Phillips, C.B., Liang, M.H., Sangha, O., Wright, E.A., Fossel, A.H., Lew, R.A., Fossel, K.K. and Shadick, N.A. (2001) Lyme disease and preventive behaviors in residents of Nantucket Island, Massachusetts. *American Journal of Preventive Medicine* 20, 219–224.

Piesman, J. (1995) Dispersal of the Lyme disease spirochete *Borrelia burgdorferi* to salivary glands of feeding nymphal *Ixodes scapularis* (Acari: Ixodidae). *Journal of Medical Entomology* 32, 519–521.

Pound, J.M., Miller, J.A. and LeMeilleur, C.A. (1994) Device and method for use as an aid in control of ticks and other ectoparasites on wildlife. US Patent #5,367,983.

Pound, J.M., Miller, J.A., George, J.E., Oehler, D.D. and Harmel, D.E. (1996) Systemic treatment of white-tailed deer with ivermectin-medicated bait to control free-living populations of lone star ticks (Acari: Ixodidae). *Journal of Medical Entomology* 33, 385–394.

Pound, J.M., Miller, J.A. and George, J.E. (2000) Efficacy of amitraz applied to white-tailed deer by the '4-poster' topical treatment device in controlling free-living lone star ticks (Acari: Ixodidae). *Journal of Medical Entomology* 37, 878–884.

Rand, P.W., Lacombe, E.H., Holman, M.S., Lubelczyk, C. and Smith, R.P. Jr. (2000) Attempt to control ticks (Acari: Ixodidae) on deer on an isolated island using ivermectin-treated corn. *Journal of Medical Entomology* 37, 126–133.

Randolph, S.E. (2000) Ticks and tick-borne disease systems in space and from space. *Advances in Parasitology* 47, 217–243.

Randolph, S.E. and Rogers, D.J. (2000) Fragile transmission cycles of tick-borne encephalitis virus may be disrupted by predicted climate change. *Proceedings of the Royal Society of London B* 267, 1741–1744.

Randolph, S.E. and Steele, G.M. (1985) An experimental evaluation of conventional control measures against the sheep tick, *Ixodes ricinus* (L.) (Acari: Ixodidae). II. The dynamics of the tick–host interaction. *Bulletin of Entomological Research* 75, 501–518.

Randolph, S.E., Green, R.M., Peacey, M.F. and Rogers, D.J. (2000) Seasonal synchrony: the key to tick-borne encephalitis foci identified by satellite data. *Parasitology* 121, 15–23.

Ribeiro, J.M.C., Mather, T.N., Piesman, J. and Spielman, A. (1987) Dissemination and salivary delivery of Lyme disease spirochetes in vector ticks (Acari: Ixodidae). *Journal of Medical Entomology* 24, 201–205.

Rogers, A.J. (1955) The abundance of *Ixodes scapularis* Say as affected by burning. *The Florida Entomologist* 38, 17–20.

Samish, M. and Alekseev, E. (2001) Arthropods as predators of ticks (Ixodidae). *Journal of Medical Entomology* 38, 1–11.

Samish, M. and Glazer, I. (1991) Killing ticks with parasitic nematodes of insects. *Journal of Invertebrate Pathology* 58, 281–282.

Samish, M. and Rehacek, J. (1999) Pathogens and predators of ticks and their potential in biological control. *Annual Review of Entomology* 44, 159–182.

Samish, M., Alekseev, E. and Glazer, I. (1999) Efficacy of entomopathogenic nematode strains against engorged *Boophilus annulatus* females (Acari: Ixodidae) under simulated field conditions. *Journal of Medical Entomology* 36, 727–732.

Samish, M., Alekseev, E. and Glazer, I. (2000) Mortality rate of adult ticks due to infection by entomopathogenic nematodes. *Journal of Parasitology* 86, 679–684.

Samsinakova, A., Kalalova, S., Daniel, M., Dusbabek, F., Honzakova, E. and Cerny, V. (1974) Entomogenous fungi associated with the tick *Ixodes ricinus* (L.). *Folia Parasitologia (Praha)* 21, 39–48.

Schmidtmann, E.T. (1994) Ecologically based strategies for controlling ticks. In: Sonenshine, D.E. and Mather, T.N. (eds) *Ecological Dynamics of Tick-borne*

Zoonoses. Oxford University Press, New York, pp. 240–280.

Schulze, T.L. and Harrison, R.L. (1995) Potential influence of leaf litter depth on effectiveness of granular carbaryl against subadult *Ixodes scapularis* (Acari: Ixodidae). *Journal of Medical Entomology* 32, 205–208.

Schulze, T.L., McDevitt, W.M., Parkin, W.E. and Shisler, J.K. (1987) Effectiveness of two insecticides in controlling *Ixodes dammini* (Acari: Ixodidae) following an outbreak of Lyme disease in New Jersey. *Journal of Medical Entomology* 24, 420–424.

Schulze, T.L., Taylor, G.C., Jordan, R.A., Bosler, E.M. and Shisler, J.K. (1991) Effectiveness of selected granular acaricide formulations in suppressing populations of *Ixodes dammini* (Acari: Ixodidae): short-term control of nymphs and larvae. *Journal of Medical Entomology* 28, 624–629.

Schulze, T.L., Jordan, R.A., Vasvary, L.M., Chomsky, M.S., Shaw, D.C., Meddis, M.A., Taylor, R.C. and Piesman, J. (1994) Suppression of *Ixodes scapularis* (Acari: Ixodidae) nymphs in a large residential community. *Journal of Medical Entomology* 31, 206–211.

Schulze, T.L., Jordan, R.A. and Hung, R.W. (1995) Suppression of subadult *Ixodes scapularis* (Acari: Ixodidae) following removal of leaf litter. *Journal of Medical Entomology* 32, 730–733.

Schulze, T.L., Jordan, R. and Hung, R. (2000) Effects of granular carbaryl application on sympatric populations of *Ixodes scapularis* and *Amblyomma americanum* (Acari: Ixodidae) nymphs. *Journal of Medical Entomology* 37, 121–125.

Schulze, T.L., Jordan, R.A., Hung, R.W., Taylor, R.C., Markowski, D. and Chomsky, M.S. (2001) Efficacy of granular deltamethrin against *Ixodes scapularis* and *Amblyomma americanum* (Acari: Ixodidae) nymphs. *Journal of Medical Entomology* 38, 344–346.

Shadick, N.A., Daltroy, L.H., Phillips, C.B., Liang, U.S. and Liang, M.H. (1997) Determinants of tick-avoidance behaviors in an endemic area for Lyme disease. *American Journal of Preventive Medicine* 13, 265–270.

Shapiro, E.D., Gerber, M.A., Holabird, N.B., Berg, A.T., Feder, H.M., Bell, G.L., Rys, P.N. and Persing, D.H. (1992) A controlled trial of antimicrobial prophylaxis for Lyme disease after deer-tick bites. *New England Journal of Medicine* 327, 1769–1773.

Sigal, L.H., Zahradnik, J.M., Lavin, P., Patella, S.J., Bryant, G., Haselby, R., Hilton, E., Kunkel, M., Adler-Klein, D., Doherty, T., Evans, J. and Malawista, S.E. (1998) A vaccine consisting of recombinant *Borrelia burgdorferi* outer-surface protein A to prevent Lyme disease. *New England Journal of Medicine* 339, 216–222.

Smith, C.N. and Cole, M.M. (1943) Studies of parasites of the American dog tick. *Journal of Economic Entomology* 36, 569–572.

Solberg, V.B., Neidhardt, K., Sardelis, M.R., Hoffman, F.J., Stevenson, R., Boobar, L.R. and Harlan, H.J. (1992) Field evaluation of two formulations of cyfluthrin for control of *Ixodes dammini* and *Amblyomma americanum* (Acari: Ixodidae). *Journal of Medical Entomology* 29, 634–638.

Sonenshine, D.E. (1991) Tick pheromones. In: *Biology of Ticks*, Vol. 1. Oxford University Press, New York, pp. 331–369.

Sonenshine, D.E. (1994) *Biology of Ticks*, Vol. 2. Oxford University Press, New York, 465 pp.

Sonenshine, D.E. and Haines, G. (1985) A convenient method for controlling populations of the American dog tick, *Dermacentor variabilis* (Acari:Ixodidae) in the natural environment. *Journal of Medical Entomology* 22, 577–583.

Sonenshine, D.E., Hamilton, J.G.C., Philips, J.S., Ellis, K.P. and Norval, R.A.I. (1992) Innovative techniques for control of tick disease vectors using pheromones. In:

Munderloh, U.G. and Kurtti, T.J. (eds) *First International Conference on Tick-borne Pathogens at the Host–Vector Interface: an Agenda for Research.* University of Minnesota, St Paul, pp. 308–313.

Sonenshine, D.E., Allan, S.A., Norval, R.A. and Burridge, M.J. (1996) A self-medicating applicator for control of ticks on deer. *Medical and Veterinary Entomology* 10, 149–154.

Sood, S.K., Salzman, M.B., Johnson, B.J., Happ, C.M., Feig, K., Carmondy, L., Rubin, L.G., Hilton, E. and Piesman, J. (1997) Duration of tick attachment as a predictor of the risk of Lyme disease in an area in which Lyme disease is endemic. *Journal of Infectious Diseases* 175, 996–999.

Stafford, K.C. III (1991) Effectiveness of carbaryl applications for the control of *Ixodes dammini* (Acari: Ixodidae) nymphs in an endemic residential area. *Journal of Medical Entomology* 28, 32–36.

Stafford, K.C. III (1992a) Oviposition and larval dispersal of *Ixodes dammini* (Acari: Ixodidae). *Journal of Medical Entomology* 29, 129–132.

Stafford, K.C. III (1992b) Third-year evaluation of host-targeted permethrin for the control of *Ixodes dammini* (Acari: Ixodidae) in southeastern Connecticut. *Journal of Medical Entomology* 29, 717–720.

Stafford, K.C. III (1993) Reduced abundance of *Ixodes scapularis* (Acari: Ixodidae) with exclusion of deer by electric fencing. *Journal of Medical Entomology* 30, 986–996.

Stafford, K.C. III (1994) Survival of immature *Ixodes scapularis* (Acari: Ixodidae) at different relative humidities. *Journal of Medical Entomology* 31, 310–314.

Stafford, K.C. III (1997) Pesticide use by licensed applicators for the control of *Ixodes scapularis* (Acari: Ixodidae) in Connecticut. *Journal of Medical Entomology* 34, 552–558.

Stafford, K.C. III (1998) Impact of controlled burns on the abundance of *Ixodes scapularis* (Acari: Ixodidae). *Journal of Medical Entomology* 35, 510–513.

Stafford, K.C. III and Magnarelli, L.A. (1993) Spatial and temporal patterns of *Ixodes scapularis* (Acari: Ixodidae) in south central Connecticut. *Journal of Medical Entomology* 30, 762–771.

Stafford, K.C. III, DeNicola, A.J. and Magnarelli, L.A. (1996) Presence of *Ixodiphagus hookeri* (Hymenoptera: Encyrtidae) in two Connecticut populations of *Ixodes scapularis* (Acari: Ixodidae). *Journal of Medical Entomology* 33, 183–188.

Stafford, K.C. III, Cartter, M.L., Magnarelli, L.A., Ertel, S. and Mshar, P.A. (1998) Temporal correlations between tick abundance and prevalence of ticks infested with *Borrelia burgdorferi* and increasing incidence of Lyme disease. *Journal of Clinical Microbiology* 36, 1240–1244.

Steere, A.C., Sikand, V.K., Meurice, F., Patenti, D.L., Fikrig, E., Schoen, R.T., Nowakowski, J., Schmid, C.H., Laukamp, S., Buscarino, C., Krause, D.S. and Group, L.D.V.S. (1998) Vaccination against Lyme disease with recombinant *Borrelia burgdorferi* outer-surface lipoprotein with adjuvant. *New England Journal of Medicine* 339, 209–215.

Taso, J., Barbour, A.G., Luke, C.J., Fikrig, E. and Fish, D. (2001) OspA immunization decreases transmission of *Borrelia burgdorferi* spirochetes from infected *Peromyscus leucopus* mice to larval *Ixodes scapularis* ticks. *Vector Borne and Zoonotic Diseases* 1, 65–74.

Tatchell, R.J. (1992) Ecology in relation to integrated tick management. *Insect Science and its Application* 13, 551–561.

Telford, S.T. III (1993) Comments in the forum: management of Lyme disease. In: Ginsberg, H.S. (ed.) *Ecology and Environmental Management of Lyme Disease.* Rutgers University Press, New Brunswick, New Jersey, pp. 164–167.

Uspensky, I. (1996) Tick-borne encephalitis prevention through vector control in Russia: an historical review. *Review of*

Medical and Veterinary Entomology 84, 679–689.

Uspensky, I. (1999) Ticks as the main target of human tick-borne disease control: Russian practical experience and its lessons. *Journal of Vector Ecology* 24, 40–53.

des Vignes, F., Piesman, J., Heffernan, R., Schulze, T.L., Stafford, K.C. III and Fish, D. (2001) Effect of tick removal on transmission of *Borrelia burgdorferi* and *Ehrlichia phagocytophila* by *Ixodes scapularis* nymphs. *Journal of Infectious Disease* 183, 773–778.

Warshafsky, S., Nowakowski, J., Nadelman, R.B., Kamer, R.S., Peterson, S.J. and Wormser, G.P. (1996) Efficacy of antibiotic prophylaxis for prevention of Lyme disease: a meta-analysis. *Journal of General Internal Medicine* 11, 329–333.

Weiser, J. and Rehacek, J. (1975) *Nosema slovaca* sp. n.: a second microsporidian of the tick *Ixodes ricinus*. *Journal of Invertebrate Pathology* 26, 411.

Wikel, S.K., Bergman, D.K. and Ramachandra, R.N. (1996) Immunological-based control of blood-feeding arthropods. In: Wikel, S.K. (ed.) *The Immunology of Host–Ectoparasitic Arthropod Relationships*. CAB International, Wallingford, UK, pp. 290–315.

Wilkinson, P.R. (1957) The spelling of pasture in cattle tick control. *Australian Journal of Agricultural Research* 8, 414–423.

Willadsen, P. and Jongejan, F. (1999) Immunology of the tick–host interaction and the control of ticks and tick-borne diseases. *Parasitology Today* 15, 258–262.

Willadsen, P., Bird, P., Cobon, G.S. and Hungerford, J. (1995) Commercialisation of a recombinant vaccine against *Boophilus microplus*. *Parasitology* 110, S43–S50.

Wilson, M.L. (1986) Reduced abundance of adult *Ixodes dammini* (Acari: Ixodidae) following destruction of vegetation. *Journal of Economic Entomology* 79, 693–696.

Wilson, M.L. and Deblinger, R.D. (1993) Vector management to reduce the risk of Lyme disease. In: Ginsberg, H.S. (ed.) *Ecology and Environmental Management of Lyme Disease*. Rutgers University Press, New Brunswick, New Jersey, pp. 126–156.

Wilson, M.L., Telford, S.R. III, Piesman, J. and Spielman, A. (1988a) Reduced abundance of immature *Ixodes dammini* (Acari: Ixodidae) following elimination of deer. *Journal of Medical Entomology* 25, 224–228.

Wilson, M.L., Litwin, T.S. and Gavin, T.A. (1988b) Microgeographic distribution of deer and *Ixodes dammini*: options for reducing the risk of Lyme disease. *Annals of the New York Academy of Sciences* 539, 437–440.

Wormser, G.P., Nadelman, R.B., Dattwyler, R.J., Dennis, D.T., Shapiro, E.D., Steere, A.C., Rush, T.J., Rahn, D.W., Coyle, P.K., Persing, D.A., Fish, D. and Luft, B.J. (2000) Practice guidelines for the treatment of Lyme disease. *Clinical Infectious Disease* 31, S1–S14.

Yoder, J.A. and Spielman, A. (1992) Differential capacity of larval deer ticks (*Ixodes dammini*) to imbibe water from subsaturated air. *Journal of Insect Physiology* 38, 863–869.

Zhioua, E., Lebrun, R.A., Ginsberg, H.S. and Aeschlimann, A. (1995) Pathogenicity of *Steinernema carpocapsae* and *S. glaseri* (Nematoda: Steinernematidae) to *Ixodes scapularis* (Acari: Ixodidae). *Journal of Medical Entomology* 32, 900–905.

Zhioua, E., Browning, M., Johnson, P.W., Ginsberg, H.S. and LeBrun, R.A. (1997) Pathogenicity of the entomopathogenic fungus *Metarhizum anisopliae* (Deuteromycetes) to *Ixodes scapularis* (Acari: Ixodidae). *Journal of Parasitology* 83, 815–818.

Zhioua, E., Ginsberg, H.S., Humber, R.A. and Lebrun, R.A. (1999) Preliminary survey for entomopathogenic fungi associated with *Ixodes scapularis* (Acari: Ixodidae) in southern New York and New England, USA. *Journal of Medical Entomology* 36, 635–637.

Index

Page numbers in **bold** refer to illustrations and tables

abundance, *definition* 36
Acari **156**
acaricides 271, 302, **309**, 313–**317**, 323
acrodermatitis chronica atrophicans (ACA)
    associated with *Borrelia afzelii* 264
    characteristic sign of LB 16
    description 1, 9–10
    first description 1
    inoculation experiment 2
    Japan 211
    treatment 19
see also Erythema (chronicum) migrans; manifestations; skin infection
adhesin 59, 293–294
Advisory Committee on Immunization Practices (ACIP) 291
African swine fever 71
animal reservoirs 12–16
animal disease models 127–129
antibiotics 2, 12, 18–20

antibodies
    against *Borrelia burgdorferi* sensu lato 12–13
    determination 16
    erythema (chronicum) migrans 201
    IgG 16, 255
    IgM 16, 255
    against OspA 57
    against OspC 293
    persistence 19
antigenic variation 59–60,122
argasid tick-borne borrelioses (ATBB) 176
argasid ticks 238–239
arthritis
    Connecticut 3, 23, 268
    Europe 11
    induction study 210
    outbreak seasonality 251
    treatment 19
    treatment-resistant 56, 284
    USA 11
see also joints; manifestations

arthropod carrier species 153, **156–157**
Asia 4, 72, 258
    East Asia 211–215
    enzootic cycles 105
avermectin 316
avian *see* bird

babesiosis 91, **92**, 223, 225
bacterial adhesion, host tissues 293–294
bacterial genomics 71–72
bacteriolysis **135**
bacteriophages 65–66
Bannwarth syndrome 10–11, 16
barrier hosts, *definition* 34–35
bears 231
behaviour
    feeding, *Ixodes* spp. 118–119, 301, 307
    human 252, 304, 306
    personal protection 304, 306
    questing 101–103, 104, 154, 203, 254
    see also human, activities
birds
    –*Ixodes pacificus* relationship 235
    competent reservoirs 253
    complement 135, 235
    hosts, *Ixodes scapularis* 234–235
    marine 311
    migrating and *Borrelia garinii* 161–162
    migratory 70, 124, 132, 139, 209–210
    passive dispersal, immature *Ixodes scapularis* 311
    reservoir hosts 70, 131, 209, 210
    seabirds 131, 162
    species 158–159, 162, 183, 184–185, 208–210
    see also host
blebs 53–54
blood-meal identification 41–42
Boophilus spp. 310

Borrelia afzelii
    association with acrodermatitis chronica atrophicans 10, 264
    association with mice 161
    association with small rodents 163, 164
    association with voles 161
    distribution 72, **178**, 181
    host specificity 161–162
    reservoir hosts 70
    resistant strains 135
    skin infection 126
    vaccine, effectiveness 290–292
Borrelia bissettii 238
Borrelia burgdorferi sensu lato
    adaptive radiation 139
    bacteriolysis **135**
    biology 47–73
    cardiomyopathy 11
    distribution, geographical **182**, 211–215
    diversity 130
    DNA molecular typing 70–71
    ecology 29–42, 223–242
    evolution 67–69, 71–72
    genome **63**, 65, 98
    identification 4, **187**, **206–207**
    membrane composition 52–53
    phylogenetic tree **213**
    prevalence in Russia **179–180**
    reservoir hosts 70, 163–164, 208–210
    species complex 67
    transformation and plating methodology **68**
Borrelia burgdorferi sensu stricto
distribution, geographical **178**
    frequency 154
    heterogeneity 72
    immunity against 283
    nomenclature modification 224, 225
    strain B31 **63**, 65, 98
    transmission risk 240
    vaccines 20

Borrelia garinii
   association with birds 161–164
   distribution, geographical **177**, 181
   flagellin gene (*flaB*) 70
   frequency 154
   heterogeneity 150, 204–205
   host specificity 161–162
   isolates from *Ixodes persulcatus*, Japan 205
   in nervous tissue 126
   migration 132–133
   reservoir hosts 16, 70–71
   vaccine, effectiveness 290–292
   variants **177**
Borrelia japonica 208, 210, 214
Borrelia lonestari 238–239
Borrelia lusitaniae 155, 163, **178**
Borrelia sinica 214
Borrelia valaisiana
   association with birds 162, 163, 164
   distribution, geographical 154–155, **178**
   reservoir hosts, 16, 70–71

C3 120, 129
carditis **7**, **8**, 19
carrier hosts, *definition* 33–34
carrier tick, *definition* 32
carriers 38–40, 150–155
   see also hosts; mammals; vectors
case definition of Lyme borreliosis for US national surveillance **260**
case reporting 259–267
cats 12, 13, 268, 315–316
cattle 12, 14–15
cattle tick (*Boophilus* spp.) 310
causative agents 3–4, 29, 149
   see also Borrelia
ceftriaxone 19, 270
Centers for Disease Control and Prevention (CDC) 290

chemoprophylaxis 19–20
chemotaxis 49–52, 126
chickens 15–16
children 3, 265, 282, 287–288
chromosomes 64, 65
chronic neuroborreliosis 11
chronic sclerodermic skin manifestations 10
ciprofloxacin 66
circular plasmids 65
classification of *Borrelia burgdorferi* 67–69, 201
clinical features 3, 5–10
   see also Acrodermatitis chronica atrophicans; erythema (chronicum) migrans; manifestations
clinical trials 284–288
co-evolution 72
co-feeding 34, 103, 160–161, 236–237
co-infection 39
coefficient of transmission 160
Colorado tick fever 223
complement 120, 133–134, 135, 235–237
complement regulatory acquiring surface proteins (CRASPs) 120, 133–134, 137
computer simulation model, LYMESIM 312–313, 322–323
conferences, international 4–5, **6**
control 295–296, 301–324
   see also prevention; vaccination
cost-effectiveness analyses, per case averted 269–270
cyst formation 53–54

dark-field microscopy (DFM) 39
DDT 302
dead-end hosts 12, 30, 158
deer
   acaricide treatment 316
   black-tailed deer 232

deer *continued*
   borreliacidal serum 235
   co-feeding transmission 161
   complement 135
   density 231–232, 312
   exclusion 311–313
   four-poster passive topical treatment 316–**317**
   hosts for *Ixodes scapularis* 231–233
   maintenance hosts 34, 257–258
   population reduction 311–313
   roe deer–*Ixodes ricinus* system 317
   repellents
   sika deer 202
   susceptibility to infection 15
   white-tailed deer 311–313
   zooprophylactic role 131
density, *definition* 36
developmental cycle, *Ixodes* spp. **100**
diagnosis 16–18
disease classification 255–256
DNA detection by PCR 17
DNA molecular typing, *Borrelia burgdorferi* sensu lato 70–71
DNA–DNA reassociation analysis 69
dogs 12–13, 268, 282, 283, 315–316
domestic animals 12–16, 127, 184, 231, 268
   see also hosts; mammals
dormice 158
doxycycline 19, 20, 269–270, 271

ecology
   Borrelia burgdorferi sensu lato 201–242
   research 29–42
   risk factors 192–194
   terminology 31–37
   tick 225–228
ehrlichiosis, granulocytic 91, **92**, 225
emergence of LB, USA 258–259
endemic regions 257–258, 259
environmental management 301–324

   see also climatic conditions; habitat
enzootic transmission cycles 104–105, 208–210
enzyme-linked immunosorbent assay (ELISA) 16, 261, 266, 270, 286–288, 289
epidemiology 70–71, 204, 251–271, 282–283
erythema (chronicum) migrans
   case definition 266
   cases 251
   description 7–10
   ecological factors 265
   Japan 211
   penicillin treatment 3
   rash 282
   relationship with tick bite and nervous disorders 2
   seasonal occurrence **267**
   skin lesions 16, 17
   symptoms 211
   treatment 19
   unusual case, Wisconsin 223
   see also acrodermatitis chronica atrophicans; manifestations; skin infection
Eurasia 5, **8**, 178
Europe
   arthritis 11
   Borrelia afzelii, geographical distribution 72, 154
   Borrelia burgdorferi s.s., geographical distribution 154, **178**
   Borrelia garinii, geographical distribution, 154, **177**, 181
   Borrelia lusitaniae, geographical distribution 155
   Borrelia valaisiana, geographical distribution 154–155
   clinical features 5, **8**
   descriptive population data 264–267
   ecological correlates 258

ecological research 30
enzootic maintenance cycle 104–105
seasonality of LB cases 103
serosurveys 13
therapy 19
European Union Concerted Action on Lyme borreliosis 5, 264–265
experimental reservoir hosts, *definition* 35
experimental vector (tick), *definition* 35
extracellular matrix (ECM) 58
extracellular vesicles (blebs) 53–54

feeding behaviour, *Ixodes* spp. 118–119, 301, 307
fibronectin-binding protein BBK32 59, 294
fipronil 315–316
flagella 49–52
focal patterns 257–258
Food and Drug Administration (FDA) 290, 291
foxes 158
fungi, entomopathogenic 319

Garin–Bujadoux–Bannwarth syndrome 10–11, 16
genetic recombination 122
genetic tools 66–67
genome and genome organization 62–67, 71–72
genospecies
    differences 129
    differentiated with molecular methods 30
    distribution, geographic **48**
    distribution in small mammals **185**
    diversity 237–238
    incidence **188**
    new 238, 208, 214
genotypic methods, molecular characterization 69–70

geographic distribution *see* distribution
goats, antibodies 15

habitat, tick 189, 228–231, 257–258, 302–304
    see also environmental management
hares 158, 183–184
hedgehogs 152, 158
hedgehog tick (*Ixodes hexagonus*) 150, 152, 158
heterogeneity 72, 150, 204–205
high-risk areas 257–258
horizontal transmission 209, 237
horses 12, 13–14
host
    abundance 324
    barrier host, *definition* 34–35
    bridging host, *definition* 35
    complement **134**, 235–237
    dead-end 12, 30, 158
    detection by ticks 99
    diversity 130–131
    immune status 159–160
    infection detection 40
    infection persistence 121–126
    dissemination 119–121
    immune status 159–160
    infectivity, *definition* 36
    maintenance 34, 104–105, 150–151, 257–258
    maintenance hosts, *definition* 33
    management 310
    non-reservoir 160–161
    of *Ixodes scapularis* 231–235
    parasitization 258
    preference, tick 226
    rodent 98
    species 131–132, 155–164
    specificity, *Borrelia* spp. 131–138, 161
    spirochaetes preferences 184–185
    to tick transmission 126–127
    tick–vertebrate cycles 104–105

host *continued*
    tissue bacterial adhesion 293–294
    tissue colonization 58–59
    see also carriers; mammals; reservoir hosts; vector
human
    clinical trials 284–288
    disease agents **92**
    infection dynamics 255–256
    infection risk 192–194
    Ixodes spp. attacks 184
    risk of disease 239–241
    risk-associated activities and behaviours 252, 258, 266, 268, 283
    tick-bite cases 201

immunoblot tests 16, 265, 270
immunofluorescence 16, 39, **50**
immunoprophylaxis 20
incubation period, LB 255
infection
    co-infection 39
    detection, vertebrate host 40
    determinants 267–269
    dynamics, human 255–256
    early disseminated 255
    early localized 255
    intensity, *definition* 36
    late persistent infection 255
    mixed 182–183
    prevalence, tick 38, 39
    prevalence, *definition* 36
    reservoir hosts 253–254
    risk, human 192–194
    skin infection, *Borrelia afzelii* 126
    transmission to humans 254–256
Insecta 97, **156, 157**
insecticides 315–316
insectivores 152, 157, 158, 183
integrated pest management (IPM) 321–323
intensity of infestation, *definition* 36
intervention strategies 323

intervention model 271
isolates, collection 181
ivermectin 316
Ixodes dammini 3–4, 226–227
    see also Ixodes scapularis
Ixodes hexagonus 150, 152, 155, 158
Ixodes jellisoni 238
Ixodes nipponensis 213
Ixodes ovatus 201–203, 204, 205, 208, 210
Ixodes pacificus
    abundance 103
    –bird relationship 235
    association with *Borrelia bissettii* 238
    association with Columbian black-tailed deer 232
    hosts 233–234
    infesting lizards 103
    life stages **101**
    principal vector 230–231
    seasonal distribution of LB cases 104
Ixodes pavlovskyi 178
Ixodes persulcatus
    adult density 192, **194**
    distribution, geographical 105, **151**
    enzootic transmission 210
    isolates from 205
    landscape preference **190, 191**
    life cycle 203
    maintenance hosts 258
    preferred hosts 209
    prevalence of borreliae **179–180,** 204
    questing activity 101–102, 104, 203
    seasonal fluctuation of tick bite **202**–203
    transmission rates 306
    vertebrate host 183–184
Ixodes ricinus
    abundance 101, 103
    co-feeding 236–237
    complement activity 135

distribution, geographical 105,
    151, 256–257
habitat 189
host range 310
infesting rodent 103
life cycle 152, 266–267
maintenance hosts 258
questing activity 154
rodent resistance to 159
transmission rates 306
vector in European Russia 176
vertebrate hosts 152, 155,
    157–159, 164
Ixodes scapularis
    -borne illness 267–268
    –*Ixodes dammini* crosses 226
    abundance 101
    abundance reduction 311–312
    alternative mammalian host 233
    bite 223
    control 313–**317**
    ecology 225–226
    habitat 228–230
    hosts 231–235, 236
    life cycle 152, 266–267
    nymphs 98, 101, 103, 240
    passive dispersal 311
    population genetics 227
    reservoir hosts 235–236
Ixodes spinipalpis 233–234
Ixodes spp.
    control methods 304–307
    developmental stages 99–102
    distribution, geographical
        150–**151**
    feeding behaviour 118–119, 301,
        307
    infection 164–165
    questing behaviour 101–103,
        104, 154, 203, 254
    reservoir hosts 186
    seasonal activity **203**
    vaccines 295
    vector competence 93, 103–104
Ixodes trianguliceps 178

Ixodes uriae
    Borrelia burgdorferi sensu lato
        circulation in Europe 152
    DNA identified in 70
    maintenance in secondary
        transmission cycles
        150–151
Borrelia garinii
    association with seabirds 162
    DNA amplified from 47
    OspA sequences **123**, 132–133
    vertebrate hosts 155, 159
ixodid tick-borne borreliosis (ITBB)
    176, 184, 185–186, 189–192,
    193, 194

joints 11, 126, 285
    see also arthritis; manifestations

Koch's postulates 14

laboratory animal species 128–129
laboratory evidence 16–17
lagomorphs 158, 183–184
landscape management 308–**310**
landscape preference, *Ixodes persulcatus*
    190, **191**
late persistent infection 255
later stage manifestations 255
life cycle, tick **101**, 138, 152, 203,
    266–267
life stages, tick 103–104
linear chromosome 64
linear plasmids 65
lipopolysaccharide (LPS) **52**–53
lipoprotein **52**, 54–56, 59–60, 62,
    121
lizards 34–35, 103, 104, 160, 234,
    235, 236
Lyme, Connecticut, *ix* 1, 3, 223,
    258–259
Lyme arthritis *see* arthritis
LYMErix 281, 286, 291–292

LYMESIM, computer simulation
    model 312–313, 322–323
lymphocytoma 1, 2, 9, 16, 19, 264

maintenance hosts 34, 104–105,
    150–151, 257–258
maintenance hosts, *definition* 33
mammals
    acaricide application 314–317
    hosts 233
    large 209
    medium-sized 157, 184, 236
    passive dispersal immature *Ixodes
        scapularis* 311
    see also carriers; host; small mammal; vector; vertebrate host
management strategies **305**
manifestations
    canine LB 12
    cardiac 11
    cattle 14–15
    characteristic signs 16
    clinical **7**–10, 149
        Russian terminology 175
    clinical definition 264–265
    development 2–3
    dynamics of infection 255–256
    early neurological 10–11
    equine LB 13–14
    eye 7
    joint 11, 12, 13
    later stage 255
    meningo-polyradiculoneuritis
        1–2
    musculoskeletal **7, 8**
    nervous system **7, 8**
    skin 7–8, 9–10
    three stages of LB 5–11
    see also acrodermatitis chronica
        atrophicans; arthritis;
        clinical features; erythema
        (chronicum) migrans; joints
mean density, *definition* 36

membrane proteins 60–62
    see also outer surface proteins
        (Osps); proteins
membrane structure, borreliae 49–62
methodological pitfalls 29–42
MHC haplotypes 129
mice
    association with *Borrelia afzelii*
        161
    Borrelia burgdorferi host 310
    C3H mouse strain 129
    density 101, 312
    immune status 159–160
    mouse models 128–129, 312
    reservoir competence 157
    sera 235–237
    white-footed mice 101, 104,
        235–237, 312
    see also rodents
migration
    bird 70, 124, 132, 139, 161–162,
        209–210
    Borrelia spp. 72, 132–133
models
    animal disease models 127–129
    intervention model 271
    LYMESIM 312–313, 322–323
    mouse models 128–129
    transmission 134–138
morbidity 176, 194
morphology, borreliae 49–62
motility, borreliae 49–52
moult, ticks 39
mouse models 128–129
Museum of Borrelia, Laboratory of
    Vectors of Infection 181

national surveillance, USA 259–264
nematode parasites, ticks 319–320
neuroborreliosis 2, 10–11, **267**
New York Times article 5
non-reservoir hosts 160–161
non-reservoir hosts, *definition* 34
non-vector tick, *definition* 31–32

non-vector tick species **96**
North America
    clinical features 5, **8**
    deer abundance 33
    descriptive population data 258–264
    ecological correlates 257–258
    ecological research 30
    ecology *Borrelia burgdorferi* sensu lato 223–242
    maintenance hosts, deer 34
    nymphal vs. female tick, importance 104

Osps *see* outer surface proteins (Osps)
outer membrane proteins 54–55
outer membrane spanning (Oms) 61
outer membrane structure 52–53
outer surface proteins (Osps)
    characterized 49
    during tick feeding 97–98
    founding members of family 58
    ligand for factor H 120
    lipoproteins 55–58
    major lipoproteins 55–58
    OspA
        function 55–56
        gene sequences **123**, **124**, 132–133, 213–214
        intraspecific heterogeneity 150
        recombinant 54
        serotypes 204–205
        typing system 4
        vaccine 20
    OspC 56–57, **125**
    OspD 57
    OspE 58, 61, 294
    OspEF-related proteins (ERP) 58
    OspF 58, 294
    selectively expressed 118–119
    targeted 121
    see also proteins

parasiticide 316
parasitization 183–184
pasture spelling 310
pathogen detection 37–42
pathogen-exposed hosts, *definition* 33
pathogen-exposed ticks, *definition* 32
pathogenesis 127–129
pathogenicity, *Borrelia* spp. 210–211
penicillin 2, 3, 19
permethrin 315
persistence, vertebrate host 121–126
personal protection 304, 306
phase contrast microscopy (PCM) 39
pheasants 16
phenotypic methods, molecular characterization 69–70
pheromones, tick 320–321
phylogenetic tree, *Borrelia burgdorferi* sensu lato **213**
plasmids 62, 65
plasminogen system 120–121
polymerase chain reaction (PCR) technology
    Borrelia burgdorferi sensu lato detection method **38**
    diagnosis support 12–13, 14, 15
    DNA detection 17
    molecular characterization 69–70
    RNA detection 39
preferred hosts, *Ixodes persulcatus* 209
pregnancy, risk for adverse outcomes 268–269
prevalence, *Borrelia burgdorferi* **179–180**
prevalence of infection, *definition* 36
prevalence of infestation, *definition* 36
prevention 270–271, 290, 302, 308, 322
    see also personal protection; vaccination; vaccine
primary vectors **95**, 99–102, 103–104
prophylaxis 19–20

proteins
    complement regulatory acquiring surface proteins (CRASPs) 120, 133–134, 137
    decorin-binding proteins (Dbps) 58–59, 294
    exported plasmid protein 62
    fibronectin-binding protein BBK32 59, 294
    host-derived complement control proteins 120
    integral membrane proteins 60–61
    lipoprotein **52**, 54–56, 59–60, 62, 121
    outer membrane proteins 54–55
    surface-exposed membrane proteins 61–62
    variable lipoproteins (Vls) 59, 60
    variable membrane proteins (Vmps) 59–60, 122, 124
    see also outer surface proteins (Osps)
pyrethrin, natural 314
pyrethroid cyfluthrin 314

qualitative terms 31–36
quality-adjusted life years (QALY) 269
quantitative terms 36–37
questing activity 101–104, 154, 203, 254

rabbit 158
rat 104, 157, 163, 233
recombinant vaccines 13, 57
relapsing fever 223
relative reservoir capacity, *definition* 37
relative vector capacity, *definition* 37
repellents 51, 306, 311
reported cases 251–252, **261**, **262**, **263**

reservoir hosts
    capacity, *definition* 37
    competence 131–132, 157, 159, 161, 235–237
    definition 29–30, 32–33, 34
    enzootic cycles 104
    identification 40–42, 155
    infection 253–254
    natural 117, 127
    relative reservoir capacity, *definition* 37
    see also birds; deer; host; mammals; mice; rodents
reservoir potential 160
reservoir potential, *definition* 37
reservoirs *see* reservoir hosts
resistance of *Borrelia burgdorferi* to, host complement **134**
resistant strains, *Borrelia afzelii* 135
restriction fragment length polymorphism analysis (RFLP) 204, 205
risk
    ecological factors 192–194
    EUCALB 264–265
    factors 267–269
    high-risk areas 257–258, 282
    human 239–241
    infection 192–194
    mapping **224**, 302–304, 323–324
    pregnancy, adverse outcomes 268–269
    transmission 239–240
Rocky Mountain spotted fever 3, 223
rodent
    borrelial isolates 211
    complement 135
    dormice 158
    enzootic transmission 210
    hosts 98, 132, 232–234
    infection reservoirs 253
    infestation, *Ixodes ricinus* 103
    principal reservoir hosts 117
    rats 104, 157, 163, 233

reservoir, habitat 257
reservoir hosts 70, 131, 163,
    236, 237, 238
squirrels 158, 163
topical treatment 315
voles 159–160, 183, 186
see also host; mice; reservoir hosts
Russian terminology, clinical manifestations 175

satellite imagery 303–304
scientific publications 4–5
sclerodermic skin manifestations 10
seasonality
    activity, ticks 102–103, **203**
    arthritis 3, 251
    Borrelia burgdorferi sensu stricto
        240–241
    disease onset 263, 265, 266–267
    distribution, human LB 104
    erythema migrans **267**
    fluctuation tick bite cases **202**
    host-seeking in ticks 30
    tick questing activity 30 101,
        102–103, **203,** 254
serological diagnostic testing cost 270
serosurveys 13
sheep 12, 15, 236–237
shrews 157, 158, 183
skin infection 7–**8**, 9–10, 16, 17, 126
    see also acrodermatitis chronica
        atrophicans; erythema
        (chronicum) migrans;
        manifestations
small mammals
    distribution of *Borrelia burgdorferi*
        sensu lato genospecies
        185
    genera 130–131
    identification *Borrelia burgdorferi*
        sensu lato isolates
        186–**187**
    parasitization by larvae and
        nymphs 183

reservoir competent 132
reservoir hosts 157, 184–185
species **182**, 184, 208–209
see also mammal
SmithKline Beecham 286
specific host infectivity, *definition* 36
specific tick infectivity, *definition* 36
spirochaetes
    biology 47–73
    host preferences 184–185
    phenotypic and genotypic heterogeneity 150
    primary bridging vectors **92**
    primary vector distribution **95**
    tick–host–spirochaete interactions
        diversity 129–130
squirrels 158, 163
    see also rodent
surface-exposed membrane proteins
    61–62
surveillance statistics 264
symptoms *see* manifestations
synthetic pyrethroid cyfluthrin 314
systematics, tick ecology 225–228

T-cell proliferation assay 17
taiga tick *see Ixodes persulcatus*
terminology 29–42, 175
tests 16–17, 270
ticks
    Amblyomma americanum (lone
        star tick) 238–239, 283,
        308
    American dog tick (*Dermacentor
        variabilis*) 233
    Argas spp. 92, **156**
    argasid ticks 238–239
    biological control 318
    black-legged tick *see Ixodes scapularis*
    Boophilus spp. 92, 310
    bite cases 202–203
    castor bean tick *see Ixodes ricinus*
    chemical control, 313–**317**

ticks *continued*
    control methods 295–296, 301, 313–320
    Dermacentor spp. 92, 176–177, 233
    ecology 102–105, 225–228
    European sheep tick *see Ixodes ricinus*
    habitat 189, 228–231, 257–258, 302–304
    Haemaphysalis spp. 92, **156,** 214
    hard 29
    hosts **100**
    hosts, *definition* 33
    Hyalomma 92
    infectivity, *definition* 36
    landscape management 308–310
    life stages 103–104
    Lone star tick, *see Amblyomma americanum*
    microenvironment modification 308–310
    natural enemies 318
    non-vector, *definition* 32
    non-vector species **96**
    off-host survival 101
    Ornithodoros 92
    pheromones 320–321
    population reduction 310–313
    population suppression 308
    questing activity 101–103
    Rhipicephalus 92, **156**
    saliva 98
    seabird tick *see Ixodes uriae*
    soft 29
    species 91–92, **94**, 176, **182**
    survival, humidity 308
    taiga tick *see Ixodes persulcatus* 177–183
    vector–vertebrate cycles 104–105
    western black-legged tick *see Ixodes pacificus*
    see also host; *Ixodes*; ixodid tick-borne borreliosis biotopes (ITBB); vectors
tick-borne encephalitis (TBE) 91, **92**, 175–176, 184, 192, 302
tick-borne zoonoses **32**, **33**
tick–borrelia relationships 97–99
transformation and plating methodology **68**
transmembrane-spanning domains 61
transmission
    co-feeding 34, 103
    coefficient of transmission 160
    current concepts 118–139
    cycle, *Borrelia burgdorferi* sensu lato model 134–138
    enzootic transmission 208–210
    horizontal 209, 237
    host to tick 126–127
    infection, transmission to humans 254–256
    model 134–138
    modes 254–255
    rates 306
    risk 239–240, 302–304
    transovarial 105, 153–154, 239
    transstadial 41, **96**, 99, 105, 138, 155
        definition 153
    zoonotic cycles 117
    see also vectors
treatment 18–20, 66, 269–270
treatment-resistant Lyme arthritis 56, 284
Treponema spp. 12, 14, 71

ungulates 160, 258
    see also deer; host
United States of America
    arthritis 3, 11, 23, 268
    Borrelia burgdorferi sensu stricto, heterogeneity 72
    canine LB 12
    emergence of LB 258–259
    equine LB 13
    national surveillance 259–264
    reported cases **261**, **262**, **263**

risk map **224**
seasonality of LB cases 103
USSR, ecological research 30

vaccination 281–296, 301, 306–308
    see also prevention; control
vaccine
    Borrelia burgdorferi sensu stricto 20
    canine LB 13
    equine LB 14
    LB 255
    recently developed 271
    recombinant 13, 57
Vaccine Adverse Events Reporting
    System (VAERS) 290
variable lipoproteins (Vls) 59, 60
variable membrane proteins (Vmps)
    59–60, 122, 124
vector
    associations 252–253
    biology 99–102
    bridging vectors 35, **92**, 93, **95**,
        102, 103–104
    capacity, *definition* 36–37
    climatic conditions 102
    competence 92–97, 103–104,
        214
    definition 31–32
    distribution **95**

experimental vector (tick),
    definition 35
identification 40–42
potential 93, 102–105
potential *definition* 37
relative vector capacity, *definition*
    37
species relationships with *Borrelia
    burgdorferi* sensu lato
    203–208
    see also *Ixodes scapularis*; tick
vector/reservoir competence, *definition*
    35–36
vegetation management 308
vertebrate host *see* host
voles 159–160, 161, 183, 186

wasp parasites, ticks 318–319
white-tailed deer 311–313
    see also deer
woodchip barrier **309**
woodrat 237, 238

xenodiagnosis 40, 41–42

zoonoses **32**, **33**
zoonotic transmission cycles 117